Anthocyanins in Fruits, Vegetables, and Grains

Anthocyanins in Fruits, Vegetables, and Grains

G. Mazza, Ph.D.
Senior Research Scientist and
Head, Food and Horticultural Sciences
Agriculture Canada Research Station
Morden, Manitoba, Canada

E. Miniati, M.Sc.
Research Associate
Istituto di Industrie Agrarie
Universitá di Perugia
Perugia, Italy

CRC Press
Taylor & Francis Group
Boca Raton London New York

CRC Press is an imprint of the
Taylor & Francis Group, an **informa** business

First published 1993 by CRC Press
Taylor & Francis Group
6000 Broken Sound Parkway NW, Suite 300
Boca Raton, FL 33487-2742

Reissued 2018 by CRC Press

© 1993 by CRC Press, Inc.
CRC Press is an imprint of Taylor & Francis Group, an Informa business

No claim to original U.S. Government works

Library of Congress Cataloging-in-Publication Data

Mazza, G., 1946-
    Anthocyanins in fruits, vegetables, and grains / G. Mazza, Enrico
  Miniati
        p.   cm.
    Includes bibliographical references and index.
    ISBN 0-8493-0172-6
    1. Anthocyanin. 2. Fruit—Composition. 3. Vegetables —
Composition. 4. Grain—Composition. I. Miniati, Enrico, 1945-.
    II. Title.
QK898.A55M39 1993
582'.019'218—dc20   92-13169

A Library of Congress record exists under LC control number: 92013169

ISBN 13: 978-1-315-89060-9 (hbk)
ISBN 13: 978-1-351-06970-0 (ebk)

Visit the Taylor & Francis Web site at http://www.taylorandfrancis.com and the
CRC Press Web site at http://www.crcpress.com

DEDICATION

To my wife Rachel and children, Joseph, Michael and Victoria, for their patience and understanding, and to my mother and father for always believing in me.

G.M.

To my wife Daniela and son Francesco, from whom the time devoted to this book has been withdrawn.

E.M.

FOREWORD

Among the natural pigments, anthocyanins are the largest water-soluble group responsible for the attractive colors ranging from salmon and pink, through scarlet, magenta, and violet to purple and blue of most fruits, juices, flower petals, and leaves. These ubiquitous compounds are fascinating in that it is now known that they can exist in many structural forms, both simple and complex, governed by certain physicochemical phenomena which have profound effects on their colors and stabilities. Studies on anthocyanins impinge into many aspects of chemistry, biochemistry, biology, botany and medicine.

In his 1963 classic book, *Naturally Occurring Oxygen Ring Compounds*, F.M. Dean speculated about the "acute frustration" of those late 19th century researchers who could see with their microscopes the brilliantly pigmented anthocyanins swimming in cell vacuoles but had little idea of their composition. If some personal reminiscence is allowable, I, too, experienced a similar sense of frustration when, as a young researcher in the early 1950s, one of my tasks was to measure the anthocyanin content of fruit juices using a pH differential method. The colors were unstable, changing with pH and time. What was going on? What were the structures present? I was aware of the basic structures of anthocyanins as benzopyrilium salts through the work of Willstätter, Robinson, and Karrer, amongst others, but it was painfully obvious that there was a great deal more to learn about anthocyanins and their behavior. I had to wait some 25 years for a truly satisfactory explanation, when Prof. Raymond Brouillard and associates published a series of papers on the structural transformations of anthocyanins. In the interim, a milestone which had a very significant effect in stimulating further research was the publication in 1967 of Prof. J.B. Harborne's book, *Comparative Biochemistry of the Flavonoids*. Knowledge of anthocyanin properties, distribution, biosynthesis and enzymology, physiological role and function is now greatly increased and anthocyanin pigmentation continues to provide a challenge to scientists in many fields because of its intricate nature and complexity.

Anthocyanins are particularly characteristic of the angiosperms or flowering plants, which themselves provide our major source of food crops. Two families predominate here, the Vitaceae (grapes) and the Rosaceae with its wide generic spectrum of pigmented fruits such as apple (*Malus*), pear (*Pyrus*), apricot, cherry, plum, peach and sloe (*Prunus*), blackberry and raspberry (*Rubus*), strawberry (*Fragaria*), and quince (*Cydonia*). Other families containing pigmented fruits include the Ericaceae (blueberry, cranberry), Saxifragaceae (black and red currants), and Caprifoliceae (elderberry). The Solanaceae contain the tamarillo, huckleberry, aubergine and potato. The natural function of pigmentation in root crops and vegetables, particularly when the colored part is hidden undergound, is not as evident as with fruits and berries where its likely function is to aid seed dispersal by animals. But amongst cultivated crops there is a conscious selection for color as a feature of quality. Colored varieties of the potato tuber may have been prized by the Peruvian peasants of pre-Inca days in much the same way that red anthocyanin color in potato skin is associated with quality today. The aesthetic pleasure which color gives to man is well illustrated by the Cruciferae, in the red roots of radish (*Raphanus sativus*) and red cabbage (*Brassica oleracea*), which appear to be grown principally for their color. Similar considerations apply to the colored pods of legumes (Leguminosae) and may be also applied to the colored seeds of cereals (Gramineae).

Color also makes an essential contribution to the attractive appearance of processed foods and beverages, and a major interest of the food industry in anthocyanins is their use as natural colorants to replace synthetic red dyes. The most abundant and historically oldest anthocyanin extracted used for this purpose is derived from grape pomace. Others, either available commercially, studied extensively with a view for industrial production or used locally as natural extracts for coloring foodstuffs, include pigments of roselle, blueberry, black currant, bilberry, cranberry, elderberry, red cabbage, purple corn, miracle fruit, sweet potato, *Viburnum* berries, morning glory, hollyhock, and *Clitoria ternatea* petals, leaves of cherry-plum and perilla, hulls of sunflower seed, black olive wastes, black cherry, black chokeberry, cowberry or lingonberry, crowberry, and saskatoon berries. Early man, whose diet was largely based on wild fruits and berries, probably consumed large amounts of anthocyanins, more than consumed today, no doubt attracted by their vivid and brilliant colors. Man is, thus, well acclimatized to ingesting anthocyanins. One could argue that the modern diet with its increasing dependence on processed foods may have now become deficient in anthocyanins. The increasing use of natural extracts to color foodstuffs could be regarded in this context as maintaining levels if not redressing the balance, which might be desirable in view of their beneficial effects.

Available evidence suggests that the anthocyanins are not only nontoxic and nonmutagenic but have beneficial therapeutic properties, particularly in ophthalmology, where they enhance sight acuteness, and for treatment of various blood circulation disorders. Toxicity tests with anthocyanins of purple corn (*Maiz morado*) and grapes (*Vitis labrusca*) fed to rats and beagles have shown little adverse effects, established no-effect levels, and satisfied regulatory agencies of the safety of their use in foodstuffs. The physiological effect of anthocyanins is related to the prevention of capillary fragility in treatment of illnesses involving tissue inflammation; indeed, they may well have replaced rutin and its derivatives in this regard. Most work in this area has been done with pharmaceutical preparations containing extract of *Vaccinium myrtillus*. This contains such a complex mixture of simple glycosides of all the common anthocyanidins except pelargonidin that it might be considered representative of the anthocyanins as a whole. Such extracts have been shown to inhibit blood platelet aggregation and to inhibit porcine elastase *in vitro*, are of interest in various aspects of diabetic microangiopathy, and exhibit anti-ulcer activity. Many pharmaceutical products are now in use containing anthocyanins.

Necessary for basic research and industrial applications, such as those outlined, is knowledge of sources and characteristics of identified anthocyanins. At the latest count, over 270 individual structures had been characterized. The object of this book is to assemble in one volume the large amount of scattered information which has been published over many years on the identification, properties, and applications of anthocyanins in edible plants and, where appropriate, to give the occurrence and significance of other selected phenolics and flavonoids. The authors are well qualified to undertake this task because of their long-standing interest in anthocyanins. The authors have not arranged the plants according to any botanical classification (as has been done in previous reviews) but rather in a more general way, based partly on their morphological characteristics. After an initial state-of-the-art summary of current anthocyanin science, six chapters are devoted to fruits (pome, stone, small, tropical, grapes and others; grapes warrant a separate chapter because of the wealth of information on this largest cultivated single fruit crop) and single chapters to cereals, legumes, roots, tubers and bulbs, cole crops, and others. There are over 2200 references, including several which are cited for the first time in an English language publication. The book will be of immense value to all those working with anthocyanins or contemplating so, but especially to those in Food and Horticultural Sciences.

C.F. Timberlake, Bristol, England

PREFACE

This book was written to serve as a primary reference text for all those interested in any aspects of anthocyanins, the most important and widespread group of plant pigments. The book deals with the chemistry, physiology, chemotaxonomy, inheritance, pharmacology, biotechnology, and food technology aspects of the anthocyanins. Types of anthocyanins, structural transformations, color stabilization and intensification factors, biosynthesis, analysis, and functions of anthocyanins are discussed first. This is followed by an in-depth review of the literature dealing with anthocyanins of all major and minor fruits, cereals, legumes, roots, tubers, bulbs, cole crops, oilseeds, herbs, spices, and several minor crops. Whenever possible, qualitative and quantitative composition of the anthocyanins is given in tabulated form and the changes associated with the development of the plant and/or with the storage/processing are presented graphically or in tables.

Some of the crops, such as grapes, blueberries, raspberries, cherries, and apples, have been studied extensively, and many aspects of their anthocyanins are well documented. The research on the anthocyanins of corn, rice, and potatoes has been focused primarily on the inheritance aspect. In many other crops, the knowledge of their red pigments is limited to the chemical nature of the anthocyanin(s), and in several cases the identification needs to be confirmed. Because of this, the depth of coverage varies from plant to plant. A reason for including plants about which little information on their anthocyanins is available is to draw attention to the gaps in our knowledge of such plants in the hope that they will receive more attention in the future.

One of the most comprehensive chapters of the book is that dealing with grapes. This chapter, which contains 17 figures and 12 tables, reviews the current knowledge of the anthocyanins in the different species of grapes; the changes that occur during growth and ripening of the fruit; the extraction, chemical, and physical changes, and role of anthocyanins in wine and juice making; the recovery and application of anthocyanins from grape pomace; and the production of anthocyanins in grape cell cultures.

A total of 2234 references are cited, with many patents and publications from eastern European countries and Japan cited for the first time in an English language publication.

We hope that this work will be useful to all those interested in anthocyanins, especially phytochemists, physiologists, food scientists, and technologists.

We shall be very obliged to readers who would call our attention to aspects that have been neglected and to errors or omissions that might appear in the publication.

G. Mazza
E. Miniati

THE AUTHORS

G. (Joe) Mazza, Ph.D., is Senior Research Scientist and Head of the Food and Horticultural Sciences Section, Agriculture Canada Research Station, Morden, MB, Canada. He is also Adjunct Professor of Food Chemistry in the Department of Food Science, University of Manitoba, Winnipeg, MB, Canada. Dr. Mazza was born in Pietrapertosa, Potenza, Italy, in 1946. He studied agricultural sciences at the University of Naples, Italy, and graduated with a B.Sc.A. degree in food science from the University of Manitoba in 1973. He received his Ph.D. degree from the University of Alberta, Edmonton, AB, and did postdoctoral studies at the Institut de Chimie, Université Louis Pasteur, Strasbourg, France.

Prior to joining the Research Branch of Agriculture Canada, Dr. Mazza worked as Food Scientist at Burns Foods Ltd., Winnipeg and Calgary, and at the Alberta Horticultural Research Center, Brooks, AB. Over the years he has served as a scientific authority for several joint government projects with industry and academia which have led to new developments in crops/products utilization, quality, and storage.

Dr. Mazza has published over 70 research papers, 15 on characterization and properties of anthocyanin pigments, and over 180 miscellaneous publications including chapters in the following books: *Engineering and Food* (Applied Science Publishers, 1984), *Quality and Preservation of Plant Foods* (CRC Press, 1989), *Potato Production, Processing and Products* (CRC Press, 1991) and *Encyclopaedia of Food Science, Food Technology and Nutrition* (Academic Press, 1992).

He has served as member of the editorial boards of the *Canadian Institute of Food Science and Technology Journal* and the *Journal of Food Quality*, and reviewed several books and many research papers for journals and granting agencies. He is a member of the Canadian Institute of Food Science and Technology, the Institute of Food Technologists, the Potato Association of America, and the Groupe Polyphénols. He also serves on several regional, national, and international committees, including the Manitoba Horticultural Council, the Prairie Potato Council, the Expert Committee on Grain Quality, and the Expert Committee on Plant Products. One of Dr. Mazza's current major research interests concerns characterization, properties, and applications of anthocyanins as natural colorants in food/pharmaceutical products.

Enrico Miniati is Technical Coordinator in the Department of Food Science and Technology and Nutrition, University of Perugia, Italy. He was born in Perugia in 1945 and graduated from the 'Alessandro Volta' Institute of Industrial Technology in 1965 with a Diploma in Industrial Chemistry. He received his M.Sc. in food chemistry from the Faculty of Pharmacy, University of Perugia for work on the copigmentation and self-association of anthocyanins in food model systems. From 1966 to 1969 he worked for the National Research Council of Italy. He joined the Department of Food Science and Technology and Nutrition at the University of Perugia, first as Technical Officer and then as Technical Coordinator. He has published 13 research papers on phenolic compounds and has been a member of the Group Polyphénols since 1976. He has served as the Italian member of the European Economic Community Expert Committee on reference methods of analysis for polyphenols in foods and has lectured on the chemistry and analysis of food colorants at the Faculty of Industrial Chemistry, University of Bologna, Italy. His current research interest concerns properties of anthocyanins and other phenolic compounds and their role on sensory and nutritional aspects of foods and beverages.

ACKNOWLEDGMENTS

We thank most sincerely Dr. Colin F. Timberlake, Bristol, England, who commented on the draft of the entire book and who wrote the foreword; Prof. Raymond Brouillard, Strasbourg, France, who commented on the introductory chapter; Dr. Tibor Fuleki, Vineland, Canada, who commented on the chapter on small fruits; Dr. T.C. Somers, Kingston Park, South Australia, who commented on the chapter on grapes; Prof. P.G. Pifferi, Bologna, Italy, who commented on the chapter on other crops; and Dr. Eiichi Idaka, Gifu, Japan, who commented on the chapter on cole crops.

Many thanks are also due to Sr. Nancy Ann Hutchinson and Mrs. Darcie Hills, who helped with the preparation of portions of the book, and several of our colleagues at Agriculture Canada who read and commented on the drafts of various chapters; in particular, we thank Dr. Walter Dedio, Dr. Dave Oomah, Dr. Sudhir Deshpande, Ms. Audra Davies, and Mr. Brian Rex. They cannot be held responsible for errors or omissions.

We express our gratitude to Enocanossa S.p.A., Ciano d'Enza, Reggio Emilia, Italy, a leading commercial producer of skin grape extract, for the financial assistance which allowed one of us (E.M.) to cover some of the expenses.

The assistance of the Art and Design and Audiovisual Group of Agriculture Canada Research Program Service with the preparation of the illustrations, including the cover page, is also acknowledged.

G.M.
E.M.

ABBREVIATIONS FOR CITED ABSTRACTS

CA	*Chemical Abstracts*
HA	*Horticultural Abstracts*
PBA	*Plant Breeding Abstracts*
CZ	*Chemisches Zentralblatt*
FSTA	*Food Sciences and Technology Abstracts*
JSFA	*Journal of the Science of Food and Agriculture Abstracts*
IAA	*Industries Alimentaires et Agricoles*
BA	*Biological Abstracts*
FCA	*Field Crop Abstracts*

TABLE OF CONTENTS

Chapter 1

INTRODUCTION

I. TYPES OF ANTHOCYANINS

The anthocyanins (Greek *anthos*, flower and *kyanos*, blue) are part of the very large and widespread group of plant constituents known collectively as flavonoids. They are glycosides of polyhydroxy and polymethoxy derivatives of 2-phenylbenzopyrylium or flavylium salts (Figure 1). Differences between individual anthocyanins are the number of hydroxyl groups in the molecule, the degree of methylation of these hydroxyl groups, the nature and number of sugars attached to the molecule and the position of the attachment, and the nature and number of aliphatic or aromatic acids attached to the sugars in the molecule. The known naturally occurring anthocyanidins or aglycones are listed in Table 1. Of these, six occur most frequently in plants. They are pelargonidin, cyanidin, peonidin, delphinidin, petunidin, and malvidin.[1,2] 6-Hydroxycyanidin has been recently found in red flowers of *Alstroemeria* (Alstroemeriaceae) where it occurs as the 3-glucoside and 3-rutinoside.[3] Several other anthocyanidins have been partly described (i.e., carajurin from *Arrabidaea*, carexidin from *Carex*, columnidin from *Columnea*, purpurinidin from *Salix*, and margicassidin from *Cassia*),[2] but these still await complete characterization.

Since each anthocyanidin may be glycosylated and acylated by different sugars and acids, at different positions, the number of anthocyanins is 15 to 20 times greater than the number of anthocyanidins. Most of the known anthocyanins are listed in Table 2. The sugars most commonly bonded to anthocyanidins are glucose, galactose, rhamnose, and arabinose. Di- and trisaccharides, formed by combinations of these four monosaccharides, may also glycosylate some anthocyanidins. The following four classes of anthocyanidin glycosides are the most common: 3-monosides, 3-biosides, 3,5-diglycosides, and 3,7-diglycosides.[1,2] Glycosylation of the 3′-, 4′-, and 5′-hydroxyl group, however, has also been demonstrated.[4-9] In many cases, the sugar residues are acylated by *p*-coumaric, caffeic, ferulic, sinapic, *p*-hydroxybenzoic, malonic, oxalic, malic, succinic, or acetic acids.[1,4-7,9] Methoxyl substituents are found at the 3′ and 5′ positions and, less frequently, at positions 7 and 5.[1,4-7] To our knowledge, however, no natural anthocyanin where all the three hydroxyl groups at the 5, 7, and 4′ positions are substituted at the same time has been reported. A free hydroxyl at one of the 5, 7, or 4′ positions is essential for generating the *in situ* colors responsible for plant pigmentation.[10] This arises mainly from the loss of an acidic hydroxyl hydrogen from the flavylium structure.

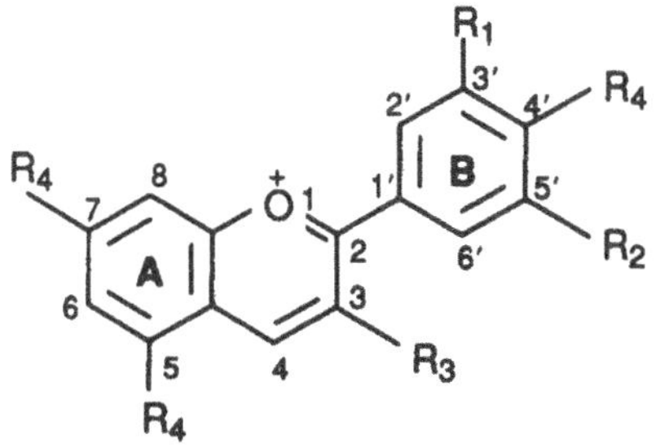

FIGURE 1.   The flavylium cation. R_1 and R_2 are H, OH, or OCH_3; R_3 is a glycosyl or H; and R_4 is OH or a glycosyl.

TABLE 1
Naturally Occurring Anthocyanidins

Name								Color

Name	3	5	6	7	3'	4'	5'	Color
Apigeninidin (Ap)	H	OH	H	OH	H	OH	H	Orange
Aurantinidin (Au)	OH	OH	OH	OH	H	OH	H	Orange
Capensinidin (Cp)	OH	OMe	H	OH	OMe	OH	OMe	Bluish-red
Cyanidin (Cy)	OH	OH	H	OH	OH	OH	H	Orange-red
Delphinidin (Dp)	OH	OH	H	OH	OH	OH	OH	Bluish-red
Europinidin (Eu)	OH	OMe	H	OH	OMe	OH	OH	Bluish-red
Hirsutidin (Hs)	OH	OH	H	OMe	OMe	OH	OMe	Bluish-red
6-Hydroxycyanidin (6 OHCy)	OH	OH	OH	OH	OH	H		Red
Luteolinidin (Lt)	H	OH	H	OH	OH	OH	H	Orange
Malvidin (Mv)	OH	OH	H	OH	OMe	OMe	OMe	Bluish-red
5-Methylcyanidin (5-Mcy)	OH	OMe	H	OH	OH	H		Orange-red
Pelargonidin (Pg)	OH	OH	H	OH	H	OH	H	Orange
Peonidin (Pn)	OH	OH	H	OH	OMe	OH	H	Orange-red
Petunidin (Pt)	OH	OH	H	OH	OMe	OH	OH	Bluish-red
Pulchellidin (Pl)	OH	OMe	H	OH	OH	OH	OH	Bluish-red
Rosinidin (Rs)	OH	OH	H	OMe	OMe	OH	H	Red
Tricetinidin (Tr)	H	OH	H	OH	OH	OH	OH	Red

The largest monomeric anthocyanin known to date is the pigment ternatin A1 from the blue petals of the butterfly pea (*Clitoria ternatea* L. [Leguminosae]). It is a polyacylated derivative of delphinidin 3,3',5'-triglucoside composed of delphinidin with seven molecules of glucose, four molecules of *p*-coumaric acid, and one molecule of malonic acid.[9,11] It has a molecular weight of 2107 and occurs with five other derivatives of delphinidin 3,3',5'-triglucoside called ternatin A2, B1, B2, D1, and D2 (Figure 2).

Other anthocyanins acylated with phenolic and aliphatic acids include malonylawobanin,[12] monardaein,[13] and cinerarin.[14] Anthocyanins acylated with two or more aromatic acids are platyconin, cinerarin, gentiodelphin, and zebrinin.[5,6]

TABLE 2
Known Anthocyanins

Apigeninidin
  5-Glucoside (gesnerin)
  Adientum [apigeninidin (?) glycoside]
  Dryopteris 1 [apigeninidin (?) glycoside]
Aurantinidin
  3-Sophoroside
  3,5-Diglucoside
Capensinidin
  3-Rhamnoside
Cyanidin
  3-Arabinoside
  3-Galactoside
  3-Glucoside
  3-Rhamnoside
  5-Glucoside
  4'-Glucoside
  3-Arabinosylgalactoside
  3-Arabinosylglucoside
  3-Gentiobioside
  3-Lathyroside
  3-Robinobioside
  3-Rutinoside
  3-Sambubioside
  3-Sophoroside
  3-Xylosylarabinoside
  3-Gentiotrioside
  2-(2^G-Glucosylrutinoside)
  3-Rhamnosyldiglucoside
  3-Xylosylglucosylgalactoside
  3-(2^G-Xylosylrutinoside)
  3-Arabinoside-5-glucoside
  3-Rhamnoside-5-glucoside
  3,5-Diglucoside
  3,7-Diglucoside
  3-Glucoside-7-rhamnoside
  3,4'-Diglucoside
  3-Rutinoside-5-glucoside
  3-Sambubioside-5-glucoside
  3-Sophoroside-5-glucoside
  3-Rhamnosylglucoside-7-xyloside
  3-Acetylglucoside
  3-*p*-Coumarylglucoside (hyacinthin)
  3-Caffeylglucoside
  3-*p*-Coumarylgentiobioside
  3-Caffeylgentiobioside
  3-*p*-Coumarylxylosylglucoside
  3-Caffeylglucosylarabinoside
  3-Ferulylglucosylgalactoside
  3-Sinapylglucosylgalactoside

  3-Caffeylrhamnosylglucoside
  3-Caffeylrhamnosyldiglucoside
  3-Ferulylxylglucosylgalactoside
  3,5-Diglucoside malonyl ester
  3,5-Diglucoside-*p*-coumaric acid ester (perillanin)
  3-(6-*p*-Coumarylglucoside)-5-glucoside
  3-(*p*-Coumarylrutinoside)-5-glucoside (cyananin)
  3-Caffeylglucoside-5-glucoside
  3-(*p*-Coumarylsophoroside)-5-glucoside
  3-Malonylsophoroside-5-glucoside
  3-Caffeylsophoroside-5-glucoside
  3-Ferulylsophoroside-5-glucoside
  3-Sinapylsophoroside-5-glucoside
  3-(di-*p*-Coumaryl)sophoroside-5-glucoside
  3-(di-Ferulyl)sophoroside-5-glucoside
  3-(di-Sinapyl)sophoroside-5-glucoside
  3-(*p*-Coumarylcaffeyl)-sophoroside-5-glucoside
  3-Rhamnosylarabinoside
  3,3'-Diglucoside
  3,5-3'-Triglucoside
  3-Rutinoside-3'-glucoside
  3-Rutinoside-5,3'-diglucoside
  3,7,3'-Triglucoside
  3-Glucuronosylglucoside
  3-(Sinapylglucoside)
  3-(Ferulylglucoside)
  3-(*p*-Coumarylferuloylglucoside)
  3-(6''-*p*-Coumarylsophoroside)
  3-(Caffeylsophoroside)
  3-(Dicaffeylsophoroside)
  3-(Sinapylxyloxylglucosylgalactoside)
  3,7,3'-(Caffeylglucoside)
  3-(6''-Malonylglucoside)
  3-(6''-Malylglucoside)
  3-(6''-Oxalylglucoside)
  3-Dimalonylglucoside
  3-Malonylglucuronosylglucoside
  3-(6''-Malonylglucoside)-5-glucoside
  3,5-di(Malonylglucoside)
  3-(6''-Succinylglucoside)-5-glucoside
  3-(*p*-Coumarylglucoside)-5-malonylglucoside
Delphinidin
  3-Arabinoside
  3-Galactoside
  3-Glucoside

TABLE 2 (continued)
Known Anthocyanins

3-Rhamnoside
3-Rutinoside
3-Sambubioside
7-Galactoside
7-Glucoside
3-Glucosylglucoside
3-Lathyroside
3-Rhamnosylgalactoside
3,5-Diglucoside
3-Rhamnoside-5-glucoside
3-Rutinoside-5-glucoside
3-Acetylglucoside
3-*p*-Coumarylglucoside
3-Caffeylglucoside
3-(di-*p*-Coumaryl)glucoside
3,5-Diglucoside malonyl ester
3-(*p*-Coumarylglucoside)-5-glucoside
  (awobanin)
3-(*p*-Coumarylrutinoside)-5-glucoside (delphanin)
3-(*p*-Coumarylsophoroside)-5-glucoside
3-Caffeylglucoside-5-glucoside
3-(di-Caffeylrutinoside)-5-glucoside
3-Rutinoside-5,3′,5′-triglucoside-caffeyl-
  ferulyl-*p*-coumaryl ester
3,7-Diglucoside
3-Rhamnosylglucoside-7-xyloside
3,5,3′-Triglucoside
3,7,3′-Triglucoside
3,3′,5′-Triglucoside
3,5-di(Malonylglucoside)
3-)*p*-Coumarylglucoside)-5-malonylgluco-
  side
Cinerarin
Ternatins A-F
Lobelin A-B
Gentiodelphin
Europinidin
  3-Glucoside
  3-Galactoside
Hirsutidin
  3,5-Diglucoside
6-Hydroxycyanidin
  3-Glucoside
  3-Rutinoside
Luteolinidin
  5-Glucoside
  5-Diglucoside
  Adientum 2 [luteolinidin (?) glycoside]
  Dryopteris 2 [luteolinidin (?) glucoside]

Malvidin
  3-Arabinoside
  3-Galactoside
  3-Glucoside
  3-Rhamnoside
  5-Glucoside
  3-Gentiobioside
  3-Laminaribioside
  3-Rutinoside
  3,5-Diglucoside
  3-Rhamnoside-5-glucoside
  3-Rutinoside-5-glucoside
  3-Gentiotrioside
  3-Arabinosylglucoside-5-glucoside
  3-Sophoroside-5-glucoside
  3-Acetylglucoside
  3-(*p*-Coumarylglucoside)
  3-Caffeylglucoside
  3-(di-*p*-Coumaryl)-glucoside
  3,5-Diglucoside acetyl ester
  3,5-Diglucoside-*p*-coumaric acid ester (ti-
    bouchinin)
  3-(*p*-Coumarylglucoside)-5-glucoside
  3-Caffeylglucoside-5-glucoside
  3-(*p*-Coumarylrutinoside)5-glucoside (ne-
    gretein)
  3-Xyloside-5-glucoside
  3,7-Diglucoside
  3-(*p*-Coumarylsambubioside)-5-glucoside
  3-(di-*p*-Coumarylxyloside)-5-glucoside
Pelargonidin
  3-Galactoside
  3-Glucoside
  3-Rhamnoside
  5-Glucoside
  7-Glucoside
  3-Gentiobioside
  3-Lathyroside
  3-Sambubioside
  3-Sophoroside
  3-Glucosylxyloside
  3-Rhamnosylgalactoside
  3-Rhamnoside-5-glucoside
  3-Robinobioside
  3-Rutinoside
  3-Gentiotrioside
  3-(2^G-Glucosylrutinoside)
  3-Galactoside-5-glucoside
  3,5-Diglucoside
  3,7-Diglucoside

TABLE 2 (continued)
Known Anthocyanins

3-Rutinoside-5-glucoside
3-Sambubioside-5-glucoside
3-Sophoroside-5-glucoside
3-Sophoroside-7-glucoside
3-(4''-*p*-Coumarylrhamnoglucoside)
3,5-Diglucoside malonyl ester
3,5-Diglucoside-*p*-coumaric acid ester (monardein)
3,5-Diglucoside-caffeic acid ester (salvi-anin)
3-(*p*-Coumarylglucoside)-5-glucoside
3-(Caffeylglucoside)5-glucoside
3-(*p*-Coumarylrutinoside)5-glucoside
3-(*p*-Coumarylsophoroside)-5-glucoside
3-(Dicaffeyldiglucoside)-5-glucoside
3-(Ferulylsophoroside)-5-glucoside
3-(*p*-Coumarylcaffeyldiglucoside)-5-gluco-side
3-(*p*-Coumarylferulylsambubioside)-5-glu-coside
3-(Dicaffeyldiglucoside)-5-glucoside
3-(*p*-Coumaryldicaffeyldiglucoside)-5-glu-coside
3-(di-*p*-Hydroxybenzoylrutinoside)-7-glu-coside
3-Diglucoside-5-glucoside-*p*-coumaric acid ester (raphanusin A)
3-Diglucoside-5-glucoside-ferulic acid ester (raphanusin B)
3-Rhamnoglucoside-5-glucoside-*p*-cou-maric acid ester (pelanin)
3-Xylosylglucoside-5-glucoside-*p*-cou-maric-ferulic acid ester (matthiolanin)
3-(6''-Malonylglucoside)
3-(6''-Malylglucoside)
3-Malonylsophoroside
3-(6''-Malonylglucoside)-5-glucoside
3,5-di(Malonylglucoside)
3-(6''-Coumarylglucoside)-5-(4'',6''-di-malonylglucoside)
Peonidin
3-Arabinoside
3-Galactoside
3-Glucoside
3-Rhamnoside
5-Glucoside
3-Gentiobioside
3-Lathyroside
3-Rutinoside
3-Galactoside-5-glucoside
3,5-Diglucoside

3-Rhamnoside-5-glucoside
3-Rutinoside-5-glucoside
3-Gentiotrioside
3-Glucosylrhamnosylglucoside (?)
3-Acetylglucoside
3-*p*-Coumarylglucoside
3-Caffeylglucoside
3-*p*-Coumarylgentiobioside
3-Caffeylgentiobioside
3-(*p*-Coumarylglucoside)-5-glucoside
3-(*p*-Coumarylrutinoside)-5-glucoside (peonanin)
3-(*p*-Coumarylsophorosides)-5-glucoside
3-(di-*p*-Coumaryl)glucoside
3-(di-Caffeylsophoroside)-5-glucoside
3-(*p*-Coumarylcaffeylsophoroside)-5-gluco-side
3-Arabinoside-5-glucoside
3-Sophoroside-5-glucoside-tri(caffeylglucose) ester (HBA)
3,5-(Malonyl-*p*-coumaryldiglucoside)
Petunidin
3-Arabinoside
3-Galactoside
3-Glucoside
3-Rhamnoside
3-Rutinoside
5-Glucoside
3-Gentiobioside
3-Sophoroside
3,5-Diglucoside
3-Rhamnoside-5-glucoside
3-Rutinoside-5-glucoside
3-Gentiotrioside
3-Acetylglucoside
3-(*p*-Coumarylglucoside)
3-Caffeylglucoside
3-(Dicaffeyl)-glucoside
3-(*p*-Coumarylglucoside)-5-glucoside
3-(*p*-Coumarylrutinoside)-5-glucoside (pe-tanin)
3-(di-*p*-Coumaryl)rutinoside (guineesin)
3-(2G-Glucosylrutinoside)-5-glucoside
Pulchellidin
3-Glucoside
Rosinidin
3,5-Diglucoside
Tricetinidin
5-(?) Glucoside
3,5-Diglucoside

TABLE 2 (continued)
Known Anthocyanins

Adapted from Hrazdina, G., *The Flavonoids, Advances in Research*, Chapman & Hall, London,
1982, 135; Harborne, J.B. and Grayer, R.J., Flavonoids, *Advances in Research Since 1980*,
Chapman & Hall, London, 1988, 1; and Kondo, T., Yamashita, J., Kawahari, K., and Goto,
T., *Tetrahedron Lett.*, 30(46), 6055, 1989.

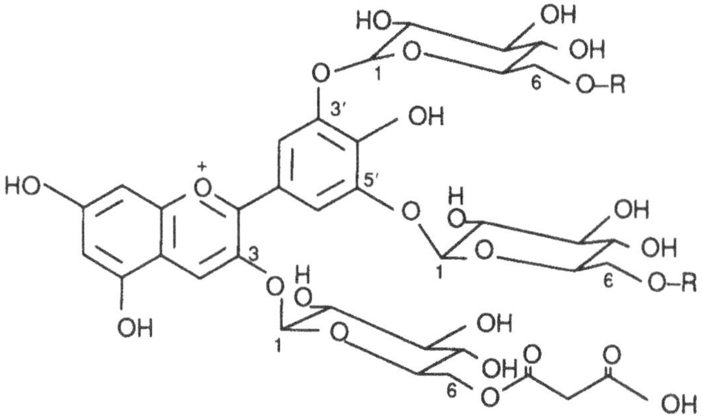

Ternatin	R
T–A1	–CGCG or –CGCG
T–A2	–CGCG or –CG
T–B1	–CGCG or –CGC
T–B2	–CGC or –CG
T–D1	–CGC or –CGC
T–D2	–CGC or –C

C: p–Coumaric acid
G: D–Glc

FIGURE 2. Structures of ternatins from blue petals of butterfly pea (*Clitoria ternatea* L.).
(From Terahara, N., Saito, N., Honda, T., Toki, K., and Osajima, Y., *Phytochemistry*, 29, 949,
1990. With permission. Copyright 1990, Pergamon Press, Ltd.)

II. STRUCTURAL TRANSFORMATIONS

In aqueous media, most of the natural anthocyanins behave like pH indicators, being red at low pH, bluish at intermediate pH, and colorless at
high pH. The nature of the chemical structures which these anthocyanins can
adopt upon changing the pH has been clarified recently.[10,15-19] It has been
demonstrated that in acidic or neutral media, four anthocyanin structures exist
in equilibrium: the flavylium cation AH$^+$, the quinonoidal base A, the carbinol
pseudobase B, and the chalcone C (Figure 3).

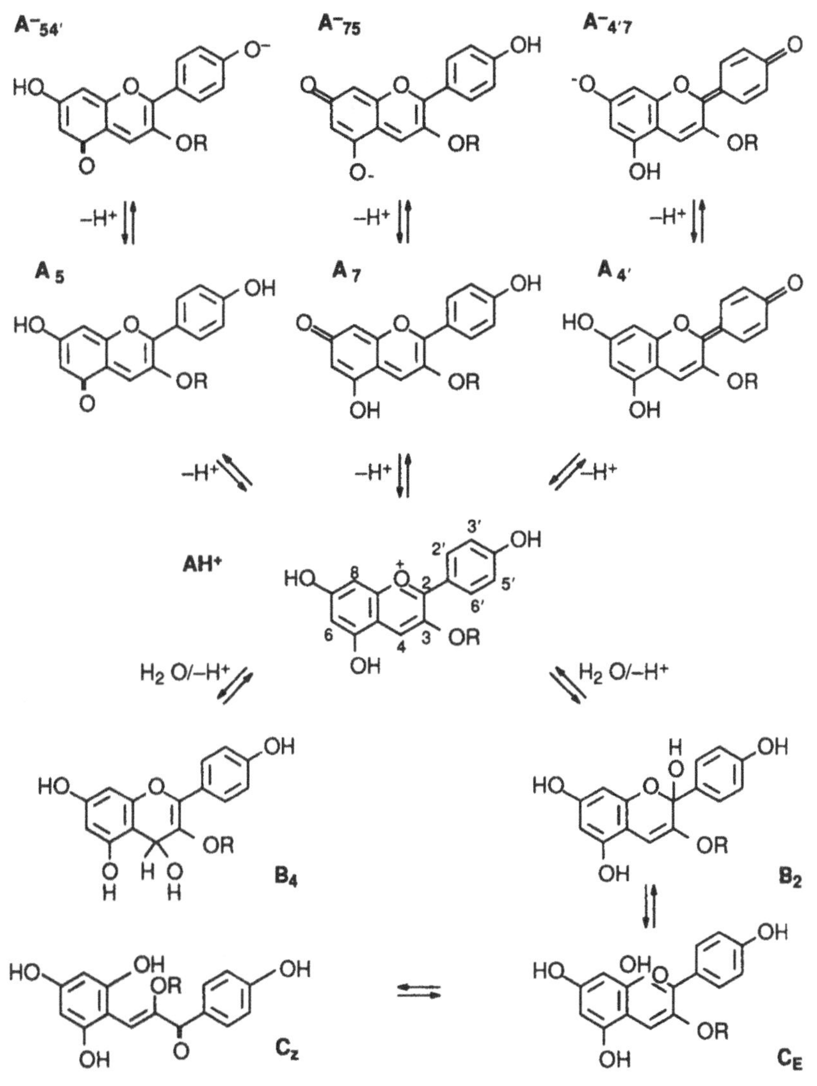

FIGURE 3. Structural transformations of anthocyanins in water. (From Cheminat, A. and Brouillard, R., *Tetrahedron Lett.* 27, 4457, 1986. With permission. Copyright 1986, Pergamon Press, Ltd.)

At pHs below 2, the anthocyanin exists primarily in the form of the red (R_3 = O-sugar) or yellow (R_3 = H) flavylium cation. As the pH is raised, a rapid proton loss occurs to yield the red or blue quinonoidal forms. The quinonoidal forms usually exist as a mixture, as the pK_a of the 4′,7-, and (if present) 5-OH groups are very similar.[10] On standing, a further reaction occurs; that is, hydration of the flavylium cation to give the colorless carbinol

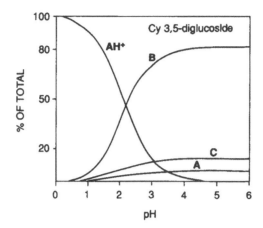

FIGURE 4. Equilibrium distribution of AH⁺, A, B, and C forms for cyanidin 3,5-diglucoside as a function of pH. (Reprinted with permission from Mazza, G. and Brouillard, R., *J. Agric. Food Chem.*, 35, 422, 1987. Copyright 1987, American Chemical Society.)

or pseudobase. This, in turn, equilibrates to the open chalcone forms, which are also colorless. The relative amounts of cation, quinonoidal forms, pseudobases, and chalcones at equilibrium vary with both pH and structure of the anthocyanin. For cyanidin 3,5-diglucoside (Figure 4), for instance, the red cation (AH⁺) is the sole structure only when the pH of the solution is less than 0.5. With increasing pH its concentration decreases as hydration to the colorless pseudobases occurs, the equilibrium being characterized by a pK_h value of 2.23 ± 0.10 when equal amounts of both forms exist. At this pH, however, small amounts of the colorless chalcones and the blue quinonoidal bases are also present, and the proportions of these and the carbinol forms increase with increasing pH at the expense of the red cationic form up to about pH 4.5. Between pH 4 and 6 very little color remains in the anthocyanin since the amounts of the colored forms AH⁺ and A are very small. The equilibrium between these species is characterized by pK_a value of 3.38 ± 0.15.

By varying the substitution pattern of the flavylium ring, anthocyanidins can be prepared that can exist primarily in the colored quinonoidal form A. This is exemplified by 4-methyl-7-hydroxyflavylium (Figure 5) and 4'-methoxy-4-carboxyl-7-hydroxyflavylium chloride (Figure 6). As can be noted, these compounds are essentially stable in the form of flavylium cation in fast equilibrium with the quinonoidal base. At equilibrium and in nearly neutral solutions (pH 6), the colored quinonoidal forms account for about 83 and 67% of the analytical concentration of the 4-methyl-7-hydroxyflavylium and 4'-methoxy-4-carboxyl-7-hydroxyflavylium chloride, respectively.

The flavylium cation of natural anthocyanins behaves as a weak acid, whereas a neutral quinonoidal base is, at the same time, a weak acid and a

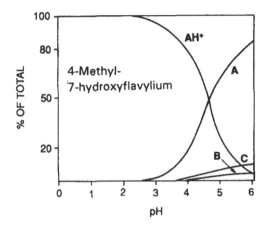

FIGURE 5. Equilibrium distribution of 4-methyl-7-hydroxy-flavylium chloride as a function of pH. (Reprinted with permission from Mazza, G. and Brouillard, R., *J. Agric. Food Chem.*, 35, 422, 1987. Copyright 1987, American Chemical Society.)

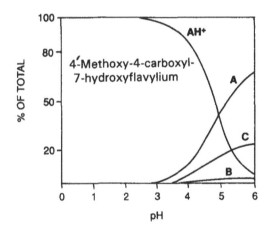

FIGURE 6. Equilibrium distribution of 4'-methoxy-4-carboxyl-7-hydroxyflavylium chloride as a function of pH. (Reprinted with permission from Mazza, G. and Brouillard, R., *J. Agric. Food Chem.*, 35, 422, 1987. Copyright 1987, American Chemical Society.)

weak base.[21] Loss of a proton can occur at any of the hydroxyl groups at C-4', C-5, or C-7. If two acidic hydroxyls are present in the flavylium cation, an ionized quinonoidal base is formed at pH values above 6. Ionized quinonoidal bases, shown on top of Figure 3, give rise to large bathochromic and hyperchromic shifts, and contribute to the diversity of fruit and flower color.

Hydration of the flavylium cation of most anthocyanins to the colorless carbinol pseudobases occurs at pH values ranging from 3 to 6. Water addition takes place at the C-2 (B_2) and to a lesser extent at the C-4 (B_4). The mechanism associated with this reaction involves proton transfer and C–O bond breaking or forming occurring at the same time. In the absence of a 3-glycosyl substituent, the hydration is less efficient and carbinol pseudobases only form at pH values higher than 4 to 5.[21]

The *cis*- (C_E) and *trans*-chalcones (C_Z), shown at the bottom of Figure 3, are formed from the carbinol pseudobase by a fast ring-opening reaction and a slow isomerization process.[18,22] Both chalcones differ from the usual chalcones[23] by having their carbonyl function next to the B-ring, whereas in the normal chalcones the carbonyl group is adjacent to the A-ring. Using nuclear magnetic resonance (NMR) spectroscopy and malvidin 3-glucoside, Cheminat and Brouillard[19] were able to detect a small amount of the water 4-adduct. At sufficiently high pH values, an anionic *cis*-chalcone appears. This slightly yellow species is believed to be in rapid equilibrium with the anionic water 2-adduct, a structure not yet identified. Both structures can also be regarded as canonical forms of a resonance hybrid, in which case identification would be impossible.[21]

III. COLOR STABILIZATION AND INTENSIFICATION

Color of anthocyanin-containing media depends on structure and concentration of the pigment, pH, temperature, presence of copigments, metallic ions, enzymes, oxygen, ascorbic acid, sugars and their degradation products, sulfur dioxide, and other factors.

A. STRUCTURE

Hydroxyl groups, methoxyl groups, sugars, and acylated sugars have a marked effect on color intensity and stability of anthocyanins.[1,10,20] As the number of hydroxyl groups on the B-ring is increased, the visible absorption maximum of the anthocyanidin is shifted to longer wavelengths, and the color changes from orange to blue; for instance, λ_{max} in 0.01% HCl-MeOH solutions are pelargonidin 520 nm (orange), cyanidin 535 nm (orange-red), and delphinidin 545 (bluish-red). Methoxyl groups replacing hydroxyl groups reverse the trend. The hydroxyl group at C3 is particularly significant because it shifts the color of the pigment from yellow-orange to red. This is exemplified by the difference in color between the majority of the anthocyanins, which are red; and the 3-deoxyanthocyanidins, apigeninidin, luteolinidin, and tricetinidin, which are yellow. The same hydroxyl, however, destabilizes the molecule, as illustrated by the fact that the 3-deoxyanthocyanidins are much more stable than the other anthocyanidins.[20,24]

Similarly, the presence of a hydroxyl group at C5 and substitution at C4 both stabilize the colored forms through the arresting of hydration reactions

TABLE 3
Effect of Structure on the Stability of Selected
Pigments at 25°C in 0.01 *M* Citric Acid, pH 2.8

Compound	λ_{max} (nm)	Half-life (d)
3-Hydroxy-4',5,5',7-tetramethoxyflavylium	512	0.04
3,4',5.5',7-Pentahydroxyflavylium	512	0.5
3,4',5.5',7-Pentamethoxyflavylium	512	6
3-Rutinose-4',5,5',7-tetramethoxyflavylium	512	12
3-Rutinose-4',5,5'7-tetrahydroxyflavylium	512	65
4',5.5',7-Tetramethoxyflavylium	488	170
4',5,7-Trihydroxyflavylium	547	400
4',7-Dihydroxyflavylium	458	400
4'-Hydroxyflavylium	436	400
4'-Methoxyflavylium	437	35
7-Hydroxyflavylium	428	300
7-Methoxyflavylium	427	8

Adapted from Iacobucci, G.A. and Sweeny, J.G., *Tetrahedron*, 39(19), 3005, 1983.

which lead to the formation of colorless species.[10,20,24] At a given pH, anthocyanin 3-glycosides are more colored than 3,5- and 5-glucosides. Optical density comparisons, carried out by many investigators,[1,2] show that 3,5- and 5-glycosides have only 50% of the absorption at 440 nm (when compared with the color maximum) as do the 3-glycosides and the free anthocyanidins. Similarly, the intensity of the shortwave maximum is lower in the case of the 3,5-diglucosides than with 3-glucosides. Glycosylation also affects the stability of the pigment. For instance, the half-life (50% reduction in absorbance at λ_{max}) of a typical anthocyanin, cyanidin 3-rutinoside, is about 65 d at room temperature in 0.01 *M* citric acid, pH 2.8. The corresponding free anthocyanidin, however, has a half-life of only 12 h.[24] Similarly, at pH 2.5 and 4.5, the stability of peonidin and malvidin is significantly lower than their corresponding 3-glucosides.[25] The slow hydrolysis of the 3-*O*-sugar unit of anthocyanins under acidic conditions is supposedly responsible for the higher stability of these pigments. Attempts to improve the stability by methylation of free phenolic hydroxyl groups have been found to reduce stability instead. The presence of either a 4'-OH or a 7-OH in the molecule significantly stabilizes the pigment while methylation of these hydroxyls decreases it (Table 3).

Anthocyanins containing two or more aromatic acyl groups (such as ternatins, platyconin, cinerarin, gentiodelphin, and zebrinin[6,7,9]) are stable in neutral or weakly acidic media, possibly as a result of hydrogen bonding between phenolic hydroxyl groups in anthocyanidins and aromatic acids. Brouillard[10,26] and Goto et al.[27,28] observed that diacylated anthocyanins are stabilized by the sandwich-type stacking caused by hydrophobic interaction between the anthocyanidin ring and the two aromatic acyl groups.

B. CONCENTRATION

The amount of anthocyanin in plant tissues may vary several fold. For instance, Green and Mazza[29] found that the concentration of cyanidin 3-galactoside and 3-glucoside in most cultivars of saskatoon berries (*Amelanchier alnifolia*) increased from about 60 mg/100 g in red berries to over 160 mg/100 g in dark purple berries. Similarly, in developing 'Montmorency' sour cherries the total anthocyanin content increased continuously during ripening from 2 to 43.6 mg/100 g fresh weight (FW).[30] Oydvin[31] reported a doubling of the total anthocyanin content during ripening of the red currant from the pink stage to the dark, fully ripe stage. For this fruit, the increase in concentration of anthocyanins was accompanied by a change in the glycoside pattern of the only aglycone present, cyanidin. In 'Meeker' raspberry, the total anthocyanin content increased fourfold in the ripe fruit. The relative proportion of the four major pigments, which were all found in unripe fruit, did not vary appreciably with maturity.[32] Increased concentration of anthocyanin(s) in these and other plant tissues intensifies their color and may enhance color stability through the phenomena of intermolecular copigmentation and self-association.

C. COPIGMENTATION

1. Intermolecular Copigmentation

Intermolecular copigmentation of anthocyanins with other flavonoids and related compounds produces an increase in color intensity (hyperchromic effect) and a shift in the wavelength of maximum absorbance toward higher wavelengths (bathochromic shift), giving purple to blue colors.[33-38] Molecules acting as copigments include a large variety of compounds, such as flavonoids, polyphenols, alkaloids, amino acids, organic acids . . . , and the anthocyanins themselves. Of these substances, however, only a few components have been investigated in some detail.[34-39] Colorless flavonoids and polyphenols are frequently found in association with anthocyanins in the vacuoles of the colored cells of higher plant organs.[4,40-42] Therefore, the copigmentation phenomenon is widespread in nature. It also occurs in fruits and vegetable products such as juices and wines.[43]

Copigmentation effect has been shown to be a molecular interaction occurring between the anthocyanin colored forms and the copigment.[38,44] The main role of a copigment is that of controlling the extent of the hydration reaction between the flavylium cation and the colorless carbinol pseudobases (Figure 3).

The intensity of the copigmentation effect has been shown to be dependent upon several factors including type and concentration of anthocyanin type and concentration of copigment, and pH and temperature of medium and of solvent.[34,37-39,44] The most efficient copigments so far discovered are flavonols, aureusidin (an aurone), and particularly *C*-glycosyl flavones such as swertisin.[34,45] Flavonolsulfonic acids are also good copigments, presumably due to the added attraction of the negative charge of the sulfonic acid groups to the flavylium cation.[46]

Under identical conditions of pH, pigment and copigment concentrations, temperature, ionic strength, and solvent, the copigment effect increases with the degree of methoxylation and glycosylation of the anthocyanin (Table 4).[38] Thus, at the copigment/pigment molar ratio of 40 (pigment concentration, 2.58×10^{-4} *M*; pH, 3.65; temperature, 20°) for instance, the incrases in absorbance at λ_{max} were 4.2-fold for malvin, 2.6-fold for cyanin, 2.5-fold for malvidin 3-glucoside, and 1.7-fold for cyanidin 3-glucoside, respectively. Anthocyanins containing an aromatic acyl group form much more stable pigment-copigment complexes with *C*-glucosylflavones than unacylated anthocyanins.[6,47] Color intensification by copigmentation increases with increasing anthocyanin concentration and increasing ratio of copigment to anthocyanins (Table 4).[38] Increasing temperature strongly reduces the color-intensifying effect produced by addition of copigment to solutions of anthocyanins (Table 5).[38] At the copigment/pigment molar ratio of 20, pigment concentration of 7.73×10^{-4} *M*, pH 3.65, and 20°, for instance, the absorbance of malvin solutions is 2.884 A/cm (Table 5). Raising the temperature of these solutions to 30, 40, 50, 60, and 70° decreases their absorbance to 2.308, 1.849, 1.469, 1.215, and 1.047 A/cm, respectively. The decreases in absorbance are larger at low temperatures than at elevated temperatures. Copigmentation occurs from pH values close to 1 to neutrality.[38,44,48-50] The pH value for maximum effect is about 3.5 and may vary slightly depending on the pigment-copigment system (Table 4).[38,49,50]

2. Intramolecular Copigmentation

Intramolecular copigmentation is responsible for the color stability of anthocyanins containing two or more aromatic acyl groups. The stability of these anthocyanins in weakly acidic or neutral solutions is attributed to a sandwich-type stacking of the aromatic residue of acyl groups with the pyrylium ring of the flavylium cation thereby decreasing hydration at C-2 and C-4 positions.[6,10,21] Anthocyanins exhibiting this unique property include ternatins,[9,11,51] platyconin,[52] lobelinin,[7] cinerarin,[13,53,54] heavenly blue anthocyanin,[6,35,55] gentiodelphin,[56] and zebrinin.[57,58] Of these, however, the ternatins A, B, and D are reportedly the most stable in neutral aqueous solution (Figure 7). Color stability appears to increase with increasing content of organic acids (cinnamic and malonic acids) and also increased substitution of the aglycone (delphinidin more stable than cyanidin or peonidin derivatives).[51] Therefore, cinerarin, a delphinidin-based anthocyanin containing two caffeic acids and one malonic acid, is more stable than rubrocinerarin, a cyanidin-based pigment also with two caffeic acids and one malonic acid. The effect

TABLE 4
Effect of pH, Anthocyanin and Chlorogenic Acid Concentration, and Type of Anthocyanin of λ_{max} and Absorbance at Visible λ_{max} and at 525 nm of Anthocyanin-Chlorogenic Acid Solutions

Pigment	Pigment conc (M)	Copigment/pigment molar ratio	pH	λ_{max}	Absorbance λ_{max}	λ_{525}	$(A-A_o)^a / A_o$	n^b
Cyanidin 3,5-diglucoside	2.58×10^{-4}	0	2.74	509.2	1.625	1.403	0.0	0.75
		5	2.74	512.0	1.858	1.677	0.20	
		10	2.73	513.6	1.987	1.849	0.32	
		20	2.73	516.8	2.287	2.201	0.57	
		40	2.75	521.2	2.736	2.700	0.92	
Cyanidin 3,5-diglucoside	2.58×10^{-4}	0	3.64	510.8	0.329	0.284	0.0	0.99
		1	3.62	512.8	0.347	0.300	0.05	
		5	3.62	514.4	0.370	0.338	0.19	
		10	3.62	515.2	0.449	0.420	0.47	
		20	3.61	518.8	0.580	0.570	1.00	
		40	3.62	523.2	0.847	0.844	1.96	
		80	3.62	528.0	1.371	1.365	3.79	
		150	3.64	531.6	2.069	2.023	6.10	
Cyanidin 3,5-diglucoside	2.58×10^{-4}	0	4.72	521.6	0.110	0.098	0.0	0.95
		5	4.70	526.0	0.126	0.119	0.21	
		10	4.71	526.4	0.138	0.127	0.30	
		20	4.73	528.8	0.160	0.155	0.58	
		40	4.71	530.0	0.229	0.225	1.30	
		80	4.72	532.4	0.338	0.330	2.37	
		150	4.69	535.2	0.603	0.578	4.89	
		200	4.72	537.2	0.710	0.675	5.89	

Compound	Concentration							
Cyanidin 3,5-diglucoside	2.58 × 10⁻⁴	0	5.74	528.8	0.097	0.090	0.0	1.06
		5	5.77	530.4	0.103	0.097	0.08	
		10	5.72	531.6	0.111	0.105	0.17	
		20	5.76	532.4	0.136	0.135	0.50	
		40	5.73	538.8	0.174	0.176	0.80	
		80	5.72	539.2	0.261	0.244	1.71	
		150	5.75	542.8	0.420	0.375	3.17	
		200	5.74	544.4	0.514	0.458	4.09	
Cyanidin 3,5-diglucoside	7.73 × 10⁻⁴	0	3.70	514.4	0.953	0.881	0.0	1.02
		1	3.70	514.4	1.045	0.971	0.10	
		5	3.70	516.8	1.351	1.305	0.48	
		10	3.69	520.8	1.821	1.793	1.04	
		20	3.69	524.8	2.829	2.813	2.19	
Cyanidin 3,5-diglucoside	8.58 × 10⁻⁵	0	2.74	511.2	0.090	0.077	0.0	1.01
		5	3.65	511.2	0.097	0.085	0.10	
		10	3.66	515.6	0.102	0.093	0.20	
		20	3.65	515.6	0.117	0.108	0.40	
		40	3.67	517.6	0.140	0.135	0.75	
		80	3.66	521.0	0.203	0.200	1.60	
		150	3.64	524.0	0.320	0.320	3.16	
		200	3.66	526.0	0.387	0.387	4.03	
Malvidin 3,5-diglucoside	2.58 × 10⁻⁴	0	3.66	523.6	0.155	0.139	0.0	1.09
		5	3.65	525.6	0.196	0.180	0.29	
		10	3.68	528.0	0.246	0.229	0.65	
		20	3.67	532.4	0.360	0.345	1.48	
		40	3.66	536.4	0.637	0.593	3.27	
		80	3.66	542.8	1.195	1.069	6.69	
		150	3.66	546.4	2.049	1.759	11.65	
Malvidin 3,5-diglucoside	7.73 × 10⁻⁴	0	3.66	524.0	0.550	0.548	0.0	1.00
		1	3.65	526.4	0.656	0.653	0.19	
		5	3.65	529.6	1.064	1.043	0.90	

TABLE 4 (continued)
Effect of pH, Anthocyanin and Chlorogenic Acid Concentration, and Type of Anthocyanin of λ_{max} and Absorbance at Visible λ_{max} and at 525 nm of Anthocyanin-Chlorogenic Acid Solutions

Pigment	Pigment conc (*M*)	Copigment/pigment molar ratio	pH	λ_{max}	A_{max}	A_{525}	$\dfrac{(A-A_o)}{A_o}$ [a]	n [b]
		10	3.67	535.2	1.650	1.583	1.89	
		20	3.66	541.2	2.884	2.685	3.90	
Cyanidin 3-glucoside	2.58 × 10^{-4}	0	3.61	510.0	1.445	1.301	0.0	0.72
		4	3.59	513.2	1.621	1.538	0.18	
		10	3.62	515.2	1.758	1.680	0.29	
		20	3.62	519.2	2.090	2.055	0.58	
		40	3.64	525.2	2.480	2.468	0.90	
Malvidin 3-glucoside	2.58 × 10^{-4}	0	3.68	520.8	0.937	0.915	0.0	0.74
		1	3.64	522.4	1.023	1.009	0.10	
		5	3.65	525.6	1.180	1.176	0.28	
		10	3.63	527.2	1.496	1.485	0.62	
		20	3.65	531.2	1.816	1.775	0.94	
		40	3.65	537.6	2.347	2.235	1.44	

Note: Solvent: aqueous H_3PO_4-NaOAc buffer; $l = 1$ cm; ionic strength $= 0.20$ *M*; T $= 20°$.

a   A and A_o are absorbance of anthocyanin solutions at 525 nm in the presence and absence of chlorogenic acid.

b   n is the stoichiometric constant, and corresponds to the slope of line obtained when plotting ln $(A-A_o)/A_o$ vs. ln$[CP]_o$.

From Mazza, G. and Brouillard, R., *Phytochemistry*, 29, 1097, 1990. With permission. Copyright 1990, Pergamon Press, Ltd.

TABLE 5
Effect of Temperature on Visible Maxima and Absorbances of Malvidin 3,5-Diglucoside and Cyanidin 3,5-Diglucoside-Chlorogenic Acid Solutions

Pigment	Pigment conc (M)	Copigment/pigment molar ratio	Temp (°)	Absorbance			$(A-A_o)^a$
				λ_{max}	λ_{max}	λ_{525}	A_o
Malvidin	7.73×10^{-4}	0	20	524.0	0.550	0.550	0.0
3,5-diglucoside		20	20	541.2	2.884	2.704	3.92
			30	537.6	2.308	2.190	2.98
			40	535.2	1.849	1.778	2.23
			50	532.8	1.469	1.436	1.61
			60	532	1.215	1.178	1.14
			70	530	1.047	1.027	0.87
Malvidin	7.73×10^{-4}	0	20	524.0	0.501	0.501	0.0
3,5-diglucoside		10	20	543.0	1.573	1.500	1.99
			30	534.0	1.290	1.253	1.50
			40	532.0	1.075	1.050	1.10
			50	529.6	0.931	0.915	0.83
			60	529.6	0.830	0.818	0.63
			70	528.4	0.753	0.743	0.48
Malvidin	7.73×10^{-4}	0	20	524.0	0.510	0.508	0.0
3,5-diglucoside		2	20	526.8	0.714	0.714	0.41
			30	526.4	0.662	0.662	0.30
			40	524.8	0.623	0.623	0.23
			50	524.4	0.596	0.596	0.17
			60	524.4	0.575	0.575	0.13
Cyanidin	7.72×10^{-4}	0	20	514.0	0.963	0.900	0.0
3,5-diglucoside		20	20	526.4	2.889	2.865	2.18
			30	526.0	2.477	2.477	1.75
			40	524.4	2.129	2.119	1.35
			50	522.4	1.827	1.819	1.02
			60	522.0	1.594	1.583	0.76
			70	520.4	1.388	1.373	0.53
			80	518.0	1.256	1.238	0.38

Note: Solvent: aqueous H_3PO_4-NaOAc buffer; $1 = 1$ cm; ionic strength $= 0.20$ M; pH 3.65.

[a] A and A_o are absorbances of anthocyanin solutions at 525 nm in the presence and absence of chlorogenic acid.

From Mazza, G. and Brouillard, R., *Phytochemistry*, 29, 1097, 1990. With permission. Copyright 1990, Pergamon Press, Ltd.

of an additional organic acid appears, however, more significant than the aglycone structure. Thus, rubrocinerarin, which possesses one organic acid (malonic acid) more than platyconin and gentiodelphin (both delphinidin based with two caffeic acids), is more stable than the two latter pigments, despite being cyanidin based. On the other hand, ternatin D appears more stable than ternatins A and B, even though all of these pigments are delphinidin based

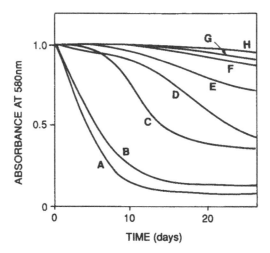

FIGURE 7. Stability of acylated anthocyanins in buffer solution (pH 6.95, ca. 30 mg/l, path length 10 mm, ca. 15 to 20°). (A) platyconin; (B) gentiodelphin; (C) rubrocinerarin; (D) cinerarin; (E) *Ipomea* anthocyanin; (F) ternatin B; (G) ternatin A; (H) ternatin D. (From Saito, N., Abe, K., Honda, T., Timberlake, C.F., and Bridle, P., *Phytochemistry*, 24, 1583, 1985. With permission. Copyright 1985, Pergamon Press, Ltd.)

with four *p*-coumaric acids and one malonic acid (see Figure 2). This indicates that the occurrence of the *p*-coumaric acid at the end of the GCGC side chains in ternatin D may enhance color stability. Therefore, as also pointed out by Brouillard,[10,21] Saito et al.,[51] and others,[27,28] polyacylation does not seem to be sufficient for intramolecular copigmentation to occur. The structure of the aglycone; the nature, number, and position of attachment of the acyl group(s) to the sugar; and the structure of the sugar and its location all seem to play a role. This renders the current explanation of the color stability of some diacylated and polyacylated anthocyanins not fully satisfactory.

D. SELF-ASSOCIATION

Self-association occurs when the color intensity of the anthocyanin increases more than linearly with an increase in pigment concentration. This deviation from Beer-Lamber's law was first reported by Asen et al.,[34] who observed that at pH 3.16 the absorbance of cyanidin 3,5-diglucoside at λ_{max} increased 300-fold when the concentration was increased from 10^{-4} to 10^{-2} *M* or 100-fold. Malvidin 3-glucoside shows similar behavior at pH 3.5 and 20°C.[59] Self-association also occurs in concentrated "neutral" solutions and is affected by the type and concentration of anthocyanin.[60-62] The exact nature of the complex formed and self-association of anthocyanins at pH 3 to 4 remains unknown, but the studies by Hoshino et al.[60-62] at pH 7 have provided strong evidence that a vertical stacking of the anthocyanin quinonoidal bases is occurring at this pH. The driving force of this stacking is attributed to the hydrophobic interaction between the aromatic nuclei rather than to hydrogen bonding.[6] It is not clear whether self-association actually leads to color stabilization, because in equilibrated solutions of pH 7, only 3 to 5% of the pigment remains in the colored form.[61] Nonetheless, a small loss in analytical pigment concentration can result in a proportionately great loss of color, especially at high anthocyanin concentrations.

E. METAL-COMPLEXING

Stable anthocyanin-metal complexes have been reported by Salt and Thomas[63] for tin, Somaatmadja et al.[64] for copper, and Jurd and Asen[65] for aluminum. Other authors have reported the presence of metal ions (Mg, Fe, K) in the structure of stable metallic anthocyanins.[6,66,67] Goto et al.[68] reported that anthocyanin quinonoidal forms as well as flavylium ions are strongly stabilized and do not form colorless pseudobases by hydration when dissolved in concentrated aqueous solutions of neutral salts such as magnesium chloride and sodium chloride. The color stabilization in NaCl solution may be due to the promotion of self-association of anthocyanins and the stabilization of $MgCl_2$ to a reduction in concentration of free water by hydration of magnesium ions.[6,61]

F. pH

pH has a marked influence on the color of anthocyanins in aqueous medias.[10,20] Their structure and thus their color and color stability vary with pH. The structural changes that occur with change of pH are shown in Figures 3 to 6 and were discussed previously under structural transformations. Below pH 2, anthocyanin solutions display their most red ($R_3 = O$-sugar) or yellow ($R_3 = pH$) coloration. As the pH of such solutions is increased their color fades, and at pH 4 to 6 most anthocyanins appear colorless. Further increases in pH give rise to solutions that are purple and blue; these, on standing or heat treatment, may change to yellow. Acidification will reverse the change although sufficient time for reequilibration is required. This time lag for conversion of chalcone to carbinol pseudobase has been exploited by Preston and Timberlake[69] to isolate malvidin 3-glucoside chalcone and malvin 3,5-diglucoside chalcone for the first time by high-performance liquid chromatography (HPLC).

G. OTHER FACTORS

In addition to pH, structure, concentration, copigmentation and metal complexing, intensity and stability of color of anthocyanins are influenced by several other factors such as temperature, light, oxygen, acetaldehyde, ascorbic acid, sugars and their degradation products, sulfur dioxide, amino acids, catechin, and some others. Reviews of literature discussing these factors have been published by Markakis,[70] Jackman et al.,[71] and Francis.[72]

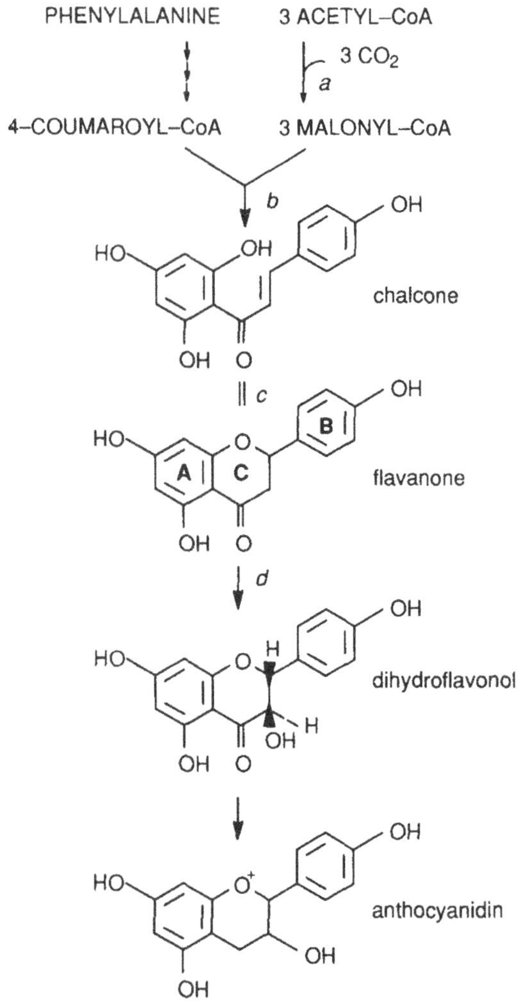

FIGURE 8. Biosynthetic pathway of anthocyanins: (a) acetyl-CoA carboxylase; (b) chalcone synthase; (c) chalcone isomerase; (d) flavanone 3-hydroxylase.

IV. BIOSYNTHESIS

Biosynthesis of anthocyanins is discussed in recent comprehensive reviews by Wong,[73] Ebel and Hahlbroch,[74] Grisebach,[75] Heller and Forkmann,[76] and Stafford.[77] It proceeds by the pathway chalcone → flavanone → dihydroflavonol → anthocyanidin → anthocyanin (Figure 8). The chalcone is synthesized in the plant from three molecules of malonyl-CoA and a suitable hydroxycinnamic acid CoA ester, ordinarily of 4-coumaroyl-CoA, by action of the enzyme chalcone synthase. The C_{15} chalcone is then isomerized into flavanone by a chalcone isomerase. Two types of chalcone isomerase are known, and their difference is based on their substrate specificity. In plants containing 5-deoxy- and 5-hydroxyflavonoids, the enzyme is able to isomerize both 2',4'-dihydroxy-(resorcinol-type) and 2',4',6'-trihydroxy-(phloroglucinol-type) chalcones. In plants devoid of 5-deoxyflavonoids, only phloroglucinol-type chalcones are substrate for the enzyme.[76,77] Dihydroflavonol is formed from flavanone by the action of flavanone 3-hydroxylase. Evidence for dihydroflavonols as intermediates in anthocyanin biosynthesis comes from experiments with inhibitors of phenylalanine ammonia-lyase (PAL)[76] and from supplementation experiments with genetically defined acyanic lines of *Matthiola incana*,[78] *Petunia hybrida*,[79,80] and *Antirrhinum majus*[81] and with aleurone tissue of maize endosperm (*Zea mays*).[81] Enzymes for the conversion of dihydroflavonols into anthocyanidin are still unknown. Leucoanthocyanidins (monomeric flavan-3,4-diols) may also be intermediates in anthocyanin biosynthesis. Evidence for this has come from supplementation experiments on genetically defined white-flowering lines of *Matthiola*.[82]

It has not yet been clearly established whether the different structures of anthocyanidins are determined at the cinnamic acid stage or at the stage of C_{15} intermediate. If the former possibility is correct, it means that 4-coumarate is the precursor of pelargonidin, caffeate is that of cyanidin, ferulate is that of peonidin, and so on; in the latter possibility, 4-coumarate would be the precursor of all anthocyanidins, and introduction of additional hydroxyl groups or methylation reactions would occur at the chalcone/flavanone stage or later in the pathway. Most accumulated evidence favors the latter route.[76,77] Both possibilities can also be envisaged for introduction of hydroxyl groups in the case of cyanidin, but the situation is less clear with methylated anthocyanidins such as peonidin and petunidin.[76] Biochemical and genetic studies of anthocyanin methylation, such as cyanidin to peonidin and delphinidin to petunidin and malvidin, reveal at least four methyltransferases, each being controlled by a separate gene.[76,77] Glycosylation is probably the last step in anthocyanin biosynthesis; however, according to Grisebach,[75] the possibility of conversion of dihydroflavonol 3-glucoside to anthocyanidin 3-glucoside exists. The enzymes involved in the glycosylation reactions are glucosyltransferases. The known enzymes involved in the acylation reactions include *O*-malonyltransferase and acyltransferase.[77]

V. ANALYTICAL METHODS

Analytical methods for anthocyanins have been reviewed in detail by Francis,[83] Jackman et al.,[84] Gross,[85] and Strack and Wray.[86] The analysis of anthocyanins is complicated because of their ability to undergo structural transformations and complexation reactions; and they are difficult to measure independently of other flavonoids because they have similar structural and reactivity characteristics. In addition, pure standard compounds are not easily available.

Qualitative analysis generally involves extraction with weakly acidified alcoholic solvent, followed by concentration under vacuum, and purification and separation of the pigments. Paper and column chromatography have been widely used for the purification and separation of anthocyanins. Paper chromatography on Whatman No. 3 is recommendable with a variety of solvent systems including butanol:acetic acid:water (BAW), chloroform:acetic acid:water (CAW), and butanol:formic acid:water (BFW).[84] Several column supports have been tested, and ion exchange resin, polyamide powders, and gel material appear to be the most useful.[4]

Identification of anthocyanins has normally been carried out by paper and/or thin layer chromatography (TLC), UV/VIS spectroscopy, and controlled hydrolysis and oxidation tests.[83,86] High-performance chromatography of anthocyanins, pioneered during the 1970s, has now become routine in most laboratories for both preparative and quantitative work.[86-91] This has proven to be a powerful tool; in *Vitis labrusca*, for instance, 18 of a 20-anthocyanin mixture could be separated in one 2-h run.[91] Nuclear magnetic resonance (NMR) spectroscopy and fast atom bombardment-mass spectrometry (FAB-MS) are now the most powerful methods for the structural elucidation of anthocyanins. Structures as large as the heavenly blue anthocyanin (Figure 2) have been determined by application of proton NMR spectroscopy methods.[92] FAB-MS has been particularly useful in the characterization of acylated anthocyanins carrying malonyl substitution.[88,93] Direct qualitative and quantitative analysis of unknown mixtures of anthocyanins still remains a difficult, expensive, and time-consuming task. At present, the most satisfactory method for mixture analysis is the multistep method of quantification, separation, and isolation by HPLC and peak identification by FAB-MS and high-field NMR.

VI. FUNCTIONS OF ANTHOCYANINS

The most significant function of anthocyanins is their ability to impart color to the plant or plant product in which they occur.[94] Color is one of the most important attributes of food and beverages. It is appreciated for its intrinsic aesthetic value and as a basis for identification and quality judgment. Food color and color of the environment in which the food is viewed can significantly increase or decrease our desire or appetite for it.

The presence of color in flowers and fruits is believed to ensure fertilization and seed dispersal by animals.[94] In leaves, the presence of anthocyanins is to act as a light screen against damaging UV radiation. In this regard, experiments by Barber[95] with natural stands of *Eucalyptus urnigera* revealed that increasing anthocyanin content in the young leaves was correlated with increasing altitude, and suggested that more anthocyanin was produced at the higher altitudes where the UV radiation from the outer atmosphere is more intense. Anthocyanins have also been associated with resistance to pathogens in species of *Brassica*,[96] sunflower,[97] pea seedlings,[98,99] and maize.[100] Anthocyanins have also been implicated in affecting the growth of *Diplodia maydis* in liquid cultures,[101] as *Rhizobium*-specific markers in lupine nodules,[102] as enhancers of photosynthesis in leaves of tropical rain forest plants,[103] and as regulators of photosynthesis in some woody plant species.[104]

Anthocyanins have also known pharmacological properties and are used by man for therapeutic purposes. Several reviews on the pharmacological and medicinal properties of phenolic compounds have been published.[105-108] Applied orally or by intravenal or intramuscular injections, pharmaceutical preparation of *Vaccinium myrtillus* anthocyanins (VMA) reduce capillary permeability and fragility.[106] This anti-inflammatory activity of VMA accounts for their significant antiedema effect and their action on diabetic microangiopathy.[106,109] It has also been reported that anthocyanins possess antiulcer activity[110] and provide protection against UV radiation.[111]

REFERENCES

1. **Harborne, J.B.,** *Comparative Biochemistry of the Flavonoids,* Academic Press, New York, 1969.
2. **Timberlake, C.F. and Bridle, P.,** The anthocyanins, in *The Flavonoids,* Harborne, J.B., Mabry, T.J., and Mabry, H., Eds., Chapman & Hall, London, 1975, 214.
3. **Saito, N., Yokoi, M., Yarnaji, M., and Honda, T.,** 6-Hydroxyanthocyanidin glycosides in the flowers of *Alstroemeria, Phytochemistry,* 24, 2125, 1985.
4. **Hrazdina, G.,** Anthocyanins, in *The Flavonoids, Advances in Research,* Harborne, J.B. and Mabry, T.J., Eds., Chapman & Hall, London, 1982, 135.
5. **Harborne, J.B. and Grayer, R.J.,** The anthocyanins, in *Flavonoids, Advances in Research Since 1980,* Harborne, J.B., Ed., Chapman & Hall, London, 1988, 1.
6. **Goto, T.,** Structure, stability and color variation of natural anthocyanins, *Prog. Chem. Organ. Nat. Prod.,* 52, 113, 1987.
7. **Kondo, T., Yamashita, J., Kawahari, K., and Goto, T.,** Structure of lobelin A and B, novel anthocyanin acylated with three and four different organic acids, respectively, *Tetrahedron Lett.,* 30, 6055, 1989.
8. **Yoshitama, K. and Abe, K.,** Chromatographic and spectral characterization of 3'glycosylation in anthocyanins, *Phytochemistry,* 16, 591, 1977.
9. **Terahara, N., Saito, N., Honda, T., Kenjiro, T., and Osajima, Y.,** Structure of ternatin A1, the largest ternatin in the major blue anthocyanins from *Clitoria ternatea* flowers, *Tetrahedron Lett.,* 31, 2920, 1990.
10. **Brouillard, R.,** Chemical structure of anthocyanins, in *Anthocyanins as Food Colors,* Markakis, P., Ed., Academic Press, New York, 1982, 1.
11. **Terahara, N., Saito, N., Honda, T., Toki, K., and Osajima, Y.,** Acylated anthocyanins of *Clitoria ternatea* flowers and their acyl moieties, *Phytochemistry,* 29, 949, 1990.
12. **Goto, T., Kondo, T., Tamara, H., and Takena, S.,** Structure of malonylawobanin, the real anthocyanin present in blue colored flower petals of *Commelina communis, Tetrahedron Lett.,* 24, 4863, 1983.

13. **Kondo, T., Nakane, Y., Tamura, H., Goto, T., and Eugester, C. H.,** Structure of monardaein, a bis-malonylated anthocyanin isolated from golden balm, *Monarda didyma, Tetrahedron Lett.,* 26(48), 5879, 1985.

14. **Goto, T., Kondo, T., Kawai, T., and Tamura, H.,** Structure of cinerarin, a tetraacylated anthocyanin isolated from the blue garden cineraria, *Senecio cruentus, Tetrahedron Lett.,* 25(5), 6021, 1984.

15. **Brouillard, R. and Dubois, J. E.,** Mechanism of the structural transformations of anthocyanins in aqueous media, *J. Am. Chem. Soc.,* 99, 1359, 1977.

16. **Brouillard, R. and Delaport, B.,** Chemistry of anthocyanin pigments. II. Kinetic and thermodynamic study of proton transfer, hydration and tautomeric reactions of malvidin 3-glucoside, *J. Am. Chem. Soc.,* 99, 8461, 1977.

17. **Brouillard, R., Delaporte, B., and Dubois, J. E.,** Chemistry of anthocyanin pigments. III. Relaxation amplitudes in pH jump experiments, *J. Am. Chem. Soc.,* 100, 6202, 1978.

18. **Brouillard, R. and Lang, J.,** The hemiacetal-*cis*-chalcone equilibrium of malvin, a natural anthocyanin, *Can. J. Chem.,* 68, 755, 1990.

19. **Cheminat, A. and Brouillard, R.,** PMR investigation of 3-*O*-(β-D-glucosyl) malvidin structural transformations in aqueous solutions, *Tetrahedron Lett.,* 27, 4457, 1986.

20. **Mazza, G. and Brouillard, R.,** Color stability and structural transformations of cyanidin 3,5-diglucoside and four 3-deoxyanthocyanins in aqueous solutions, *J. Agric. Food Chem.,* 35, 422, 1987.

21. **Brouillard, R.,** Flavonoids and flower colour, in *The Flavonoids — Advances in Research,* Harborne, J.B., Ed., Chapman & Hall, London, 1988, 526.

22. **McClelland, R.D., Devine, D.B., and Sorensen, P.E.,** Hemiacetal formation with a phenol nucleophile: simple proton transfers as rate-limiting steps, *J. Am. Chem. Soc.,* 107, 5459, 1985.

23. **Bohm, B. A.,** The minor flavonoids, in *Flavonoids, Advances in Research,* Harborne, J.B. and Mabry, T.J., Eds., Chapman & Hall, London, 1982, 313.

24. **Iacobucci, G.A. and Sweeny, J.G.,** The chemistry of anthocyanins, anthocyanidins and related flavylium salts, *Tetrahedron,* 39(19), 3005, 1983.

25. **Ohta, H., Akuta, S., and Osajima, Y.,** Stabilization of anthocyanin pigments and their utilization. II. Stability of anthocyanin pigments and related compounds in acidic solutions, *Nippon Shokuhin-Kogyo Gakkai-Shi,* 27, 81, 1980 (CA 93, 24763).

26. **Brouillard, R.,** Origin of the exceptional colour stability of the *Zebrina* anthocyanin, *Phytochemistry,* 20, 143, 1981.

27. **Goto, T., Kondo, T., Tamura, H., Imagawa, H., Iino, A., and Takeda, K.,** Structure of gentiodelphin, an acylated anthocyanin isolated from *Gentiana makinoi,* that is stable in dilute aqueous solution, *Tetrahedron Lett.,* 23, 3695, 1982.

28. **Goto, T., Kondo, T., Tamura, H., and Kawahori, K.,** Structure of platyconin, a diacylated anthocyanin isolated from the Chinese bellflower *Platycodon grandiflorum, Tetrahedron Lett.,* 24, 2181, 1983.

29. **Green, R. and Mazza, G.,** Relationships between anthocyanins, total phenolics, carbohydrates, acidity and colour of saskatoon berries, *Can. Inst. Food Sci. Technol. J.,* 19, 107, 1986.

30. **Dekazos, E. D.,** Anthocyanin pigments in red tart cherries, *J. Food Sci.,* 35, 237, 1970.

31. **Øydvin, J.,** Inheritance of four cyanidin 3-glucosides in red currant, *Hortic. Res.,* 14, 1, 1974.

32. **Barrit, B.H. and Torre, L.C.,** Cellulose thin-layer chromatographic separation of *Rubus* fruit anthocyanins, *J. Chromatogr.,* 75, 151, 1973.

33. **Robinson, G.M. and Robinson, R.,** A survey of anthocyanins, *Biochem. J.,* 25, 1687, 1931.

34. **Asen, S., Stewart, R.N., and Norris, K.H.,** Co-pigmentation of anthocyanins in plant tissues and its effect on color, *Phytochemistry,* 11, 1139, 1972.

35. **Asen, S., Stewart, R.N., and Norris, K.H.**, Anthocyanin and pH involved in the colour of 'Heavenly Blue' morning glory, *Phytochemistry*, 16, 1118, 1977.

36. **Scheffeldt, P. and Hrazdina, G.**, Co-pigmentation of anthocyanins under physiological conditions, *J. Food Sci.*, 43, 517, 1978.

37. **Osawa, Y.**, Copigmentation of anthocyanins, in *Anthocyanins as Food Colors*, Markakis, P., Ed., Academic Press, New York, 1982, 41.

38. **Mazza, G. and Brouillard, R.**, The mechanism of co-pigmentation of anthocyanins in aqueous solutions, *Phytochemistry*, 29, 1097, 1990.

39. **Chen, L.J. and Hrazdina, G.**, Structural aspects of anthocyanin-flavonoid complex formation and its role in plant colour, *Phytochemistry*, 20, 297, 1981.

40. **McClure, W.**, The physiology of phenolic compounds in plants, in *Biochemistry of Plant Phenolics*, Swain, T., Harborne, J.B., and Van Sumere, C.F., Eds., *Rec. Adv. Phytochemistry*, 12, Plenum Press, New York, 1979, 525.

41. **Pecket, R.C. and Small, C.J.**, Occurrence, location and development of anthocyanoplasts, *Phytochemistry*, 19, 2571, 1980.

42. **Small, C.J. and Pecket, R.C.**, The ultrastructure of anthocyanoplasts in red cabbage, *Planta*, 154, 97, 1982.

43. **Mazza, G. and Brouillard, R.**, Recent developments in the stabilization of anthocyanins in food products, *Food Chem.*, 25, 207, 1987.

44. **Brouillard, R., Mazza, G., Saad, Z., Albrecht-Gary, A.M., and Cheminat, A.**, The copigmentation reaction of anthocyanins: a microprobe for the structural study of aqueous solutions, *J. Am. Chem. Soc.*, 111, 1604, 1989.

45. **Somers, T.C. and Evans, M.E.**, Spectral evaluation of young wines: anthocyanin equilibria, total phenolics, free and molecular sulfur dioxide 'chemical age', *J. Sci. Food Agric.*, 28, 279, 1977.

46. **Sweeny, J.G., Wilkinson, M.M., and Iacobucci, G.A.**, Effect of flavonoid sulfonates on the photobleaching of anthocyanins in acid solution, *J. Agric. Food Chem.*, 29, 563, 1981.

47. **Hoshino, T., Matsumoto, U., and Goto, T.**, The stabilizing effect of the acyl group on the co-pigmentation of acylated anthocyanins with C-glucosylflavones, *Phytochemistry*, 19, 663, 1980.

48. **Asen, S., Stewart, R.N., Norrish, K.H., and Massie, D.R.**, A stable blue non-metallic copigment complex of delphanin and C-glucosylflavones in Prof. Blaauw iris, *Phytochemistry*, 9, 619, 1970.

49. **Brouillard, R., Wigand, M.C., Dangles, O., and Cheminat, A.**, pH and solvent effects on the copigmentation reaction of malvin by polyphenols, purine and pyrimidine derivatives, *J. Chem. Soc., Perkin Trans. 2*, 8, 1235, 1991.

50. **Williams, M. and Hrazdina, G.**, Anthocyanins as food colorants: effect of pH on the formation of anthocyanin-rutin complexes, *J. Food Sci.*, 44, 66, 1979.

51. **Saito, N., Abe, K., Honda, T., Timberlake, C.F., and Bridle, P.**, Acylated delphinidin glucosides and flavonols from *Clitoria ternatea*, *Phytochemistry*, 24, 1583, 1985.

52. **Saito, N., Osawa, Y., and Hayashi, K.**, Platyconin, a new acylated anthocyanin in Chinese Bell-Flower, *Platycodon grandiflorum*, *Phytochemistry*, 10, 445, 1971.

53. **Yoshitama, K. and Hayashi, K.**, Concerning the structure of cinerarin, a blue anthocyanin from garden *Cineraria*. Studies on anthocyanins. LXVI, *Bot. Mag. (Tokyo)*, 87, 33, 1974.

54. **Yoshitama, K., Hayashi, K., Abe, K., and Kakisawa, H.**, Further evidence of the glycoside structure of cinerarin. Studies on anthocyanins. LXVII, *Bot. Mag. (Tokyo)*, 99, 213, 1975.

55. **Goto, T., Kondo, T., Imagawa, H., Takase, S., Atobe, M., and Miura, I.**, Structure confirmation of tris-deacyl heavenly blue anthocyanin obtained from flower of Morning Glory "Heavenly Blue", *Chem. Lett.*, 883, 1981.

56. **Goto, T., Kondo, T., Tamura, H., Imagawa, H., Iino, A., and Takeda, K.,** Structure of gentiodelphin, an acylated anthocyanin isolated from *Gentiana makinoi,* that is stable in diluted aqueous solution, *Tetrahedron Lett.,* 23(36), 3695, 1982.

57. **Stirton, J.I. and Harborne, J.B.,** Two distinctive anthocyanin patterns in the Commelianaceae, *Biochem. Syst. Ecol.,* 8, 285, 1980.

58. **Idaka, E., Ohashi, Y., Ogawa, T., Kondo, T., and Goto, T.,** Structure of zebrinin, a novel acylated anthocyanin isolated from *Zebrina pendula, Tetrahedron Lett.,* 28, 1901, 1987.

59. **Timberlake, C.F.,** Anthocyanins — occurrence, extraction and chemistry, *Food Chem.,* 5, 69, 1980.

60. **Hoshino, T., Matsumoto, U., and Goto, T.,** Self-association of some anthocyanins in neutral aqueous solution, *Phytochemistry,* 20, 1971, 1981.

61. **Hoshino, T., Matsumoto, U., Goto, T., and Harada, N.,** Evidence for the self-association of anthocyanins. IV. PMR spectroscopic evidence for the vertical stacking of anthocyanin molecules, *Tetrahedron Lett.,* 23, 433, 1982.

62. **Hoshino, T.,** An approximate estimate of self-association constants and the self-stacking conformation of malvin quinonoidal boxes studied by HNMR, *Phytochemistry,* 30, 2049, 1991.

63. **Salt, F.W. and Thomas, J.G.,** Aerobic corrosion of tin in anthocyanin solutions and fruit syrups, *J. Appl. Chem.,* 7, 231, 1957.

64. **Somaatmadja, D., Powers, J.J., and Hamdy, M.,** Anthocyanins. VI. Chelation studies on anthocyanins and other related compounds, *J. Food Sci.,* 29, 655, 1964.

65. **Jurd, L. and Asen, S.,** The formation of metal and co-pigment complexes of cyanidin 3-glucoside, *Phytochemistry,* 5, 1263, 1966.

66. **Takeda, K.,** Metallo-anthocyanins. II. Further experiments of synthesizing crystalline blue metallo-anthocyanins using various kinds of bivalent metals, *Proc. Jpn. Acad.,* 53, B, 257, 1977.

67. **Takeda, K. and Hayashi, K.,** Metallo-anthocyanins. I. Reconstruction of commelinin from its components, awobanin, flavocommelin and magnesium, *Proc. Jpn. Acad.,* 53, B, 1, 1977.

68. **Goto, T., Hoshino, T., and Ohba, M.,** Stabilization effect of neutral salts on anthocyanins, flavylium salts, anhydro bases and genuine anthocyanins, *Agric. Biol. Chem.,* 40, 1593, 1976.

69. **Preston, N.W. and Timberlake, C.F.,** Separation of anthocyanin chalcones by high-performance liquid chromatography, *J. Chromatogr.,* 214, 222, 1981.

70. **Markakis, P.,** Stability of anthocyanins in food, in *Anthocyanins as Food Colors,* Markakis, P., Ed., Academic Press, New York, 1982, 163.

71. **Jackman, R.L., Yada, R. Y., Tung, M. A., and Speers, R.A.,** Anthocyanins as food colorants — a review, *J. Food Biochem.,* 11, 201, 1987.

72. **Francis, F.J.,** Food colorants: anthocyanins, *Crit. Rev. Food Sci. Nutr.,* 28, 273, 1989.

73. **Wong, E.,** Biosynthesis of flavonoids, in *Chemistry and Biochemistry of Plant Pigments,* Goodwin, T.W., Ed., Academic Press, New York, 1976, 464.

74. **Ebel, J. and Hahlbroch, K.,** Biosynthesis, in *The Flavonoids: Advances in Research,* Harborne, J.B. and Mabry, T.J., Eds., Chapman & Hall, London, 1982, 641.

75. **Grisebach, H.,** Biosynthesis of anthocyanins, in *Anthocyanins as Food Colors,* Markakis, P., Ed., Academic Press, New York, 1982, 41.

76. **Heller, W. and Forkmann, G.,** Biosynthesis, in *Flavonoids, Advances in Research since 1980,* Harborne, J.B., Ed., Chapman & Hall, London, 1988, 399.

77. **Stafford, H.A.,** *Flavonoid Metabolism,* CRC Press, Boca Raton, FL, 1990, 1.

78. **Forumann, G.,** Precursors and genetic control of anthocyanin synthesis in *Matthiola incana* R.Br., *Planta,* 137, 159, 1977.

79. **Kho, K.F.F., Bessnick, G.J.H., and Wiering, H.,** Anthocyanin synthesis in a white flowering mutant of *Petunia hybrida* by a complementation technique, *Planta,* 127, 271, 1975.

80. **Kho, K.F.F., Bolsman-Louwen, A.C., Vuik, J.C., and Bessnink, G.J.H.,** Anthocyanin synthesis in white flowering mutant of *Petunia hybrida.* II. Accumulation of mutant dihydroflavonol intermediate, *Planta,* 135, 139, 1977.

81. **Harrison, B.J. and Stickland, R.G.,** Precursors and genetic control of pigmentation. II. Genotype analysis of pigment controlling genes in acyanic phenotypes in *Antirrhinum majus, Heredity,* 33, 112, 1974.

82. **Heller, W., Britsch, L., Forkmann, G., and Grisebach, H.,** Leucoanthocyanidins as intermediates in anthocyanidin biosynthesis in flowers of *Matthiola incana* R.Br., *Planta,* 163, 191, 1985.

83. **Francis, F.J.,** Analysis of anthocyanins, in *Anthocyanins as Food Colors,* Markakis, P., Ed., Academic Press, New York, 1982, 182.

84. **Jackman, R., Yada, R.Y., and Tung, M.A.,** A review: separation and chemical properties of anthocyanins used for their qualitative and quantitative analysis, *J. Food Biochem.,* 11, 279, 1987.

85. **Gross, J.,** *Pigments in Fruits,* Academic Press, London, 1987, 58.

86. **Strack, D. and Wray, V.,** Anthocyanins, in *Methods in Plant Biochemistry,* Vol. 1, Academic Press, New York, 1989, 326.

87. **Mazza, G.,** Anthocyanins and other phenolic compounds of saskatoon berries (*Amelanchier alnifolia* Nutt.), *J. Food Sci.,* 51, 1260, 1986.

88. **Takeda, K., Harborne, J.B., and Self, R.,** Identification and distribution of malonated anthocyanins in plants of the Compositae, *Phytochemistry,* 25, 1337, 1986.

89. **Anderson, O.M.,** Anthocyanins in fruits of *Vaccinium japonicun, Phytochemistry,* 26, 1220, 1987.

90. **Hong, V. and Wrolstad, R.E.,** Characterization of anthocyanin-containing colorants and fruit juices by HPLC/photodiode array detector, *J. Agric. Food Chem.,* 38, 698, 1990.

91. **Williams, M., Hrazdina, G., Wilkinson, M. M., Sweeny, J.G., and Iacobucci, G.A.,** High performance liquid chromatographic separation of 3-glucosides, 3,5-diglucosides, 3-(6-*O*-*p*-coumaryl) glucosides and 3-(6-*O*-*p*-coumarylglucoside)-5-glucosides of anthocyanins, *J. Chromatogr.,* 155, 389, 1978.

92. **Kondo, T., Kawai, T., Tamura, H., and Goto, T.,** Structure determination of Heavenly Blue anthocyanin, a complex monomeric anthocyanin from the Morning Glory *Ipomoea tricolor,* by means of the negative NOE method, *Tetrahedron Lett.,* 28, 2273, 1987.

93. **Bridle, P., Loeffler, R.S.T., Timberlake, C.F., and Self, R.,** Cyanidin 3-malonylglucoside in *Cichorium intybus, Phytochemistry,* 23, 2968, 1984.

94. **Harborne, J.B.,** Functions of flavonoids in plants, in *Chemistry and Biochemistry of Plant Pigments,* Goodwin, T.W., Ed., Academic Press, New York, 1976, 736.

95. **Barber, H.N.,** Selection in natural populations, *Heredity,* 20, 551, 1965.

96. **Weisaeth, G.,** Quality problems in breeding for disease resistance in cabbage and cauliflower, *Qual. Plant. Plant Foods Hum. Nutr.,* 26, 167, 1976.

97. **Burlov, V.V.,** Inheritance of the resistance to the local race of broomrape (*Orobanche cumana* Wallr.) in sunflower, *Genetika,* 12, 44, 1976.

98. **Kraft, J.M.,** The role of delphinidin and sugars in the resistance of pea seedlings to *Fusarium solani* F. Sp. *pisi* root rot, *Phytopathology,* 8, 1057, 1977.

99. **Muehlbauer, F.J. and Kraft, J.M.,** Effect of pea seed genotype on preemergence damping-off and resistance to *Fusarium solani* and *Pythium ultimum* root rot, *Crop Sci.,* 18, 321, 1978.

100. **Hammerschmidt, R. and Nicholson, R.L.,** Resistance of maize to anthracnose: effect of light intensity on lesion development, *Phytopathology,* 67, 247, 1977.

101. **Larson, R.L.,** Influence of phenolic compounds on growth of *Diplodia maydis* in liquid cultures, *Z. Pflanzenkr. Pflanzenschutz,* 86, 247, 1979.

102. **Caradus, J.R. and Silvester, W.B.,** Rhizobium specific anthocyanin-like marker in lupine (*Lupinus arboreus*) nodules, *Plant Soil,* 51, 437, 1979.

103. **Lee, D.W., Loury, J.B., and Stone, B.C.,** Abaxial anthocyanin layer in leaves of tropical rain forest plants: enhancer of light capture in deep shade, *Biotropica,* 11, 70, 1979.
104. **Chernyshiev, V.D.,** Spectral reflection of light by leaves of certain woody species in the far east USSR, *Lesovedenie,* 6, 63, 1975.
105. **Wagner, H.,** Phenolic compounds in plants of pharmaceutical interest, in *Biochemistry of Plant Phenolics,* Swain, T., Harborne, J.B., and Van Sumere, C.F., Eds., *Rec. Adv. Phytochemistry,* 12, Plenum Press, New York, 1979.
106. **Wagner, H.,** New plant phenolics of pharmaceutical interest, in *Ann. Proc. Phytochem. Soc. Eur.,* Vol. 15, Van Sumere, C.F. and Lea, P.J., Eds., Clarendon Press, Oxford, 1985, 409.
107. **Mabry, T.J. and Ulubelen, A.,** Chemistry and utilization of phenylpropanoids including flavonoids, coumarins and lignans, *J. Agric. Food Chem.,* 28, 188, 1980.
108. **Beretz, A. and Cazahave, J.-P.,** The effect of flavonoids on blood vessel wall inter-actions, in *Plant Flavonoids in Biology and Medicine II,* Cody, V., Middleton, E., Harborne, J.B., and Beretz, A., Eds., Alan R. Liss, New York, 1988, 187.
109. **Boniface, R., Miskulin, M., Robert, L., and Robert, A.M.,** Pharmacological prop-erties of *Myrtillus* anthocyanosides: correlation with results of treatment of diabetic mi-croangiopathy, in *Flavonoids and Bioflavonoids 1985,* Farkas, L., Gabor, M., and Kallay, F., Eds., Elsevier, Amsterdam, 1986, 193.
110. **Cristoni, A. and Magistretti, M. J.,** Antiulcer and healing activities of *Vaccinium myrtillus* anthocyanosides, *Farmaco, Ed. Prat.,* 42, 29, 1987.
111. **Kano, E. and Miyakoshi, J.,** UV protection effect of keracyanin an anthocyanin de-rivative on cultured mouse fibroblast L cells, *J. Radiat. Res.,* 17, 55, 1976.

Chapter 2

POME FRUITS

I. APPLE

Skin color of apples, particularly in red cultivars, is an important factor in consumer acceptance. It is also important in establishing government grades and standards, because a particular grade must have a certain proportion of the apple skin colored, and proportions vary depending on the grade and variety of apples considered.[1] In most markets, red-skinned apples are preferred to others, and within a cultivar better colored fruits generally earn higher prices. This consumer preference is not only a matter of attractive exterior appearance. Consumers generally know from experience that brighter colored apples often taste better than green apples.[2] This is because of the higher sugar and aromatic compounds levels.

A. ANTHOCYANINS IN APPLE

The pigments causing red coloration in apple skin — *Malus pumila* L., Rosaceae (syn. *Malus domestica* Borkh.) — are mainly anthocyanins, although colorless phenolic compounds (flavonols, flavan 3-ols, dihydrochalcones, phenolic acids, and tannins) also aid in the intensification of color through the copigmentation reaction.[2-4]

Duncan and Dustman[5] and Sando[6] characterized the major anthocyanin of 'Winesap', 'Grimes Golden', and 'Jonathan' apples as cyanidin 3-galactoside or idaein. Later Fouassin[7] confirmed the presence of a monoglucoside of cyanidin in extracts of cultivated and wild apple skins, and Walker[8] found cyanidin 3-galactoside and quercetin 3-galactoside in numerous apple varieties grown in New Zealand. Pais and Gombkötö,[9] using gel filtration on a Sephadex-G column, identified cyanidin 3-galactoside and cyanidin 3-glucoside and their corresponding acylated forms in the skin of 'Jonathan' apples. Sun and Francis[10] identified cyanidin 3-galactoside as the major pigment, and cyanidin 3-arabinoside and 7-arabinoside as the minor anthocyanins of American 'Red Delicious' apples of the Richared strain and of several other varieties. The occurrence of these three anthocyanins was also reported in the skin of 'Jonathan' and 'Wagner'' apples from the U.S.S.R.[11] Cyanidin 7-arabinoside was, however, not found in strains of 'Red Delicious' apples from England and France and in 'Stoke Red', 'Jonathan', 'Tremlett's Bitter', 'Ingrid Marie', and 'Cox's Orange Pippin' fruit analyzed by Timberlake and Bridle.[12] The latter authors used three extracting solvents (methanol-hydrochloric acid, acetone-sulfur dioxide, and acetone alone); three purification procedures (ion exchange, gel filtration and adsorption chromatography); and hydrolysis, spectral analysis, and paper chromatography to characterize the

TABLE 1
Anthocyanins in Peel of Different Cultivars of Apples
(*Malus pumila* L.)

Anthocyanins	Cultivar and relative conc (%)					
	Red Delicious (Starkrimson)	Stoke Red	Jonathan	Tremletts	Cox's Orange Pippin	Ingrid Marie
Cyanidin 3-galactoside	85	94	94	90	92	90
Cyanidin 3-arabinoside	10	5	4	8	6	6
Cyanidin 3-glucoside	5	1	3	2	2	4
Cyanidin 3-xyloside	t	t	t	t	t	t
Their acylated derivatives	t	t	t	t	t	t

Note: t = Trace amounts.

Adapted from Timberlake, C.F. and Bridle, P., *J. Sci. Food Agric.*, 22, 509, 1971.

anthocyanins of six apple cultivars. They confirmed the presence of cyanidin 3-galactoside as the major anthocyanin of apples and identified the minor components as cyanidin 3-glucoside, cyanidin 3-arabinoside, and the hitherto undescribed cyanidin 3-xyloside. In addition, the acylated derivatives of all four of these pigments were detected for the first time. They also observed a loose association between anthocyanins and proanthocyanins, and demonstrated instability of the latter during purification by gel filtration with an acidified solvent. In the cultivars analyzed by Timberlake and Bridle,[12] cyanidin 3-galactoside accounted for 83 to 94%, cyanidin 3-arabinoside for 4 to 13%, cyanidin 3-glucoside for 1 to 8%, and cyanidin 3-xyloside for less than 1% of the total anthocyanins.

Samorodova-Bianki and Bazarova[13-14] reported the presence of three cyanidin glycosides in the skin of three apple varieties from eastern Europe, and the occurrence of cyanidin 3-arabinosylgalactoside in the pulp of one of the varieties analyzed. Mazza and Velioglu[15] characterized cyanidin 3-galactoside, 3-glucoside, 3-arabinoside, and 3-xyloside in red-flesh 'Scugog' apples (*M. pumila* var. *niedzwetzkyana*). Qualitatively the anthocyanin composition of these apples is very similar to that of other apples,[12] but quantitatively the red flesh apples contain much more cyanidin 3-glucoside, 3-arabinoside, and 3-xyloside and less cyanidin 3-galactoside (Tables 1 and 2). Also, although using a highly sophisticated HPLC system with a diode array detector and a computer system, Mazza and Velioglu[15] were unable to detect the presence of acylated anthocyanins in red-flesh apples. At 95 to 100 mg/kg of anthocyanin content, 'Scugog' apples are, however, a rich source of the relatively rare cyanidin 3-xyloside.

TABLE 2
Anthocyanin Composition of Red-Flesh Apples
(*M. pumila* var. *niedzuetzkyana*)

Anthocyanins	R_t^a	k'	α	Area (%)
Cyanidin 3-galactoside	10.6	4.9	—	39.1
Cyanidin 3-glucoside	11.3	5.3	1.08	27.0
Cyanidin 3-arabinoside	13.1	6.3	1.19	23.3
Cyanidin 3-xyloside	18.5	9.3	1.48	10.5

[a] R_t = retention time (s × 100); k' = capacity factor (T_r − t_o)/t_o (t_o = 1.8 s); α = separation factor k_2'/k_1'; flow rate = 1 ml/min; concentration of total anthocyanins 100 ± 5 mg/kg of fresh fruit.

From Mazza, G. and Velioglu, Y.S., *Food Chem.*, 43, 113, 1992. With permission.

Anthocyanins are also found in the bark and leaves of the apple tree. According to Okuse and Naoki,[16] the bark contains cyanidin 3-galactoside, cyanidin 3-arabinoside, and cyanidin 7-arabinoside. The pigments are located in the hypodermal cells and are apparently present in a higher concentration in winter than in summer.[17-18] The reddening which accompanies infection of the tree by the mosaic virus is not associated with an increase in content or type of anthocyanins normally present.[19] Anthocyanins in the leaves of apples increase in concentration under water stress conditions.[20]

In addition to anthocyanins, apples also contain the flavonols quercetin 3-galactoside, 3-arabinoside, 3-rhamnoside, 3-xyloside, 3-rutinoside, 3-glucoside, and 3-diglucoside;[21-25] the flavan 3-ols (+)-catechin, (−)-epicatechin, (+)-gallocatechin, and (−)-epigallocatechin;[26-28] the dihydrochalcones phloridzin and phloretin xyloglucoside;[21,24,29,30] and the phenolic acids caffeoylquinic acids, *n*-chlorogenic (5^1-CQA), neochlorogenic (3^1-CQA), cryptochlorogenic (4^1-CQA), and isochlorogenic (diCQA) acids; *p*-coumaroylquinic acids (3^{1}pCQA, 4^1-pCQA, 5^1-pCQA); and glucose derivatives of *p*-coumaric, ferulic, caffeic, and sinapic acids.[15,26,28,31-35] 5′-Feruloylquinic acid has also been reported in cell suspensions from apple fruits.[36]

These colorless phenolic compounds may form intensely colored complexes with anthocyanins[3,4] and/or play a role in taste characteristics such as bitterness and astringency; formation of yellow and brown pigments in bruised fruit, apple juice, or wine; and formation of hazes and sediments in juices.[37-40] In addition the quercetin 3-glycosides are known to have inhibitory properties toward a β-galactosidase and toward softening of apples during cold storage.[41-43]

B. FACTORS AFFECTING ANTHOCYANIN LEVELS

The formation of anthocyanin in apples depends on a number of internal and external factors. Knowledge of the role of these factors in the control of anthocyanin biosynthesis is particularly valuable because they can be used for practical control of fruit quality and market value.

1. Genetics

Distribution and heritability of anthocyanins in apples are controlled by several genes, and the presence of the red color appears to be dominant.[44,45] The anthocyanins are most frequently located in the external tissue (epiderm and a few subepidermal cell layers), but they may be distributed throughout the whole fruit as in 'Scugog' apples.[15] The vacuole generally forms the major cell compartment in which anthocyanins and other soluble phenolic compounds accumulate.[46,47] The accumulation level is generally controlled and thus can vary between cultivars.[48,49] In highly pigmented cultivars such as Cherry Cox, skin pigmentation is transmitted from parents to their progeny.[50] The anthocyanins are regulated by a single dominant gene, R; the environment, however, strongly influences its expression.[45,51] In the cultivar Richared 31.6% of the total skin cells contain anthocyanins; in the cultivars Jonathan and Cox's Orange Pippin the percentage of pigmented cells is 30.4 and 13.0%, respectively. The order in which the distribution of anthocyanins decreases is subepidermal to epidermal to the third layer of cells.[52] The cellular distribution of anthocyanin in the skin is a distinctive varietal characteristic.[53] Several authors have reported on the genetic variation of red pigmentation in some of the better known apple varieties such as McIntosh,[54] Delicious and Jonathan,[55] and Northern Spy.[56] Schmidt[57] studied the pigmentation in the fruit and leaves of crosses between *M. niedtzwetzkyana* species in an attempt to obtain a homozygous *RR* by crossing with the variety Jonathan. The presence of anthocyanins as genetic markers in the apple allows apomixis in *Malus* spp. to be expressed.[58] Isaev and Maksimova[59] observed a significant correlation between leaf pigmentation and the quantity of anthocyanins in the fruit. Nybom and Bergendal[60] reported an apparent relationship between the presence of anthocyanins in the petiole and the pigmentation in the fruit, but this has been negated by others.[61] The inheritance of anthocyanins in apple fruit skin is discussed in more detail in a recent review by Schmidt.[45]

2. Light

Anthocyanin synthesis in apple skins is markedly affected by light intensity and quality. Siegelman and Hendricks[48,49] found a linear increase of anthocyanin concentration with light intensity above a certain threshold value. The requirement for long irradiation times at high intensities for appreciable anthocyanin synthesis has been confirmed by Downs et al.[62] Similarly, Proctor and Creasy[63] found a close correlation between the distance of the fruit from fluorescent lamps and the quantity of anthocyanin produced when natural light within the tree canopy was supplemented by continuous irradiation for 48 h. The increase of anthocyanin synthesis was also nearly linear with fluence rate or energy flux in experiments performed by Arakawa et al.[64] with detached whole apples and isolated apple skin. In these, anthocyanin synthesis was considerably higher in continuous light than under intermittent light cycles of 14 h light and 10 h dark.[65]

The most effective light for maximum anthocyanin formation is blue-violet with wavebands around 650 nm and 430 to 480 nm.[49] These spectral characteristics are consistent with the requirement for photosynthetic activity in the expression of the response, as demonstrated by the fact that an inhibitor of photosynthesis — dichlorophenyl dimethyl urea (DCMU) — is very effective in reducing the rate of anthocyanin synthesis.[62] Anthocyanin synthesis in apple fruit is also stimulated by light in the UV region.[66,67] It has been shown that light with wavelengths from 280 to 320 nm is highly effective in stimulating anthocyanin synthesis and exerts a synergistic effect when applied simultaneously with red light.[68]

The maximum energy requirement for anthocyanin synthesis varies considerably with the cultivar and changes during the season.[69,70] Anthocyanin accumulation during ripening of apples is directly correlated with PAL activity; and in the apple skin, PAL activity has been observed only in the red parts of the peel.[71,72] In 'Jonathan' apples anthocyanin synthesis appeared to be regulated by the level of phenylalanine ammonia-lyase (PAL) activity, which in turn was initiated by light. PAL, although not the critical enzyme in the anthocyanin synthesis in apple skin, was found to be rate limiting.[73] A close relationship between PAL activity and anthocyanin synthesis was also found in 'Red Spy' apples by Tan.[74,75] In the absence of light, anthocyanin synthesis does not occur. The regulation of PAL levels seems to be due to a phenylalanine ammonia-lyase inactivity system (PAL-IS) that has been isolated from different plants including apples.[76]

3. Temperature

Temperature is a major factor in anthocyanin accumulation in apples. Low temperatures have long been considered to promote anthocyanin synthesis; and high temperatures in autumn, to inhibit it. The effect of temperature differs for fruit on the tree or fruit after harvest, because the level of ripeness and light are also involved.[65] It also varies with cultivar and with the stage of fruit development.[77] 'Jonathan' apples require a night temperature of 10°C and high daytime temperatures for maximum coloration, whereas 'Ontario' apples develop less color during warm days. Diener and Naumann[78] obtained the highest anthocyanin accumulation in immature apples on and off the tree at a combination of the lowest daytime and the lowest night temperature. However, during ripening a change occurred from promotion to inhibition of anthocyanin formation at low temperatures, both in attached and detached fruit. Investigations by Tan[74] on the relationship between PAL and anthocyanin at different temperature regimes showed that low temperatures of 6°C promoted both anthocyanin synthesis and PAL activity. Fruits held at alternating temperatures of 6 and 18°C in light produced twice the amount of anthocyanin as fruits at constant 18°C.

Several explanations for the effect of temperature on anthocyanin synthesis have been proposed. Uota[79] suggested that at higher temperature a greater amount of energy is required for the synthesis of the pigment and high levels of pigmentation can be achieved by increasing the light intensity. Creasy[80] explained the increase in anthocyanin formation in apple skin at low temperature on the basis of the carbohydrate metabolism in the skin. Faragher[65] and Faragher and Brohier[81] noted that the increase in anthocyanin in autumn may be more directly related to the ripening process, as indicated by the rise in ethylene, than to a fall in temperature. Saure[2] concluded that low temperature may contribute to color formation by directly reducing gibberellin activity.

4. Agronomic Factors

Generally, excessive nitrogen fertilization is associated with a reduction in the percentage of well-colored fruits at harvest, although the total of well-colored fruit may be higher.[82] Lekhova[83,84] found that a low level of nitrogen fertilization increased anthocyanic pigmentation without affecting the size and specific weight of the fruit or the suitability for conservation. The threshold value above which nitrogen has a negative effect upon anthocyanin accumulation is about 90 kg/ha.[85] It is uncertain whether the effect of nitrogen is direct, as more nitrogen-containing substances such as amino acids and protein are synthesized, interfering with the biosynthesis of sugars and anthocyanins; or indirect, producing increasing vegetative growth which hinders the penetration of light. Most authors seem to support the direct effect of nitrogen.[2]

Kaether[86] observed that potassium per se had only a small positive effect on anthocyanin synthesis, but a high potassium supply favored the positive effect of a low nitrogen level on anthocyanin formation. Lüdders and Bünemann[87] increased anthocyanin synthesis by high potassium fertilization in summer.

Soil moisture generally promotes anthocyanin formation, especially in dry areas or dry seasons as along as it is adequate for normal fruit development. Both excessive and deficient soil moisture that hamper normal fruit development impair anthocyanin synthesis.[88,89]

The reported effect of pruning on anthocyanin formation is diverse. Severe dormant pruning has been reported to improve, but also to reduce fruit color.[89] Summer pruning usually results in improved anthocyanin synthesis, but is not always effective.[90] If performed early in the season and in poorly colored cultivars such as Jonagold apples, summer pruning may have detrimental effects.[90] Root pruning improves the percentage but often reduces total yield of well-colored fruit.[91-93] Fruit thinning also increases fruit color, but the amount of thinning required for maximum color development is still disputed and depends on many factors inside and outside the fruit.[2]

TABLE 3
Anthocyanin Synthesis in 7 Apple Cultivars
Irradiated with White + UV312 (5.3 wm^{-2}
in Total) Light for 96 h

Cultivar	Anthocyanin (gcm^{-2})	
	Nonbagged	Bagged[a]
Starking Delicious	20.84	21.49
Jonathan	25.24	77.69
McIntosh	16.43	18.26
Fuji	7.67	6.72
Ralls Janet	6.83	10.03
Mutsu	1.98	6.47
Golden Delicious	1.21	4.41

[a] Bagged fruit had been covered during fruit development
with paper bags.

Adapted from Arakawa, O., *J. Jpn. Soc. Hortic. Sci.*, 57,
373, 1988.

Covering the fruit with paper bags about 1 month after full bloom, followed by removal of the bags after several months, under appropriate conditions promotes anthocyanin synthesis.[94,95] Although it is a very laborious
task, this practice is widely used in Japan as an effective practice for inducing
color formation in most apple cultivars, including those that usually do not
show red color upon maturation.[96] Arakawa[97] noted that, under white +
UV312 light, fruit which had been covered with paper bags for about a month
from flowering produced a much higher anthocyanin content at immature and
mature stages than nonbagged fruit, regardless of cultivar (Table 3). This
anthocyanin accumulation, however, decreased markedly during fruit ripening, with a parallel increase in ethylene concentration in the cortical tissue
of apples. Color formation in bagged fruit may still be poor after light exposure
if the fruit is picked too early, or if the fruit is insufficiently developed.[94,97]

Chemicals used to promote anthocyanin formation in apples include thi-
ocyanates;[98] ethephon;[99] daminocide;[99,100] paclobutrazol;[101,102] auxins,[103] cytokinins;[104,105] gibberellins;[103,104] growth regulators such as 2,4,5-T, NAA,
and NAD;[106,107] and the fungicide Tuzet.[108] Thiocyanate sprays to leaves and/
or fruits result in an increase in anthocyanin formation and a brightening of
the red color of various apple cultivars.[98] The sprays, however, also cause
injuries to foliage, and the color responses vary among cultivars and among
neighboring trees of the same cultivar.

Ethephon, also known as Ethrel and CEPA, is an ethylene-releasing product that may promote color formation primarily in some cultivars if applied
up to 2 weeks before the normal harvesting date. In 'McIntosh' apples,

Looney[99] reported an increase in color in almost all parts of the tree and over the entire fruit surface, whereas 'Fuji' apples gave no response at all in experiments by Fortes.[109]

Some authors hold that ethephon acts primarily by accelerating the ripening process, and that it is incapable of inducing red color formation without sufficient light or in nonred cultivars such as Golden Delicious.[110,111] Other authors, however, have found that when applied in combination with daminocide or if applied up to 60 d before harvest, ethephon may promote color formation without substantially hastening the ripening process.[112]

Daminocide, SADH or Alar is a growth retardant well known for its ability to promote anthocyanin synthesis in most red cultivars.[99,100,113,114] Daminocide does not, however, induce red coloration in nonred cultivars and may have only a limited effect in red cultivars such as Jonagold. Tree physiology at the time of application also influences the effectiveness of daminocide application. Paclobutrazol, PP333 or Cultar is another growth retardant, similar to daminocide, that may increase red color formation in some apple cultivars when applied under favorable conditions.[101,102]

Applications of auxins and gibberellins also may promote anthocyanin formation, but it remains unclear whether this is a direct effect or a consequence of advanced/delayed fruit maturity.[104,115] Many other chemicals have been tested in recent years; however, most of them have not been introduced for color promotion because of inconsistent or insufficient effects or because of negative side effects.[2]

5. Composition Factors

The presence of sugars is essential for the synthesis of anthocyanins, and it has a triggering effect on the pigment accumulation.[116,117] Fructose, glucose, lactose, maltose, and sucrose all stimulate anthocyanic synthesis in apple skin.[118] Catabolism of glucose via the pentose phosphate pathway (PPP) has been associated with anthocyanin production in apple fruit by several workers.[119,120] Indeed, PPP is believed to be at the origin of the formation of erythrose-4-phosphate which undergoes a condensation reaction with phosphoenolpyruvate, enabling entry to the shikimate pathway.[121]

Phenylalanine ammonia-lyase (PAL) is recognized as a critical enzyme in the regulation of anthocyanin synthesis. Other enzymes governing the biosynthesis of anthocyanins include: chalcone synthase, chalcone isomerase, flavanone 3-hydroxylase, flavanone 3'-hydroxylase, several glycosyltransferases, and acyltransferase.[122-125] The role of these enzymes is further discussed in Chapter 1.

Phytochrome, a pigment bound to protein, is the photoreceptor in light-dependent anthocyanin synthesis. It consists of two physiological forms, one active far red absorbing (Pfr) and the other inactive red absorbing (Pr). Red light converts Pr to Pfr, and Pfr in turn is converted to Pr, both by far red light (FR) and in the dark. Phytochrome involvement in anthocyanin synthesis of the apple is supported by the small but significant R-FR reversible response in immature apple skin disks after long-term irradiation with red light observed by Arakawa.[97]

Ethylene biosynthesis is coupled with fruit ripening and color formation.[99,109-111] Ethephon or Ethrel is frequently used to accelerate the ripening of fruit, and accumulation of anthocyanins is normally associated with the maturation process of apples. The increase in anthocyanin content caused by endogenous or exogenous ethylene is due to increased activity of enzymes involved in the anthocyanin biosynthesis, especially PAL.[67,81,112,126] The activity of PAL is stimulated by ethylene, light, and fruit ripening.[127] However, ethylene can only stimulate PAL activity which is light induced. When this condition is not fulfilled, the anthocyanin synthesis is hindered.[127]

Calcium ions enhance synthesis of anthocyanins in disks of apple peel, but the magnitude of the increase depends on the stage of fruit maturity and on the form in which the calcium is supplied.[118] Anthocyanin levels in percentage of control (100) were 152, 145, and 187 for calcium carbonate, calcium chelate, and calcium succinate, respectively.

trans-Cinnamic acid and L-phenylalanine, presumed precursors in anthocyanin synthesis, stimulated red color development in skin disks of 'McIntosh' apples at relatively low concentrations, but acted as inhibitors at concentrations above 0.0025 M.[118]

Endogenous growth regulators, such as auxins, cytokinins, gibberellins, and abscisic acid, are known to affect color development in apples. However, most available data deal with the effect of exogenous applications of these compounds to apples and apple trees.[2,121]

6. Storage and Processing

Apple anthocyanins are relatively stable during storage at 2°C and 70 to 90% relative humidity.[128,129] In a low oxygen and high carbon dioxide atmosphere, however, anthocyanin concentrations generally decrease and undesirable browning may occur.[129] Lin et al.[129] compared the stability of anthocyanins in the skin of 'Starkrimson' apples stored for 0, 7, 23, and 30 weeks under heat shrinkable wrap, in packaged modified atmosphere, and unpackaged. They found that the concentration of cyanidin 3-galactoside, cyanidin 3-arabinoside, and an unidentified cyanidin arabinoside remained practically unchanged in apples stored in cardboard boxes at 2°C and 73% relative humidity for 23 weeks. However, the flavonoids decreased by 0 to 36% in apples stored under shrink wrap and by about 80% in fruit stored in packaged atmosphere at 2°C and 73% relative humidity. The modified atmosphere packages contained 5% oxygen, 10% carbon dioxide, and 85% nitrogen at the beginning of the storage period, but about 0.5% oxygen, 73% carbon dioxide, and 27% nitrogen after 23 weeks of storage. These conditions are, however, well outside the recommended oxygen, carbon dioxide, and nitrogen levels for controlled atmosphere storage of apples. Thus, it may very well be that storage of red-skinned apples in atmosphere of 0.5 to 0.6% carbon dioxide, 1.5 to 1.6% oxygen, and about 98% nitrogen may not decrease the anthocyanin content of apples.

Juice and cider made from red apples are pink in color and rich in anthocyanin(s) when processed but turn brown during storage.[120] The degree of browning is, however, dependent on storage, temperature, and concentration of solids in the juice.[130] Thus, while juice stored at 1°C gave no visible change after 48 h, storage at 37°C caused an initial increase in absorbance at 500 to 520 nm followed by progressive development of a peak at 446 nm and a yellow-brown coloration. Storage of ruby-red 70° Brix concentrate at 37°C resulted in the loss of an absorbance peak at 527 to 528 nm and an increasing absorbance at wavelengths below 500 nm. The color of the concentrate also changed from ruby-red to orange after 167 h and to brown after 648 h at 37°C.[130] Juice blends containing apple and grape, cherry, or raspberry juices developed haze, lost the red color, and browned in storage at 25°C.[131] Also, as length of storage time increased, polymeric color and percent of color due to tannin increased while total anthocyanin decreased in all of the anthocyanin pigmented juices. The results of these studies are consistent with other literature reports demonstrating that a higher temperature shifts the equilibrium between anthocyanin structures toward the yellow and colorless chalcone structures[132,133] and that ascorbic acid and sugars and their degradation products have a detrimental effect on anthocyanins.[134]

Cyanidin 3-galactoside and other anthocyanins of apple may also be subject to enzymatic degradation by an endogenic anthocyanase of the fruit.[134] The activity of this enzyme can be separated into two groups: a glycosidase activity thought to release the aglycone from the anthocyanin, followed by spontaneous transformation of the aglycone to colorless derivatives, and an activity resulting from the action of polyphenoloxidase (PPO) on anthocyanins in the presence of *O*-diphenols.[134] Anthocyanins can also be destroyed directly by PPO but they remain fairly poor substrates for the enzyme.[134,135] Apple anthocyanins are also degraded by light and enzymatic preparations obtained from eggplant, potatoes, and mushrooms.[136] pH, metal ions, copigments, and ascorbic acid contents of the fruit or fruit products also affect the color intensity and stability of the anthocyanins.[134]

II. PEAR

Fruit of the pear, *Pyrus communis* L. (Rosaceae), is very similar to the apple in that it has many common characteristics and is widely cultivated. Pear skin can be rough but is usually smooth and generally yellow at ripeness. Some varieties have red coloration which may influence consumer preference for the fruit.[137]

The red pigments are anthocyanins and their concentration in the peel is 5 to 10 mg/100 g fresh weight (FW).[138] Harborne and Hall[139] described the presence of cyanidin 3-galactoside in the flesh and in the bark of some *Pyrus* varieties. Using chromatographic and spectrophotometric techniques, Francis[140] identified cyanidin 3-galactoside and cyanidin 3-arabinoside in the newly introduced colored mutants 'Stark Crimson' and 'Max-Red Bartlett', and in the red mutants of 'Clapps Favorite' and 'Bartlett'. Timberlake and Bridle[12] characterized cyanidin 3-galactoside in 'Painted Lady' and 'Blakeney Red' pears using sulfur dioxide in acetone to prevent hydrolysis of acylated pigments. Minor pigments were separated by paper chromatography, but their concentration was too small for complete characterization. In contrast to apple, in which there is a notable presence of acylated pigments, these pigments are practically irrelevant in the pear. Redelinghuys et al.[141] identified cyanidin 3-galactoside and peonidin 3-galactoside in the epidermis of the red cultivar, Williams Bon Chrètien.

The genetics of anthocyanic pigmentation in the pear was studied by Dayton[142] and Fregoni and Roversi.[143] Dayton's investigations revealed that the general pattern of anthocyanin distribution in pear fruits involves a non-pigmented epidermis and one or two additional nonpigmented layers, lying above two to five layers containing anthocyanins (Table 4). In most pear cultivars, the gene or genes responsible for anthocyanin development are carried in cells of the second histogenic layer. The cultivar Starkrimson is a major exception, because its anthocyanin development is confined to the epidermis. Dayton[142] also reported the presence of anthocyanins in young leaves and shoots of all red-fruited seedlings of 'Farmingdale' x 'Max-Red Bartlett' pears and noted that in a red-fruited plant all tissues should be capable of anthocyanin development. Fregoni and Roversi[143] observed the phenomenon of reversion, or returning to the ancestral yellow pigmentation, in the red-fruited 'Bartlett' pears.

Pears are rich in flavan 3-ols, phenolic acids, and flavonols. (−)-Epicatechin and (+)-catechin contents are reported to be 1.03 and 0.23 mg/100 g, respectively.[26] Chlorogenic acid and total phenolic contents of normal 'd'Anjou' pear skin were 7.2 and 28.3 mg/100 g FW.[144,145] Chlorogenic acid decreased 70% and total phenolic content decreased by 50% during browning of the skin; however, both components increased during the first 7 to 9 weeks of storage at 1°C. The flavonols of pear include: quercetin 3-glucoside, 3-diglucoside, and 7-xyloside; isorhamnetin 3-rutinoside, 3-glucoside, 3-rhamnosylgalactoside, and 3-galactoside; and malonylglucosides of kaempferol, quercetin, and isorhamnetin.[146-149] As in apples, these compounds may co-pigment with anthocyanins;[3,4] and play a role in taste characteristics, formation of yellow and brown pigments in processed and bruised fruit, and formation of hazes and sediments in juices.[131,150] Nortje and Koeppen[148] also reported traces of leucoanthocyanins in 'Bon Chrètien' pear with arbutin in the seeds

TABLE 4
Distribution of Cell Layers Containing Anthocyanin in Red Pear Varieties and Selections of Known Parentage

Variety	Number of nonpigmented cell layers including epidermis	Number of cell layers containing anthocyanin	Estimated % of surface with red color	Estimated intensity of color
Nonmutated Varieties				
Seckel	1–2	4–7	20–40	Medium-high
Early Seckel	1–2	3–4	20–40	Medium-high
Anjou	1–2	1–4	10–20	Low
Forelle	2	3	10–20	Low
Guerre Giffard	2	3–4	10–20	Medium
Fondante de Moulins-Lille	2	3	10–20	Low
Eureka (Seckel × Kieffer)	0–3	3	20–40	Medium-high
Reimer No. 80 (Farmingdale × Seckel)	1–3	2–3	20–40	Medium-high
Early Seckel × Farmingdale Selection 1	1	2–4	50–60	Medium-high
Early Seckel × Farmingdale Selection 2	1–2	3–5	30–40	Medium
Bud-sport Varieties				
Max-Red Bartlett	2–3	2–7	60–80	Medium-high
Rosi-Red Bartlett	1–2	3–4	70–80	Low-medium
Red Bartlett, P.I. 258948	3	3–4	70–80	Medium-high
Royal Red Hardy	2–3	3–4	20–50	Low
	Epidermal cells containing anthocyanins, %	1st hypodermal layer cells containing anthocyanins, %		
Starkrimson	99+	ca. 50	99	Very high

From Dayton, D.F., *Proc. Am. Soc. Hortic. Sci.*, 89, 110, 1966. With permission.

only. The leucoanthocyanins of pome fruits are complex, and their exact configuration and degree of polymerization are uncertain.

If pear juice is mixed with other anthocyanin-containing juices such as grape, raspberry, or cherry, it may become turbid and unstable. Spayd et al.[131] attributed this phenomenon to the formation of polymers.

III. OTHER POMES

A. CHOKEBERRY

Black chokeberry, *Aronia melanocarpa* Elliott (Rosaceae), is a native North American plant introduced in Europe in the late 18th century. The plant is 1 to 4 m tall, produces clusters of black berries, is winter hardy, and is used as an ornamental shrub in Canada and northern Europe.

Recently there has been renewed interest in chokeberry as a commerical plant, particularly in Poland, where the berries are used to make jams, juices, wines, and natural food colors.[151,152] The fruit is rich in anthocyanins and other biologically active substances.[153,154] Yuditskaite[155] observed that grafting of *A. melanocarpa* onto *Sorbus aucuparia* (mountain ash) favors the accumulation of anthocyanins in the fruit. Also, application of chlorocholine chloride (CCC) increased the anthocyanin content of the fruit by 30%,[156] while foliar applications of microelements (particularly boron and zinc) increase anthocyanins by 20%.[157] According to Martynov et al.,[158] the average anthocyanin content of fruits ranges from 725 to 800 mg/100 g fresh fruit and is made up of three pigments which were not identified. Kalemba et al.[159] reported an anthocyanin content of 1050 mg/100 g FW (at 77% moisture content) compared to a total polyphenol content of 4500 mg/100 g.

Using HPLC, Oszmianski and Sapis[160] characterized the two major anthocyanins of *A. melanocarpa* fruit as cyanidin 3-galactoside (64.5%) and cyanidin 3-arabinoside (28.9%), and two minor constituents as cyanidin 3-glucoside (2.4%) and cyanidin 3-xyloside (4.2%) (Figure 1). Szepczinska[161] reported cyanidin 3-glucoside and the 3,5-diglucosides of cyanidin and pelargonidin as well as flavonols and phenolic acids. These findings, however, suggest that the plant material used by this author may have not been *A. melanocarpa,* as the anthocyanin 3,5-diglucosides are not characteristic of the *Aronia* species.

Aronia fruit is used to improve the organoleptic quality of apple products. According to Rosa and Krugly[162] *Aronia* fruits give 10% more juice than black currants. When added to apple juice in a 30% concentration, it gives a high quality product. They also observed that concentrated chokeberry juice is more stable during storage than single-strength juice. The fruit can also be preserved by drying, and Shapiro et al.[163] optimized the process for keeping the loss of anthocyanins and other flavonoids to a minimum.

Aronia fruit is also considered a potential source of natural colorants. Experimental trials for extracting the anthocyanins with water acidified with citric acids[164,165] and by direct pressing of the fruit followed by extraction of the pressed pomace with 72% aqueous ethanol have been reported.[166] The high content of anthocyanins and the ease of extraction indicate that chokeberry may be considered an excellent source of natural colorants.

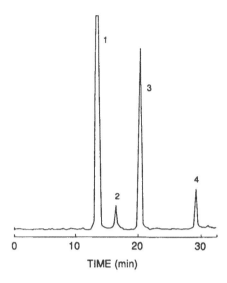

TIME (min)

FIGURE 1.   HPLC separation of *A. melanocarpa* (chokeberry) anthocyanin. (1) Cyanidin 3-galactoside; (2) cyanidin 3-glucoside; (3) cyanidin 3-arabinoside; (4) cyanidin 3-xyloside. (From Oszmianski, J. and Sapis, J.C., *J. Food Sci.*, 53, 1241, 1988. With permission.)

B. COTONEASTER

Cotoneaster (*Cotoneaster* Medic. spp.) is a deciduous or evergreen shrub of the temperate regions of Europe, North Africa, and Asia (except Japan) that is grown primarily for its attractive fruits and flowers. There are about 50 species and many subspecies of *Cotoneaster*.[167] Few, however, have been studied for their anthocyanin content and composition. According to Kiselev et al.[168] the anthocyanins protect *Cotoneaster* from cold injuries.

The black-berried variety found in the mountainous zones of Altai contains between 6.6 and 9.5% anthocyanins.[169] The major pigment of these berries is cyanidin 3-rutinoside which decreases rapidly with ripening. Shrubs growing in rocky areas with a southern exposure apparently contain higher pigment concentrations.[170]

Demina[171] studied the anthocyanin composition of *C. uniflora* and *C. melanocarpa* and reported values of 8 and 2.5%, respectively. The major pigment of these species of *Cotoneaster* is cyanidin 3-glucoside. The same pigment is present almost exclusively in *C. pyracantha*,[172] which indicates that this fruit could be a possible source of cyanidin 3-glucoside for pharmaceutical use.

C. divaricata contains cyanidin 3-galactoside and pelargonidin 3-galactoside.[173] Kruegel and Krainhoefner[174] studied the distribution of these two pigments in 53 taxa of *Cotoneaster*. Cyanidin 3-galactoside was present in all of them, while pelargonidin 3-galactoside was only present in 32 of the taxa studied. The colors of the berries varied from yellow-orange to red to black, depending on the different proportions of these two anthocyanins and the presence of the corresponding flavonol, quercetin 3-galactoside. The study also showed that the subgenera *Cotoneaster* and *Chaenopetalum* can be distinguished from the anthocyanin composition.

According to Robinson and Robinson[175] berries of *C. simmonsii* contain a pentose glucoside of pelargonidin as the principal pigment while the major pigment of the pink berries of *C. frigida* is a pentose glucoside of cyanidin. These pigments, however, are most likely the 3-galactosides of pelargonidin and cyanidin, as also reported by Du and Francis[172] and Kruegel and Krainhoefner.[174]

C. HAWTHORN

The fruit of hawthorn, *Crataegus* spp. (Rosaceae), contains different biologically active substances including ascorbic acid (>800 ppm) and other vitamins (B1, B2, and E).[176] This fruit also contains carotene and various polyphenols such as catechin, flavonols, and anthocyanins. In some varieties, the anthocyanin content is characteristically low, varying from 50 to 250 mg/100 g FW, while the procyanidin content ranges from 400 to 1500 mg/100 g FW.[177] The anthocyanin content of mature fruits of *C. pinnatifida* fruit is between 11 and 16 mg/100 g.[178] In 60 *Crataegus* spp. the content of polyphenols, catechins, and flavonols varied noticeably, while the anthocyanin concentration remained fairly constant at 358 to 407 mg/100 g FW basis.[179]

Chekalinskaya and Dovnar[180] suggested that hawthorn berries could be used for the production of pharmaceutical products if fruit variety and stage of maturity were considered, because these factors influence the content of biologically active constituents. It would also seem, for instance, that (−)-epicatechin is the predominant phenolic of fruits of selected species of hawthorn[181] and these species could serve as possible sources of this compound, were it to be used for pharmaceutical purposes.[182,183]

Vecher and Benkovic[184] found that the anthocyanin content in some varieties of hawthorn increased 10-fold during fruit maturity while the procyanidin level decreased considerably. Using paper chromatography, these authors identified the anthocyanidins, cyanidin and peonidin, in seven *Crataegus* varieties.[185] In nine species of *Crataegus*, Benkovic[186] reported derivatives of cyanidin, and hyperin and vitexin as the flavonoids present in greatest concentration in the skin.

The leaves are also red in *Crataegus oxyacantha* and appear to contain an enzymatic system capable of discoloring the anthocyanins.[187] Freudenberg and Weinges[188] presented the structure of the two procyanidins of *C. oxyacantha* fruit. After acidic hydrolysis one procyanidin yielded (−)-epicatechin and cyanidin, and the other yielded cyanidin and an unidentified anthocyanidin.

Chumbalov and Nurgalieva[189] found cyanidin 3-galactoside in *C. alma-atensis.* Santagati et al.[190] identified cyanidin 3-glucoside and cyanidin 3-arabinoside in the skin of *C. azarolus.* This fruit has small yellow or red-orange pomes and is cultivated in the Middle East, Northern Africa, and Southern Europe.[191] Hawthorn fruits are useful for the preparation of jams and jellies, particularly those of the species *C. oxyacantha* and *C. monogyna.*[192] The fruit of hawthorn, however, loses more than 50% of its vitamin C content when packaged in paper cartons. The absence of light and air noticeably reduces this negative effect.[193] In canned fruit, the pigments present in the epidermal layer of the fruit pass into the syrup. In air, the liquid turns purple in metal cans, while they remain unaltered in glass containers.[194]

Robinson and Robinson[175,195] reported that a monoside of cyanidin was found in *C. coccinea* (American hawthorn), *C. oxyacantha* L., *C. macroacanthae* Lodd, and *C. orientalis* Pall. Fouassin[7] examined three varieties and identified the pigment common to all three as cyanidin 3-galactoside.

D. MOUNTAIN ASH BERRY

The chemical composition of the fruit of *Sorbus* spp. (Rosaceae) has been the subject of numerous investigations. The berries of *S. aucuparia* L., or mountain ash are rich in vitamins, especially vitamin C, which constitutes 40 to 60 mg/100 g fresh berries.[196] They also contain anthoxanthin, the bitter precursor of sorbic acid; *trans*-3-β-D-glucopyranosyloxy-5-hexanolide;[197] and the phenolics cyanidin 3-galactoside and/or cyanidin 3,5-diglucoside, quercetin, isoquercitrin, and rutin.[195,198] *S. aria* Crantz. also contains cyanidin 3-β-D-glucopyranoside.[199]

Pyysalo and Kuusi[200] identified 18 phenolic compounds in fruit of *S. aucuparia* including cyanidin 3-galactoside, 3-glucoside, 3,5-diglucoside, and a fourth anthocyanin which on hydrolysis gave cyanidin and arabinose. Other phenolic compounds included: *trans*- and *cis*-chlorogenic acid, *trans*- and *cis*-neochlorogenic acid, isochlorogenic acid, *trans*-caffeic acid, *trans*-ferulic acid, feruloylquinic acid, *trans*-coumaric acid, *p*-coumaroylquinic acid, quercetin, rutin, hyperoside, and leucoanthocyanidin. The total amount of polyphenols obtained from fresh berries was 0.45%, which includes 0.15% anthocyanins. Kolesnik and Elizarova[201] found that the cyanidin aglycone is common to most *Sorbus* species, and the anthocyanin concentration varies with cultivar and place of origin. According to these authors, storage of fruit in a carbon dioxide-rich environment favors pigment retention more than refrigeration.

There is considerable interest in the industrial use of this fruit in eastern Europe where it is widely distributed. Proposed uses include wine,[202] juice,[203] natural food colorants,[204-207] and therapeutic products.[208] Nakhmedov et al.[204] patented a process for producing food colorants from fresh berries and berry by-products. The process involves extraction of the pigments with boiling aqueous solution of citric acid (0.5%) and shaking the mixture of 1 h, followed by pressing, filtration, and concentration of the pigments under vacuum. The resulting pigment extract is a dark red syrup (25 to 30° Brix) containing 5 to 7% anthocyanins.[204-207] This colorant is stable at room temperature for 12 months or longer. Improved color stability was achieved by pasteurization of the extract at 85°C.[204]

The presence of about 2600 mg of anthoxanthin per 100 g of dry fruit together with the high concentration of anthocyanins and other biologically active phenolic compounds has led several authors to suggest consumption of berries of mountain ash to reduce blood cholesterol and fat content of the liver and to inhibit the increase in capillary permeability[208] which is the initial sign of capillary inflammation.[209]

E. QUINCE

Quince, *Cydonia oblonga* Mill. (Rosaceae), is a small fruit-bearing tree native to the area from Iran to Turkestan. The pear-shaped fruit, 7 to 10 cm in diameter, is fragrant and fuzzy and contains many seeds. It is hard and acidic, but cooking with sugar turns it into delicious preserves and candied fruit.[210]

Markh and Kozenko[211] examined 40 varieties, some of which had red anthocyanin pigmented skin. The anthocyanins and flavonoids identified were cyanidin 3-glucoside, cyanidin 3,5-diglucoside, and quercetin 3-glucoside. The unripe fruit of quince also contains proanthocyanidins which on hydrolysis yield (−)-epicatechin and (+)-catechin.[212]

Shariova and Illarionova[213] evaluated quince varieties for content of sugar, acid, pectin, and biologically active substances such as anthocyanins, flavonols, and catechin. Gribovskaya et al.[214] studied the effect of microelements in the soil and the accumulation of biologically active substances in the fruit; they found that pigment intensity and tannin content are directly correlated with levels of Cu, Co, V, and Ba in the soil. Other microelements — Fe, Mo, Cu, and V — were found to influence the ascorbic acid content.

Quince leaves contain the flavonols kaempferol and quercetin, as well as unknown cyanidin derivatives.[215]

F. SASKATOON BERRY

The saskatoon (*Amelanchier alnifolia* Nutt.) is a fruit-bearing shrub native to the southern Yukon and Northwest Territories, and to the prairies of Canada and the northern United States.[216] It is extremely adaptable and grows under a wide range of environmental conditions. Saskatoon plants begin to bear fruit when they are 2 to 4 years old and with adequate management can yield 8 to 10 tonnes of fruit per hectare. The fruit is usually referred to as a berry but it is actually a pome.[216,217] Saskatoon berries were originally a major food source for the native people and early settlers of the North American prairies;

until recently, they could be picked only in the wild. In the last decade there has been increasing interest in the commercial cultivation and utilization of this fruit, and currently there are approximately 200 ha of saskatoon berries in production in western Canada.

Saskatoon berries contain 78 to 81% moisture; up to 19% sugar; small amounts of protein and fat; a fair amount of fiber; and relatively large amounts of potassium, iron, magnesium, and aluminum.[218] The predominant aroma component is benzaldehyde,[219] and the predominant acids are malic and citric.[230]

There are at least four anthocyanins in ripe saskatoon berries, of which cyanidin 3-galactoside and 3-glucoside account for about 61 and 21% of the total anthocyanins, respectively. Other phenolic compounds characterized include cyanidin 3-xyloside, chlorogenic acid, and rutin.[221] Anthocyanin content of saskatoon berries ranges from 25 to 179 mg/100 g of berries, and total phenolics range from 0.17 to 0.52%.[222] Cultivar, maturity level, and year of production, however, affect content of total anthocyanin and total phenolics of fruit.[222]

The influence of acetaldehyde and catechin on aqueous solutions of cyanidin 3-glucoside and aqueous and alcohol extracts of saskatoon berries was investigated by Green and Mazza.[223] Presence of acetaldehyde in the fruit extracts and both acetaldehyde and catechin in the cyanidin 3-glucoside model system caused a marked increase in color intensity during storage. Color intensification was attributed to molecular condensation involving catechin, acetaldehyde, and anthocyanin as well as other phenolic compounds in the berry extracts.

REFERENCES

1. **Anon.,** *U.S. Standards for Apples,* Fresh Products Standardization and Inspection Branch, Fruits and Vegetables Division, U.S. Department of Agriculture/PMA, Washington, DC, 1966.
2. **Saure, M.C.,** External control of anthocyanin formation in apple, *Sci. Hortic.,* 42, 181, 1990.
3. **Asen, S., Stewart, R.N., and Norris, K.H.,** Copigmentation of anthocyanins in plant tissues and its effect on color, *Phytochemistry,* 11, 1139, 1972.
4. **Mazza, G. and Brouillard, R.,** The mechanism of copigmentation of anthocyanins in aqueous solutions, *Phytochemistry,* 29, 1097, 1990.
5. **Duncan, I.J. and Dustman, R.B.,** The anthocyanin pigment of the Winesap apple, *J. Am. Chem. Soc.,* 58, 1511, 1936.
6. **Sando, C.E.,** Coloring matters of Grimes Goden, Jonathan and Stayman Winesap apples, *J. Biol. Chem.,* 117, 45, 1937.
7. **Fouassin, A.,** Identification par chromatographie des pigments de fruits et des legumes, *Rev. Ferment. Ind. Aliment,* 11, 175, 1956.
8. **Walker, J.R.L.,** Flavonoid pigments in the skin of New Zealand apples, *N.Z. J. Sci.,* 7, 585, 1964 (CA 62, 9461a).

9. **Pais, I. and Gombkoto, G.**, Chemical analysis of Hungarian types of apples. I. Anthocyanin pigments of Jonathan apple, *Kert. Szolesz. Foiskola Kozl.*, 31, 71, 1967 (CA 68, 84949).

10. **Sun, B.H. and Francis, F.J.**, Apple anthocyanins: identification of cyanidin 7-arabinoside, *J. Food Sci.*, 32, 647, 1967.

11. **Posokhlyarova, N.S.**, Effect of tree top formation on the dynamics of anthocyanins and sugars in apple skins, *Fiziol. Rast.*, 22, 859, 1976 (CA 83, 175642).

12. **Timberlake, C.F. and Bridle, P.**, Anthocyanins of apples and pears. Occurrence of acylated derivatives, *J. Sci. Food Agric.*, 22, 509, 1971.

13. **Samorodova-Bianki, G.B. and Bazarova, V.I.**, Extraction and identification of anthocyanins in apples, *Fiziol. Rast.*, 17, 189, 1970 (CA 72, 131136).

14. **Bazarova, V.I. and Samorodova-Bianki, G.B.**, Comparative study of apple anthocyanins, *Izv. Vyssh. Uchebn. Zaved., Pishch. Tekhnol.*, 5, 46, 1971 (CA 76, 44780).

15. **Mazza, G. and Velioglu, Y.S.**, Anthocyanins and other phenolic compounds in fruits of red-flesh apples, *Food Chem.*, 43, 113, 1992.

16. **Okuse, I. and Naoki, S.**, Anthocyanins in the bark of tissue of apple tree branches, *Hirosaki Daigaku Nogakubu Gakujutsu Nokaku*, 25, 56, 1974 (CA 84, 14690).

17. **Protsenko, D.P., Bogomaz, K.I., and Korshuk, T.P.**, Anthocyanins of the bark of cultivated fruit trees and their dynamics during the course of the year, *Visn. Kiiv. Univ. No. 3, Ser. Biol.*, 1, 56, 1960 (CA 60, 4459d).

18. **Leonchenko, V.G. and Khanina, N.P.**, Dynamics of cyanidins in the bark of one-year old apple branches in relation to frost resistance, *Byull. Nauchn. Inf. Tsentr. Genet. Lab*, 42, 15, 1985 (CA 105, 206448).

19. **Baker, E.A. and Campbell, A.I.**, Effect of viruses on the pigment composition of apple bark and pear leaf, *Annu. Rep. Long Ashton Agric. Hortic. Res. Stn.*, p. 141, 1967 (HA 38, 659).

20. **Andersen, P.C., Lombard, P.B., and Westwood, M.N.**, Leaf conductance, growth and survival in willow and deciduous fruit tree under flooding soil conditions, *J. Am. Soc. Hortic. Sci.*, 109, 132, 1984.

21. **Dick, A.J., Redden, P.R., DeMarco, A.C., Lidster, P.D., and Brindley, T.B.**, Flavonoid glycosides of Spartan apple peel, *J. Agric. Food Chem.*, 35, 529, 1987.

22. **Fisher, D.J.**, Phenolic compounds of the apple fruit cuticle, *Annu. Rep. Long Ashton Agric. Hortic. Res. Stn.*, p. 255, 1966–1967.

23. **Harborne, J.B.**, *Comparative Biochemistry of the Flavonoids*, Academic Press, London, 1967.

24. **Oleszek, W., Lee, C.Y., Jaworski, A.W., and Price, K.R.**, Identification of some phenolic compounds in apples, *J. Agric. Food Chem.*, 36, 430, 1988.

25. **Teuber, H. and Herrmann, K.**, Flavonolglykoside der Apfel (*Malus silvestris* Mill.), *Z. Lebensm. Unters. Forsch.*, 166, 80, 1978.

26. **Mosel, H.D. and Herrmann, K.**, The phenolics of fruits. III. The contents of catechins and hydroxycinnamic acids in pome and stone fruits, *Z. Lebensm. Unters. Forsch.*, 154, 6, 1974.

27. **Lea, A.G.H. and Timberlake, C.G.**, The phenolics of ciders. I. Procyanidins, *J. Sci. Food Agric.*, 25, 1537, 1974.

28. **Risch, B. and Hermann, K.**, Die Gehalte an Hydroxyzimtsäure-Verbindungen und Catechinen in Kern und Steinobst, *Z. Lebensm. Unters. Forsch.*, 186, 225, 1988.

29. **Hulme, A.C. and Rhodes, M.J.C.**, Pome fruits, in *The Biochemistry of Fruits and Their Products*, Vol. 2, Hulme, A.C., Ed., Academic Press, London, 1971, 363.

30. **Dick, A.J.**, Apple fruit polyphenols and fruit softening, *Bull. Liaison Groupe Polyphènols*, 13, 358, 1986.

31. **Herrman, K.**, Über Oxydation Ferment under phenolische Substrate in Gemüse und Obst. Catechin, Oxyzimtsaüre und *O*-phenol Oxydase, *Z. Lebensm. Unters. Forsch.*, 108, 152, 1953.

32. **Möller, B. and Herrmann, K.**, Ester des Hydroxyzimtsäuren mit Chinasäure als Inhaltsstoffe des Stein und Kernobstes, *Lebensmittelchem. Gerichtl. Chem.*, 36, 135, 1982.

33. **Möller, B. and Herrmann, K.**, Quinic acid esters of hydroxycinnamic acids in stone and pome fruit, *Phytochemistry*, 22, 477, 1983.

34. **Macheix, J.J., Rateau, J., Fleuriet, A., and Bureau, D.**, Les esters hydroxycinnamiques des fruits, *Fruits*, 32, 397, 1977.

35. **Macheix, J.J. and Fleuriet, A.**, Les dèrivès hydroxycinnamiques des fruits, *Bull. Liaison Groupe Polyphènols*, 13, 337, 1986.

36. **Koumba-Koumba, D. and Macheix, J.J.**, Biosynthesis of hydroxycinnamic derivatives in apple fruit cell suspension culture, *Physiol. Veg.*, 20, 137, 1982.

37. **Lea, A.G.H. and Arnold, G.M.**, The phenolics of ciders: bitterness and astringency, *J. Sci. Food Agric.*, 29, 478, 1978.

38. **Lea, A.G.H.**, Tannin and colour in English cider apples, *Fluess. Obst.*, 8, 356, 1984.

39. **Oleszek, W., Lee, C.Y., and Price, K.R.**, Apple phenolics and their contribution to enzymatic browning reactions, *Acta Soc. Bot. Pol.*, 58, 273, 1989.

40. **Spanos, G.A., Wrolstad, R.E., and Heatherbell, D.A.**, Influence of processing and storage on the phenolic composition of apple juice, *J. Agric. Food Chem.*, 38, 1572, 1990.

41. **Dick, A.J., Williams, R., Bearne, S.C., and Lidster, P.D.**, Quercetin glycosides and chlorogenic acid: inhibitors of apple β-galactosidase and of apple softening, *J. Agric. Food Chem.*, 33, 798, 1985.

42. **Lidster, P.D. and McRae, K.B.**, Effects of vacuum infusion of a partially purified β-galactosidase inhibitor on apple quality, *HortScience*, 20, 80, 1985.

43. **Lidster, P.D., Dick, A.J., DeMarco, A., and McRae, K.B.**, Application of flavonoid glycosides and phenolic acids to suppress firmness loss in apples, *J. Am. Soc. Hortic. Sci.*, 111, 892, 1986.

44. **Wellington, R.**, An experiment in breeding apples. II, *N.Y. State Agric. Exp. Stn., Tech. Bull.*, 106, 1924.

45. **Schmidt, H.**, The inheritance of anthocyanin in apple fruit skin, *Acta Hortic.*, 224, 89, 1988.

46. **Pecket, R.C. and Small, C.J.**, Occurrence, location and development of anthocyanoplasts, *Phytochemistry*, 19, 2571, 1980.

47. **Yamaki, S.**, Isolation of vacuoles from immature apple fruit flesh and compartmentation of sugars, organic acids, phenolic compounds and amino acids, *Plant Cell Physiol.*, 25, 151, 1984.

48. **Siegelman, H.W. and Hendricks, S.B.**, Photocontrol of alcohol, aldehyde and anthocyanin production in apple skin, *Plant Physiol.*, 33, 409, 1958.

49. **Siegelman, H.W. and Hendricks, S.B.**, Photocontrol of anthocyanin synthesis in apple skin, *Plant Physiol.*, 33, 185, 1958.

50. **Misic, P.D. and Tesovic, Z.V.**, Anthocyanin color distribution in the skin of Cox's Orange Pippin apple and its dark-red sports, *Hortic. Res.*, 11, 161, 1971.

51. **Schmid, P.**, Idaein-decomposing enzymes in apple peel, *Z. Lebensm. Unters. Forsch.*, 146, 198, 1971.

52. **Tesovic, Z.V.**, A histological investigation of cell contents and the distribution of anthocyanins in the skin of apples from parental varieties and F1 hybrid seedlings, *Jugosl. Vocarstvo*, 10, 275, 1976 (CA 88, 186222).

53. **Dayton, D.F.**, Red color distribution in apple skin, *Proc. Am. Soc. Hortic. Sci.*, 74, 72, 1959.

54. **Dayton, D.F.**, The distribution of red color in the skin of apple varieties of McIntosh parentage, *Proc. Am. Soc. Hortic. Sci.*, 82, 51, 1963.

55. **Dayton, D.F.**, Variation in the pattern of red color distribution in the skin of Delicious and Starking bud sport varieties, *Proc. Am. Soc. Hortic. Sci.*, 84, 44, 1964.

56. **Pratt, C., Way, R.D., and Einset, J.**, Chimeral structure of red sport of Northern spy apple, *J. Am. Soc. Hortic. Sci.*, 110, 419, 1975.

57. **Schmidt, M.**, Contributions to research on apple breeding. II. Morphological-pomological studies on the F1 progeny of a cross between a cultivated variety and *M. niedzwetzkyana, Zuechter,* 23, 327, 1953 (PBA 1954, 1389).

58. **Sampson, D.R.**, Use of a leaf color market gene to detect apomixis in *Malus* species and observations on the variability of the apomictic seedlings, *Can. J. Plant Sci.*, 49, 409, 1969.

59. **Isaev, S.I. and Maksimova, I.P.**, The inheritance of the character of leaf coloration and quantity of anthocyanins in reciprocal apple hybrids, *Nauchn. Dokl. Vyssh. Shk., Biol. Nauk.*, 1, 109, 1968 (PBA 1970, 8313).

60. **Nybom, N. and Bergendal, P.O.**, Pome fruits and bush fruits, *Inst. Vaxtfor. Frukt Baer, Balsgard,* 1960, 8, 1960 (PBA 1962, 1073).

61. **Anon.**, Horticultural research, *Annu. Rep., s'Gravenhage, The Netherlands*, 242, 1965 (PBA 1968, 1201).

62. **Downs, R.J., Siegelman, W.L., Butler, W.L., and Hendricks, S.B.**, Photoreceptive pigments for anthocyanin synthesis in apple skin, *Nature (London),* 205, 909, 1962.

63. **Proctor, J.T.A. and Creasy, L.L.**, Effect of supplementary light on anthocyanin synthesis in McIntosh apples, *J. Am. Soc. Hortic. Sci.,* 96, 523, 1971.

64. **Arakawa, O., Hori, Y., and Ogata, R.**, Characteristics of color development and relation between anthocyanin synthesis and PAL activity in Starking Delicious, Fuji and Mutus apple fruits, *J. Jpn. Soc. Hortic. Sci.,* 54, 424, 1986 (CA 105, 23313).

65. **Faragher, J.D.**, Temperature regulation of anthocyanin accumulation in apple skin, *J. Exp. Bot.,* 34, 1291, 1983.

66. **Arthur, J.M.**, Red pigment production in apples by means of artificial light sources, *Contrib. Boyce Thompson Inst.*, 4, 1, 1932.

67. **Chalmers, D.J. and Faragher, J.D.**, Regulation of anthocyanin synthesis in apple skin. I. Comparison of the effect of cycloheximide, ultraviolet light, wounding and maturity, *Aust. J. Plant Physiol.,* 4, 111, 1977.

68. **Arakawa, O., Hori, Y., and Ogata, R.**, Relative effectiveness and interaction of ultraviolet-B, red and blue light in anthocyanin synthesis of apple fruit, *Physiol. Plant,* 64, 323, 1985.

69. **Proctor, J.T.A.**, The color of plants, *Span,* 13, 165, 1970.

70. **Proctor, J.T.A.**, Color stimulation in attached apples with supplementary light, *Can. J. Plant Sci.,* 54, 499, 1974.

71. **Aoki, S., Araki, C., Kaneo, K., and Katayama, O.**, L-Phenylalanine ammonia-lyase activities in Japanese chestnut, strawberry, apple fruit and bracken, *J. Food Sci. Technol.,* 17, 507, 1970.

72. **Hyodo, H.**, Phenylalanine ammonia-lyase in strawberry fruits, *Plant Cell Physiol.,* 12, 989, 1971.

73. **Faragher, J.D. and Chalmers, D.J.**, Regulation of anthocyanin synthesis in apple skin. III. Involvement of PAL, *Aust. J. Plant Physiol.,* 4, 133, 1977.

74. **Tan, S.C.**, Relationship and interactions between phenylalanine ammonia-lyase, phenylalanine ammonia-lyase inactivating system, and anthocyanin in apples, *J. Am. Soc. Hortic. Sci.,* 104, 581, 1979.

75. **Tan, S.C.**, Phenylalanine ammonia-lyase and phenylalanine ammonia-lyase inactivating system: effects of light, temperature, and mineral deficiencies, *Aust. J. Plant Physiol.,* 7, 159, 1980.

76. **Creasy, L.L.**, Phenylalanine ammonia-lyase inactivating system in sunflower leaves, *Phytochemistry,* 15, 673, 1976.

77. **Naumann, W.D.**, Untersuchungen über den Einfluß der Temperatur auf die Nachreife von Äpfeln der Sorten 'Jonathan' und 'Ontario', *Gartenbauwissenschaft,* 29, 523, 1964.

78. **Diener, H.-A and Naumann, W.D.**, Der einfluß von Tag-und Nachttemperaturen auf die Anthocyansynthese in der Apfelschale, *Gartenbauwissenschaft*, 46, 125, 1981.

79. **Uota, M.**, Temperature studies on the development of anthocyanin in McIntosh apples, *Proc. Am. Soc. Hortic. Sci.*, 59, 231, 1952.

80. **Creasy, L.L.**, The role of low temperature in anthocyanin synthesis in McIntosh apples, *Proc. Am. Soc. Hortic. Sci.*, 93, 716, 1968.

81. **Faragher, J.D. and Brohier, R.L.**, Anthocyanin accumulation in apple skin during ripening: regulation by ethylene and phenylalanine ammonia-lyase, *Sci. Hortic.*, 22, 89, 1984.

82. **Beattie, J.M.**, The effect of differential nitrogen fertilization on some of the physical and chemical factors affecting the quality of Baldwin apples, *Proc. Am. Soc. Hortic. Sci.*, 63, 1, 1954.

83. **Lekhova, E.**, Effect of N fertilization on some apple fruit qualities, *Gradinar. Lozar. Nauka*, 9, 9, 1971 (CA 78, 158237).

84. **Lekhova, E.**, Changes in apple fruit quality as affected by N fertilization, *Bulgarski Plodove, Zelench. Konservi*, 6, 20, 1974 (HA 1975, 9142).

85. **Ulyanova, D.A. and Ulyanov, A.M.**, Effect of N fertilizers on the pigment content in apples, *Dokl. Vses. Akad. Skh. Nauk V.I. Lenina*, 8, 19, 1977 (HA 1978, 2028).

86. **Kaether, K.E.**, Def Einfluß der mineralischen Stickstoffernährung auf Inhaltsstoffe des Apfels, insbesondere auf die Pigmente der Fruchtschale, *Gartenbauwissenschaft*, 30, 361, 1965.

87. **Lüdders, P. and Bünemann, G.**, Die Wirkung Jahreszeitlich unterschiedlicher Kaliumverfügbarkeit auf Apfelbäume. III. Einfluß auf des generative Wachstum, *Gartenbauwissenschaft*, 39, 69, 1974.

88. **Walter, T.E.**, Factors affecting fruit colour in apples: a review of world literature, *Rep. East Malling Res. Stn.*, p. 70, 1967.

89. **Clerinx, P.**, Evaluation de l'aptitude du sol á la culture de la variètè Jonagold, *Rev. Agric.*, 40, 25, 1987.

90. **Saure, M.C.**, Summer pruning effects in apple — a review, *Sci. Hortic.*, 30, 253, 1987.

91. **Brunner, T. and Droba, B.**, Korrelációs gèpi gyümölcsfametszès (gèpi metszès èlettani alapon), *Bot. Kozl.*, 657, 145, 1980.

92. **Schumacher, R., Frankhauser, F., and Stadler, W.**, Influence of growth regulators, ringing and root cutting on apple quality and physiological disorders, *Acta Hortic.*, 179, 731, 1986.

93. **Schupp, J.R. and Ferree, D.C.**, Effects of root pruning at four levels of severity on growth and yield of 'Melrose'/M. 26 apple trees, *J. Am. Soc. Hortic. Sci.*, 113, 194, 1988.

94. **Fletcher, L.A.**, A preliminary study of the factors affecting red color of apples, *Proc. Am. Soc. Hortic. Sci.*, 265, 191, 1928.

95. **Proctor, J.T.A. and Lougheed, E.C.**, The effect of covering apples during development, *HortScience*, 11, 108, 1976.

96. **Kikuchi, T.**, Influence of fruit-bag practice on colouration process in apples of different varieties, *Bull. Fac. Agric., Hirosaki Univ.*, 10, 89, 1964.

97. **Arakawa, O.**, Characteristics of color development in some apple cultivars: changes in anthocyanin synthesis during maturation as affected by bagging and light quality, *J. Jpn. Soc. Hortic. Sci.*, 57, 373, 1988 (HA 1990, 7900).

98. **Dustman, R.B. and Duncan, I.J.**, Effect of certain thiocyanate sprays on foliage and fruit in apples, *Plant Physiol.*, 15, 343, 1940.

99. **Looney, N.E.**, Interaction of ethylene, auxin, and succinic acid-2,2-dimethylhydrazine in apple fruit ripening control, *J. Am. Soc. Hortic. Sci.*, 96, 350, 1971.

100. **Edgerton, L.J. and Hoffman, M.B.**, Inhibition of fruit drop and colour stimulation with *N*-dimethylaminosuccinamic acid, *Nature (London)*, 209, 314, 1966.

101. **Williams, M.W.**, Use of bioregulators to control vegetative growth of fruit trees and improve fruiting efficiency, *Acta Hortic.*, 146, 97, 1984.
102. **Lever, B.G.**, 'Cultar' — a technical overview, *Acta Hortic.*, 179, 459, 1986.
103. **Bhartiya, S.P., Thakur, D.R., and Kar, P.L.**, Effect of auxin, gibberellic acid and NU-Spartin on anthocyanin and ascorbic acid in apple fruit cv. Starking Delicious, *Prog. Hortic.*, 15, 69, 1983 (HA 1985, 4173).
104. **Hegazi, E.S. and Plich, H.**, The effect of gibberellin, auxin, cytokinin and abscisic acid on ethylene production and some other symptoms of fruit ripening in apples, *Bull. Acad. Pol. Sci., Ser. Sci. Biol.*, 28, 253, 1980.
105. **Oota, S., Masuda, T., and Tamura, T.**, Apple flesh tissue culture and anthocyanin formation in derived callus tissues, *J. Jpn. Soc. Hortic. Sci.*, 52, 117, 1980.
106. **Pollard, J.E.**, Effect of SADH, ethephon and 2,4,5 T on color and storage quality of McIntosh apples, *Proc. Am. Soc. Hortic. Sci.*, 99, 341, 1974.
107. **Kabil, M.T., Fayek, M.A., Moursy, H.A., and Khalil, F.H.**, Effect of some growth regulators on quality and coloration of Barkher apple fruits, *Egypt. J. Hortic.*, 9, 71, 1982 (FSTA 15, 2J160).
108. **Böemeke, H.**, Der Tuzeteinsatz im Obstbau unter Beruecksichtigung der Anthozyanbildung, *Mitt. Ovr. Jork.*, 14, 139, 1959 (HA 1960, 249).
109. **Fortes, G.R. de L.**, Avaliaçao da coloraçao vermelha e anticipaçao da colheita do frutos de macieira (*Malus domestica* Borckh.) cv. Fuji com o emprego de ethephon, *Empresa Catarinese Pesq. Agropecu. S.A.*, 3, 741, 1984.
110. **Graf, H.**, Die Anwendungsbereiche des Wachstumsregulators Alar im Apfelanbau an der Niederelbe, *Mitt. Ovr. Jork.*, 41, 186, 1986.
111. **Graf, H.**, Möglichkeiten und Ergebnisse zur Förderung der Fruchtfarbe bei Apfelfrüchten durch Wachstumsregulatoren, *Mitt. Ovr. Jork.*, 41, 399, 1986.
112. **Blanpied, G.D., Forshey, G.C., Styles, W.C., Green, D.W., Lord, W.J., and Bramlage, W.J.**, Use of ethephon to stimulate red color without hastening ripening of McIntosh apples, *J. Am. Soc. Hortic. Sci.*, 100, 379, 1975.
113. **Castro, H.R., Rodriguez, R., Barria, J., Benitez, C.E., and Francile, S.**, Efectos del Alar sobre la calidad de la produccion de manzanas Red Delicious Spur, *Estac. Exp. Reg. Agropecu., Alto Valle*, 34, 1984.
114. **Castro, H.R., Rodriguez, R., Barria, J., Benitez, C.E., and Francile, S.**, Effecto de algunos regulardores del crecimiento sobre la coloracion y conservaccion frigorifica de los frutos de manzana cv. Red Delicious. *Estac. Exp. Reg. Agropecu., Alto Valle*, 37, 1984.
115. **Alfermann, W. and Reinhard, E.**, Isolierung anthocyanhaltiger und anthocyanfreier Gewebestämme von *Daucus carota:* Einfluß von Auxinen auf die Anthocyanbildung, *Experientia*, 27, 353, 1971.
116. **Thiman, K.V., Edmonson, Y.H., and Radner, B.S.**, The biogenesis of anthocyanins. III. The role of sugars in anthocyanin formation, *Arch. Biochem. Biophys.*, 34, 305, 1951.
117. **Pirie, A.J. and Mullins, M.G.**, Changes in anthocyanin and phenolics content of grape leaf and fruit tissues treated with sucrose, nitrate and abscisic acid, *Plant Physiol.*, 58, 468, 1976.
118. **Vestrheim, S.**, Effects of chemical compounds on anthocyanin formation in 'McIntosh' apple skin, *J. Am. Soc. Hortic. Sci.*, 95, 712, 1970.
119. **Faust, M.**, Physiology of anthocyanin development in McIntosh apple. I. Participation of pentose phosphate pathway in anthocyanin development *Proc. Am. Soc. Hortic. Sci.*, 87, 1, 1965.
120. **Gianfagna, T.J. and Berkowitz, G.A.**, Glucose catabolism and anthocyanin production in apple fruit, *Phytochemistry*, 25, 607, 1986.
121. **Macheix, J.J., Fleuriet, A., and Billot, J.**, *Fruit Phenolics*, CRC Press, Boca Raton, FL, 1990, chap. 3.

122. **Grisebach, H.,** Recent developments in flavonoids biosynthesis, in *Pigments in Plants,* Czygan, F.-C., Ed., Fischer, Stuttgart, Germany, 1980, 187.

123. **Ebel, J. and Hahlbroch, K.,** Biosynthesis, in *The Flavonoids: Advances in Research,* Harborne, J.B. and Mabry, T.J., Eds., Chapman & Hall, London, 1982, 641.

124. **Grisebach, H.,** Biosynthesis of anthocyanins, in *Anthocyanins as Food Colors,* Markakis, P., Ed., Academic Press, New York, 1982, 41.

125. **Heller, W. and Forkmann, G.,** Biosynthesis, in *Flavonoids, Advances in Research Since 1980,* Harborne, J.B., Ed., Chapman & Hall, London, 1988, 399.

126. **Murphey, A.S. and Dilley, D.R.,** Anthocyanin synthesis and maturity of McIntosh apples as influenced by ethylene-releasing compounds, *J. Am. Soc. Hortic. Sci.,* 113, 718, 1988.

127. **Camm, E.L. and Towers, G.H.N.,** Review article: phenylalanine ammonia lyase, *Phytochemistry,* 12, 961, 1973.

128. **Schulz, H.,** Veranderungen der Farbstoffe in der Apfelfruchte wahrend der Entwiklung und Lagerung, in *Physiologische Probleme im Obstbau* — Dtsch. Akad. Landwirtsch., Martin Luther Universität, Berlin, 1970.

129. **Lin, T.Y., Koehler, P.E., and Shewfelt, R.L.,** Stability of anthocyanins in the skin of Starkrimson apples stored unpackaged, under heat-shrinkable wrap and in-package modified atmosphere, *J. Food Sci.,* 54, 405, 1989.

130. **Beveridge, T., Franz, K., and Harrison, J.E.,** Clarified natural apple juice: production and storage stability of juice and concentrate, *J. Food Sci.,* 51, 411, 1986.

131. **Spayd, S.E., Nagel, C.W., Hayrynen, L.D., and Drake, S.R.,** Color stability of apple and pear juices blended with fruit juices containing anthocyanins, *J. Food Sci.,* 49, 411, 1984.

132. **Brouillard, R. and Delaporte, B.,** Degradation termique des anthocyanes, *Bull. Liaison Groupe Polyphènols,* 8, 25, 1978.

133. **Timberlake, C.F.,** Anthocyanins — occurrence, extraction and chemistry, *Food Chem.,* 5, 69, 1980.

134. **Markakis, P.,** Stability of anthocyanins in foods, in *Anthocyanins as Food Colors,* Markakis, P., Ed., Academic Press, New York, 1982, 163.

135. **Walker, J.R.L.,** Enzymic browning of apples, *Trien. Rep. Cawton Inst.,* 1966, 238 pp (HA 1967, 6385).

136. **Casoli, U., Dall'Aglio, G., and Leoni, C.,** Azione di estratti enzimatici vegetali sui pigmenti antocianici delle melanzane, delle mele e del ribes, *Ind. Conserve,* 44, 193, 1969.

137. **Snapyan, G.G., Palazan, T.N., and Melkonyan, L.T.,** Biochemical and marketing characteristics of autumn and winter varieties of Armenian pears, *Biol. Zh. Arm.,* 37, 241, 1984 (FSTA 18, 2J79).

138. **Garcìa-Mina, M.C., Fuertes-Lasala, M.E., Fernandez, M., Vega, F.A., and Martinez-Valls, L.,** Componentes fenolicos en variedades cultivades de *Pyrus communis* L. H. Estimaciones quantitativas, *An. Bromatol.,* 26, 177, 1974.

139. **Harborne, J.B. and Hall, E.,** Plant polyphenols. XIII. The systematic distribution and origin of anthocyanins containing branched trisaccharides, *Phytochemistry,* 3, 453, 1964.

140. **Francis, F.J.,** Anthocyanins in pears, *HortScience,* 5, 42, 1970.

141. **Redelinghuys, H.J.P., Koeppen, B.H., and De Swardt, G.H.,** The anthocyanin in a South African red sport of *Pyrus communis* L. cv. Williams Bon Chrètien, *Agrochemophysica,* 5, 21, 1973 (HA 1974, 5401).

142. **Dayton, D.F.,** The pattern and inheritance of anthocyanin distribution in red pears, *Proc. Am. Soc. Hortic. Sci.,* 89, 110, 1966.

143. **Fregoni, M. and Roversi, A.,** Le reversioni cromatiche sui frutti di Max Red Bartlett, *Ann. Fac. Agrar. Univ. Bari,* 8, 96, 1968.

144. **Wang, C.Y. and Mellenthin, W.M.,** Chlorogenic acid levels, ethylene production and respiration of d'Anjou pears affected with cork spot, *HortScience,* 8, 180, 1973.

145. **Wang, C.Y. and Mellenthin, W.M.,** Relationship of friction discoloration to phenolic compounds in d'Anjou Pears, *HortScience,* 8, 321, 1973.

146. **Duggan, M.B.,** Identity and occurrence of certain flavonol glycosides in four varieties of pears, *J. Agric. Food Chem.,* 17, 1098, 1969.

147. **Van Buren, J.,** Fruit phenolics, in *The Biochemistry of Fruits and Their Products,* Vol. 1, Hulme, A.C., Ed., Academic Press, London, 1970, 269.

148. **Nortje, B.K. and Koeppen, B.H.,** The flavonol glycosides in the fruit of *Pyrus communis* L. cultivars Bon Chrètien, *Biochem. J.,* 97, 209, 1965.

149. **Wald, B., Wray, W., Galensa, R., and Hermann, K.,** Malonated flavonol glycosides and 3,5-dicaffeoylquinic acid acid from pears, *Phytochemistry,* 28, 663, 1989.

150. **Beveridge, T. and Harrison, J.E.,** Nonenzymatic browning in pear juice concentrate at elevated temperatures, *J. Food Sci.,* 49, 1335, 1984.

151. **Plocharski, W., Zbrosczyk, J., and Lenartowicz, W.,** Aronia fruit (*Aronia melanocarpa* Elliott) as a natural source of anthocyanin colorants. II. The stability of the color of aronia juices and extracts, *Fruit Sci. Rep.,* 16, 41, 1989.

152. **Wilszka-Jeszka, J., Anders, B., Los, J., and Urbaniak-Olszewska, M.,** Influence des techniques de prètraitement des fruits d'*Aronia melanocarpa* sur la composition polyphènolique des vins, *Bull. Liaison Groupe Polyphènols,* 13, 43, 1986.

153. **Shapiro, D.K., Chekalinskaya, I.I., Dovnar, T.V., and Yankovskaya, A.K.,** Biologically active substances of preserves from black chokeberry (*Aronia melanocarpa*), *Vestsi Akad. Navuk BSSR, Ser. Biyal. Navuk,* 6, 37, 1973 (CA 80, 106968).

154. **Chekalinskaya, I.I.,** Biologically active substances of the fruit of some hawthorn and black-fruited chokeberry species, *Tezisy Dokl. Konf. BSSR Biokhim. Ova,* 2, 1974, 145, 1974 (CA 85, 139751).

155. **Yuditskaite, S.V.,** Some data on the contents of vitamin C, carotene and anthocyanins in the fruit of *Arona (Amelanchier) melanocarpa* grafted on *Sorbus aucuparia* in Lithuania, *Vitam. Rastit, Resur. Ikh. Ispolz., Moskow, Russia, Moskow Univ.,* 158, 1977 (HA 1978, 6451).

156. **Stroev, E.A. and Martynov, E.G.,** Effect of chlorocoline chloride on anthocyanin content in *Aronia melanocarpa* fruit, *Khim. Prir. Soedin.,* 409, 1979 (CA 91, 187862).

157. **Martynov, E.G.,** Effect of microelement on anthocyanin accumulated in *Aronia (Amelanchier) melanocarpa* fruit, *Khim. Prir. Soedin.,* 4, 526, 1978 (HA 1979, 2444).

158. **Martynov, E.G., Suprunov, N.I., Frolova, V.A., Promakhova, S.S., and Skokova, I.G.,** Anthocyanins of *Amelachier vulgaris* fruits, *Khim. Prir. Soedin.,* 5, 725, 1980 (CA 94, 99787).

159. **Kalemba, D., Kurowski, A., and Gora, J.,** Polyphenols in *Aronia melanocarpa* fruits, *Bull. Liaison Groupe Polyphènols,* 12, 132, 1984.

160. **Oszmianski, J. and Sapis, J.C.,** Anthocyanins in fruits of *Aronia melanocarpa* (chokeberry), *J. Food Sci.,* 53, 1241, 1988.

161. **Szepczinska, K.,** Polyphenolic compounds in *Aronia melanocarpa*, *Acta Pol. Pharm.,* 46, 404, 1989 (CA 113, 74808).

162. **Rosa, J. and Krugly, G.,** Black rowanberries as a raw material for manufacture of fruit juices, juice concentrates and wines, *Przem. Ferment. Owocowo Warzywny,* 31, 25, 1987 (FSTA 21, 5H75).

163. **Shapiro, D.K., Bereskovskii, V.V., Dashkevich, L.E., and Dovnar, T.V.,** Effect of the drying process and conditions on the stability of bioflavonoids of the black chokeberry (*Aronia melanocarpa*), *Vestsi Akad. Navuk BSSR, Ser. Biyal. Navuk,* 6, 28, 1974 (CA 82, 96615).

164. **Nakhmedov, F.G., Frumkin, M.L., Svitsunova, V.A., and Myachin, V.M.,** Colouring matter from the pomace of black rowanberries and blackcurrants, *Konservn. Ovoshchesush. Promst.,* 4, 15, 1975 (FSTA 8, 6J903).

165. **Nakhmedov, F.G., Frumkin, M.L., Bolshakova, T.I., and Soloveva, K.V.**, Experience in the industrial production of natural food dyes from black chokeberry and black currants residues, *Konserv. Ovoshchesush Promst.*, 1, 3, 1980 (CA 92, 127192).

166. **Plocharski, W. and Zbrosczyk, J.**, Aronia fruit (*Aronia melanocarpa* Elliott) as a natural source of anthocyanin colorants. I. Recovery of anthocyanins from aronia fruit by using alcoholic extraction, *Fruit Sci. Rep.*, 16, 33, 1989.

167. **Bailey, L.H.**, *Manual of Cultivated Plants*, Macmillan, New York, 1977.

168. **Kiselev, V.E., Demina, T.G., Azovtsev, G.R., Polyakova, L.V., and Vysochina, G.I.**, Flavonoids of plants of Siberian flora, *Rastit. Bogatstva Sib.*, 1971, 93 (CA 77, 85643).

169. **Sobolevskaya, K.A. and Demina, T.G.**, Characteristics of the accumulation of anthocyanins in the bush fruits of Gorny Altai, *Izv. Sib. Otd. Akad. Nauk SSSR, Ser. Biol. Nauk*, 3, 62, 1970 (CA 75, 59777).

170. **Sobolevskaya, K.A. and Demina, T.G.**, Anthocyanins of the ripe fruits of several wild bush plants, *Tr. Vses. Semin. Biol. Aktiv. (Lech.), Veshchestvam Plodov Yagod*, 1970, 129 (CA 81, 74951).

171. **Demina, T.G.**, Anthocyanins of the genus *Cotoneaster* growing in the Gorny Altai, *Biol. Akt. Soedin. Rastit. Sib. Flory*, 1974, 27 (CA 82, 40689).

172. **Piferri, P.G. Vaccari, A., and Bisognani, M.**, La sostanza colorante del *Cotoneaster pyracantha* (agazzino), *Ind. Conserve*, 50, 216, 1975.

173. **Du, C.T. and Francis, F.J.**, Anthocyanins of *Cotoneaster* and barberry, *HortScience*, 9, 40, 1974.

174. **Kruegel, T. and Krainhoefner, A.**, Flavonoide Farbstoffe der Gattung *Cotoneaster* Medicus (Rosaceae, Maloidae), *Wiss. Z. Friedrich-Schiller-Univ. Jena Naturwiss. Reihe*, 33, 801, 1984.

175. **Robinson, G.M. and Robinson, R.**, A survey of anthocyanins. II. *Biochem. J.*, 26, 1647, 1932.

176. **Han, Y.**, Nutrients of hawthorn (*Crataegus* species) products, *Shipin Kexue (Beijing)*, 125, 56, 1990 (CA 113, 229952).

177. **Chekalinskaya, I.I.**, Biologically active substances of the fruit of some hawthorn and black-fruited chokeberry speices, *Tezisy Dokl. Konf. BSSR, Biokhim. Ova*, 2, 245, 1974 (CA 85, 139751).

178. **Lin, D., Li, W., and Lui, X.**, Color and pigments in hawthorn fruit and juice, *Shipin Fajao Gongye*, 4, 59, 1988 (CA 100, 74109).

179. **Chekalinskaya, I.I. and Dovnar, T.V.**, Studies on the biochemical characteristics of the fruit of hawthorns introduced into Byelorussia, *Intr. Rastit. Optimiz. Okruzhayushch. Sredy Sredstvanii Ozelenyia. Minsk, BSSR, Nauk. Tekhn.*, 1977, 187 (HA 1978, 6352).

180. **Chekalinskaya, I.I. and Dovnar, T.V.**, Biologically active substance of several species of hawthorn introduced into Belorussia, *Polez. Rastit. Priblat. Respub. BSSR Mater. Nauch. Konf.*, 1973, 273 (CA 81, 88018).

181. **Weinges, K.**, The occurrence of catechins in fruits, *Phytochemistry*, 3, 263, 1984.

182. **Ullsperger, R.**, The mode of action of a new active body from hawthorn and of several pure anthocyanins, *Pharmazie*, 8, 923, 1953 (CA 50, 17161e).

183. **Seel, H.**, *Crataegus* active principles in heart therapy, *Planta Med.*, 2, 117, 1954 (CA 50, 10265h).

184. **Vecher, A.S. and Benkovic, E.I.**, Level of flavonoid compounds in ripening fruits of the hawthorn, *Vestsi Akad. Nauk, BSSR, Ser. Biyal. Navuk*, 2, 16, 1971 (CA 75, 31214).

185. **Vecher, A.S. and Benkovic, E.I.**, Flavonoids of some species of the genus *Crataegus*, *Vestsi Akad. Navuk BSSR, Ser. Biyal. Navuk*, 6, 107, 1970 (CA 74, 95463).

186. **Benkovic, E.I.**, Biologically active substances in hawthorn fruits and changes in them during ripening, *Biol. Akt. Vesch. Plodov Yagod, Moskow*, 1976, 170 (HA 1977, 11778).

187. **Paech, K. and Eberhardt, F.**, Research on the biosynthesis of anthocyanins, *Z. Naturforsch.*, 76, 666, 1952.

188. **Freudenberg, K. and Weinges, K.,** Pro-anthocyanidin aus den Fruechten des Weiss-dorns, *Tetrahedron Lett.*, 17, 267, 1961.

189. **Chumbalov, T.C. and Nurgalieva, G.M.,** Anthocyanins of *Malus niedzwetzkyana* Mill., and *Crataegus almaatensis*, *Khim. Khim. Tekhnol.*, 17, 153, 1975.

190. **Santagati, N.A., Duro, R., and Duro, F.,** Ricerche sui pigmenti naturali colorati presenti nelle azzeruole, *Riv. Merceol.*, 23, 97, 1984.

191. **Benk, E.,** Weniger bekannte tropische und subtropische Fruechte, *Fluess. Obst.*, 53, 61, 1986.

192. **Benk, E.,** Wildfruechte als Rohware fuer Konfituren und Gelees, *Gordian*, 85, 7/8, 152, 1985.

193. **Kobodziejski, J. and Gill, S.,** Ascorbic acid content in raw materials containing anthocyanin dyes depending on manner of storage and method of determination, *Acta Pol. Pharm.*, 15, 1985 (JSFA 1959, 170i).

194. **Culpepper, C.W. and Caldwell, J.S.,** The behavior of the anthocyan pigments in canning, *J. Agric. Res.*, 35, 107, 1927.

195. **Robinson, G. and Robinson, R.,** A survey of anthocyanins, I. *Biochem. J.*, 25, 1687, 1931.

196. **Kuusi, T.,** On the chemical composition and characteristic indexes of some Finnish berries, *The State Institute for Technical Research, Series IV-Chemistry*, Helsinki, Finland, 1969, 100.

197. **Pyysalo, H. and Kuusi, T.,** The taste in the berries of mountain ash. The glycoside precursor of sorbic acid, *Suom. Kemistil. B*, 44, 393, 1971.

198. **Shnaidman, L., Kushinskaja, I., Mitelman, M., Efumov, A., Klementeva, I., and Alekseieva, S .,** Biologically active substances in mountain ash berries, *Rastit. Resur.*, 7, 68, 1971.

199. **Vatulina, G.G.,** Chemical study of anthocyanin of fruits of the wild mountain ash, *Tr. Vses. Semin. Biol. Aktiv. (Lech.), Veshchestvam Plodov Yagod*, 4, 162, 1970 (CA 81, 150664).

200. **Pyysalo, H. and Kuusi, T.,** Phenolic compounds from the berries of mountain ash, *Sorbus aucuparia*, *J. Food Sci.*, 39, 636, 1974.

201. **Kolesnik, A.D. and Elizarova, L.G.,** Anthocyanin pigments in the fruit of the mountain ash, *Fenolnye Soedin. Ikh. Biol. Funkts., Mater. Vses. Simp.*, 1, 233, 1966 (CA 71, 10265).

202. **Lenchine, V.S.,** Le sorbe norie, matière première de vinification, *Vinodel. Vinograd. SSSR*, 18, 53, 1958 (IAA 52, 20).

203. **Fan-Yung, A.F., Rechits, M.A., and Dobrosok, L.P.,** Changes in polyphenols of rowanberries during manufacture of concentrated juice, *Izv. Vyssh. Uchebn. Zaved., Pishch. Tekhnol.*, 3, 20, 1979 (FSTA 12, 4H700).

204. **Nakhmedov, F.G., Frumkin, M.L., Stroitelev, G.M., Pushkhareva, P.I., and Svistunova, V.A.,** Method of obtaining a food colorant from vegetable material, U.S.S.R. Patent 522,216, 1976 (FSTA 9, 1T18).

205. **Nakhmedov, F.G., Frumkin, M.L., Bolshakova, T.I., and Soloveva, K.V.,** Industrial manufacture of natural pigments from press cake of black rowanberries and blackcurrants, *Konservn. Ovoshchesush. Prom.*, 1, 3, 1980.

206. **Nakhmedov, F.G., Frumkin, M.L., Malykhina, E.S., Myatkovskii, O.N., and Soloveva, K.V.,** Production of colorants from black rowanberry wastes, *Konservn. Ovoshchesush. Prom.*, 9, 34, 1981 (CA 95, 218985).

207. **Nakhmedov, F.G., Frumkin, M.L., Svitsunova, V.A., and Myachin, V.M.,** Colouring matter from the pomace of black rowanberry and blackcurrant, *Konservn. Ovoshchesush. Prom.*, 4, 15, 1975.

208. **Shnaidman, L., Kushinskaya, I., Mitelman, M., Efumov, A., Klementeva, I., and Alekseieva, S .,** Biologically active substances in mountain ash berries, *Rastit. Resur.*, 7, 68, 1971.

209. **Mabry, T.J. and Ulubelen, A.**, Chemistry and utilization of phenylpropanoids including flavonoids, coumarins and lignans, *J. Agric. Food Chem.*, 28, 188, 1980.

210. **Anon.**, Quinces, *Chocolate, Confect. Bakery Rev.*, 5, 22, 1980.

211. **Markh, A.T. and Kozenko, S.I.**, Biochemical characteristics of quince fruits, *Prikl. Biokhim. Mikrobiol.*, 380, 1965 (CA 64, 1014d).

212. **Foo, L.Y. and Porter, L.J.**, The structure of tannins of some edible fruits, *J. Sci. Food Agric.*, 32, 711, 1981.

213. **Shariova, N.I. and Illarionova, N.P.**, Results of evaluating quince varieties for content of nutrients and biologically active substances, *Byull. Vses. Ordena Lenina Ordena Druz. Narod. Nauchnoissled. Inst. Rast. Im. N.I. Vavilova*, 121, 41, 1982 (HA 1985, 8466).

214. **Gribovskaya, I.F., Ugulava, N.A., and Karyakin, A.V.**, Effect of microelements in the soil and fruit of quinces on the accumulation of biologically active substances in the fruit, *Agrokhimiya*, 12, 87, 1977 (CA 88, 101412).

215. **Krstic-Pavlovic, N., Lukic, P., Jokanovic, M., and Gorunovic, M.**, Components of quince leaves, *Jugosl. Vocarstvo*, 17, 27, 1983 (CA 101, 69376).

216. **Harris, R.E.**, The saskatoon, *Agriculture Canada Publication 1246*, 1972.

217. **Olson, A.R. and Steeves, T.A.**, Structural changes in the developing fruit wall of *Amelanchier alnifolia, Can. J. Bot.*, 60, 1880, 1982.

218. **Mazza, G.**, Chemical composition of Saskatoon berries (*Amelanchier alnifolia* Nutt.), *J. Food Sci.*, 47, 1730, 1982.

219. **Mazza, G. and Hodgins, M.W.**, Benzaldehyde, a major aroma component of saskatoon berries, *HortScience*, 20, 742, 1985.

220. **Wolfe, F.H. and Wood, F.W.**, Non-volatile organic acid and sugar composition of saskatoon berries during ripening, *Can. Inst. Food Sci. Technol. J.*, 4, 29, 1971.

221. **Mazza, G.**, Anthocyanins and other phenolic compounds of saskatoon berries (*Amelanchier alnifolia* Nutt.), *J. Food Sci.*, 51, 1260, 1986.

222. **Green, R.C. and Mazza, G.**, Relationships between anthocyanins, total phenolics, carbohydrates, acidity and colour of saskatoon berries, *Can. Inst. Food Sci. Technol. J.*, 19, 107, 1986.

223. **Green, R.C. and Mazza, G.**, Effect of catechin and acetaldehyde on colour of saskatoon berry pigments in aqueous and alcoholic solutions, *Can. Inst. Food Sci. Technol.*, 21, 537, 1988.

Chapter 3

STONE FRUITS

I. APRICOT

The botanical name for apricot, *Prunus armeniaca* L. (Rosaceae), implies an Armenian origin. However, it is now generally agreed that apricots came from western China and were brought to Rome at about the time of Christ.[1] The fruit is round, 6 to 8 cm in diameter, yellow, and sometimes flushed with red.

The apricot is important for the fresh fruit market industry, as well as the canning and dehydration industries. Joshi et al.[2] reviewed the data on the chemical composition of apricot fruit and the evolution of its components during ripening, preservation, and storage.

Sun-dried and dehydrated apricots are deep orange with a faint purple hue. The orange color is due to α- and β-carotene, phytofluene, phytoene, *cis*-β- and *cis*-α-carotene, cryptoxanthin, lutein, and other carotenoids;[3] and the red color is due primarily to cyanidin 3-glucoside. Red color development of maturing apricots is accentuated when the fruit is exposed to pollution by fluoride compounds.[4] According to Quinche et al.,[4] the fluoride had general corrosive effects, but also enhanced the production of anthocyanin in the fruit. The presence of cyanidin was also observed by Fouassin[5] in a peach-apricot hybrid and by Lafuente et al.[6] in apricots treated with 200 ppm of 2,4,5-T.

Also containing anthocyanin pigments is the Japanese apricot (*Prunus mume*), in which Ishikura[7] identified the 3-glucoside and 3-rutinoside of cyanidin.

Other phenolics of apricots include: quercetin 3-rutinoside and 3-glucoside; kaempferol 3-rutinoside and 3-glucoside;[8] (+)-catechin; (−)-epicatechin;[9,10] chlorogenic, neochlorogenic, and cryptochlorogenic acids; *cis*- and *trans*-3-, 4-, and 5-*p*-coumaroylquinic acids; *cis*- and *trans*-3-, 4, and 5-feruloylquinic acids; *p*-coumaric acid glucoside; ferulic acid glucoside; coumarin; and scopoletin.[9,10] The presence and concentration of these flavonols, flavan 3-ols, and phenolic acids contribute to the color and flavor characteristics of apricots.

II. CHERRIES

A. SWEET CHERRY

The sweet cherry, *Prunus avium* L. (Rosaceae), is imported commercially as a table fruit and as an ingredient for fruit cocktails and maraschino cherries. Color is the most important indicator of maturity and quality for both fresh and processed cherries. In cherries for processing, color has a direct influence

TABLE 1
Anthocyanins of Sweet Cherries (*Prunus avium* L.)

Anthocyanin	Bing[a]	Burlat Moreau[b]	Lambert + other cvs[c]	Mora[d]
Cyanidin 3-glucoside	+ +	+	+ +	+ +
Cyanidin 3-rutinoside	+ +	+ +	+ +	+ +
Cyanidin 3-sophoroside	—	—	+	—
Peonidin 3-glucoside	+	—	—	—
Peonidin 3-rutinoside	+	—	+	—

Note: +, + + = Relative quantity.

[a] From Lynn, D.Y. and Luh, B.S., *J. Food Sci.*, 29, 735, 1964.
[b] From Ryugo, K., *Proc. Am. Soc. Hortic. Sci.*, 88, 160, 1966.
[c] From Tanchev, S.S., Ioncheva, N., Vasilev, V., and Tanev, T., *Nauchn. Tr. Vyssh. Inst. Khranit. Vkusova Prom. Plodiv*, 18, 379, 1971.
[d] From Pifferi, P.G. and Cultrera, R., *J. Food Sci.*, 39, 786, 1974.

on quality of the finished product.[11] Quality is good when fully ripe, dark sweet cherries become purple or even black. Development of red color in dark sweet cherries is an index of maturity and can be used to predict processing grades. Light-colored cultivars of sweet cherries develop a red blush, but intensity of the red color is not a good index of maturity for them.[11]

Ripe sweet cherries contain two readily detectable anthocyanins first characterized by Willstätter and Zollinger[12] and Robinson and Robinson[13] as 3-rhamnoglucoside and cyanidin 3-glucoside. These identifications were confirmed by Li and Wagenknecht,[14] Lynn and Luh,[15] and others.[5,16] Lynn and Luh[15] also reported the presence of peonidin and two of its glycosidic derivatives in 'Bing' cherries (Table 1), and Okombi[16] identified peonidin 3-rutinoside as the main pigment of 'Bigarreau Napoléon' cherries. However, Harborne and Hall[17] and Olden and Nybom[18] found only cyanidin derivatives and no peonidin glycosides in varieties of *P. avium*. Similarly, Fouassin[5] reported the presence of the two monosides of cyanidin in different varieties of *P. avium* analyzed by paper chromatography.

Casoli et al.,[19] using two-dimensional chromatography, separated nine pigments from which they identified a 3-monoglucoside of cyanidin, a 3-diglucoside of peonidin, and a diglucoside of cyanidin which appeared to be acylated with coumaric acid. Tanchev et al.[20] and Tanchev[21] identified cyanidin 3-glucoside; cyanidin 3-rutinoside; peonidin 3-rutinoside; and, for the first time, cyanidin 3-sophoroside in 'Lambert', 'Helmsdorf', 'Somaya', 'Kozerskia', and 'Bing' cherries.

The concentration of anthocyanins in sweet cherries ranges from 4 mg/100 g peel in blush-yellow cultivars, such as Bigarreau Napoléon, to 350–

450 mg/100 g fresh weight (FW) in dark-fruited cultivars such as Bigarreau Burlat and Guignes.[16-20] Other major phenolic compounds of sweet cherries include: (−)-epicatechin at 15 mg/100 g FW, + (−)-catechin at 6.8 mg/100 g FW,[22] kaempferol 3-rutinoside and 3-galactosyl-7-diglucoside, quercetin 3-rutinoside and 3-rutinosyl-4'-diglucoside, and neochlorogenic acid.[8-10,22-25]

Eaton et al.[26] noted that γ-irradiation produced a uniform color distribution in the skin. Pigment degradation was proportional to the dose of irradiation, although doses between 400 and 600 krd aided in pigment regeneration.[27] Drake et al.[28-31] found that the anthocyanin content in canned cherries was directly related to the color measured by reflectance even if it was artificially increased by growth regulators such as daminozide.

Aminozide,[32] Alar,[33-34] and Stim-80[35] are widely used to enhance ripening of commercially grown cherries. Their use also stimulates synthesis of anthocyanins.[32] In fruits of 'Hative Burlat' and 'Bigarreau Moreau' (treated with Alar) the reported ratios between the two main pigments, cyanidin 3-rutinoside and cyanidin 3-glucoside, are 1.6:1 and 1.7:1, respectively; while in the control fruits the pigment ratio for 'Burlat' is 3.3:1 and 4.3:1 for 'Moreau'. In 'Bing' cherries, the ratios remain unchanged.[34]

The inheritance of fruit color in sweet cherries has been studied by several authors.[36-38] Lamb[36] reported that fruit color can be segregated for dark and light fruits by simple monohybrid ratios. Fogle[37] concluded that fruit color was determined by two factors: a major factor, *A*, producing dark skin and flesh color and a minor factor, *B*, intensifying skin color and expressing itself as a blush in the absence of the major factor. Toyama,[38] however, concluded that the color of sweet cherries is determined by two major genes, *A* for dark vs. *a* for light fruit and *B* for blushed vs. *b* for nonblushed fruit. Therefore, the genotype of the blushed yellow cultivars Emperor Francis and Napoleon is *aaBB* and the genotype of the dark-fruited cultivars Badacsoner Reisenkirsche, 'Bing', 'Lambert', and 'Ulster' is *AaBB*. The genotypes of the dark- and light-fruited crosses are *AaBb* and *aaBb*, respectively.[38]

Postharvest enzymatic activity and a combination of bruising and time-temperature effects related to picking, transportation, and storage cause migration of anthocyanins from the skin to the pulp and loss of the fruit color. This phenomenon, known as scald, is believed to be influenced by the activity of anthocyanase, which is apparently very high in sweet cherries.[39] Pifferi and Cultrera,[40] however, demonstrated the involvement of polyphenol oxidase in decoloration of cyanidin 3-glucoside and cyanidin 3-rutinoside from 'Mora' sweet cherries. According to these authors, the reaction leading to loss of color requires mediating phenolics and follows a scheme of sequential reactions which involves formation of *o*-quinone prior to oxidation of the anthocyanins by the *o*-quinone to a colorless product. Ascorbic acid protects the anthocyanins from oxidation by the *o*-quinone; however, as soon as all ascorbic acid is used, oxidation of the anthocyanins proceeds rapidly. In sweet cherries, color loss occurs at a pH higher than that optimal for the enzyme. This appears to contribute to a greater stability of the *o*-quinone from pH 4.7 to 5.3 and to a greater reactivity of the predominant anthocyanin.[41] According to Scheiner,[42] the enzyme system capable of decolorizing anthocyanins contains Cu in the molecule and is inhibited by heavy metals and carbonyl compounds. Yankov[43] noted that during mashing of cherries, the action of the oxidase systems of the pulp was particularly strong resulting in more than a 50% loss of the anthocyanins.

In storage, cherries gradually lose their brightness in color, and this is accompanied by a decline in anthocyanin content. These losses can be reduced by storing the fruit in an inert gas atmosphere (CO_2, N_2).[44] The combined use of controlled atmosphere and low temperatures ($-1°C$) allows the cherries to be preserved for 4 weeks without appreciable color and anthocyanin losses.[45]

Losses of anthocyanin are minimized when cherries are frozen;[46] however, dark-fruited cultivars are the most suitable for freezing.[47] 'Bing', 'Lambert', and 'Black Republican' are the most popular dark sweet cherries for freezing. Light sweet cherries ('Napoleon', 'Rainier') can also be frozen, but darkening of color due to oxidation is more evident in those cultivars. The exclusion of oxygen by addition of sugar syrups helps to prevent darkening.

Cherries are well adapted for either commercial or home canning. Sweet cherries are canned in sugar syrups ranging in concentration from 0 to 56° Brix[48] at a processing temperature above 91°C in the center of the can.[48] At this temperature, structural transformation and loss of color by the cherries occur. The color loss by the anthocyanins of sweet cherries is a first order reaction with the rate constant dependent on the fruit cultivar.[49] The half-life of anthocyanins (430 to 680 min at 78°C; 30 to 68 min at 108°C) together with the enthalpy and entropy of activation was determined by Ruskov and Tanchev.[50]

Color changes also occur during juicing of sweet cherries. The methods of juice processing influence product color intensity and stability. Beschia et al.[51] used pectolytic enzymes during mashing of the fruit to obtain a juice with higher flavonol and anthocyanin contents. Pretreatment of the juice at 80°C for 1 min before enzymatic treatment is useful to reduce anthocyanin loss caused by the fruit polyphenol oxidase. The greatest loss of anthocyanins occurs in the presence of oxygen. Addition of an antioxidant, such as cystein, prevents the color loss during sterilization and concentration of cherry juice.[52,53]

B. SOUR CHERRY

The first research on the anthocyanins of sour cherry, *Prunus cerasus* L. (Rosaceae), dates back to the work of Rochleder[54] in 1870. Willstätter and Zollinger[12] isolated a pigment from the skin of the fruit which they call keracyanin. From the fruit of the 'Montmorency' cherries, Li and Wagenknecht[55] isolated and characterized two anthocyanins, cyanidin 3-rhamnoglucoside (keracyanin) and cyanidin 3-gentiobioside (mecocyanin).

The presence of these two pigments was confirmed by Markakis[56] who used zone electrophoresis to separate the two anthocyanins. He also reported their isoelectric points as pH 6.9 for keracyanin and pH 6.0 for mecocyanin.

Dekazos[57] separated and identified additional pigments in sour cherries. In addition to cyanidin 3-rhamnoglucoside and cyanidin 3-sophoroside, he identified the triglucoside cyanidin 3-glucosylrhamnoglucoside (as well as cyanidin 3-glucoside and peonidin 3-rhamnoglucoside) as the predominant pigment of 'Montmorency' cherries. This triglucoside was also found in seven other cultivars of sour cherries by Harborne and Hall.[17]

The presence of cyanidin 3-glucoside as a minor pigment of 'Montmorency' cherries was confirmed by Schaller and Von Elbe.[58] These authors also identified cyanidin 3-rhamnoglucoside and cyanidin 3-glucosylrhamnoglucoside.[59,60]

The pigment that was previously described as mecocyanin was characterized as cyanidin 3-(2^G-glucosyl)-rhamnoglucoside by Fischer and Von Elbe.[61] Fouassin[5] had hypothesized the occurrence of this anthocyanin from the chromatographic behavior of one of the major pigments of *Prunus cerasus*. The cultivar Precocious N.2, analyzed by Timberlake and Tanchev,[62] contained cyanidin 3-(2^G-glucosyl)-rhamnoglucoside, cyanidin 3-sophoroside, cyanidin 3-rutinoside, and cyanidin 3-glucoside, while the cultivar Plovdiv lacked cyanidin 3-(2^G-glucosyl)-rhamnoglucoside.

Working with 'Morello', 'Early Richmond', and 'Meteor' cherries, Shrikhande and Francis[63] identified cyanidin 3-glucosylrutinoside; cyanidin 3-sophoroside; cyanidin 3-rutinoside; peonidin 3-rutinoside; and, for the first time, cyanidin 3-sambubioside or 3-xylosylglucoside. Free cyanidin and peonidin were, however, not found. Cyanidin 3-gentiobioside and cyanidin 3-rutinoside were found as the major anthocyanin pigments and cyanidin 3-glucoside as the minor pigment in eight sour cherry varieties analyzed by electrophoresis and paper chromatography.[64] A summary of the known anthocyanins of sour cherries is given in Table 2. Cyanidin 3-gentiobioside is not listed in Table 2, because reports of this anthocyanin in sour cherries are believed to be incorrect.[65]

Dekazos[66] presented data on the content of seven anthocyanins in ripening and mature 'Montmorency' cherries (Table 3). In 100 g of mature, pitted cherries there were 1.3 mg peonidin 3-rutinoside, 8.2 mg cyanidin 3-glucoside, 11.6 mg cyanidin 3-rutinoside, 7.1 mg cyanidin 3-sophoroside, 11.0 mg cyanidin 3-(2^Gglucosylrutinoside), and small quantities of free cyanidin and peonidin.

Regarding the other polyphenolic components of 'Montmorency' cherries, Schaller and Von Elbe[60] reported the presence of six isomers of caffeoylquinic acid, four isomers of *p*-coumaroylquinic acid, and two free phenolic acids. In addition, two flavonols were identified, kaempferol 3-rhamnoglucoside and kaempferol 3-glucoside.

TABLE 2
Anthocyanins of Sour Cherries (*Prunus cerasus* L.)

Anthocyanin	Cultivar			
	Morello 240[a]	Morello 240D[a]	Morello 348[a]	Montmorency[b]
Cyanidin (Cy)	—	—	—	7
Peonidin (Pn)	—	—	—	3
Cy 3-glucoside	15[d]	3	4	19
Cy 3-sambubioside	—	—	—	t[c]
Cy 3-rutinoside	15	11	16	27
Cy 3-sophoroside	1	3	3	16
Cy 3-(2^G-glucosylrutinoside	69	77	77	25
Pn 3-glucoside	—	—	—	t
Pn 3-rutinoside	—	—	—	3

Note: t = Trace amount.

[a]   Recalculated from Hong, V. and Wrolstad, R.E., *J. Agric. Food Chem.*, 38, 698, 1990.
[b]   Recalculated from Dekazos, E.D., *J. Food Sci.*, 35, 242, 1970.
[c]   Found by Shrikhande, A.J. and Francis, F.J., *J. Food Sci.*, 38, 649, 1973.
[d]   Percent of total anthocyanin.

TABLE 3
Changes in Anthocyanins during Maturation and Ripening of
'Montmorency' Cherries

Anthocyanin	Stage of maturity			
	Very immature	Immature	Partially mature	Mature
Cyanidin (Cy)	11.8	5.1	6.8	7.0
Peonidin (Pn)	4.3	5.9	4.7	3.2
Cy 3-glucoside	16.5	19.5	16.6	18.7
Cy 3-rutinoside	28.0	21.2	27.1	26.6
Pn 3-rutinoside	6.6	4.0	4.0	3.0
Cy 3-sophoroside	11.5	18.3	14.6	16.2
Cy 3-(2^G-glucosylrutinoside)	21.2	26.0	26.2	25..2
Total anthocyanins (mg/100 g)	2.0	10.3	29.0	43.6

Recalculated from Dekazos, E.D., *J. Food Sci.*, 35, 242, 1970.

The average content of total anthocyanin in sour cherries was determined by Savic,[67] and in six varieties it ranged from 35 to 82 mg/100 g. The HPLC pattern of anthocyanin pigment was dependent on variety only and did not change with harvesting season.[68] Ben and Gaweda[69] followed anthocyanin content during fruit development. A high anthocyanin content and improved fruit quality are found in 'Schattenmorelle' sour cherries grafted on *P. avium*.[70,71]

Dekazos and Worley[72] studied the effects of Alar (2,2-dimethylhydrazide of succinic acid) treatments on the biosynthesis of anthocyanins in 'Montmorency' cherries. They found an increase in anthocyanin content and a noticeable change in the proportions of individual pigments. Peonidin 3-rutinoside, however, remained unchanged. Poll[73] reported the relationship between anthocyanins, total polyphenols, and stage of fruit maturity.

Sour cherries are preserved frozen, canned, or as juice. Von Elbe[74] studied the storage stability of the two main pigments of sour cherries and found that the highest degradation occurs at pH 3.5 and increases with raising the temperature from 4.4 to 15.5 and 37.8°C. Dalal and Salunkhe[75] confirmed that for maintenance of quality of canned sour cherries, the product should be stored at a temperature not higher than 4.4°C; otherwise degradation of the anthocyanins occurs. According to these authors degradation is due in part to hydroxymethylfurfural (HMF) which forms from the added sugars. These results were confirmed in 'Marasca' cherries by Lovric et al.[76] and Lovric[77] who reported that blanching the fruit prior to processing improved product quality and stability by enhancing extraction of pigments and inactivating the oxidizing and hydrolytic enzymes. Also, a more effective blanching temperature is 80 to 85°C for 5 min.[78] Thermal treatments such as pasteurization, sterilization, and concentration cause degradation of the anthocyanins; and in the absence of oxygen, formation of dark brown polymers. This author also reported that the use of pectolytic enzymes did not cause color loss. Siegel et al.[79] confirmed that initial blanching of the fruit for 45 to 60 s at 100°C reduces enzymatic degradation of the anthocyanins in frozen sour cherries. According to Lovric[77] the addition of sucrose to sour cherry juice has a positive effect upon the stability of the anthocyanins during storage. Tanchev,[80] however, found no significant differences in color stability of juices with or without added sucrose. A mathematical model for color evaluation in sour cherry products was proposed by Milkov and Atanasov.[81]

The scald of cherries or the discoloration of the fruit due to migration of anthocyanins from the skin to the less colored pulp and subsequent browning has been studied in detail by Dekazos.[82] This author also summarized previous results of Wagenknecht et al.[39] and Van Buren et al.[83] who isolated an enzyme from sour cherries that was able to decolorize the anthocyanins in the presence of catechol and oxygen, and was inhibited by thiourea. The occurrence of catechol as an intermediate substrate in this degradation mechanism of anthocyanins is also in agreement with the results of Peng and Markakis,[84] who demonstrated that anthocyanins alone are poor substrates for phenolase. Yeatman et al.[85] evaluated the degree of scald damage by optical difference spectrometry at 780 and 695 nm.

Cherries are also preserved by freeze-drying, a process which has been found to influence anthocyanin pigmentation. Do et al.[86] observed that cyanidin 3-(2^G-xylosylrutinoside) was the least stable pigment during storage of freeze-dehydrated sour cherries. Urbanyi[87] also found that the presence of oxygen played an important role in anthocyanin breakdown. Thus, frozen dehydrated sour cherries should be packaged in a material impermeable to oxygen and light.

C. OTHER CHERRIES

In the pulp of the black cherry, *P. serotina* Ehrh. var. Szomolya, Gombkötö[88] characterized the two principal anthocyanins as cyanidin 3-*p*-coumarylglucoside (70%) and cyanidin 3-*p*-coumarylrhamnoglucoside (23%). The remaining 7% is made up of two unidentified pigments.

Du et al.[89] studied anthocyanins of the ornamental cherry (*P. sargentii*, Rehd.), and identified cyanidin 3-glucoside and 3-rutinoside as the major anthocyanins and pelargonidin 3-glucoside and 3-rutinoside as the minor pigments. They also reported the presence of two other minor anthocyanins tentatively identified as cyanidin 3-glycoside and cyanidin 3-diglucoside, most likely cyanidin 3-gentiobioside.

The tomentosa cherry (*P. tomentosa* Thunb.) is a small, hardy tree or very large shrub grown for ornamental purposes and for its globular, light red, slightly hairy fruit.[90,91] The fruit anthocyanins are pelargonidin and cyanidin 3-rutinosides.[7]

The cherry laurel — *P. laurocerasus* L. (*Laurocerasus officinalis*, Roem.) — is an evergreen bush, seldom a small tree, native to southeastern Europe and Iran, and grown for its ornamental value and its black-purple fruit. The intensely flavored, somewhat bitter fruit is used in Europe in specialty soft drinks. The fruit and leaves are also used as a source of an aromatic oil to flavor liqueurs. The anthocyanins of the fruit are peonidin 3-arabinoside and cyanidin 3-arabinoside.[92-94]

The European dwarf or ground cherry (*P. fruticosa* Pall.) is a low-spreading bush, 10 to 50 cm high, grown for its moundlike shape and profuse flowers, and often grafted on tall stocks to make a drooping compact head. The fruit is the size of a large pea, globular, deep reddish-purple, and sour in taste. Only one anthocyanin was found in this fruit by Olden,[95] but its structure has not yet been elucidated.

III. OLIVE

The olive tree (*Olea europaea* L.) is an evergreen cultivated since ancient times for its fruit, which is an oval, green drupe when immature and purple to black when completely ripe. The olive fruit has long been used as a food and source of edible oil which is used as a frying fat; in salad dressings; and in soaps, perfumes, and medicines.[96] The world production of olive oil is estimated at 1.7 million tonnes per annum, with Italy and Spain accounting for about 60% of the total production. In contrast to other edible oils, olive oil is consumed unrefined, thus containing polyphenols which are usually removed from other edible oils in the various refining stages. The polyphenolic compounds identified in olive oils include caffeic, vanillic, *p*-coumaric, syringic, ferulic, protocatechuic, and *p*-hydroxybenzoic acids; 3-hydroxyphenyl-ethanol; and 3,4-dihydroxyphenyl-ethanol.[97-99] These compounds have a strong influence on the flavor of the oil and seem to be responsible for the high oxidation resistance of olive oil.

That olives contain anthocyanins has probably been known since the last century;[100] however, the identification of cyanidin 3-rutinoside as the major pigment in *O. europaea* is attributed to Musajo[101] in 1940.

Cantarelli,[102] using paper chromatography, separated a monoglucoside and diglucoside of cyanidin as the major anthocyanins and four other pigments present in low concentration in olives from Italy. The relationship between pigment quality and quantity differs in the various cultivars examined with clear genetic relationships.[103]

Vazquez Roncero and Maestro Duran[104] identified cyanidin 3-glucoside and cyanidin 3-rutinoside as the two main pigments and three triosides believed to be cyanidin 3-(2^G-glucosylrutinoside), cyanidin 3-(2^G-xylosylrutinoside), and cyanidin 3-glucosylrutinoside. In ripe 'Manzanillo' olives, Luh and Mahecha[105] identified cyanidin 3-glucoside, 3-caffeylrhamnosylglucoside, 3-caffeylrhamnosyldiglucoside, 3-*p*-coumarylglucoside; and peonidin 3-*p*-coumarylglucoside and 3-rhamnosyldiglucoside.

A study on the same variety by Maestro Duran and Vazquez Roncero[106] confirmed the presence of the monoglucosides, diglucosides, and triglucosides of cyanidin, some of which are acylated with caffeic acid. These authors, however, provided no confirmation of the occurrence of peonidin derivatives for those acylated with *p*-coumaric acid (Table 4). Tanchev et al.[107,108] reported delphinidin 3-rhamnoglucoside-7-xyloside in the fruit of two Greek olive cultivars. This, however, is the only report of a triglycoside with glycosylation in the 7 position and remains to be confirmed.

The pigments of olives are concentrated, for the most part, in the skin of the fruit. The same anthocyanins are found in the pulp but their distribution is cultivar dependent.[109] Vazquez Roncero et al.[110] found that the pulp, in addition to being rich in anthocyanins, also contains hydroxyphenols, phenolic acids, and anthoxanthins. This information has generated an interest in chemotaxonomic classification of olives.[109]

The pattern of anthocyanic biosynthesis in olives is a rapid increase from the onset phase until a maximum is reached, followed by a decrease of anthocyanins in overripe fruit.[111] The ratio of monoglucoside to diglucosides decreases to about 1:1 around the first of November and then remains fairly

TABLE 4
Anthocyanins of Olive (*Olea europea*)
cv. Manzanillo

	Total %
Cyanidin 3-glucoside	15
Cyanidin 3-rutinoside	t
Cyanidin 3-caffeoylrutinoside	60
Cyanidin 3-glucosylrutinoside	t
Cyanidin 3-(2^G-glucosylrutinoside)	25
Cyanidin 3-(2^G-glucosylrutinoside) caffeic acid ester	t

Note: t = Trace amount.

Adapted from Maestro Durán, R. and Vázquez Roncero, A., *Grasas Aceites (Seville)*, 27, 237, 1976.

constant. The principal factor that promotes the formation of pigments is light, and fruit ripened in the dark has been found to contain one tenth the anthocyanin concentration of fruit ripened under normal conditions.[111] The cultivar also affects the pattern of biosynthesis.[112] In ripe, late-maturing cultivars, for instance, the content of anthocyanins may range from 42 to 228 mg/100 g dry weight. In intermediate and early cultivars, the anthocyanin content is much lower.[112,113] There is also an increase in polymeric anthocyanins with ripening of olives. Maximum levels of these compounds are, however, reached at different times depending on the cultivar.[114] While in 'Grossa di Cassano' and 'Leccino' olives maximum levels of these compounds are reached at the end of November, in 'Caroleo' olives maximum content of polymeric anthocyanins is achieved at the end of January. Vlahov[114] also observed that the ratio between the contents of the two major pigments, cyanidin 3-glucoside and 3-rutinoside, is a varietal characteristic and remains constant during ripening.

Treatment of picked green mature olives with exogenous cytokinins (kinetin, benzyladenine, and zeatin) caused a marked enhancement in anthocyanin accumulation without affecting other maturation parameters, CO_2, or fruit softening.[115,116] Immersion of green olives for 1 h in a cytokinin solution considerably increased the anthocyanin content of fruits after 10 d. At 3.5 ppm, zeatin produced the maximum effect, while the optimum concentration of kinetin and benzyladenine was 50 ppm. Adenine and uracil did not induce pigmentation when applied separately and had only minimal effect when applied as a mixture. Kinetin had only a weak effect on accumulation of anthocyanins in half-black olives.[117]

In 'Manzanillo' olives application of auxins, NAA, 2,4-D, or 2,4,5-TP at the pit-hardening stage of maturity caused a delay in fruit ripening, but late application can prevent postharvest drop and slightly enhance coloration.[116,117] Immersion of green olives for 1 h in NAA or 2,4,5-TP solutions caused a moderate increase in anthocyanin content after 10 d, particularly at concentrations of 200 and 300 ppm.[116]

Reports on the effect of ethephon on anthocyanin accumulation of olives are somewhat contradictory. Shulman et al.[118] reported that ethylene or ethephon treatment of green or half-black 'Manzanillo' olives resulted in delayed maturation and inhibition of anthocyanin accumulation. Rugini et al.[119] found, however, that spraying of 1500 and 2000 ppm of ethephon on trees of 'Giarraffa' olives midway between the onset of pit hardening and the maximum respiratory rate induced early pigmentation of the fruit and significantly increased anthocyanin content.

Processing of green or black olives to render the fruit edible involves removal of bitterness by lye treatment, water washing, and storage with or without fermentation and sterilization. These processes lead to a progressive loss of total anthocyanins and to a considerable increase in the polymeric anthocyanin content of the fruit.[120] According to Vlahov,[120] the polymeric anthocyanins of 'Grossa di Cassano' and 'Leccino' olives increased from 25% prior to processing to 55 and 75% of the total anthocyanin content of the processed olives, respectively.

Olives are rich in many other phenolic compounds including: oleuropein, flavonol glycosides, rutin, and luteolin 7-glucoside, in particular; and derivatives of hydroxycinnamic acids, especially verbascoside, and the heterosidic ester of caffeic acid and dihydroxyphenyl-ethanol.[121] Considerable quantities of these phenolic compounds and anthocyanins are discharged in the effluents of olive processing facilities each year.[122-124] Attempts to recover these compounds for use as antioxidants and natural colorants have been made,[122-124] but to our knowledge no recovery process has yet been commercialized.

IV. PEACH

The peach, *Prunus persica* L., originated in China and is grown commercially around the world between 25 and 45° latitude above and below the equator. These limits may be extended somewhat by warm ocean currents, large lakes, or altitude.[1] The leading peach producing countries are the United States, Italy, France, Japan, Spain, Greece, and Argentina. Wherever grown, peaches are an important fruit in the fresh fruit and the canning industries. Small quantities are also frozen and dried. If the fruit is to be canned, it should have a yellow flesh (preferably a clingstone), a small nonsplitting pit, and a good symmetrical size; and be firm, resistant to browning, and low in red pigment content.[1] The fresh fruit market prefers a yellow-flesh freestone peach, relatively free from fuzz, firm, of good size and roundish, and bright in color.

Color intensity and anthocyanin content of peaches depend on variety, ripeness of fruit, and environmental factors during growth and storage of the fruit.[125-128] "Freestone" cultivars, for instance, contain more red pigment in the pulp near the pit than "clingstone" cultivars.[125] Likewise, high-density planting systems which increase the shading of fruit decrease red pigment development, and a high correlation coefficient between light intensity and skin color has been reported.[126,127] Temporary exposure at the beginning of the color-change period is almost as effective as exposure just before picking since response can be rapid after only 8 d of light treatment.[126] Exposure of fruit to low night and day temperatures and ultraviolet light also favor color development.[128]

Culpepper and Caldwell[129] reported on characteristics of anthocyanins and their relationship to the pink discoloration which occurs in canned peaches. Johnson et al.[130] found that the peach pigment was a 3-monoglucoside of cyanidin. Hayashi et al.[131] identified the pigment as cyanidin 3-glucoside. Fouassin[5] examined five peach varieties and found three cyanidin derivatives. One was predominant in all the varieties, a second varied according to the variety, and the third occurred in trace amounts in all varieties. All three appeared to be monosides.

Luh et al.[132] studied the phenomenon of the pink discoloration in canned freestone peaches and reported that it is directly related to the anthocyanin content in the pit cavity. The pigmented skin was not involved since it had been removed before canning. These authors[133] demonstrated that the phenomenon was of a nonenzymatic nature because heat treatment of the fruit inactivated the enzymes. The discoloration was due to the anthocyanins in the pit cavity reacting with the tin in the packaging material. In a subsequent study of the problem, Hsia et al.[125,133] used chromatographic, UV-VIS and IR spectrophotometric techniques as well as acid and alkaline hydrolyses, and found that cyanidin 3-glucoside was the anthocyanin responsible for the pink discoloration in 'Elberta' peaches. They also examined the anthocyanin content in relationship to the fruit origin to identify conditions most suitable for commercial processing of peaches.

Miyakawa and Takehana[134] used model systems to study the discoloration of canned peaches and confirmed that the phenomenon was due to the complexing of cyanidin with the tin of the can. Ascorbic and isoascorbic acids were the most effective additives for improving product quality. Polyphosphates were not effective, and saccharose accentuated the negative effect caused by the tin. Above pH 3.5, the discoloration phenomenon is favored, as in storage, the anthocyanin colored structure, the flavylium cation, is converted to the blue quinonoidal base and the colorless carbinol pseudobase and chalcones.[135] Therefore, in addition to selecting varieties low in anthocyanins, extreme pH values in the can should be avoided.[135-137] Uciyama[138] proposed removing the anthocyanins from the pit cavity by using anthocyanases extracted from various molds.

A survey of nine peach varieties by Van Blaricom and Senn[127] did not reveal any qualitative differences in pigmentation. Ueno et al.[139] confirmed that cyanidin 3-glucoside was the main anthocyanin in the peach. More recently, Ishikura[7] reported the presence of cyanidin 3-rutinoside in a 10% concentration and cyanidin 3-glucoside accounting for 90% of the total anthocyanin concentration. According to Monet,[140] a single gene controls the anthocyanin-deficient character in yellow peaches, and the presence of anthocyanins in the pit cavity may be associated with plant vigor.[141]

The concentration of anthocyanins in the fruit progressively increases with ripening. In the early cultivar Flordasun, it apparently proceeds in two distinct phases.[142] El-Ashwah et al.[143] found that unripe peaches had low contents of tannins and anthocyanins which progressively increased with ripening. High temperatures during ripening favor anthocyanin formation; however, for harvested fruit, there is no correlation between anthocyanin content and temperature. Nonetheless, peaches destined for canning should be stored at temperatures below 30°C to prevent excessive pigmentation.[144]

An improved pulp quality expressed by better anthocyanin pigmentation in the skin can be obtained with various Alar treatments after flowering.[145] A similar effect can be obtained with ethephon and 2,4,5-TP.[146]

Peach flowers contain the same major pigment as the fruit, cyanidin 3-glucoside.[147] A pentosylhexoside of cyanidin[147] and monohexoside and a pentosylhexoside of peonidin have also been reported.[148]

Peaches are also rich in hydroxycinnamic acid derivatives, especially chlorogenic and neochlorogenic acids, flavan-3-ols, and flavonols.[8,9] The concentration of total phenolic compounds ranges from 28 to 180 mg/100 g FW,[149] hydroxycinnamic acid derivatives from 8.1 to 75 mg/100 g FW, and flavan 3-ols from 5.3 to 14.1 and flavonols from 0.4 to 1 mg/100 FW.[8,9,10] The concentration of anthocyanins in the peel varies from 5.3 to 13.7 g/100 g dry weight.[150]

V. PLUMS

The plum tree grows in temperate regions. It may be low and shrubby, or it may grow to 9 m high. Plums have many food uses. They are eaten fresh, canned, or frozen; used to make jelly, preserves, plum butter, and jam; and dried to make prunes. Over 2000 varieties of plums from some 15 species have been grown in North America alone. The important varieties are from five main types of plums: European, damson, Japanese, American, and ornamental.

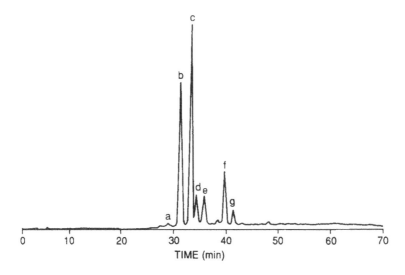

FIGURE 1. HPLC chromatogram of plum anthocyanins. Peak area percentages: (a) (unidentified), less than 1% (b) (cyanidin 3-glucoside), 37%; (c) (cyanidin 3-rutinoside), 45%; (d) (unidentified), 3%; (e) (unidentified), 3%; (f) (unidentified), 12%; (g) (unidentified), trace. (Reprinted with permission from Hong, V. and Wrolstad, R.E., *J. Agric. Food Chem.*, 38, 698, 1990. Copyright 1990, American Chemical Society.)

A. EUROPEAN PLUM

The European plum (*Prunus domestica* L.), also called garden or common plum, is a moderately vigorous tree with green, yellow, red, or blue fruit borne largely on spurs. The fruit size and shape are variable, and the stone clings or separates from the flesh.

In ripe European plums the color is located in the skin and the outer cells of the flesh. The red pigments are cyanidin 3-glucoside, cyanidin 3-rutinoside, peonidin 3-glucoside, and peonidin 3-rutinoside.[17,65,149,151] Varietal differences and separation-characterization methodology may account for differences in relative concentration and for the absence of some of these anthocyanins in certain plums. Hong and Wrolstad,[65] using HPLC with photodiode array detection, separated seven anthocyanins from a plum juice concentrate and found the 3-rutinoside and 3-glucoside of cyanidin to form 45 and 37% of the total anthocyanins, respectively (Figure 1). These authors, however, failed to find peonidin 3-rutinoside and did not fully characterize four of the seven HPLC peaks. Raynal et al.,[151] on the other hand, found the rutinoside derivatives of cyanidin and peonidin to be the major anthocyanins of 'Agen' plums from Casseneuil, France. The total anthocyanin content of this fruit was 6.34 mg/g of dry weight; and cyanidin 3-rutinoside and peonidin 3-rutinoside accounted for 44 and 42%, respectively, of the total anthocyanins. Popov[152] failed to find peonidin 3-glucoside in studies of two varieties, and

Blundstone and Crean[153] did not report this anthocyanin in 'Victoria' plums.

According to Rees,[154] in Victoria plums the anthocyanin pigments appear about 1 week after the onset of the ripening stage. Hillis and Swain[155,156] found no significant changes in the phenolic compounds of plum skin through August 27, when the total concentration is about 1.2 mg/g dry weight and formation of anthocyanins begins. The phenolic content increased to 3.0 mg/g on September 12, and to 8.4 mg/g dry matter at fruit maturity on September 26. Subsequent to publication of these findings by Swain and Hillis,[156] numerous investigations have confirmed that phenolic concentrations are generally higher in young fruits.[157] During ripening, the concentration of phenolics increases in fruits in which anthocyanins or other flavonoids accumulate, but decreases in late-maturing species and varieties.[157] In 'Ruth Gerstetter' and 'Pozegaca' plums a steady increase in the anthocyanin content is observed during ripening.[158] This pattern is similar to that observed in grapes, cherries, strawberries, olives, and many other red fruits.[157]

Several studies have dealt with the proper maturity of plums, and estimation of anthocyanin content by measurement of absorbance at 520 nm has been proposed as a maturity index for canned fruit.[159] This chemical procedure is quick and simple, requiring only limited analysis; however, it was developed long before widespread use of reflectance color instruments, which have been found to be highly reliable in predicting anthocyanin content in several red fruit,[31] and may also be used to predict color of plums. György[160,161] reported that the ideal time for harvesting 'Besztercei' plums is when the anthocyanin pigments in the skin are at their maximum level and the chlorophylls have almost completely disappeared.

Tanchev[162] and Ruskov and Tanchev[163] studied the kinetics of anthocyanin degradation during sterilization and storage of juice from 'Afuski' plums; they found that it was a first order reaction, dependent only on temperature. The storage time for halving the anthocyanin content was 143, 50, and 7 d at 10, 30, and 40°C, respectively. Anthocyanin concentration is also a reliable quality index of 'Maria Claudia' plum products. Tanchev et al.[164] described a thermal sterilization process in which degradation of anthocyanins is minimized and consistency of the final product is improved. Ichas et al.[165] showed that storage at 5°C impedes anthocyanin degradation and subsequent browning in concentrated plum juice.

In canned plums, anthocyanins diffuse from the skin into the flesh of the plum and into the surrounding liquid, thereby diluting the pigment and decreasing the overall color quality of the canned fruit.[166,167] The reported loss of pigment content during heat treatment of 'Fellenberg' plums is 17% when fresh fruit is processed and 37% when frozen raw material is used.[168] Addition of Ca^{2+} ions to the 30% sucrose syrup does not influence the redistribution of the pigment between skin flesh and surrounding liquid. Pigment content and color intensity were higher in products stored at 4°C than in those stored at 30°C. Heat processing and storage under various conditions show that full

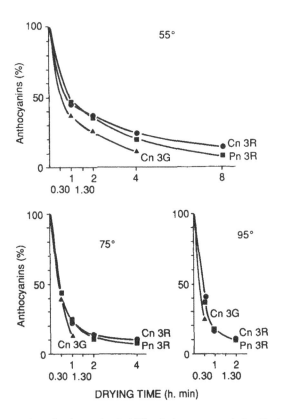

FIGURE 2. Evolution of anthocyanins in 'd'Ente' plum exocarp during the initial hours of drying at different temperatures: Cn3G, cyanidin 3-glucoside; Cn3R, cyanidin 3-rutinoside; Pn3R, peonidin 3-rutinoside. (Reprinted with permission from Raynal, J., Moutounet, M. and Souquet, J.-M., *J. Agric. Food Chem.*, 37, 1046, 1989. Copyright 1989, American Chemical Society.)

degradation with irreversible polymerization of the anthocyanins is an important factor in pigment loss, whereas changes in polymer concentration play only a small part.[169] Without the possibility of diffusion, plum skins retained approximately 70% of the anthocyanin monomers directly after pasteurization, which either increased to approximately 98% retention (no added sugar) or remained constant (in the added sugar) during the following 7d. During storage at 20°C for 8.5 weeks, the anthocyanin monomers were then reduced by 40 to 59% of the initial concentration.[169]

Comparable losses of anthocyanins also occur during drying of plums.[151] Thus, in prunes, anthocyanins disappear very rapidly with the rate of disappearance increasing with temperature. After 1 h at 95°C, about 15% cyanidin 3-rutinoside and 17% peonidin 3-rutinoside remain, whereas 46 and 48% remain after 1 h of drying at 55°C (Figure 2). Under the same conditions, a quasilinear relationship was observed for degradation of rutin as a function

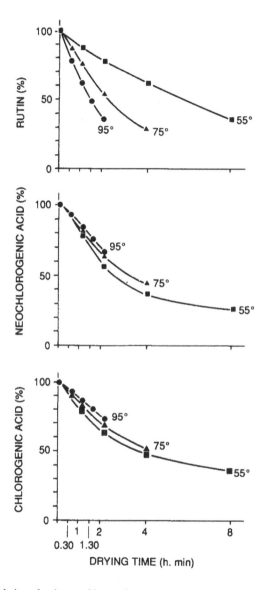

FIGURE 3. Evolution of rutin, neochlorogenic, and chlorogenic acids in 'd'Ente' plum pulp during the initial hours of drying at several temperatures. (Reprinted with permission from Raynal, J., Moutounet, M., and Souquet, J.-M., *J. Agric. Food Chem.*, 37, 1046, 1989. Copyright 1989, American Chemical Society.)

of drying time, particularly at 55°C (Figure 3). As for the anthocyanins, however, the rate of degradation increased as the drying temperature increased; and after 2 h at 95°C, 36% of the original rutin concentration remained. In contrast, the degradation of hydroxycinnamic acid derivatives and, in particular, neochlorogenic acid slowed down when the temperature was increased from 55 to 95°C, which corresponded directly with the progressive inactivation of *o*-diphenoloxydases.[151,170] Anthocyanins and other flavonols are therefore not degraded by the same mechanism as phenolic acid, because these compounds are not direct substrates of the oxidases. In fact, degradation of anthocyanins during thermal processing of plums is probably associated with two factors, enzymatic and thermic. The enzymatic degradation is dependent on the presence of quinones,[40-43,151,170] but a large part of the apparent disappearance of anthocyanins can be attributed to a decrease in concentration of flavylium cations and an increase in chalcones with heating.[135,170-172]

B. DAMSON PLUM

Damson plums (*Prunus insititia* L.) are found in home orchards, where the tart, blue fruit is limited to culinary purposes, primarily for jellies and jams. The trees resemble the European types, but are smaller, more resistant to low temperatures, relatively free from diseases, and require little care. There are several types, including 'Bullaces', 'St. Juliens', and 'Mirabelles'.[1] Robinson and Robinson[13] characterized a pentoseglucoside of cyanidin in damson and California blue plums. Using chromatographic techniques, Fouassin[5] found three pigments, of which two appeared to be monosides of cyanidin while the third appeared to be a glucoside of peonidin. This third pigment was characterized as a derivative of malvidin by Reznik.[173]

C. JAPANESE PLUM

Japanese or Salicena plums (*Prunus salicina* Lindl.) are eaten fresh, cooked, and canned. Prunes are rarely made from these plums. The fruits are quite variable but easily distinguishable from other types of plums by their large size; oblate to heart (but rarely oblong) shape; and their bright yellow, red, or purplish-red color.[1] Flesh color is yellow, amber, or red in the so-called blood plums.[1]

Hunter[174] observed that red pigmentation is dominant over yellow flesh color. Ishikura[7] found only two cyanidin derivatives in Japanese plums, cyanidin 3-glucoside and cyanidin 3-rutinoside. Ahn[175] also found only cyanidin derivatives in the variety Santa Rosa. These pigments, however, were characterized as 3-glucosides and 3-xylosylglucosides (sambubioside) of cyanidin. Tsuji et al.[176] identified the diglucoside of *P. salicina* cv. Sordum as cyanidin 3-rhamnoglucoside (23%) and the major pigment as cyanidin 3-glucoside (68%). Itoo et al.[177] analyzed the cultivar Karari and confirmed two anthocyanins, cyanidin 3-glucoside and cyanidin 3-rutinoside, in about a 1.2:1 ratio. These two pigments are also present in *P. jamasakura* Sieb. fruit.[178]

Draetta et al.[179] also characterized cyanidin 3-glucoside and cyanidin 3-sambubioside as the major anthocyanins of *P. salicina* cv. Carmesim. Two other pigments were detected in these plums, but their concentration was too low to permit identification. Anthocyanin content of 'Carmesim' plums was 29.5 mg/100 g FW.

Tsuji et al.[180] studied the relationship between biosynthesis of anthocyanins, PAL activity, and storage temperature of the fruit and found that between 20 and 30°C there was a maximum relationship between enzymatic activity and anthocyanin biosynthesis. Anthocyanin production in Japanese plums can be stimulated by pre- and postharvest treatments with ethephon.[181,182] There is also an excellent relationship between sucrose content, growth hormones (IAA, ABA, and BA), and anthocyanin development in 'Santa Rosa' plums.[183]

D. AMERICAN PLUM

Plums native to America include several species used for fresh or culinary purposes, in breeding, or as a root stock; the more important ones are *P. angustifolia, P. hortulana,* and *P. maritima.*

P. angustifolia, or Chickawa plum, is a twiggy tree about 6 m high or a bush with zigzag hard reddish glabrous branches, producing a 1.0- to 1.5-cm diameter, nearly globular, cherrylike, red or yellow fruit. Two anthocyanins have been characterized in this plum; they are cyanidin 3-glucoside and 3-rutinoside, accounting for 43.7 and 56.3% of the total anthocyanin content, respectively.[184]

P. hortulana (or Hortulan plum) is a thorny, bushy, vigorous tree yielding globose fruit, 2.0 to 2.5 cm in diameter, and red or yellow in color. Its anthocyanins are cyanidin 3-glucoside and 3-rutinoside, accounting for 48.0 and 52.0% of the total anthocyanin content, respectively.[184]

P. maritima (or beach plum) grows well on beaches and sand dunes from Virginia to New Brunswick, and is used primarily in the production of jams and jellies. It also contains cyanidin 3-glucoside and 3-rutinoside, and the respective concentrations are 54.1 and 45.9% of the total anthocyanin content.[184]

E. OTHER PLUMS
1. Sloe Plum

Sloe or blackthorn (*P. spinosa* L.) is a thick-topped, thorny, suckering hardwood tree or shrub; is 3 to 5 m high; and produces a harsh, globular blue fruit about 1 cm in diameter. It is frequently planted for its profuse early bloom, its dense head of foliage, and its abundant ornamental flowers. Werner et al.[184] reported cyanidin 3-glucoside (37.8%), cyanidin 3-rutinoside (48.0%), peonidin 3-glucoside (2.8%), and peonidin 3-rutinoside (11.2%) as the major pigments of blackthorn fruit. Popov[152] reported cyanidin 3-glucoside, cyanidin 3-rhamnoglucoside, and peonidin 3-rhamnoglucoside to be typical of all *P. spinosa* varieties, with cyanidin 3-rutinoside being the most predominant.

Rozmyslova and Samorodova-Bianki[185] found that the anthocyanin content in 30 forms of *P. spinosa* varied from 365 to 2154 mg/100 g FW. Caffeoyl-3'-quinic acid was the most abundant hydroxycinnamic derivative in sloe plums, in which were also detected some quercetin glycosides together with the known 3-glycosides and 3-rutinosides of cyanidin and peonidin.[186]

2. Cherry Plum

The cherry plum or myrobalan plum (*P. cerasifera* Ehrh.) is used widely as rootstock for European and Japanese plums. Seedlings are hardy, vigorous, and tolerant to wet soil.[1] Fruit of the cherry plum is small, round or oval, yellow or red in color and has a rather insipid flesh. A form with red foliage is called 'Pissardi'.

Werner et al.[184] found six anthocyanins in fruit of *P. cerasifera*. They are cyanidin 3-glucoside (9.7%), 3-rutinoside (64.7%), peonidin 3-glucoside (6.5%), 3-rutinoside (8.8%), and two unknowns accounting for the remaining 11.3% of the total anthocyanin content. Other authors[187,188] have also characterized the cyanidin 3-glucoside and cyanidin 3-rhamnoglucoside in the fruit of some varieties of *P. cerasifera*, while the leaves contain peonidin 3-glucoside, cyanidin 3-gentiobioside, and cyanidin 3-glucoside.[187]

REFERENCES

1. **Childers, N.F.,** *Modern Fruit Science,* Horticultural Publications, Rutgers University, New Brunswick, NJ, 1976.
2. **Joshi, S., Srivastava, R.K., and Nath, D.,** The chemistry of *Prunus armeniaca, Br. Food J.,* 88(932), 74, 1986.
3. **Curl, A.L.,** The carotenoids of apricot, *Food Res.,* 25, 190, 1960.
4. **Quinche, J.P., Balet-Buron, A., and Vogel, F.,** La pollution par les composés fluorés et ses effets sur les arbres fruitiers du Valais, *Bull. Murithienne,* 91, 3, 1974.
5. **Fouassin, A.,** Identification par chromatographie des pigments anthocyaniques des fruits et des legumes, *Rev. Ferment. Ind. Aliment.,* 11, 173, 1956.
6. **Lafuente, B., Primo, E., and Cunat, P.,** Accelerated ripening of apricots. II. Tests with apricots of the Canino variety, *Rev. Agroquim. Technol. Aliment.,* 4, 96, 1964 (JSFA 1964, 258ii).
7. **Ishikura, N.,** A survey of anthocyanins in some angiosperms. I. *Bot. Mag. (Tokyo),* 88, 41, 1975.
8. **Henning, W. and Hermann, K.,** Flavonolglykoside der Aprikosen (*Prunus armeniaca* L.) und der Pfirsiche (*Prunus persica* Batsch), *Z. Lebensm. Unters. Forsch.,* 171, 183, 1980.
9. **Mosel, H.D. and Herrmann, K.,** The phenolics of fruits. III. The contents of catechins and hydroxycinnamic acids in pome and stone fruits, *Z. Lebensm. Unters. Forsch.,* 154, 6, 1974.
10. **Möller, B. and Herrmann, K.,** Quinic acid esters of hydroxycinnamic acids in stone and pome fruit, *Phytochemistry,* 22, 477, 1983.

11. **Drake, S.R., Proebsting, E.L., Jr., and Spayd, S.E.,** Maturity index for the color grade of canned dark sweet cherries, *J. Am. Soc. Hortic. Sci.,* 107, 180, 1982.
12. **Willstätter, R. and Zollinger, E.H.,** Untersuchung ueber die Anthocyane. XIV. Ueber die Farbstoffe der Kirsche und der Schlehe, *Justus Liebigs Ann. Chem.,* 412, 1654, 1916.
13. **Robinson, G.M. and Robinson, R.,** A survey of anthocyanins, *Biochem. J.,* 25, 1687, 1931.
14. **Li, K.C. and Wagenknecht, A.C.,** A minor anthocyanin pigment of sweet cherry, *Nature (London),* 182, 657, 1958.
15. **Lynn, D.Y. and Luh, B.S.,** Anthocyanin pigments in Bing cherries, *J. Food Sci.,* 29, 735, 1964.
16. **Okombi, G.,** Les pigments de la Cerise, *Prunus avium* (L.), variète "Bigarreau Napolèon": variations au cours de la croissance, de la maturation et de la conservation, Thèse Doctorate 3ème Cycle, Universitè d'Orlèans, France, 1979.
17. **Harborne, J.B. and Hall, E.,** Plant polyphenols. XIII. The systematic distribution and origin of anthocyanins containing branched trisaccharides, *Phytochemistry,* 3, 453, 1964.
18. **Olden, E.J. and Nybom, N.,** On the origin of *Prunus cerasus* L., *Hereditas,* 59, 327, 1968.
19. **Casoli, U., Cultrera, R., and Gherardi, S.,** Ricerche sugli antociani di ciliegia, di ribes nero e di mora di rovo, *Ind. Conserve,* 42, 255, 1967.
20. **Tanchev, S.S., Ioncheva, N., Vasilev, V., and Tanev, T.,** Identification of the anthocyanins in the fruits of some cherry varieties, *Nauchn. Tr. Vyssh. Inst. Khranit. Vkusova Prom. Plovdiv,* 18, 379, 1971 (CA 79, 63533).
21. **Tanchev, S.,** Identification of the anthocyanins in sweet cherries, *Novi Tekhnol. Chereshoprozvod., Kyustendil, Bulgaria,* 116, 1975 (PBA 1977, 5787).
22. **Melin, C.,** Les composès phènoliques au cours de la croissance et de le maturation de la cerise, *Prunus avium* (L.) variètè 'Bigarreau Napolèon', Thèse Doctorate 3ème Cycle, Universitè d'Orlèans, France, 1976.
23. **Melin, C., Billot, J., and Dupin, J.F.,** Les composès phènoliques de la Cerise Bigarreau Napolèon I. Les dèrives hydroxycinnamiques, *Physiol. Veg.,* 17, 557, 1979.
24. **Melin, C., Billot, J., and Dupin, J.F.,** Les composès phènoliques de la cerise Bigarreau Napolèon. II. Les flavonoides, *Physiol. Veg.,* 17, 573, 1979.
25. **Stohr, H., Mosel, H.D., and Herrmann, K.,** The phenolics of fruits. VII. The phenolics of cherries and plums and the changes in catechins and hydroxycinnamic acid derivatives during the development of fruits, *Z. Lebensm. Unters. Forsch.,* 159, 85, 1975.
26. **Eaton, G., Meehan, C., and Turner, N.,** Some physical effects of post-harvest gamma radiation on the fruit of sweet cherry, blueberry and cranberry, *Can. Inst. Food Sci. Technol. J.,* 3, 152, 1970.
27. **Andreotti, R., Generali, P., and Tomasicchio, M.,** Effetto delle radiazioni gamma sui pigmenti antocianici del lampone e della ciliegia, *Ind. Conserve,* 50, 209, 1975.
28. **Drake, S.R., Proebsting, E.L., Jr., Carter, G.H., and Nelson, J.W.,** Effect of growth regulators on ascorbic acid content, drained weight and color of fresh and processed Rainer cherries, *J. Am. Soc. Hortic. Sci.,* 103, 162, 1978.
29. **Drake, S.R., Proebsting, E.L., Jr., and Nelson, J.W.,** Influence of growth regulators on the quality of fresh and processed Bing cherries, *J. Food Sci.,* 43, 1695, 1978.
30. **Drake, S.R., Proebsting, E.L., Jr., Thompson, J.B., and Nelson, J.W.,** Effects of diaminozide, maturity and cultivar on the color grade and character of sweet cherries, *J. Am. Soc. Hortic. Sci.,* 105, 668, 1980.
31. **Drake, S.R., Proebsting, E.L., Jr., and Spayd, S.E.,** Maturity index for the color grade of canned dark sweet cherries, *J. Am. Soc. Hortic. Sci.,* 107, 180, 1982.
32. **Chaplin, M.H. and Kenworthy, A.L.,** Influence of succinic acid mono-(2,2-dimethyl)hydrazide (Aminozide) on fruit ripening of Windsor sweet cherry, *J. Am. Soc. Hortic. Sci.,* 95, 532, 1970.

33. **Ystaas, J.,** Effect of Alar on shoot growth, fruit ripening and the occurrence of Alar residues in sweet cherries, *Forsk. Fors. Landbruket,* 24, 451, 1973 (FSTA 6, 4J479).

34. **Ryugo, K.,** Persistance and mobility of Alar (B-995) and its effects on anthocyanin metabolism in sweet cherry, *Prunus avium, Proc. Am. Soc. Hortic. Sci.,* 88, 160, 1966.

35. **Gyuro, F. and Toth, M.,** Stim-80 on the shoot growth and flower numbers of sweet and sour cherry, *Kert. Egyetem Kozl.,* 40, 127, 1976.

36. **Lamb, R.C.,** Notes on the inheritance of some characters in the sweet cherry *Prunus avium, Proc. Am. Soc. Hortic. Sci.,* 61, 293, 1953.

37. **Fogle, H.W.,** Inheritance of fruit color in sweet cherries *(Prunus avium), J. Hered.,* 49, 294, 1958.

38. **Toyama, T.K.,** Inheritance of fruit color in the sweet cherry, *HortScience,* 13, 155, 1978.

39. **Wagenknecht, A.D., Scheiner, S.M., and Van Buren, J.P.,** Anthocyanase activity and its possible relation to scald in sour cherries, *Food Technol.,* 14, 47, 1960.

40. **Pifferi, P.G. and Cultrera, R.,** Enzymatic degradation of anthocyanins: the role of sweet cherry polyphenoloxidase, *J. Food Sci.,* 39, 786, 1974.

41. **Pifferi, P.G., Cultrera, R., Mazzocco, F., Iori, R., and Manenti, I.,** Struttura degli antociani e loro degradazione enzimatica in rapporto al pH ed al fenolo substrato della polifenolossidasi, *Ind. Conserve,* 54, 105, 1979.

42. **Scheiner, D.M.,** Enzymic decolorisation of anthocyanin pigments, *Diss. Abstr. Int. B,* 21, 2882, 1961.

43. **Yankov, S.I.,** Oxidation des substances polyphenoliques des cerises au cours du broyage., *Ann. Technol. Agric.,* 21, 123, 1972.

44. **Kalitka, V.V. and Skrypnik, V.V.,** Changes in the anthocyanin content of cherries during controlled atmosphere storage, *Konservn. Ovoshchesush. Prom.,* 11, 46, 1983 (CA 100, 21744).

45. **Skrypnik, V.V., Kalitka, V.V., and Naichenko, V.M.,** Dynamics of anthocyanin pigments in storage of cherry fruits, *Fiziol. Biokhim. Kult. Rast.,* 17, 182, 1985 (CA 102, 202853).

46. **Polesello, A. and Bonzini, C.,** Observations on pigments of sweet cherries and on pigment stability during frozen storage. I. Anthocyanin composition, *Confructa,* 22, 170, 1977.

47. **Pizzocaro, F., Crotti, C., and Polesello, A.,** Observations on the pigmentation of cherries and the stability of this pigmentation during frozen storage: influence of variety, *Bull. Inst. Int. Froid,* 59, 1717, 1979.

48. **Drake, S.R.,** The cherry, in *Quality and Preservation of Fruits,* Eskin, N.A.M., Ed., CRC Press, Boca Raton, FL, 1991, 169.

49. **Tanchev, S.S. and Ioncheva, N.,** Kinetics of destruction of anthocyanins in cherry fruits, *Izv. Vyssh. Uchebn. Zaved., Pishch. Tekhnol.,* 4, 58, 1972 (CA 78, 41764).

50. **Ruskov, P. and Tanchev, S.,** Kinetics of degradation of anthocyanins in cherries, *Nauchn. Tr., Vyssh. Inst. Khranit. Vkusova Prom. Plovdiv,* 18, 215, 1971 (CA 77, 60288).

51. **Beschia, M., Segal, B., and Toma, G.,** Effect of pectolytic enzymic preparations on the flavones and anthocyanins of sweet cherries, *Ind. Aliment. (Bucharest),* 23, 683, 197 (CA 78, 122846).

52. **Chytra, M., Curda, D., and Kyzlink, V.,** Effect of processing on the stability of some fruit anthocyanins, *Sb. Pr. Vys. Sk. Chem.-Technol., Potravin Technol.,* 1960, 57, 1962 (JSFA 1962, 58ii).

53. **Savic, M.,** Changes of anthocyanin content in the production of clear concentrates of mixed varieties of cherry, *Rad. Poljopr. Fak. Univ. Sarajevu,* 32, 149, 1984 (CA 103, 36360).

54. **Rochleder, F.,** Ueber einiger Bestandtheile der Fruechte von *Cerasus acida* Borckh. *Chem. Ber.,* 3, 238, 1870.

55. **Li, K.C. and Wagenknecht, A.C.**, The anthocyanin pigments of sour cherries, *J. Am. Chem. Soc.*, 78, 979, 1956.

56. **Markakis, P.**, Zone electrophoresis of anthocyanins, *Nature (London)*, 187, 1092, 1960.

57. **Dekazos, E.D.**, Anthocyanin pigments in red tart cherries, *J. Food Sci.*, 35, 237, 1970.

58. **Schaller, D.R. and Von Elbe, J.H.**, The minor pigment component of Montmorency cherries, *J. Food Sci.*, 33, 442, 1968.

59. **Von Elbe, J.H. and Schaller, D.R.**, Hydrochloric acid in isolating anthocyanin pigments from Montmorency cherries, *J. Food Sci.*, 33, 439, 1968.

60. **Schaller, D.R. and Von Elbe, J.H.**, Polyphenols in Montmorency cherries, *J. Food Sci.*, 35, 762, 1970.

61. **Fischer, R.R. and Von Elbe, J.H.**, Identification of cyanidin 3-(2^G-glucosyl)-rutinoside in Montmorency cherries, *J. Milk Food Technol.*, 33, 481, 1970 (CA 75, 724423).

62. **Timberlake, C.F. and Tanchev, S.S.**, Qualitative identification of anthocyanins in some fruits, *Nauchn. Tr. Vyssh. Inst. Khranit. Vkusova Prom. Plovdiv*, 16, 71, 1969 (CA 78, 2975).

63. **Shrikhande, A.J. and Francis, F.J.**, Anthocyanin pigments of sour cherry, *J. Food Sci.*, 38, 649, 1973.

64. **Von Elbe, J.H., Bixby, D.G., and Moore, J.D.**, Electrophoretic comparison of anthocyanin pigments in eight varieties of sour cherries, *J. Food Sci.*, 34, 113, 1969.

65. **Hong, V. and Wrolstad, R.E.**, Characterization of anthocyanin-containing colorant and fruit juices by HPLC/photodiode array detection, *J. Agric. Food Chem.*, 38, 698, 1990.

66. **Dekazos, E.D.**, Quantitative determination of anthocyanin pigments during the maturation and ripening of red tart cherries, *J. Food Sci.*, 35, 242, 1970.

67. **Savic, M.**, Quantitative analysis of anthocyanins in sour cherry cultivars, *Hrana Ishrana*, 26, 135, 1985 (CA 107, 235108).

68. **Wander, M.**, Anthocyanins in sour cherries, *Gartenbauwissenschaft*, 50, 183, 1985 (CA 103, 175446).

69. **Ben, J. and Gaweda, M.**, Changes in the quantity of pigments in the developing tart cherries of North Star and Lutowka cultivars. II. Growth of fruit sets and amount of anthocyanins, *Fruit Sci. Rep.*, 134, 171, 1987 (CA 109, 20333).

70. **Schmid, P.P.S., Scherf, H., and Schmidt, K.**, Composition of the sour cherry cultivars Schattenmorelle in relation to various rootstocks, *Erwebsobstbau*, 24, 175, 1982.

71. **Schmid, P.P.S., Schmidt, K., Kraus, A., and Schimmelpfeng, H.**, The composition of sour cherries in relation to cultivar and rootstocks, *Erwebsobstbau*, 25, 204, 1983 (PBA 1984, 8415).

72. **Dekazos, E.D. and Worley, J.P.**, Effect of succinic acid diethylhydrazide (Alar) on anthocyanin metabolism and cell wall carbohydrates in red tart cherries, *Prunus cerasus*, *J. Am. Soc. Hortic. Sci.*, 95, 703, 1970.

73. **Poll, L.**, Studies on the quality of sour cherry juices (*Prunus cerasus* L.). I. Influence of harvesting date on chemical composition, *Acta Agric. Scand.*, 36, 205, 1986 (CA 105, 76145).

74. **Von Elbe, J.H.**, Factors Affecting the Colour Stability of Cherry Pigments and Cherry Juice, Ph.D. thesis, University of Wisconsin, Madison, 1964 (JSFA 1964, 204ii).

75. **Dalal, K.B. and Salunkhe, D.K.**, Thermal degradation of pigments and relative biochemical changes in canned apricots and cherries, *Food Technol.*, 18, 1198, 1964.

76. **Lovric, T., Debicki, J., Sevcik, D., and Skitarelic, G.**, The influence of HMF (hydroxymethylfurfural) and furfural on the degradation of anthocyanin in cherries (*Kem. Ind.*, 17, 799, 1968 (JSFA 1969, 45ii).

77. **Lovric, T.**, Studio dei fattori che determinano la stabilità del colore dei succhi di amarena e di mora di rovo, *Ind. Conserve*, 40, 208, 1965.

78. **Savic, M.**, Influence of blanching temperature on the stability of cherry anthocyanins, *Rad. Poljopr. Fak. Univ. Sarajevu*, 32, 141, 1984 (CA 103, 36359).

79. **Siegel, A., Markakis, P., and Bedford, C.L.,** Stabilization of anthocyanins in frozen tart cherries by blanching, *J. Food Sci.,* 36, 962, 1971.

80. **Tanchev, S.S.,** Kinetics of the degradation of anthocyanins during storage of juice from cherries and mazzards, *Nauchn. Tr. Vyssh. Inst. Khranit. Vkusova Prom. Plovdiv,* 20, 1643, 1973 (CA 83, 162348).

81. **Milkov, I. and Atanasov, B.,** Mathematical modelling of colour in concentrated products with sour cherry, *Khranit. Prom.,* 38, 27, 1989 (FSTA 22, 7A99).

82. **Dekazos, E.D.,** Anthocyanin in red tart cherries as related to anaerobiosis and scald, *J. Food Sci.,* 31, 226, 1966.

83. **Van Buren, J.P., Scheiner, D.M., and Wagenknecht, A.C.,** An anthocyanin-decolorizing system in sour cherries, *Nature (London),* 185, 1655, 1960.

84. **Peng, C.Y. and Markakis, P.,** Effect of phenolase on anthocyanins, *Nature (London),* 199, 597, 1963.

85. **Yeatman, J.N., Birth, G.S., Ernest, J.V., Bender, R.W., and Sidwell, A.P.,** Spectrophotometric evaluation of anthocyanin pigment development and scald damage in intact red tart cherries, *Food Technol.,* 15, 521, 1961.

86. **Do, J.Y., Potewiratananond, S., Salunkhe, D.K., and Rahman, A.R.,** Freeze dehydrated sour cherries. II. Stability of anthocyanins during storage, *J. Food Technol.,* 11, 265, 1976.

87. **Urbanyi, G.,** Changes in colour and anthocyanin content of quick-frozen sour cherries during freeze-drying and subsequent storage, *Hutoipar,* 33, 118, 1987 (FSTA 21, 5J18).

88. **Gombköto, G.,** Anthocyanin pigments of the black cherry, *Acta Aliment.,* 9, 335, 1980 (CA 95, 3387).

89. **Du, C.T., Wang, P.L., and Francis, F.J.,** Anthocyanins of ornamental cherries, *J. Food Sci.,* 40, 1142, 1975.

90. **Roversi, A.,** Anche il *Prunus tomentosa* tra i piccoli frutti, in *Lampone, Mirtillo ed altri Piccoli Frutti,* Atti, Trento, Italia, 1987, 117.

91. **Kask, K.,** The tomentosa cherry, *Fruit Var. J.,* 43, 50, 1989 (PBA 1989, 9678).

92. **Gogolishvili, Z.M.,** Polyphenols in the fruit of the forest laurel cherry, *Tr. Vses. Semin. Biol. Aktiv. (Lech.), Veshchestvam Plodov Yagod,* 4th, 1970, 192 (CA 81, 88019).

93. **Gogolishvili, Z.M.,** Fruits of the laurel cherry. *Tr., Gruz. Nauchno-Issled. Inst. Pishch. Prom.,* 5, 133, 1977 (CA 78, 108188).

94. **Tsiklauri, G.C.,** Anthocyanins of cherry laurel (*Laurocerasus officinalis* Roem.) fruit, *Soobshch. Akad. Nauk Gruz. SSR,* 79, 177, 1975 (CA 83, 190350).

95. **Olden, E.J.,** Stone fruits, rootstocks, strawberries and hazel, *Inst. Fruit Berry Breed, Rep. Activ., Balsgard,* 16, 1960 (PBA 1962, 1074).

96. **Vaughan, J.G.,** Oleaceae in *The Structure and Utilization of Oilseeds,* Chapman & Hall, London, 1970, 1979.

97. **Gutfinger, T.,** Polyphenols in olive oils, *J. Am. Oil Chem. Soc.,* 58, 966, 1981.

98. **Nergiz, C. and Unal, K.,** Determination of phenolic acids in virgin olive oil, *Food Chem.,* 39, 237, 1991.

99. **Montedoro, G.F., Servili, M., Baldioli, M., and Miniati, E.,** Simple and hydrolizable phenolic compounds in virgin olive oil. Note 1. Phenolic extracts, quantitative and semiquantitative separation and evaluation by HPLC, *J. Agric. Food Chem.,* 40, 1571, 1992.

100. **Laenderer, X.,** Vierteljahreschrift, *Prakt. Pharm.,* 13, 1370, 1864.

101. **Musajo, L.,** L'antociano delle olive, oleocianina, *Gazz. Chim. Ital.,* 70, 293, 1940.

102. **Cantarelli, C.,** Sui polifenoli presenti nelle drupe e nell'olio di oliva, *Riv. Tal. Sostanze Grasse,* 38, 69, 1961.

103. **Cantarelli, C.,** Cultvar e pigmentazione dell olive: esame analitico dei polifenoli delle drupe di diverse cultivar dell'Italia centrale, Accad. Naz. Olivo, 1° Convegno Olivicolo-Oleario, Spoleto, Italy, June 1–3, 1962.

104. **Vázquez Roncero, A. and Maestro Durán, R.,** Los colorantes antocianicos de la aceituna madura. I. Estudio cualitativo, *Grasas Aceites (Seville),* 21, 208, 1970.

105. **Luh, B.S. and Mahecha, G.,** Anthocyanins of *Olea europaea* (Manzanillo olives), *Chung Kuo Nung Yeh Hua Hsueh Hui Chih* (Special issue), 1, 1971 (CA 77, 60298).

106. **Maestro Durán, R. and Vázquez Roncero, A.,** Los pigmentos antocianicos de la aceituna Manzanillas maduras, *Grasas Aceites (Seville),* 27, 237, 1976.

107. **Tanchev, S.S., Ioncheva, N., Genov, N., and Coudounis, M.,** Identification of anthocyanins in olives, *Bull. Liaison Groupe Polyphénols,* 9, 114, 1980.

108. **Tanchev, S.S., Ioncheva, N., Genov, N., and Coudounis, M.,** Identification of anthocyanins contained in olives, *Georgike Ereuna,* 4, 5, 1980.

109. **Ragazzi, E. and Veronese, G.,** Ricerche sugli antociani nella polpa delle olive mature coltivate in Italia, *Atti Mem. Accad. Patavina Sci. Lett. Arti,* 85, 251, 1977.

110. **Vázquez Roncero, A., Graciani Constante, E., and Maestro Durán, R.,** Componentes fenolicos de la aceituna madura. I. Polifenoles de la pulpa, *Grasas Aceites (Seville),* 25, 269, 1974.

111. **Vázquez Roncero, A., Maestro Durán, R., and Janer der Valle, M.L.,** Los colorantes de la ceituna madura. II. Variaciones durante la maduracion, *Grasas Aceites (Seville),* 21, 337, 1970.

112. **Golubov, V.N., Gusar, Z.D., and Mamedov, E.S.,** Cultivar-specific characteristics of olives grown in Azerbaidzhan, *Subtrop. Kult.,* 6, 86, 1987 (CA 109, 51732s).

113. **Vlahov, G. and Solinas, M.,** Studio della composizione quali-quantitativa mediante GLC degli antociani delle olive nere. Loro evoluzione nel corso della maturazione, *Ann. Ist. Sper. Elaiotecnica,* Pescara, Italy, 1984–87, 10.

114. **Vlahov, G.,** Gli antociani polimeri delle olive nere. Loro determinazione in funzione del processo di maturazione, *Ind. Aliment.,* 29, 11, 1990.

115. **Shulman, Y. and Lavee, S.,** Effect of kinetin on anthocyanin formation in green harvested olive fruits, *J. Am. Soc. Hortic. Sci.,* 96, 808, 1971.

116. **Shulman, Y. and Lavee, S.,** Effect of cytokinins and auxins on anthocyanin accumulation in green Manzanilla olives, *J. Exp. Bot.,* 24, 655, 1973.

117. **Lavee, S.,** Abscission studies of olive fruit — physiological and horticultural aspects, *Olea,* 35, 1976.

118. **Shulman, Y., Erez, A., and Lavee, S.,** Delay in ripening of picked olives due to ethylene treatments, *Sci. Hortic.,* 2, 21, 1974.

119. **Rugini, E., Bongi, G., and Fontanazza, G.,** Effect of ethephon on olive ripening, *J. Am. Soc. Hortic. Sci.,* 107, 835, 1982.

120. **Vlahov, G.,** La polimerizzaione degli antociani nelle olive nere da mensa. Ruolo delle tecnologie, *Atti Semin. Int. Olio d'Oliva Olive da Tavola: Tecnologia e Qualità,* Pescara, Italy, 1990, 357.

121. **Amiot, M.J., Fleuriet, A., and Macheix, J.J.,** Importance and evolution of phenolic compounds in olive during growth and maturation, *J. Agric. Food Chem.,* 34, 828, 1986.

122. **Cantarelli, C. and Montedoro, G.F.,** Extraction des antioxidants naturelles des olives: etude experimentale, in *Natuerliche und Synthetische Zusatstoffe in der Nahrung der Menschen,* Amonn, R., Hollo, J., Eds., D. Steinkopf Verlag, Darmstadt, Germany, 1974, 84.

123. **Coudounis, M., Katsaboxakis, K., and Papanicolaou, D.,** Progress in the extraction and purification of anthocyanin pigments from the effluents of olive oil extracting plants, *LWT (Prog. Food Eng.),* 7, 567, 1983.

124. **Coudounis, M. and Tanchev, S.,** Recovery of anthocyanin pigments present in olive oil extraction plant effluents, *Bull. Liaison Groupe Polyphénols,* 9, 56, 1980.

125. **Hsia, C.L., Luh, B.S., and Chichester, C.O.,** Chemistry of anthocyanin in freestone peaches, *J. Food Sci.,* 30, 5, 1965.

126. **Erez, A. and Flone, J.A.,** The quantitative effect of solar radiation on 'Redhaven' peach fruit skin color, *HortScience,* 21, 1424, 1986.

127. **Van Blaricom, L.O. and Senn, T.L.,** Anthocyanin pigments in freestone peaches in the southeast, *Proc. Am. Soc. Hortic. Sci.,* 90, 541, 1967.

128. **Basiouny, F.M. and Buchanan, D.W.,** Fruit quality of Sungold nectarine as influenced by shade and sprinkling, *Soil Crop Sci. Fla. Proc.,* 36, 130, 1977.

129. **Culpepper, G.W. and Caldwell, J.S.,** The behavior of anthocyanin pigments in canning, *J. Agric. Res.,* 35, 107, 1929.

130. **Johnson, G., Mayer, M.M., and Johnson, D.K.,** Isolation and characterization of peach tannins, *Food Res.,* 16, 169, 1951.

131. **Hayashi, K., Abe, Y., Noguchi, T., and Suzushino, G.,** Anthocyanins. XXII. Analyses by paper chromatography of natural anthocyanin and its application to the investigation of dyes of the red *Impatiens* flowers and the blood-red peach fruit, *Pharm. Bull.,* 1, 130, 1963 (CA 50, 14035e).

132. **Luh, B.S., Chichester, C.O., and Leonard, S.J.,** Brown discoloration in canned freestone peaches, Proc. Congr. Food Sci. Technol., London, 1962, 401.

133. **Hsia, C.L., Luh, B.S., and Chichester, C.O.,** Chemistry of anthocyanin in Elberta freestone peaches, Proc. Congr. Food Sci. Technol., London, 1962, 429.

134. **Miyakawa, H. and Takehana, H.,** Quality of canned peaches. I. Preventing the change of the anthocyanin color to purple, *Chiba Daigaku Engeigakubu Gakujutsu Hokoku,* 14, 43, 1966 (CA 58, 21029).

135. **Mazza, G. and Brouillard, R.,** Color stability and structural transformations of cyanidin 3,5-diglucoside and four 3-deoxyanthocyanins in aqueous solutions, *J. Agric. Food Chem.,* 35, 422, 1987.

136. **Yuan, S., Tung, N-F., and Sheng, C-Y.,** Main causes of pink discoloration of the canned white peaches in syrup, *Shih Pin Yu Fa Hsiao Kung Yeh,* 2, 57, 1981 (CA 95, 5219).

137. **Denny, E.G., Coston, D.C., and Ballard, R.E.,** Peach skin discoloration, *J. Am. Soc. Hortic. Sci.,* 111, 439, 1986.

138. **Uciyama, Y.,** Anthocyanin decolorizing enzyme from moulds. I. Screening test for decoloration of anthocyanins in reddish white peach, *Agric. Biol. Chem.,* 33, 1342, 1969.

139. **Ueno, N., Takemura, E., and Hayashi, K.,** Additional data for the paper chromatographic survey of anthocyanins in the flora of Japan. IV. Studies on anthocyanins, *Bot. Mag. (Tokyo),* 82, 155, 1969 (HA 4237).

140. **Monet, R.,** Contribution a l'etude genetique du pécher, *Ann. Amelior. Plant,* 17, 5, 1967.

141. **Borzakovskaya, I.V., Shitan, I.M., Kolomiets, N.P., and Chuprina, L.M.,** Changes in the pigment complex in tissues of new peach hybrids in relation to their winter hardiness, *Introd. Aklim. Rosl. Ukr.,* 16, 64, 1980 (PBA 1981, 3562).

142. **Sandhu, S.S. and Dhillon, B.S.,** The relation between growth pattern and endogenous metabolites in the developing fruit of early maturing Flordasun peach, *J. Food Sci. Technol.,* 18, 135, 1981.

143. **El-Ashwah, F.A., Awaad, K.A., Sahran, M.A., and Keikal, H.A.,** Quality attributes of peach fruits during ripening and peeling, *Agric. Res. Rev.,* 55, 67, 1977.

144. **Manabe, M., Nakamichi, K., Shingai, R., and Tarutani, T.,** Effect of temperature in post-harvest ripening and storage on quantitative changes of anthocyanin and polyphenol in white peaches, *Nippon Shokuhin Kogyo Gakkai-Shi,* 26, 175, 1979 (CA 91, 73331).

145. **Sims, E.T., Jr., Gambrell, C.E., and McClary, J.T.,** Influence of succinic acid 2,2-dimethylhydrazide on peach quality, *J. Am. Soc. Hortic. Sci.,* 96, 527, 1971.

146. **Morini, S., Vitagliano, G., and Xiloyannis, C.,** Azione dell'Ethephon e del 2,4,5 TP sulla cascola, pezzatura e maturazione dei frutti di pesco, *Riv. Ortoflorofruttic. Ital.,* 58, 235, 1974.

147. **Maekaji, K., Sunagawa, M., and Imai, H.,** The anthocyanins in the petals of white peach (*Prunus persica* var. *vulgaris*), *Agric. Biol. Chem.,* 27, 165, 1963.

148. **Abe, Y.,** Anthocyanin pigments of the petals of peach tree, *Kenkyu Nenpo — Nihin Daigaku Buringakubu (Mishima), Shizen Kagaku Hen,* 27, 17, 1978 (CA 91, 2548).

149. **Van Buren, J.,** Fruit phenolics, in *Biochemistry of Fruits and Their Products,* Vol. 1, Hulme, A.C., Ed., Academic Press, London, 1970, 269.

150. **Sadykov, A.A., Isaev, K.T., and Vdovtseva, T.A.,** Anthocyanins from *Persica vulgaris* fruit skins, *Khim. Prir. Soedin.,* 3, 405, 1978 (CA 89, 126165).

151. **Raynal, J., Moutounet, M., and Souquet, J.-M.,** Intervention of phenolic compounds in plum technology. I. Changes during drying, *J. Agric. Food Chem.,* 37, 1046, 1989.

152. **Popov, K.H.R.,** Anthocyanin pigments in some varieties of raspberry and plum, *Nauchn. Tr., Nauchno-Issled, Inst. Konserv. Prom. Plovdiv.,* 5, 69, 1967 (CA 71, 10290).

153. **Blundstone, H.A.W. and Crean, D.E.C.,** *The Pigments of Red Fruits,* Rep. Fruit Veg. Preserv. Res. Assoc., Chipping Campden, U.K., 1966.

154. **Rees, D.I.,** Chemical constituents of Victoria plums: changes during growth on the tree, *J. Sci. Food Agric.,* 9, 404, 1958.

155. **Hillis, W.E. and Swain, T.,** The phenolic constituents of *Prunus domestica.* II. The analysis of tissues of the Victoria plum tree, *J. Sci. Food Agric.,* 10, 135, 1959.

156. **Swain, T. and Hillis, W.E.,** Phenolic constituents of *Prunus domestica.* I. Quantitative analysis of phenolic constituents, *J. Sci. Food Agric.,* 10, 63, 1959.

157. **Macheix, J.-J., Fleuriet, A., and Billot, J.,** *Fruit Phenolics,* CRC Press, Boca Raton, FL, 1991, 149.

158. **Bulatovic, M.S.,** Analysis of the anthocyanin pigment content in the plum cultivars with different time of maturity, *Acta Hortic.,* 74, 165, 1978.

159. **Ingalsbe, R.W., Carter, G.H., and Neubert, A.M.,** Anthocyanin pigments as a maturity index for processing dark sweet cherries and purple plums, *J. Agric. Food Chem.,* 13, 3580, 1965.

160. **György, K.,** The effect of plum pox on the composition of the Besztercei plum, *Gyumolcstermeszetes,* 3, 161, 1976 (HA 1977, 8274).

161. **György, K.,** Biochemical investigations on plum ripening, *Kiserlet. Kozl.,* 61C(1/3), 21, 1968 (HA 1970, 3211).

162. **Tanchev, S.S.,** Kinetics of destruction of anthocyanins during sterilization and storage of Afuski plum juice, *Nauchn. Tr. Vyssh. Inst. Khranit. Vkusova Prom. Plovdiv,* 20, 197, 1973 (CA 83, 162343).

163. **Ruskov, P. and Tanchev, S.S.,** Kinetics of the anthocyanin degradation during plum juice sterilization and storage, *Proc. Congr. Food Sci. Technol., Madrid,* 1B, 40, 1974 (FSTA 7, 5H708).

164. **Tanchev, S., Frulinski, I., and Videv, K.,** Effect of sterilization conditions on some quality indices of greengages compotes, *Nauchni Tr. Vissh. Inst. Khranit. Vkusova Prom., Plovdiv,* 265, 11, 1978 (FSTA 7, 9J1443).

165. **Ichas, H., Golebiowski, T., Pieczonka, W., and Satora, Z.,** Colour changes in concentrated plum juice on storage, *Przem. Ferment. Roln.,* 20, 17, 1976 (FSTA 9, 6H992).

166. **Weinert, I.A.G., Solms, J., and Escher, F.,** Quality of canned plums with varying degrees of ripeness. I. Chemical characterization of color changes, *Lebensm. Wiss. Technol.,* 22, 307, 1990.

167. **Weinert, I.A.G., Solms, J., and Escher, F.,** Quality of canned plums with varying degrees of ripeness. II. Texture measurement and sensory evaluation of texture and color, *Lebensm. Wiss. Technol.,* 23, 117, 1990.

168. **Weinert, I.A.G., Solms, J., and Escher, F.,** Diffusion of anthocyanins during processing and storage of canned plums, *Lebensm. Wiss. Technol.,* 23, 396, 1990.

169. **Weinert, I.A.G., Solms, J., and Escher, F.,** Polymerization of anthocyanins during processing and storage of canned plums, *Lebensm. Wiss. Technol.,* 23, 445, 1990.

170. **Raynal, J. and Moutounet, M.,** Intervention of phenolic compounds in plum technology. II. Mechanism of anthocyanin degradation, *J. Agric. Food Chem.,* 37, 1051, 1989.

171. **Brouillard, R.,** Chemical structure of anthocyanins, in *Anthocyanins as Food Colors,* Markakis, P., Ed., Academic Press, New York, 1981, 1.

172. **Mazza, G. and Brouillard, R.,** The mechanism of co-pigmentation of anthocyanins in aqueous solutions, *Phytochemistry,* 29, 1097, 1990.

173. **Reznik, H.,** Untersuchungen über die physiologische Bedeutung der chymochromen Farbstoffe, *Stiz. Ber. Heidelberg Akad. Wiss., Math. Naturwiss.,* 1956, 2abh.

174. **Hunter, N.,** Inheritance of flesh colour in the fruit of the Japanese plum, *Prunus salicina, J. Agric. Sci.,* 5, 673, 1962.

175. **Ahn, S.Y.,** Studies on the identification of anthocyanins in plums, *J. Korean Agric. Chem. Soc.,* 16, 53, 1973 (FSTA 7, 7J1079).

176. **Tsuji, M., Harakawa, M., and Komiyama, Y.,** Main anthocyanin pigments in plum fruit *(Prunus salicina)* variety Sordum, *Nippon Shokuhin Kogyo Gakkai-Shi,* 28, 517, 1981 (CA 96, 33560).

177. **Itoo, S., Matsuo, T., Noguchi, K., and Kodama, K.,** The qualities of subtropical fruits. III. Anthocyanin pigment of Japanese plum *(Prunus salicina* Lindl.) cv. Karari, *Kagoshima Daigaku Nogakubu Gakujutsu Hokoku,* 32, 35, 1982 (CA 97, 90672).

178. **Takeda, K. and Sano, S.,** Anthocyanin in the purple-black fruits of *Prunus jamasakura* Sieb., *Tokyo Gakugei Daigaku Kiyo, Dai-4-Bu,* 30, 251, 1978 (CA 90, 200314).

179. **Draetta, I.S., Iaderoza, M., Baldini, V.L.S., and Franic, F.J.,** Antocianinas de ameixa *(Prunus salicina* L.), *Cien. Tec. Aliment,* 5, 31, 1985.

180. **Tsuji, M., Harakawa, M., and Komiyama, Y.,** Inhibition of increase of pulp color and PAL activity in plum fruit at high temperature (30°C), *J. Jpn. Soc. Hortic. Sci.,* 30, 688, 1983.

181. **Lee, J.C. and Lee, Y.B.,** Physiological study on the coloration in plum fruits. I. Effect of ethephon on the fruit composition and anthocyanin development in Santa Rosa plum *(Prunus salicina), Hanguk Wonye Hakhoe Chi,* 21, 36, 1980 (CA 94, 97853).

182. **Blommaert, K.L.J., Theron, T., and Steenkamp, J.,** Earlier and more uniform ripening of Santa Rosa plums using ethephon, *Fruit Grower,* 25, 267, 1975.

183. **Lee, J.C. and Lee, Y.B.,** Physiological study on the coloration in plum fruits. II. Effect of sucrose and plant growth regulators on the anthocyanin development in Santa Rosa plum *(Prunus salicina)* fruits, *Hanguk Wonye Hakhoe Chi,* 21, 170, 1980 (CA 94, 80366).

184. **Werner, D.J., Maness, E., and Ballinger, W.E.,** Fruit anthocyanins in three sections of *Prunus* (Rosaceae), *HortScience,* 24, 488, 1989.

185. **Rozmyslova, A.G. and Samorodova-Bianki, G.B.,** Polyphenols in sloe, *Byull. Vses. Ordena Lenina Ordena Druzhby Nard. Nauchnoissled. Inst. Rastit. N.I. Vavilova,* 73, 63, 1977 (PBA 1979, 7428).

186. **Ramos, T. and Macheix, J.J.,** Phenols in blackthorn *(Prunus spinosa* L.) fruit. *Plant. Med. Phytother.,* 24, 14, 1990 (CA 114, 139742).

187. **Tanchev, S.S. and Vasilev, V.N.,** Identification of the anthocyanins in fruits of some varieties of *Prunus cerasifera, Gradinar. Lozar. Nauka,* 10, 23, 1973 (CA 79, 144166).

188. **Rahman, A.U., Frontera, M.A., and Tomas, M.A.,** Antocianinas en hojas de *Prunus cerasifera* (Forma *atropurpurea), An. Asoc. Quim. Argent.,* 63, 91, 1975.

Chapter 4

SMALL FRUITS

I. BLACKBERRIES

Blackberries have been used by man for a very long time. They were used medicinally up to the 16th century, their juice being recommended for infections of the mouth and eye.[1] There are many species of blackberries, varying in weight and size, including European and American blackberries. The European blackberries, in turn, can be divided into two large groups: the true blackberries or *Moriferi veri* with eleven subsections, and *Rubus caesius* with all its polymorphous relatives.[1] The designation of *R. fruticosus* L. agg. is sometimes used to denote the aggregate of blackberries of the *Moriferi* complex. In North America, there are at least five primary diploid species, which are *R. allegheniensis* Porter, *R. argutus* Link, *R. cuneifolius* Pursh, *R. setosus* Bigel, and *R. trivialis* L.[1]

Torre and Barritt[2] analyzed by TLC and densitometry 16 species and three cultivars of blackberries for individual anthocyanin pigments. They found that blackberry fruits were characterized by one major pigment — cyanidin 3-glucoside — and, in some cases, a second pigment — cyanidin 3-rutinoside — in lower concentrations (Table 1). Bate-Smith,[3] in his classical chromatographic study of anthocyanins, reported a monoglycoside of cyanidin and an unidentified pigment in *R. fruticosus*. Fouassin[4] and Blundstone and Crean[5] also found cyanidin 3-glucoside and other unidentified pigments. The second major pigment was first identified as cyanidin 3-rutinoside by Harborne and Hall.[6] Nybom[7] and Jennings and Carmichael[8] confirmed the occurrence of the 3-rutinoside of cyanidin and reported on the inheritance of anthocyanins in blackberries. Thirteen pigments, two of which were monoglycosides and two biosides of cyanidin, were reported by Casoli et al.[9] In addition, a bioside of malvidin and a monoglycoside of delphinidin were also found. 'Marion' blackberries are unusual in that the major pigment is cyanidin 3-rutinoside.[10] Recently, Sapers et al.,[11] using HPLC showed that in addition to cyanidin 3-glucoside and cyanidin 3-rutinoside, thornless blackberries contain cyanidin 3-xyloside along with several cyanidin derivatives acylated with a dicarboxylic acid.

During ripening, cyanidin 3-glucoside increases to a greater extent than the other pigment, and the accumulation of individual anthocyanins is linearly related to the total anthocyanin content and to the soluble solid:acidity ratio.[11] Bilyk and Sapers[12] studied the distribution of flavonols in thornless blackberry fruits and found no correlation between the amounts of flavonols and the total anthocyanin content. During ripening in warm weather conditions, antho-

TABLE 1
Anthocyanins of Blackberries (*Rubus* spp.)

Species or cultivar	Anthocyanins[a]			No. of unidentified pigments	Total conc
	Cyanidin 3-glucoside	Cyanidin 3-rutinoside	Pelargonidin 3-glucoside		
Rubus allegheniensis	129.2	51.0			180.2
R. amabilis	137.5			1	137.5
R. caesius	75.6	6.9			82.5
R. canadensis	95.7				95.7
R. caucasicus	92.9		4.9		97.8
R. dregeri	221.6	75.6		3	325.9
R. laciniatus	148.7				148.7
R. picetorum	171.1				171.1
R. plicatus	171.4	41.5		1	234.2
R. procerus	143.6			1	143.6
R. radula	145.6			1	145.6
R. scanicus	138.1	9.6			147.7
R. scheutzii	161.9			2	161.9
R. shankii	159.9			2	159.9
R. wahlbergii	186.7	25.3		3	244.4
R. ursinus	64.0	45.0			109.0
'OR-US 1122'	121.4	17.6		2	150.7
'Black Satin'	142.1			2	154.8
'Dirksen Thornless'	131.4			1	131.4

[a] Conc in mg/100 g of fruit.

Adapted from Torre, L. C. and Barritt, B. H., *J. Food Sci.*, 42, 488, 1977.

cyanin concentration in fruits reaches highest levels between the third and the fifth day of the ripening period.[13] Fruit ripening at low temperature has fewer anthocyanins and little dry matter.[13]

Juices and fermented drinks made from blackberries are particularly rich in anthocyanins — so much so that at times they precipitate.[14] Removal of excess pigments can be achieved by enzymatic treatment with anthocyanase.[15] In a blackberry juice concentrate, Hong and Wrolstad[16] found cyanidin 3-glucoside, 3-rutinoside, 3-sophoroside, and 3-glucosylrutinoside (Table 2). Neither cyanidin 3-sophoroside nor cyanidin 3-glucosylrutinoside has been reported in blackberry fruit. Thus, the presence of these unexpected pigments may have been caused by contamination of the sample with red raspberries which are known to contain these pigments.[17]

Color changes during processing and storage of blackberries have been studied by a number of authors.[18-21] Tanchev[18] evaluated the effect of sterilization and storage temperature on the thermal stability of anthocyanins in blackberry juice. Jennings and Carmichael[19] investigated the effect of freezing on berries of varying maturity levels and observed that in unripe fruit color

TABLE 2
Anthocyanins of Blackberry Juice
Concentrate

Anthocyanin	Conc (% of total)
Cyanidin 3-sophoroside	9
Cyanidin 3-glycosylrutinoside	4
Cyanidin 3-glucoside	57
Cyanidin 3-rutinoside	30

Data from Hong, V. and Wrolstad, R. E., *J. Agric. Food Chem.*, 38, 698, 1990.

varied from blue to red. They attributed these changes in color to the lowering of fruit pH caused by mixing of cellular fluids caused by cell plasmolysis. Crivelli and Rosati[20] studied the suitability of 'Smoothstem' and 'Thornfree' blackberries for freezing but did not report a color change during frozen storage. Sapers et al.[21] compared ripe fruit of 40 thornless blackberry cultivars and selections and juices obtained from them for differences in color and composition after freezing, thawing, and heating. Color changes (reddening) during frozen storage were associated with sample variability in degree of ripeness. Red subsamples of frozen berries were lower than black subsamples in soluble solids and total anthocyanin contents and higher in titratable acidity and anthocyanin in the pressed juice. When juice samples were standardized to compensate for differences in pH and anthocyanin concentration, ripeness and cultivar effects on juice tristimulus parameters were small. Rapid thawing of frozen fruit resulted in less anthocyanin loss than did slow thawing. Heating darkened blackberry juice samples and increased the value of A_{440}/A_{513}. The increase in A_{440}/A_{513} was greater in juice from black than from red fruit, which indicates that in the juice from black samples more anthocyanins are in the colorless chalcone forms because of the higher pH.

II. BLUEBERRIES

Blueberries, *Vaccinium* spp., belong to the heath family (Ericaceae). They grow wild in many parts of the world; however, Canada and the United States supply about 95% of the fruit used by the food industry. North American farmers harvest about 47 million kilograms of blueberries annually. Nearly a third of the crop is marketed as fresh fruit; another third is frozen; and the rest is canned or goes into bakery products, ice cream, and other food products. The food industry uses two main kinds of blueberries, highbush and lowbush. The rabbiteye blueberry is cultivated on a limited scale in the southern U.S. This hardy species grows to a height of 1 to 2 m and produces blue fruit which turns red when crushed. All blueberries grow best in acid soil. To blossom normally, they require a cold, inactive period during the winter. However, most blueberry plants cannot survive temperatures below $-29°C$.

A. HIGHBUSH BLUEBERRY

Highbush blueberry (*Vaccinium corymbosum* L.), also called black blueberry or swamp blueberry, is a bushy shrub which can reach a height of 3 m. It originated in North America but is now cultivated throughout the world, and Bluecrop is one of the most important commercial cultivars.[22,23]

Woodruff et al.[24] followed the chemical changes in blueberry fruit with ripening and found that it requires 16 d for the completion of the ripening process and the anthocyanin pigmentation increases only in the first 6 d. The anthocyanins which appear first during ripening are those which are the most abundant in the mature fruit.[25] Dekazos and Birth[26] described a light transmittance technique for the nondestructive assessment of degree of ripeness in intact blueberries. The method involves the measurement of the optical density of the intact fruit at two wavelengths and computation of the optical density difference which is significantly correlated with anthocyanin content of the fruit. Ballinger and Kushman[27] found that as the fruit develops, anthocyanin content, berry weight, pH, soluble solids/acid ratio (SS/A), soluble solids, and sugar (percent of fresh weight [FW] and mg per berry) increase; and only percent acid decreases. Anthocyanin content (mg/cm^2 of berry surface) was found to be significantly correlated with sugar/acid ratio and SS/A ratio. This served as the basis for development of photoelectronically based sorting systems for mechanically harvested blueberries.[28,29] In a later study, Kushman and Ballinger[30] concluded that each cultivar had its own characteristic index of quality which could be determined from the measurement of anthocyanin, acidity, and sugar contents.

In ripe blueberries of the Croatan variety, Ballinger et al.[31] characterized 14 anthocyanins. The major anthocyanins, in order of decreasing amounts, are malvidin 3-galactoside, delphinidin 3-galactoside, delphinidin 3-arabinoside, petunidin 3-galactoside, petunidin 3-arabinoside, and malvidin 3-arabinoside. Also present in lesser amounts are the 3-monoglucosides of delphinidin, petunidin, malvidin, peonidin, and cyanidin; peonidin 3-arabinoside; and cyanidin 3-galactoside (Table 3). In a study on the anthocyanins of ripe fruit of pink-fruited and blue-fruited blueberries, Ballinger et al.[32] confirmed the occurrence of five aglycones and three sugars in the pigments of blueberries, with the pink fruit containing a lower content of anthocyanins.

Makus and Ballinger[33] characterized the anthocyanins of ripening 'Wolcott' blueberries. In unripe red fruit, cyanidin 3-glucoside and 3-galactoside comprised 40% of the total pigment; however, all combinations of the five aglycones with glucose, galactose, and arabinose (15 anthocyanins) were detected. In ripe and overripe fruit, all anthocyanins were present, except peonidin 3-galactoside. Malvidin 3-glucoside and 3-galactoside constitute about

TABLE 3
Anthocyanins of Blueberries
(*Vaccinium* **spp.**)

Anthocyanin	Relative conc
Cyanidin 3-glucoside	+
Cyanidin 3-galactoside	+
Cyanidin 3-arabinoside	+
Delphinidin 3-glucoside	+
Delphinidin 3-galactoside	+ + +
Delphinidin 3-arabinoside	+ + +
Malvidin 3-glucoside	+
Malvidin 3-galactoside	+ + +
Malvidin 3-arabinoside	+ + +
Peonidin 3-glucoside	+
Peonidin 3-galactoside	+
Peonidin 3-arabinoside	+
Petunidin 3-glucoside	+
Petunidin 3-galactoside	+ + +
Petunidin 3-arabinoside	+ + +

Data from Francis, F. J., Harborne, J. B., and Barker, W. G., *J. Food Sci.*, 31, 583, 1966; and Ballinger, W. E., Maness, E. P., and Kushman, L. J., *J. Am. Soc. Hortic. Sci.*, 95, 283, 1970.

60% of the anthocyanin content of ripe fruit of the 'Wolcott' blueberry. Vereshkovskii and Shapiro[34] separated 16 anthocyanins in blueberries from the former U.S.S.R. Eight of these compounds were identified as the 3-galactoside and 3-arabinoside of delphinidin, 3-glucoside and 3-galactoside of cyanidin, 3-arabinoside and 3-galactoside of petunidin, and 3-galactoside and 3-arabinoside of malvidin. In pink blueberries with varying degrees of pigmentation, the anthocyanin composition is not qualitatively different from that of dark-colored cultivars.

The total anthocyanin content varies from 25 to 495 mg/100 g; because of this wide range in its content, it is believed that anthocyanin expression in blueberries is dependent on one or two major genes and that anthocyanin content may be quantitatively inherited.[32,35]

Recently, Sapers et al.,[36] in a study on color and composition of highbush blueberry cultivars, separated as many as 16 anthocyanins by HPLC. Three distinct anthocyanin patterns were found for the 11 different cultivars studied (Figure 1). 'Bluecrop', 'Blueray', 'Bluetta', 'Earliblue', and 'Weymouth' blueberries have moderate to large peaks corresponding to pigments 1, 2, 4–10, 14, and 16 as well as small peaks 11, 13, and 15; 'Collins' and 'Jersey' resemble the first group but lack peaks 11, 13–16. 'Berkeley', 'Burlington,'

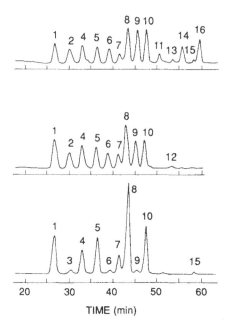

FIGURE 1.   HPLC chromatograms of anthocyanins in 'Earliblue' (top), 'Jersey' (middle), and 'Berkeley' (bottom) highbush blueberries. (From Sapers, G. M., Burgher, A. M., Phillips, J. G., and Jones, S. B., *J. Am. Soc. Hortic. Sci.*, 109, 105, 1984. With permission.)

'Colville,' and 'Elliott' resemble group 2, but are high in peak 8, deficient in peaks 2, 6, and 9, and have small peaks 3 and 15. The color of juice made from fruit of these cultivars was dependent upon the total anthocyanin content, pH, and extent of browning of the juice, but not on the pattern of individual anthocyanins. The fruit of highbush blueberries also contains small amounts of quercetin (24 to 29 mg/kg FW);[12] *p*-hydroxybenzoic, protocatechuic, and chlorogenic acids; and condensed tannins.[37]

The colored leaves of the highbush blueberry are pigmented by a smaller number of anthocyanins with only the 3-glucosides of cyanidin, delphinidin, petunidin, and malvidin present.[38]

B. LOWBUSH BLUEBERRY

The lowbush blueberry, *Vaccinium angustifolium* Ait., is grown commercially in eastern Canada (New Brunswick, Nova Scotia, and Quebec) and in the northeastern U.S. (Maine). The annual North American production of this berry varies from 23,000 to 47,000 t;[39] and it is sold in frozen, canned, and fresh markets. The plant of lowbush blueberry is characterized by the presence of anthocyanins in the branches,[40] in the leaves where they increase with increased temperature,[41] and with noticeable variations of concentration in the fruit.[42,43] In *V. angustifolium*, Francis et al.[43] identified the 3-glucosides

and 3-galactosides of delphinidin, malvidin, petunidin, cyanidin, and peonidin, plus small quantities of 3-arabinosides of the same anthocyanidins.
Several diglycosides were also present, but in quantities too small for identification. No acylation was apparent.

Fuleki and Hope[44] experimented with the enzymatic treatment of the
crushed berries and were able to improve juice yield and color intensity. The
pigments in the juice of lowbush blueberries are believed to possess a strong
affinity for the papilla on the tongue and the teeth, and it also forms deposits
on the internal wall of the containers.[45] According to Fuleki,[46] lowbush blueberries are suitable for wine production, and deeper colored wines are obtained
with thermovinification or fermenting of the berries "on the skin". With cold
pressing only about one quarter of the extractable color is obtained. The
absorption spectrum of wines resembles that of the juice but the A_{520}/A_{430}
ratio is lower, which indicates a slight degradation of the anthocyanins in
blueberry wine.

Simard et al.[47] investigated some of the factors influencing color stability
during storage of lowbush blueberry juice at the original pH, with addition
of tannin or cupric ions and with simultaneous modification of light, pH, and
storage temperature. Predictably, pH and temperature were the two most
important factors affecting color of blueberry juice. Similarly, Yang and
Yang[48] observed that lowbush blueberry puree became darker and showed a
bluish-purple color when pH shifted from low to higher values, a finding
consistent with the structural transformation of anthocyanins as a function of
pH of the medium[49] (see Chapters 1 and 6).

C. RABBITEYE BLUEBERRY

Spiers[50] reviewed the culture of rabbiteye blueberry (*Vaccinium ashei*
Reade) in the southern U.S. with reference to its agronomic aspects, diseases,
harvesting, and market potential of the fruit. As for other berries, the color
of rabbiteye blueberries is a most important parameter of quality, particularly
when the berries are harvested mechanically.[51]

The anthocyanins found in *V. ashei* are the 3-arabinoside, 3-galactoside,
and 3-glucoside of peonidin, cyanidin, malvidin, delphinidin, and petunidin.[52]
Anthocyanin content of fruits increases after harvesting. According to Basiouny and Chen,[53] during a 45-d storage period following harvest, there is a
55 and 71% increase in anthocyanin content in 'Tifblue' and 'Bluegem'
berries. At harvesting time, on June 15, the corresponding total anthocyanin
contents are 210 mg/100 g of 'Tifblue' berries and 272 mg/100 g of 'Bluegem'
berries.

D. CAUCASIAN BLUEBERRY

According to Mzhavanadze,[54] the following anthocyanins have been separated and identified in the fruit of Caucasian blueberry (*Vaccinium arctostaphylos*) by paper chromatography: delphinidin 3-arabinoside, petunidin

3-arabinoside, cyanidin 3-arabinoside, malvidin 3-arabinoside, and peonidin 3-arabinoside. Lebedeva[55] reported to have characterized different *Vaccinium* species (*V. myrtillus, V. uliginosum, V. vitis-idaea, V. arctostaphylos, V. oxycoccus,* and *V. macrocarpum*) from their contents of anthocyanins, flavons, and catechins. Phenolic acids and esters of caffeic and *p*-coumaric acids are present in *V. arctostaphylos* and were identified by Mzhavanadze et al.[56] Rutin, isoquercetin, and hyperin are the flavonol glycosides identified in the leaves of the Caucasian blueberry.[57]

E. OTHER BLUEBERRIES

In *V. arboreum* — commonly called sparkleberry, forkberry, tree huckleberry, or winter huckleberry — Ballinger et al.[58] isolated 13 of the 15 potential anthocyanins: 3-arabinosides and 3-galactosides of cyanidin, malvidin, delphinidin, petunidin, and peonidin; and the 3-glucosides of delphinidin, petunidin, and peonidin. The anthocyanin pattern of the fruit of sparkleberries is, thus, very similar to those reported for the fruit of highbush and lowbush blueberries, although the morphological appearance and the geographical range of sparkleberries are similar to those of deerberries, or southern gooseberries, (*V. stamineum*). *V. stamineum* (sectin *Polycodium*), however, contains primarily cyanidin 3-galactoside, 3-arabinoside, and 3-glucoside;[58-60] and only small amounts of the galactosides, arabinosides, and glucosides of peonidin and malvidin.[60] The total anthocyanin content varies from 7 to 101 mg/100 g, and it is located primarily in the skin of the berry.[59]

Ishikura and Sugahara[61] found glycosides of cyanidin, delphinidin, and petunidin in the fruits of *V. bracteatum* and glycosides of delphinidin, petunidin, and malvidin in *V. smallii.*

Martinod et al.[62] isolated six anthocyanins in *V. floribundum,* using TLC. They identified cyanidin in three of the pigments, while they were unable to identify the aglycone of the other three.

In *V. japonicum,* Andersen[63] identifies cyanidin 3-arabinoside (54%) and pelargonidin 3-arabinoside (39%) as two main pigments, together with small quantities of the corresponding 3-galactosides representing only 7% of the total 113 mg/100 g of fresh fruit. This is the first report of pelargonidin glycosides in the genus *Vaccinium.*

III. CRANBERRIES

American cranberry (*Vaccinium macrocarpon* Ait.) and European or small cranberry (*Vaccinium oxycoccus* L., Ericaceae) are slender, creeping bushes up to 1.20 m long. The fruit which becomes red in early autumn can stay on the plant all winter if protected by the snow.

In the fruit of *V. oxycoccus* L., or small cranberry, Andersen[64] identified peonidin 3-glucoside and cyanidin 3-glucoside as the main pigments; they accounted for 41.9 and 38.3% of the total anthocyanin content, respectively. Small amounts of 3-monogalactosides and 3-monoarabinosides of peonidin and cyanidin; and 3-monoglucosides of delphinidin, petunidin, and malvidin are also present. This anthocyanin pattern is rather different from that of the American cranberry, which is rich in the 3-galactosides and the 3-arabinosides of peonidin and cyanidin.[65,66] The total anthocyanin content of the fruit is about 78 mg/100 g fresh fruit.

The commercial value of the cranberry depends upon the red color in the fruit. Generally, harvesting takes place in late autumn when the berries are dark red. In this system, however, the fruit is often frost-damaged and a large percentage of the berries is overripe, which decreases the storability of the crop. Studies were aimed at promoting an early color development of the berry in the field by foliar application of the insecticide malathion.[67-69] Other authors experimented with the use of fungicides upon the development of pigmentation in cranberry. Francis and Zuckerman[70] noted that application of a Maneb, Zineb, Fulsan, Folpet, COCS, and Bordeaux mixture reduced anthocyanin production. However, treatment with Dichlobenil increased pigmentation,[71,72] and CEPA had no effect on the accumulation of anthocyanins[73] in cranberries.

Eck[74] reported that ethephon applied before harvest stimulated pigmentation. Bramlage et al.[75] confirmed this observation and noted that production of anthocyanins was correlated with the production of ethylene by the fruit. Cracker[76] applied ethylene under illumination to just-harvested fruit and observed that the presence of light greatly increased anthocyanin synthesis by stimulating production of ethylene in the fruit. Devlin and Demoranville[77] used CGA-15821 and R-27969 to stimulate ethylene production and obtained up to a 150% increase in anthocyanin production. Francis and Atwood[78] obtained higher levels of pigmentation by fertilizing cranberries with low levels of P and K. A lack of N had no effect, while an excess reduced the quantity of anthocyanins in the leaves.[79] The depth to which the plants are submerged, 30 to 54 cm, has no effect on the anthocyanin content or fruit size.[80]

The optimal anthocyanin content in quality cranberries is one higher than 67 mg/100 g. This level is best attained by allowing the berries to remain on the plant for 15 d after apparent ripeness, rather than by storing the fruit at room temperature for the same length of time.[81] An increase of 35 to 60% in anthocyanin content can be obtained by enhancing fruit wettability with various compounds when spraying with Ethrel.[82]

Production of anthocyanin in cranberries has been correlated with various agronomic and physiological factors which influence the physical and chemical characteristics of the fruit.[83,84] Cansfield and Francis[85] experimented with the application of several products (malathion, indoleacetic acid [IAA], sugars, etc.) aimed at stimulating pigment production in unripe cranberries after harvest. The results, however, were generally not positive. Hall and Stark[86] obtained more favorable results by lowering the temperature before and after harvest. Polyethyleneglycoldodecylether (DEPEG) was found to enhance anthocyanin development and slightly improve the storability of 'McFarlin' cranberries.[87] Lees and Francis[88] obtained an increase in anthocyanin level when treating the fruit with γ-radiation, but the maturity level of the fruit influences the intensity of the response to the radiation treatment; and this has no effect on the proportion of individual anthocyanins.

The anthocyanin content, calculated from the content in 100 berries, may be used to characterize a given variety.[89] However, within a variety, the anthocyanin content varies greatly between berries. This is due to ecological and agronomic factors, as well as berry size.[90] Vorsa and Welker[91] reported an inverse relationship between berry size and anthocyanin extractability. Therefore, a genetic selection for larger berries for increased juice production could cause a lower yield of anthocyanins. Sapers et al.[92] suggested that it is more economical to increase the quantity of anthocyanins by selecting clones with many pigment-rich berries rather than trying to increase anthocyanin synthesis or waiting for simultaneous ripening of the berries.[93]

Only slight differences have been found in the anthocyanin pattern between different clones. The proportions of individual anthocyanins in fruits of different coloration were similar, indicating a constant biosynthetic rate of each pigment during color development.[94] PAL plays an important role in anthocyanin biosynthesis.[95]

In the American cranberry, Grove and Robinson[96] identified the pigment as peonidin 3-glucoside, calling it oxycoccicyanin after the old botanical name of the fruit, *Oxycoccus macrocarpus* Pers. Sakamura and Francis[97] studied the anthocyanin composition of 'Early Black' cranberries and isolated and identified four pigments: cyanidin 3-galactoside, peonidin 3-galactoside, and monoglycosides of peonidin. Zapsalis and Francis[65] reanalyzed the pigments and identified the four major anthocyanins as cyanidin 3-galactoside, peonidin 3-galactoside, cyanidin 3-arabinoside, and peonidin 3-arabinoside. Fuleki and Francis[98] identified two other pigments, present in small quantities, as cyanidin 3-glucoside and peonidin 3-glucoside.

The developments in HPLC techniques in the early 1970s opened new possibilities for the separation and characterization of anthocyanins and allowed Camire and Clydesdale[99] to confirm the presence of the 3-galactosides and 3-arabinosides of cyanidin and peonidin as the four major anthocyanins of cranberries. Traces of two additional pigments were also detected. The separation of these pigments was carried out on a reverse phase column with methanol:acetic acid:water (90:5:5). More recently, Fuleki[100] using HPLC separated the anthocyanins in fruits of *V. macrocarpon* and characterized nine of them. The relative concentrations of the major anthocyanins varies with the cultivar (Table 4) and ranges from 17 to 25% for cyanidin 3-galactoside, 13 to 25% for cyanidin 3-arabinoside, 23 to 39% for peonidin 3-galactoside, and 11 to 19% for peonidin 3-arabinoside.

The flavonoid composition of the cranberry includes various flavonols, the most important of which is quercetin 3-galactoside.[12,101,102] Puski and

TABLE 4
Distribution of Major Anthocyanins in Cranberry Cultivars

	Anthocyanin					
Cultivar	Cyanidin 3-galactoside (%)	Cyanidin 3-arabinoside (%)	Peonidin 3-galactoside (%)	Peonidin 3-arabinoside (%)	Galactoside/ arabinoside (%)	Cyanidin/ peonidin (%)
Early Black	24	19	23	13	1.5	1.2
Franklin	25	23	39	11	1.9	1.0
Bergman	23	17	34	12	2.0	0.9
Beckwith	23	17	29	11	1.8	1.0
McFarlin	20	24	23	21	0.9	1.0
Stevens	17	17	30	13	1.6	0.8
Howes	16	16	26	14	1.4	0.8
Wilcox	21	25	26	19	1.1	1.0
CN	21	19	31	14	1.6	0.9
Ben Lear	21	13	36	13	2.2	0.7

Adapted from Fuleki, T., *Bull. Liaison Groupe Polyphénols*, 13, 374, 1986.

Francis[101] first isolated flavonol glycosides from 'Early Black' cranberries; and using chromatographic and spectral analysis identified them as 3-galactoside, 3-rhamnoside, and 3-arabinoside of quercetin and 3-arabinoside and 3-galactoside of myricetin. Using an electrophoretic method, Cansfield and Francis[102] found quercetin 3-galactoside, quercetin 3-arabinoside, quercetin 3-rhamnoside, and myricetin 3-arabinoside in an extract of 'Early Black' berries. Bilyk and Sapers[12] used an HPLC method to determine the distribution of quercetin, myricetin, and kaempferol in dark and medium-colored samples of six cranberry varieties. The range for dark-colored fruit is 112 to 250 mg/kg for quercetin, 11 to 24 mg/kg for myricetin, and 0 to 3 mg/kg for kaempferol. Within varieties, the flavonol contents are smaller in the less pigmented berries. Cranberries also contain a variety of phenolic acids: *p*-coumaric, ferulic, sinapic, and caffeic; and their derivatives,[103] flavan-3-ols[104] and condensed tannins.[105]

Cranberries are not consumed raw. They are processed into a number of popular consumer products such as cranberry sauce, cranberry juice cocktail, cranapple and crangrape juice, and others. The retention of the bright red color of the fruit during processing and storage of these products has been extensively studied.[100,106-111] Starr and Francis[106] investigated the effect of oxygen and ascorbic acid on the stability of color in cranberry juice cocktail. Their findings show that an increase in either factor causes a proportional increase in the disappearance of the anthocyanins in the juice. In particular, the arabinosides of peonidin and of cyanidin are more sensitive to oxidative changes than the respective galactosides.[106] A detailed description of the methods for color measurement in cranberry-based products is given by Fran-

TABLE 5
Effects of Freeze-Thaw Treatment or Yield and Anthocyanin Content of Cranberry Juice Prepared by Single Pressing

Cultivar	Trial	Juice yield (ml/100 g)		Juice anthocyanin (mg/100 ml)	
		Fresh	Freeze-thaw	Fresh	Freeze-thaw
Early Black	1	71	81	7.8	51.1
	2	72	81	8.7	51.2
Pilgrim	1	68	79	3.7	37.2
	2	69	80	3.5	38.5
Searles	1	54	80	1.8	34.9
	2	54	80	1.8	39.3

Adapted from Sapers, G. M., Jones, S. B., and Maher, G. T., *J. Am. Soc. Hortic. Sci.*, 108, 246, 1983.

cis and Clydesdale.[107] Canned cranberries can be stored for extended periods without marked color changes.[108] Sometimes, however, the pigments may precipitate because of their complexation with other components of the fruit.[109] Some metallic ions — Ca, Al, Sn, and Fe — have been shown to have a protective effect on color stability and anthocyanin content in cranberry juice, but the blue and brown discoloration caused by the metal-anthocyanin complexes often overshadows the protection; the net result is not beneficial.[110] Shrikhande and Francis[111] reported that 3 to 9 mg/100 g of total flavonoids (quercetin 3-galactoside) in cranberry juice retards the oxidation of ascorbic acid, thus indirectly protecting the anthocyanins.

Kuznetsova[112] obtained superior quality cranberry juices from fruit harvested at full ripeness and attributed reduced color stability in juice made from immature fruit to the higher concentration of procyanidins in immature berries. In addition, blanching the whole fruit inactivates the oxidizing enzymes, and this leads to a higher anthocyanin concentration in juices made from blanched berries. Servadio and Francis[113] evaluated the relationship between anthocyanin composition in cranberries with the color characteristics of gelatins and sauce products made from the fruit. Their findings show that, with the use of regression analysis, it is possible to predict the color of the finished product based on the anthocyanin composition data of the raw material. The ratio between the optical density of 535 and 415 nm provides an excellent index of anthocyanin degradation.

Sapers et al.[114] studied the factors that influence juice yield and anthocyanin recovery. Among these, freezing-thawing of the berries is particularly effective because it increases juice yield by as much as 50% and juice anthocyanin content by up to 15-fold (Table 5). Anthocyanin recovery can also be increased by double squeezing and by tissue homogenization, although excessive pigment degradation due to oxidation may limit the application of the latter systems.

Given the value and uniqueness of cranberry juice, recent works have focused on the detection of adulteration of this product, particularly its anthocyanin content. Adulteration is often made by addition of the less expensive grape anthocyanin extract, enocyanin. Francis[115] used a colorimetric method for detecting the addition of enocyanin to cranberry juice cocktail which contained less than 12% pigments, instead of the required 25% level. Confirmation of the addition of grape anthocyanins can more accurately be made from determination of the anthocyanin pattern of the juice by a chromatographic method of analysis. Hale et al.[116] detected adulteration of cranberry juice with 5% or more enocyanin by comparing HPLC chromatographic profiles which revealed the presence of delphinidin and petunidin 3-glucosides (characteristic components of enocyanin), but not found in cranberries. Also using HPLC analysis, Hong and Wrolstad[117] were able to show slight adulterations with enocyanin from the presence of delphinidin and malvidin peaks in the chromatograms which are practically nonexistent in genuine cranberry juice. In addition to the anthocyanin pattern, authenticity of cranberry juice can be established and/or confirmed from the organic acids and sugar profiles.[118]

Cranberries have also been suggested for use as a source of anthocyanins.[119-126] The raw material most suited for this application is the residue or pomace that remains after juice extraction. Chiriboga and Francis[120,121] described a process for the extraction and purification of the pigment from cranberry pomace. This process involves extraction of the pigment from the pomace with methanol containing 0.03% HCl; distillation of the methanol under vacuum; adsorption of the aqueous anthocyanin solution on Amberlite CG-50 resin; elution of the pigments with 0.0001% HCl in ethanol; and final concentration of the pigment for fortification of cranberry juice. Volpe[122] used concentrated cranberry juice to color pie fillings. In the former U.S.S.R., Usova et al.[123] did likewise to color soft drinks. Stahl et al.[124] patented a process involving microfiltration and reverse osmosis to purify cranberry anthocyanins. Woo et al.[125] experimented with membrane technology for the recovery of anthocyanins from pulp wastes. They used ultrafiltration for the purification and reverse osmosis for their concentration of the pigments. Clydesdale et al.[126] used methanolic extraction and concentration by spray-drying to produce an anthocyanin-rich product with a slightly astringent taste, but otherwise highly suitable as a natural colorant for beverages and gelatin desserts.

IV. CURRANTS

Currants belong to the saxifrage family, Saxifragaceae, genus *Ribes*. The red currant is *R. rubrum* L.; the black currant is *R. nigrum* L. Both types grow wild in northern Europe and are easily grown in cool, moist regions of North America. The red currant is used to make jellies, jams, wines, and pies. The black currant is more popular in Europe and is used primarily in the juice processing industry. Robinson and Robinson[127] reported a 3-bioside of cyanidin and a glycoside of peonidin or pelargonidin in red currants. Using chromatographic techniques, Fouassin[4] separated two derivatives of cyanidin, one with the property of a monoside and the other, a bioside. Mehlitz and Matzik[128] isolated three major and four minor anthocyanins in red currants, all cyanidin derivatives. Chandler and Harper[129] separated and identified two glycosides of cyanidin and two glycosides of delphinidin in black currants, but only cyanidin derivatives in red currants.

Harborne and Hall[6] reported mono-, di-, and trisaccharides of cyanidin in cultivars of *R. rubrum* L. The sugars are glucose, rhamnose, and xylose. Cyanidin 3-glucoside is the only monoglycoside present. The diglycosides are 3-rutinoside (glucose + rhamnose), 3-sambubioside (glucose + xylose), and 3-sophoroside (glucose + glucose); the triglycosides are 3-(2^G-xylosyl-rutinoside) (glucose + rhamnose + xylose) and 3-(2^G-glucosylrutinoside). Nybom[130] recognized two groups of red currants: the *R. petraeum* group containing six glycosides, and the *R. sativum* group with less than four glycosides of cyanidin and lacking the sophoroside and the 2^G-glucosylrutinoside. A single dominant gene is believed to control the presence of cyanidin, making red dominant over white.[131,132]

In red currant cv. Red Dutch, the relative amounts and the total contents of anthocyanins in fruit at the unripe pink stage and at the dark red stage are significantly different.[131] The total anthocyanin content almost doubles during ripening; whereas in unripe fruit cyanidin 3-sambubioside is the main pigment (39%), in ripe fruit it decreases to 22% of the total pigments. Also, in contrast to the findings of Harborne and Hall[6] (Table 6), in ripe fruit of 'Red Lake' currants from Norway, Øydvin[131] found large amounts (73%) of cyanidin 3(2^G-xylosylrutinoside) and smaller amounts of cyanidin 3-glucoside, 3-sambubioside, and 3-rutinoside. Cyanidin 3-sophoroside was lacking. The cultivar 'Rondom' contained all six glycosides, as did the cultivars 'Earliest of the Fourlands' and 'Red Dutch' (Table 6).

In black currant, *R. nigrum*, Green[133] found no difference in the preferential accumulation of cyanidin or delphinidin in fruits harvested over a 14-week period. Although the appearance of anthocyanins characterizes the ripening phase, the biochemical indices on the presence of cyanidin and delphinidin in relationship to the acidity and sugar level are not, in themselves, sufficient to accurately determine the optimum harvest date.[134] At maturity, black currants contain about 250 mg of anthocyanin per 100 g of fresh fruit.[5,135,136] According to Koeppen and Herrmann[136] the pigments account for at least 2% of the skin fresh weight. These authors, however, found only four anthocyanins in fully ripe *R. nigrum:* cyanidin 3-rutinoside (36%), delphinidin 3-rutinoside (34%), and minor concentrations of cyanidin 3-glucoside

TABLE 6
Anthocyanin Distribution in Five Red Currant Cultivars
Grown in Norway

| | % of Total anthocyanins | | | | |
Anthocyanin	Red Lake	Jonkheer van Tets	Earliest of Fourlands	Red Dutch	Rondom
Cyanidin 3-glucoside	2	3	5	10	10
Cyanidin 3-rutinoside	16	17	16	9	8
Cyanidin 3-sambubioside	9	11	10	29	31
Cyanidin 3-sophoroside	—	—	4	9	8
Cyanidin 3-(2^G-xylosylrutinoside)	73	69	37	28	29
Cyanidin 3-(2^G-glucosylrutinoside)	—	—	28	15	14
Total anthocyanins (mg/100 g FW)	18.0	16.9	18.6	14.0	11.9

Adapted from Øydvin, J., *Hortic. Res.*, 14, 1, 1974.

(15%) and delphinidin 3-glucoside (15%). Chandler and Harper[137] identified the same four pigments but also reported the presence of other aglycones, which were probably artifacts.

Using column chromatography, Reichel and Reichwald[138] were able to separate and identify cyanidin 3-rutinoside, delphinidin 3-rutinoside, and a delphinidin glycoside in black currants. In ten varieties from the former U.S.S.R., Demina[139] found four anthocyanins amounting to 3.9 to 5.8% of the dry weight of the fruit. Cyanidin 3-rutinoside was the major pigment at 1.4 to 2.3%, cyanidin 3-glucoside 0.5 to 1.3%, delphinidin 3-rutinoside 1.1 to 1.9%, and delphinidin 3-glucoside 0.2 to 0.9% dry weight basis. LeLous et al.[140] were the first to identify and quantitate pelargonidin 3-rutinoside, cyanidin 3-sophoroside, and delphinidin 3-sophoroside in *R. nigrum* in addition to delphinidin and cyanidin 3-glucosides and 3-rutinosides (Table 7). More recently, Skrede,[141] using HPLC analysis, found only the 3-glucosides and 3-rutinosides of cyanidin and delphinidin in black currants grown in Norway.

In black currant juice, all anthocyanins disappear in 9 weeks at 37°C and pH 2, whereas 60% remain at 4°C. Variation related to duration of storage corresponds to a second degree polynomial relationship.[142] In pasteurized, vacuum-packed juice (from 40 black currant genotypes), stored at approximately 20°C, Taylor[143] found that the concentration of delphinidins and cyanidins decreased throughout a 20-week storage period (Figure 2). There is a significant variation in the rate of loss between the major and minor pigments and between the major pigments themselves. Delphinidin 3-rutinoside is more stable than cyanidin 3-rutinoside and both major pigments are more stable than the minor pigments.[143]

TABLE 7
Anthocyanins of Black Currants
(*Ribes nigrum* L.)

Anthocyanin	% of Total
Cyanidin 3-rutinoside	35
Delphinidin 3-rutinoside	30
Cyanidin 3-glucoside	17
Delphinidin 3-glucoside	13
Pelargonidin 3-rutinoside	3
Cyanidin 3-sophoroside	1
Delphinidin 3-sophoroside	1

Adapted from LeLous, J., Majoie, B., Moriniére, J.-L. and Wulfert, E., *Ann. Pharm. Fr.*, 33, 393, 1975.

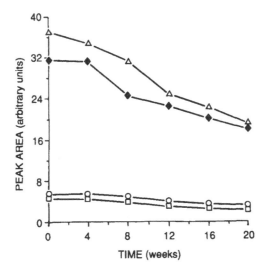

FIGURE 2. Changes in anthocyanins of black currant juice during storage at room temperature. △, Cyanidin 3-rutinoside; ◆, delphinidin 3-rutinoside; ○, delphinidin 3-glucoside; □, cyanidin 3-glucoside. (From Taylor, J., *J. Sci. Food Agric.*, 49, 487, 1989. With permission.)

Skrede[144] assessed color stability of black currant syrup packed in four different packaging materials and stored at 20°C in the dark and in artificial light (400 lx Osram white). The results show that total anthocyanin content and Hunter "a" value decreases follow first-order reaction kinetics during storage. Syrups from various cultivars differ in color stability; degassing of syrup prior to storage does not enhance color stability; daylight storage of syrup lowers half-life of Hunter "a" values by 10 to 30% compared with the

TABLE 8

Half-Life of Anthocyanin Content and Hunter "a" Values of Blackcurrant Syrups from Different Cultivars, Stored in Four Package Types at 20°C under Artificial Light

	Half-life (months)	
	Anthocyanin content	Hunter "a" values
Cultivar		
Silvergieter	5.9	23.8
Ben Nevis	6.8	24.0
Ben Lomond	5.0	14.8
An industrial syrup	4.7	16.2
Packaging material		
Glass	4.7	16.2
Polyester	3.4	12.5
Polyvinyl chloride	2.7	8.6
High-density polyethylene	0.9	1.9
$\bar{D}_{5\%}$	1.4	3.7

Adapted from Skrede, G., *J. Food Sci.*, 50, 514, 1985.

dark; packaging material greatly influences anthocyanin content and color stability (Table 8). According to Skrede et al.,[145] anthocyanin pigments of black currant syrup are more stable than anthocyanins of unfortified strawberry syrup. However, this difference is due to total anthocyanin concentration rather than qualitative pigment composition, and color stability of strawberry syrup fortified with equal anthocyanin concentration is similar to black currant syrup.

Recovery of anthocyanins from black currant press cake residue after juice extraction was evaluated by Rosa and Ichas[146] and Rosa.[147-149] In these studies, the pigments were extracted using a 6:1 mixture of 45% ethanol and dried skins for 8 to 10 h at room temperature. The extract was then concentrated six times in a Centritherm evaporator with simultaneous recovery of the alcohol. The storage of this high-quality concentrate created some problems of degradation due to the presence of hydroxymethylfurfural (HMF). The coloring potential was half after 10 months of storage at 4°C, and practically zero after 3 months of 20°C.[147] Of the four anthocyanins, the two cyanidin derivatives degrade slightly faster than the delphinidins.[148] Addition of glucose (50 to 100 g/l), starch (100 g/l), and citric acid (10 g/l) increases the stability of the concentrate, but addition of ascorbic acid decreases the stability.[149]

V. GOOSEBERRIES

The gooseberry is an oval, tart fruit or berry that is closely related to the currant. The gooseberry shrub grows both in Europe and North America. The American type is classified as *Ribes hirtellum* Pursh., and the European type is classified as *R. grossularia* L. The plant grows well in shady, cool, moist regions. People often plant them under fruit trees in gardens. Gooseberries are commonly used in preserves and pies; the fruit may be white, yellow, green, or red and may have a prickly, hairy, or smooth surface. The skins of the red berries contain the 3-rutinoside and 3-glucoside of cyanidin.[6] According to Nillson and Traikovski,[151] some varieties of red gooseberries also contain glycosides of delphinidin; and Saburov and Ulyanova[152] found cyanidin 3-glucoside and a glycoside of delphinidin in gooseberry juice. The presence of red pigments in gooseberries is a dominant character controlled by a single dominant gene.[132] Gooseberries harvested at commercial maturity apparently contain more anthocyanins and less tannins than fruit harvested when fully ripe.[153] This is unusual, since the anthocyanin content of most fruits reaches the maximum at the fully ripe stage of maturity.

Gooseberries contain many other phenolic compounds, including: (+)-catechin; (−)-epicatechin;[154] protocatechuic, 5' and 3' caffeoylquinic, 3'-*p*-coumaroylquinic, and 3'-feruloylquinic acids; and several sugar derivatives of hydroxycinnamic acids.[155]

VI. RASPBERRIES

Raspberries belong to the rose family, Rosaceae, genus *Rubus,* and subgenus *Idaeobatus.* The subgenus consists of approximately 200 species distinguishable by the ability of their mature fruits to separate from the receptacle. Several of these species have been domesticated, but red and black raspberries are the only ones grown on a large scale.[1] The fruit of *Rubus* spp. is an aggregate fruit consisting of a collection of drupelets, which are small fruits with seeds contained in hard pits surrounded by pulpy flesh. Red raspberries are widely distributed in all temperate regions of Europe, Asia, and North America. They are variable, but two well-defined types are most important: *Rubus idaeus* subsp. *vulgatus* Arrhen. or European red raspberry and *R. idaeus* subsp. *strigosus* Michx or North American red raspberry.[1] Most cultivars grown commercially in North America were derived from the above two *R. idaeus* subspecies. About 75% of the raspberries grown in the United States and nearly 100% of the raspberries grown in Canada are red.

Cultivars of the black raspberry are domesticated forms of *Rubus occidentalis* L., a species indigenous only to North America, where it is distributed widely though slightly more southerly than *R. idaeus* subsp. *strigosus.* The black raspberry is adapted to moderate climate. It is not as winter hardy as the red raspberry, and it is not adapted to the south, where it does not receive adequate chilling.[156]

In red raspberries (*R. idaeus*), Nybom[157] and Harborne and Hall[6] identified cyanidin 3-rutinoside, cyanidin 3-glucoside, cyanidin 3-sophoroside, and cyanidin 3-glucosylrutinoside. One cultivar, September, contained these four cyanidin pigments, as well as the four related pelargonidin derivatives.[6] Nybom[157] also reported that black raspberry contained a xylose in the glucosidic residues of the anthocyanins. Daravingas and Cain[158] used chromatographic and spectrophotometric techniques to identify the anthocyanin pigments in 'Monger' black raspberries and reported the presence of cyanidin 3-glucoside, cyanidin 3,5-diglucoside, cyanidin 3-diglucoside, and cyanidin 3-rhamnoglucoside-5-glucoside.

In 'Durham' red raspberries, Francis[159] found cyanidin 3-glucoside, cyanidin 3-sophoroside, cyanidin 3-rutinoside, and cyanidin 3-glucosylrutinoside; while fruit of the cultivar Heritage contained only the 3-glucoside and 3-sophoroside of cyanidin. Misic[160] identified cyanidin 3-glucoside and cyanidin 3-sophoroside as the only anthocyanins present in all 29 cultivars of red raspberry examined. Two other pigments, cyanidin 3-rutinoside and cyanidin 3-glucosylrutinoside, were found in the fruit of 25 of the 29 cultivars studied. Barritt and Torre[161] also examined 37 cultivars of *R. idaeus*, and found that the two most representative pigments were cyanidin 3-glucoside and cyanidin 3-sophoroside. Two other important pigments found in the majority of the cultivars were cyanidin 3-rutinoside and cyanidin 3-glucosylrutinoside. A few cultivars also contained cyanidin 3,5-diglucoside and/or the 3-glucoside, 3-rutinoside, 3-sophoroside, and 3-glucosylrutinoside of pelargonidin; one selection contained cyanidin 3-sambubioside and cyanidin 3-xylosylrutinoside (Table 9).

Quantitative data on the anthocyanin content of raspberries were first reported by Blundstone and Crean[5] on three cultivars of *R. idaeus*. Torre and Barritt[2] determined by TLC and densitometry the concentration of individual anthocyanin pigments in the fruits of 43 *Rubus*, including 8 cultivars of red and 3 cultivars of black raspberries. Their results show that red raspberries are characterized by the presence of cyanidin 3-sophoroside as a major pigment and by total anthocyanin concentrations in the range between 20 and 60 mg/100 g of fruit. Fruits of the black raspberry are characterized by two xylose-containing pigments, cyanidin 3-sambubioside and cyanidin 3-xylosylrutinoside; and a total anthocyanin content of more than 200 mg/100 g of fruit (Table 10). More recently, Hong and Wrolstad[162] used HPLC to confirm the presence of the 3-glucoside, 3-rutinoside, 3-sambubioside, and 3-xylosylrutinoside of cyanidin in black raspberries and detected a fifth unidentified pigment present in trace amounts.

Other raspberry species of limited economic importance in Eastern Europe include *R. caesius* and *R. caucasicus*, and *R. anatolicus*. The fruits of Caucasian raspberries (*R. caucasicus* Focke), like other raspberries, are characterized by the presence of cyanidin 3-glucoside and cyanidin 3-rutinoside as the major anthocyanins, and a total content that increases during ripening from 0.28–0.30 to 0.88–1.20% dry weight.[163]

TABLE 9
Distribution (%) of Anthocyanins in Fruits of 37 Red Raspberry Cultivar and Selections

Cultivar/selection[a]	Cyanidin					Pelargonidin			
	3-Gl	3-Ru	3-Sop	3-GlRu	3,5-diGl	3-Gl	3-Ru	3-Sop	3-GlRu
Willamette[b]	25		72		3			t	
Malling Orion	30		70					t	
Matsqui	30		70		t	t		t	
Newburgh[b]	24	6	70	t				t	
WSU 550	26		69		5			t	
Malling Exploit	33		67			t		t	
Washington	34		66			t		t	
WSU 553[b]	32		62		6	t		t	
Heritage[b]	38		62			t		t	
Fairview	38		62			t		t	
Malling Admiral	21	6	59	14				t	
Meeker	17	5	58	20	t			t	
Heritage[c]	45		55			t		t	
Glen Clova	17	7	50	26	t			t	
Sumner[b]	16	9	50	25				t	
WSU 404	13	10	50	24	3			t	
Milton	13	10	48	27	2			t	
Sunrise	14	7	47	32				t	t
Canby[b]	26	10	46	18	t	t		t	
Haida	15	13	46	26				t	t
Durham	24	9	45	22				t	
SHRI 6531/42	27	11	45	17				t	
SHRI 676/4 (M 30)	25	12	45	18	t				
Cuthbert	24	12	43	21		t		t	t
Latham	14	12	41	33				t	t
Lloyd George	25	12	40	23	t	t		t	t
SHRI 6531/84[b]	24	16	40	20				t	
Cherokee	16	18	38	28				t	t
Madawaska	11	16	35	35	3				
Taylor	16	20	35	29					
Carnival	14	22	32	32				t	t
Puyallup[b]	18	21	31	30	t			t	t
Rideau	25	26	25	24					
Chief	12	27	23	38					t
Newman	19	25	22	34			t	t	t
September[c]	15	32	21	32		t	t	t	t
WSU 455	20	32	20	28			t	t	t
SHRI 6626/41 (M 28)[d]	15	7	43	9					

t = Trace amounts; less than 2% of total pigment composition.

[a] WSU refers to Washington State University and SHRI to Scottish Horticultural Research Institute advanced red raspberry selections.

[b] Percentages are means of 2 evaluation years.

[c] Fruit harvested in September.

[d] Also contained 20% cyanidin 3-sambubioside and 6% cyanidin 3-xylosylrutinoside.

Adapted from Barritt, B. H. and Torre, L. C., *J. Am. Soc. Hortic. Sci.*, 100, 98, 1975.

TABLE 10
Anthocyanin Distribution in Black Raspberries

Anthocyanin	Cultivar (mg/100 g FW)		
	New Hampshire	Munger	Cumberland
Cyanidin 3-glucoside	35.9	76.3	89.0
Cyanidin 3-rutinoside	49.8	85.7	167.7
Cyanidin 3-sambubioside	43.6	33.5	55.2
Cyanidin 3-xylosylrutinoside	84.5	59.1	115.8
Total	213.8	254.6	427.7

Adapted from Torre, L. C. and Barritt, B. H., *J. Food Sci.*, 42, 488, 1977.

The concentration of anthocyanins in raspberries is determined by a series of genes; gene T is the most important of them, its recessive allele t giving a very low concentration of anthocyanins.[1] The R, Xy, and So genes control the formation of rhamnoside, xyloside, and sophoroside-containing glycosides, respectively.[1] The i gene controls the pigmentation in the thorns, hairs, and leaves of the raspberry plants.[164] Dimitrov et al.[165] found a significant correlation between fruit color (red or yellow) and color of thorns, soft hairs, and leaves.

The synthesis of anthocyanins in fruits of raspberry depends on ecological and physiological factors,[166] but certain quantitative differences within a species are more or less characteristics of the cultivar. Sági et al.[167] reported that in some cultivars, anthocyanin synthesis proceeds uniformly with ripening, while in others there is a maximum in the first or second phase of ripening. In addition, overripening in raspberry causes an undesirable dark color.

Postharvest storage of raspberry fruits generally produces an increase in anthocyanin concentration and a loss in titratable acidity, but the increase in anthocyanin content is influenced by the degree of fruit ripeness at harvest.[168,169] In 'Meeker' red raspberry fruit harvested at the inception stage of maturity (red color present but 25 to 75% of the fruit surface still green) and at the red ripe stage, the rate of increase in anthocyanin concentration is considerably higher than in fruit harvested when 100% of the surface is red to purplish-red (processing ripe) (Figure 3).

Many studies have been conducted on color, appearance, and pigment composition of processed raspberry products such as preserves and juices. Diemair and Schormueller[170] studied the effect of addition of water on color and its thermal stability in raspberry juice. Guadagni et al.[171] published the results of a study on the kinetics of color degradation in frozen raspberry sauces. The thermal degradation of anthocyanin pigments of black raspberries as influenced by pH, oxygen, and sugars and their degradation products was studied by Daravingas and Cain.[172,173] It was found that factors which posi-

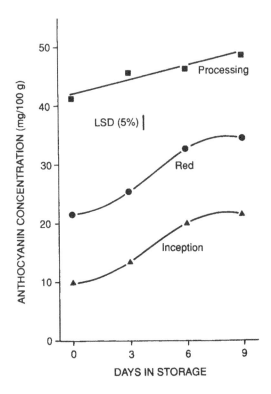

FIGURE 3.   Influence of maturity and storage time on anthocyanin concentration on 'Meeker' red raspberry fruit. (From Sjulin, T. M. and Robbins, J., *J. Am. Soc. Hortic. Sci.*, 112, 481, 1987. With permission.)

tively influence stability of anthocyanins are absence of oxygen, low pH, and low processing and storage temperatures. Temperature also plays a role in the stability of anthocyanins of frozen raspberry products. The losses in anthocyanins during 12 months storage were lower when juice was stored at −18°C than at −2°C. Coffey et al.[175] investigated the possibility of stabilizing the raspberry anthocyanins by the formation of complexes with Sn^{2+} and Al^{3+} ions. Their results indicate that complex formation occurs with cyanidin 3-glucoside and Al^{3+} at pH 2 and 3 and with Sn^{2+} at pH 3 and 4; however, these complexes are of limited practical value.

Pectolytic enzyme treatment of the fruit during juice extraction, used to increase juice yield and improve clarification, reduces total content of anthocyanins in raspberry juice.[176,177] According to Jiang et al.[177] treatments of 'Malling Jewel' raspberries with commercial pectinase enzyme preparations reduce the total anthocyanin content of juice by up to 32.5%. Treatments with one preparation, Pectinex, also resulted in an increase of cyanidin 3-glucoside and cyanidin 3-rutinoside and a decrease of cyanidin 3-sophoroside and cyanidin 3-glucosylrutinoside contents. The yield of the two former and the degradation of the two latter increased with increasing enzyme concentration and reaction time. Thus, it appears that the enzyme(s) convert cyanidin 3-sophoroside into cyanidin 3-glucoside and cyanidin 3-glucosylrutinoside into cyanidin 3-rutinoside, suggesting that the (β1 → 2) glucosidic bonds can be hydrolyzed by enzymes in the crude pectinase preparation.

Fermentation of raspberry juice also degrades most anthocyanin pigments, and the loss in red color of bottled wine (stored in the dark at 20°C for 6 months) is well over 50%.[178] Cyanidin 3-glucoside is the most unstable anthocyanin, disappearing completely during fermentation; while cyanidin 3-sophoroside, the major pigment of raspberries, was the most stable.[145] In addition to anthocyanin, red raspberries also contain high contents of flavonols (7.2 to 10.2 mg/100 g FW),[179] flavan-3-ols (3.2 to 4.9 mg/100 g FW),[180] and phenolic acids[155] which are known to interact with anthocyanins to produce an increase in color intensity and a bathochromic shift in the spectrum of the anthocyanin to give purple to blue colors.[49,119,181]

VII. STRAWBERRIES

Botanically, the strawberry (*Fragaria* × *ananassa* Duch.) is an aggregate fruit formed by the ripening together of several ovaries, all belonging to a single flower and adhering as a unit on a common, pulpy receptacle. Thus, strawberry is not a true berry, because true berries (such as blueberries and cranberries) have seeds within their fleshy tissue. The fruit is greenish white first and ripens to a bright red. It is a good source of vitamin C, and it is often consumed fresh. Strawberries are also frozen or canned and are used in making jam, jelly, and wine.

Chlorophyll and carotenoids are synthesized up to the 28th day after petal fall; between the 25th and 28th day, synthesis of anthocyanins begins.[182] Total pectin and HCl-soluble pectin decrease with ripening, while water-soluble pectin increases in relationship to anthocyanin production.[183] There is no relationship between pulp color and ascorbic acid content.[184] Maclachan[185] studied the hereditary characters and the genetic factors which govern skin and pulp color. It is believed that a single gene controls the white or red fruit color.[186] Anthocyanin synthesis in strawberry is directly related to PAL activity during ripening.[187] When the fruit is green, the PAL activity is very low but it is significantly higher at the fully ripe stage.[188]

In a study on the influence of location and cultivar on color and chemical composition of strawberry fruit, Nelson et al.[189] found that the anthocyanin content varied widely from location to location. A cold, humid climate promotes anthocyanin formation.[190] Guttridge and Nunns[191] reported that the fungus *Fusarium sambucinum* causes anthocyanins to form in immature strawberries. Foliar applications of CaCl$_2$ delays ripening of strawberries and increases anthocyanin content.[192] Anthocyanins were also produced in leaf disks of *Fragaria* spp. when immersed in sugar solution of varying concentrations and then exposed to light.[193]

The main anthocyanin pigment in cultivated strawberries was identified as pelargonidin 3-glucoside by Robinson and Robinson.[194] The report of pelargonidin 3-galactoside[195] in strawberries was found incorrect.[196,197] Sondheimer and Kertesz[198] confirmed the presence of pelargonidin 3-glucoside as the major pigment of cultivated strawberries. This identification was also confirmed by Akuta and Koda[196] and by Robinson and Smith.[197] Lukton et al.[199] described a second, minor pigment to be cyanidin 3-glucoside in *F. virginiana*.

Using column chromatography, Sondheimer and Karash[200] separated three anthocyanins in wild strawberries (*F. vesca*). The major pigments were identified as pelargonidin 3-glucoside and cyanidin 3-glucoside. The difference between cultivated and wild strawberries was found to be in the relative proportions of the pigments. In cultivated species, the ratio of pelargonidin 3-glucoside to cyanidin 3-glucoside is 20:1, whereas in *F. vesca* it is closer to 1:1. Sondheimer and Karash[200] and Lamort[201] reported a third pigment in wild strawberries identified as a glycoside of pelargonidin. Fouassin[4] also reported that both cultivated and wild strawberries had monoglycosides of pelargonidin and cyanidin.

In a study on the flavonoid compounds of strawberry, Co and Markakis[202] confirmed the presence of cyanidin 3-glucoside and pelargonidin 3-glucoside, and reported quercetin 3-glucoside and kaempferol 3-glucoside among the flavonols. Recently, Poei-Langston and Wrolstad[203] showed that these flavonols play an important role in improving anthocyanin stability. Wrolstad et al.,[204] in a study of 40 lots of strawberries consisting of 13 selections and 5 cultivars, noted the presence of a third pigment not previously identified in most samples. Quantitatively, this new anthocyanin was present at up to 5% of the total anthocyanin content and was composed of pelargonidin and glucose in a 1:1 ratio. From its chromatographic and spectral properties, it was suggested that this third anthocyanin is either the furanose form or an anomeric form of pelargonidin 3-glucoside.

Fuleki[205] reported the presence of cyanidin 3-glucoside, pelargonidin 3-glucoside, and four derivatives of pelargonidin (Table 11) in the cultivated strawberry. These included a diglucoside and a trioside of pelargonidin, as well as a glycoside which was less hydrophylic than pelargonidin 3-glucoside. A diglucoside of pelargonidin was also noted by Akuta and Koda,[196] and Polesello and Rampilli[206] who concluded that it was pelargonidin 3,5-diglucoside.

Loss of color in strawberries can occur after thawing of frozen fruit; and during processing and storage of strawberry juices, concentrates, wines, preserves, and purees.[207] Juice is the strawberry product most susceptible to color loss. In products kept at 0°C for 180 d, the anthocyanin level decreases by

TABLE 11
Anthocyanins Identified in Strawberries

Cultivar	Anthocyanin	Ref.
Unnamed	Pelargonidin 3-glucoside	194
Bailey	Pelargonidin 3-glucoside	198
Unnamed	Pelargonidin 3-glucoside	196
	Pelargonidin diglycoside	
Shasta, Marshall	Cyanidin 3-glucoside	199
Culver	Pelargonidin 3-glucoside	200
	Cyanidin 3-glucoside	
	Pelargonidin glycoside	
Sovereign	Pelargonidin 3-glucoside	201
	Cyanidin 3-glucoside	
	Pelargonidin glycoside	
Redcoat, Senga Sengana	Pelargonidin monoside	205
	Pelargonidin 3-glucoside	
	Pelargonidin monoside	
	Cyanidin 3-glucoside	
	Pelargonidin bioside	
	Pelargonidin trioside	
Tioga and others	Pelargonidin 3-glucoside	204
	Cyanidin 3-glucoside	
	Cyanidin 3-glucoside isomer	

26 to 50%; at 20°C, there is a 75 to 90% loss.[207] In a model system, Markakis et al.[208] found that thermal degradation is a first order reaction with a linear phase between 45 and 110°C. The presence of oxygen, ascorbic acid, metal ions, and HMF increases the negative effect of temperature on color stability. Some natural components of the juice, such as ascorbic acid and riboflavin[209] or additives such as cystein, have a synergic effect on loss of color by strawberry anthocyanins.[210] Addition of acetaldehyde, however, increases color intensity through the formation of acetaldehyde-bridged catechin-anthocyanin complexes.[211] Lukton et al.[212] were among the first to show the importance of pH on the breakdown of strawberry anthocyanins. They found that the rate of destruction of pelargonidin 3-glucoside, both in buffer solutions and in strawberry juice at 45°C, was virtually pH-independent in the pH range of 2.0 to 4.5 and in the absence of oxygen. In the presence of oxygen, however, anthocyanin degradation increased dramatically with pH. A number of authors[213-215] noticed the accelerating effect of sugars on the degradation of the major anthocyanin of strawberries. Fructose, arabinose, lactose, and sorbose were found to be more deleterious to the pigment than sucrose, glucose, and maltose. Oxygen aggravated the destructive effect of the sugars. Meschter[213] related the effect of sugars on anthocyanins to sugar degradation products, namely, furfural and 5-hydroxymethylfurfural, which are formed when sugars are heated with acids. Sondheimer and Lee[215] found that under

certain conditions glucose caused a violet shade in strawberries, due to the formation of a stable complex between pelargonidin and the sugar. Recently, Wrolstad et al.[216] showed that sugar addition has a slight, but statistically significant, protective effect on monomeric anthocyanin pigments in frozen strawberries. The cause of the protective effect is believed to be due to the inhibition of pigment-degrading enzymes (i.e., β-glucosidase and polyphenoloxidase); steric interference with condensation reactions (i.e., anthocyanin-phenolic polymerization); anthocyanin-ascorbate interaction, or acting as partial oxygen barrier.[216]

Strawberry polyphenoloxidase (PPO) is known to play a role in the degradation of anthocyanin pigments.[217] However, direct oxidation of anthocyanins by PPO does not appear to be the main route of decolorization. Quinones and intermediary compounds formed during the oxidation of flavan-3-ols by PPO may be responsible for the destruction of anthocyanin either through oxidation or copolymerization.[217] The detrimental effect of endogenous enzymes on strawberry juice color is eliminated by blanching the product. Wrolstad et al.[218] reported that the anthocyanins in juice concentrate do not undergo degradation and that the color stability during storage is maintained when the fruit is blanched with microwave energy before pressing.

Enzymatic treatment of the crushed berries before pressing increases the yield of extracted anthocyanins by up to 50%. The quantity of extracted pigments depends on the type of enzyme used. The most efficient is the simultaneous addition of pectinase and protease.[219] Rwbahizi and Wrolstad,[220] on the other hand, observed an acceleration in the loss of anthocyanins in juices produced from moldy fruit or concentrated by ultrafiltration.

In strawberry preserves, Decareau et al.[221] reported a direct relationship between temperature and pigment loss. Any temperature increase above 60°C causes a proportional color loss. Similar results were obtained by Tinsley and Bockian.[214] Abers and Wrolstad[222] found that catechin and leucoanthocyanins played an important role in color degradation in strawberry preserves. During storage, color deterioration occurred at a higher rate in preserve rich in leucoanthocyanins, flavonols, and total phenolics than in preserves made from strawberries low in these constituents. These findings indicate that color deterioration during storage in strawberry preserves is due to the development of brown pigments rather than the actual anthocyanin pigment loss.

Sistrunk and Cash[223] used Sn and Al salts to stabilize the color of strawberry puree. They found that pH was the parameter having the greatest influence on the color. Stannic chloride was the most effective chemical in stabilizing color in the 3.0 to 3.8 pH range. Wrolstad and Erlandson,[224] however, showed that the stabilized red color was not due to a tin complex of the strawberry anthocyanin but rather to the formation of reddish leucoanthocyanin-metal complex. Spayd and Morris[225] determined the influence of immature fruits, thaw time, and acid level on color stability during storage of jam prepared from cultivars suited to once-over mechanical harvest. Their results revealed that although the addition of immature fruits reduced jam color, anthocyanin levels in the jam were more important than the percentage of ripe fruit. The presence of immature fruits in jam did not influence color loss and discoloration during storage of 2, 25, and 35°C. Of the experimental variables, storage temperature had the greatest influence on anthocyanin content and discoloration. In order to retain acceptable jam color after 6 and 12 months at 25°C, initial anthocyanin levels should be in the range of 4.8 to 6.0 OD_{520} units/g of jam, and initial discoloration index value should be in the range of 1.3 to 1.6, respectively.[225]

Austin et al.[226] determined the color development in immature strawberries and reported that in fruit harvested while still green, the red color will completely develop after 48 h at 28°C in the dark or after 96 h at 18°C. Below 13°C, no color will develop. Landfald[227] observed an increase in anthocyanin in harvested ripe and especially in semiripe berries stored at 4, 12, or 22°C. The average storage life (less than 10% incidence of rot) was 13 d at 4°C, 7 d at 12°C, and 3 d at 22°C. Storage life was 2 to 3 d longer for semiripe than ripe strawberries. Storage does not seem to cause a qualitative variation in anthocyanin composition but, in some varieties, there is a quantitative increase in pigments.[228]

In frozen strawberries, Kyzlink and Vit[229] noted that pelargonidin 3-glucoside tended to increase during storage while the other pigments decreased; after 6 months of storage it was the only anthocyanin present. Color loss was more evident in the darker varieties; total pigment loss varied from 11 to 27%. Wrolstad et al.,[230] however, showed that in order to have a frozen product with an acceptable color, the pH should not exceed 3.51 and the total anthocyanin content should be between 450 and 700 mg/kg.

Color losses also occur in freeze-dried strawberries, and Carballido and Rubio[231] found that if dark red strawberries are used, the rehydrated product has an appearance similar to (but lighter than) the original fruit although the taste and fragrance deteriorate. Zhukhova[232] determined the stability of anthocyanins during freeze-drying of four cultivars of strawberries and concluded that for maximum pigment and aroma retention, the fruit should be subjected to a rapid freezing step prior to freeze-drying. The degradation of anthocyanins in freeze-dried strawberries also depends on the water activity of the product.[233] For maximum color retention, the water activity should be below 0.11.

Strawberries are also canned and processed into wine and liqueurs. For the production of alcoholic beverages, color retention is most important. Production of quality products has apparently been achieved by using pectolytic enzymes,[234] 10% sucrose,[235] and well-ripened fruit free from molds.[236,237] In canned strawberries, the most important factor for maximum color retention is storage temperature.[238-240] Crushing of the fruit and heat processing of the canned product have only limited influence on color retention. Adams and Ongley[239,240] recommended a storage temperature of 0 to 5°C, and showed that omitting thermal processing at high temperature and using various additives had no effect on color stability. Blom,[241] however, reported that the most critical processing steps for color retention during strawberry jam manufacturing are crushing of the fruit, heat processing at 90°C which should not exceed 4 min, and cooling of the product from 90 to 52°C which appears to have the most negative effect on color.[242] Blom and Thomassen[243] studied the kinetic characteristics of a thermostable anthocyanin-β-glycosidase from *Aspergillus niger* on strawberry anthocyanins. Their results show that the pH for maximum decolorizing activity is 3.9 to 4.0 and the optimum temperature is 68°C.

An extensively studied method of preservation for strawberries is ionizing radiation. This process has been evaluated for its effect on the stability of anthocyanins themselves and for its role on fruit ripenings.[244-258] Markakis et al.[245] showed that 465 krd of γ-rays destroyed 63% of the anthocyanins in strawberry juice. Cooper and Salunkhe[246] found that 20 krd prolonged the shelf-life by 15 d and controlled fungal infection, but had no effect on pigment development. Much higher doses (300 to 800 krd) of γ-rays, however, produced a consistent color loss.[247] Johnson et al.[248] reported that the red color in the peripheral tissues of strawberries seemed to move toward the internal white tissue of the fruit. It was suggested that this apparent diffusion of anthocyanins toward the inside of the berries may be due to an increase in their synthesis as a result of irradiation. Herregods and De Proost[249] observed that the anthocyanins in irradiated strawberries increased with the dose applied and decreased with the duration of the application. Mahmoud et al.[250] reported a pigment content in ripe strawberry inversely proportional to the radiation dose. Irradiation at 250 krd caused no color loss in preserved fruit stored in open containers or PVC bags.[251]

VIII. WHORTLEBERRIES

Whortleberries (bilberry, cowberry, lingonberry, bog whortleberry) include several *Vaccinium* species native to parts of Europe and northern regions of Asia and America.

A. BILBERRY

The bilberries (*Vaccinium myrtillus* L.) are small, edible, bluish-black berries, whose major anthocyanin was first detected and named mirtillin by Willstätter and Zollinger in 1915.[252] In 1927, Karrer and Widmer[253] reported that mirtillin was a mixture of glycosides of delphinidin and its methylated derivatives, malvidin, and petunidin. Delphinidin appeared glycosylated by galactose, the other two anthocyanidins by glucose. Fouassin[4] used chromatographic analysis to detect the presence of two derivatives of delphinidin, a derivative of petunidin, and two derivatives of both malvidin and cyanidin. Casoli et al.[254] isolated 11 pigments, glycosides of delphinidin, cyanidin,

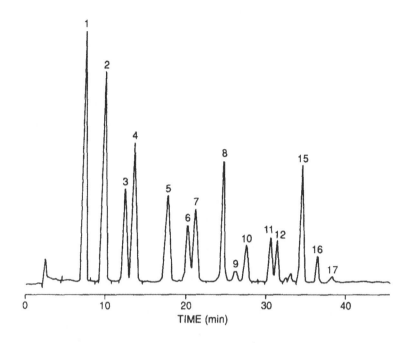

FIGURE 4. HPLC separation of the anthocyanins of *V. myrtillus*. 1: Delphinidin 3-galactoside; 2: delphinidin 3-glucoside; 3: cyanidin 3-galactoside; 4: delphinidin 3-arabinoside; 5: cyanidin 3-glucoside; 6: petunidin 3-galactoside; 7: cyanidin 3-arabinoside; 8: petunidin 3-glucoside; 9: peonidin 3-galactoside 10: petunidin 3-arabinoside; 11: peonidin 3-glucoside; 12: malvidin 3-galactoside; 13: peonidin 3-arabinoside; 14: cyanidin; 15: malvidin 3-glucoside; 16: malvidin 3-arabinoside; 17: petunidin. (From Baj, A., Bombardelli, E., Gabetta, B., and Martinelli, E. M., *J. Chromatogr.*, 279, 365, 1983. With permission.)

pelargonidin; and, in lesser amounts, malvidin, petunidin, and peonidin. Some with acyl moieties were also isolated. Later work by Bombardelli et al.[255] and Baj et al.[256] using both high-performance liquid chromatography and gas chromatography found all possible combinations of cyanidin, delphinidin, petunidin, peonidin, and malvidin 3-galactosides, 3-glucosides, and 3-arabinosides in bilberry. Quantitatively, the delphinidin glycosides were present in the largest quantities, and the peonidin glycosides were the least abundant (Figure 4). Recently, Hong and Wrolstad[16] by using HPLC separated 13 anthocyanins in a bilberry colorant. The major pigments in this colorant were malvidin 3-glucoside (22%), delphinidin 3-galactoside (19%), cyanidin 3-galactoside (15%), petunidin 3-galactoside (14%), cyanidin 3-glucoside (9%), and delphinidin 3-glucoside (9%).

Senchuk and Borukh[257] reported an anthocyanin content of 648 to 698 mg/100 g in *V. myrtillus* from Byelorussia, and noted that the greatest concentration of pigments is in the skin of the fruit. Fruit of bilberries from the Piedmont Alps contains 300 to 320 mg of anthocyanins per 100 g of fresh fruit.[258] During fruit ripening, the anthocyanin content increases, and flavan 3-ols and procyanidins decrease.[259]

In fruit from five species of *Vaccinium* section *Myrtillus*, Ballington et al.[260] identified 3-arabinosides, 3-galactosides, and 3-glucosides of cyanidin, delphinidin, malvidin, peonidin, and petunidin.

In addition to anthocyanins, flavan 3-ols, and procyanidins, bilberries contain a large number of phenolic compounds including flavonol glycosides, hydroxycinnamic acids, and hydroxybenzoic acid derivatives.[261-264] Recently, Azar et al.[262] found 3 flavonoids (quercetin 3-rhamnoside or quercitrin, quercetin 3-galactoside or hyperoside, and quercetin 3-glucoside or isoquercitrin) and 12 phenolic acids (caffeic, chlorogenic, ferulic, syringic, gallic, protocatechuic, *p*-hydroxybenzoic, *m*-hydroxybenzoic, vanillic, *m*-coumaric, *o*-coumaric, and *p*-coumaric) in the ethyl acetate-extractable fraction of bilberry juice.

Pharmaceutical preparations of *V. myrtillus* extracts are widely used to treat various microcirculatory diseases, and their pharmacological activities are attributed to the unique anthocyanin composition of the prepared product.[256,265-267] Thus, the production of these pharmaceutical products has to be conducted under highly controlled conditions. Pourrat et al.[267] described a novel method for the preparation of bilberry anthocyanin extracts that uses cellulase enzymes. The phenolic compounds extracted from bilberry juice after fermentation constitute the active principle in a French drug (Difrarel) used for its vascular properties.[268]

B. COWBERRY

The cowberry or lingonberry (*Vaccinium vitis-idaea* L.) is a low shrub bearing red, globose fruit less than 0.75 cm in diameter, used for making excellent jams.

The pigment of this minor fruit was first identified by Willstätter and Mallison[269] as cyanidin 3-galactoside. Fouassin[4] confirmed the identity of cyanidin 3-galactoside and noted the presence of a second glycoside of cyanidin. Troyan and Borukh[270] reported malvidin 3,5-diglucoside and cyanidin 3-xylosylglucoside in cowberries from the Carpathians. More recently, Andersen[271] identified the 3-galactoside (88.0%), 3-arabinoside (10.6%), and 3-glucoside (1.4%) of cyanidin and delphinidin 3-glucoside (less than 0.1%) as the anthocyanins of Norwegian cowberries. The total anthocyanin content was found to be 174 mg/100 g fresh fruit. In fruit of *Gaylussacia brachycera* (section *Vitis-idaea*), Ballington et al.[260] found delphinidin 3-galactoside (12.6%), 3-glucoside (17.4%), and 3-arabinoside (13.6%); malvidin 3-glucoside (16.4%), 3-galactoside (5.6%), and 3-arabinoside (5.4%); petunidin 3-glucoside (11.4%), 3-arabinoside (3.3%), and 3-galactoside (2.8%); cyanidin 3-glucoside (2%), 3-galactoside (1.7%), and 3-arabinoside (1.8%); and peonidin 3-galactoside (4.7%) and 3-glucoside (0.9%).

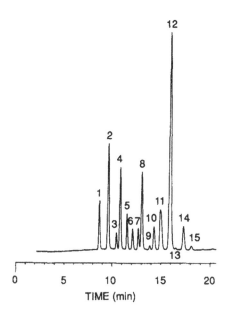

FIGURE 5. HPLC separation of the anthocyanins of *V. uliginosum*. 1: Delphinidin 3-galactoside; 2: delphinidin 3-glucoside; 3: cyanidin 3-galactoside; 4: delphinidin 3-arabinoside; 5: cyanidin 3-glucoside; 6: petunidin 3-galactoside; 7: cyanidin 3-arabinoside; 8: petunidin 3-glucoside; 9: peonidin 3-galactoside; 10: petunidin 3-arabinoside; 11: peonidin 3-glucoside and malvidin 3-galactoside; 12: malvidin 3-glucoside; 13: peonidin 3-arabinoside; 14: malvidin 3-arabinoside; 15: unknown. (From Andersen, O. M., *J. Food Sci.*, 52, 665, 1987. With permission.)

According to Bandaztiene and Butkus,[272] the anthocyanin content of *V. vitis-idaea* increases from 3–30 mg to 180–330 mg/100 g during ripening, but this final level is still considerably lower than that of *V. myrtillus* (650 to 700 mg/100 g).[257] Cowberries, however, have higher contents of catechin (430 mg/100 g), flavonols, and leucoanthocyanins than bilberries have.[273]

C. BOG WHORTLEBERRY

Bog whortleberry (*Vaccinium uliginosum* L.) is a deciduous, dwarf shrub with a circumboreral distribution. Its blue-black edible fruit is an important crop in some northern European countries, especially in the former U.S.S.R. Hayashi[274,275] concluded that the principal anthocyanin pigment in *V. uliginosum* was malvidin 3-galactoside, while Fouassin[4] reported that the anthocyanin composition of bog whortleberries appeared to be qualitatively the same as *V. myrtillus*. Recently, Andersen[276] used combined droplet countercurrent chromatography and semipreparative HPLC to isolate 15 anthocyanins (Figure 5). The major 13 of these pigments were identified as 3-glucoside (35.9%), 3-arabinoside (3.4%), and 3-glucoside (3.2%) of malvidin; 3-glucoside (14.0%), 3-arabinoside (10.2%), and 3-galactoside (6.5%) of

3-arabinoside (1.8%), and 3-galactoside (1.9%) of cyanidin; 3-glucoside (9.1%), 3-arabinoside (2.9%), and 3-galactoside (2.1%) of petunidin; and 3-glucoside (4.5%), 3-arabinoside (0.1%), and 3-galactoside (0.4%) of peonidin. According to Andersen,[276] the total anthocyanin content of bog whortleberry is 256 mg/100 g fresh fruit. Butkus et al.[277] found the total anthocyanins to be 114 to 224 mg/100 g fresh fruit.

IX. OTHERS

A. ARCTIC BRAMBLES

Arctic brambles are species of the genus *Rubus* occurring in north circumpolar or alpine regions.[1] Of these species *R. chamaemorus* L. (commonly known as baked-apple berry, cloudberry, or mountain bramble) is the most common of the arctic *Rubi*. Other species are *R. arcticus* L. diffused in northern latitudes higher than 62° N and *R. stellatus* Sm. which grows spontaneously in northwest Alaska, the Yukon, the Aleutian Islands, and Kamtchatka. In Scandinavia, the fruits of *R. arcticus* L. (Arctic raspberry) with those of *R. chamaemorus* L. (cloudberry) are harvested by local people and eaten fresh or processed in jams and jellies.[1,278]

Willstaedt[279] first noted the presence of small quantities of anthocyanins in immature *R. chamaemorus* berries. Keränen and Suomalainen[280] identified a cyanidin 3-glucoside and a bioside of cyanidin which was different from 3,5-diglucoside or gentiobioside in its chromatographic behavior.

B. BOYSENBERRY

The boysenberry is very likely a cross between the loganberry (*Rubus loganbaccus*) and a trailing blackberry (*Rubus baileyanus* Britt.) such as 'Lucretia' or 'Austin Mayes'. The fruits are large, deep purple in color and have an acid taste. They ripen distinctly earlier than blackberries and loganberries and tend to be rather soft for long-distance transport.[1] Boysenberries are grown extensively in the states of Oregon and Washington and in New Zealand, and are processed into juices and syrups.[1] Luh et al.[281] studied the color stability of boysenberry products and found that puree stored in plastic containers lined with aluminum retain more red color than products stored in containers without the aluminum lining.[282]

According to Luh et al.,[283] the anthocyanin pigments of boysenberries are cyanidin 3-glucoside, cyanidin 3-diglucoside, cyanidin 3-rhamno-glucoside, and cyanidin 3-rhamnoside 5-glucoside. Torre and Barritt,[2] however, characterized anthocyanins of the boysenberry as cyanidin 3-sophoroside (44.5%), cyanidin 3-glucoside (26.4%), cyanidin 3-glycosylrutinoside (25.8%), and cyanidin 3-rutinoside (3.3%). The total anthocyanin content is over 160 mg/100 g fresh fruit.[2]

According to Monro and Sedcole[284] and Monro and Lee,[285] anthocyanins are a good index of maturity for boysenberries because it is the fruit component that changes most during fruit ripening, when anthocyanins and sugars (glucose and fructose) increase and citric, malic, and isocitric acids show a decrease in concentration. Of particular interest is the apparent relationship between anthocyanin content and mechanical harvesting, where higher levels of anthocyanins coincide with an easier fruit removal from the plant and, thus, a higher yield of mechanically harvested fruit.[286,287]

C. LOGANBERRY

The loganberry, *Rubus loganbaccus,* is a most important hybrid between blackberry and raspberry, but it is often considered a red variety of blackberry. The fruit has an attractive conical shape, a large size, and a characteristic flavor. Its color is reddish-purple, and it has to be very ripe before its acidity is low enough for eating fresh.[1] Because of their high acidity, loganberries are usually processed into jams and syrups.

Loganberry fruits are characterized by the presence of cyanidin 3-sophoroside as the major pigment (48.1%) and cyanidin 3-glucosylrutinoside (23.6%), 3-glucoside (21.6%), and 3-rutinoside (6.2%).[2] The total anthocyanin content is about 77 mg/100 g of fresh fruit.[2] Thus, loganberry possesses an anthocyanin pattern typical of red raspberry and a total anthocyanin content and morphological characteristics more typical of blackberry.[2]

D. MYRTLE

Myrtle (*Myrtus communis* L., Myrtaceae) is one of the most typical plants of the Mediterranean basin. Myrtle berries have the shape and size of blueberries and can be utilized as fresh fruit, as medicinal plants, and for the production of essential oils. Myrtle berries are rarely eaten fresh, but are used as spice for meat and sausages. The ancient Roman *mortatum* (from which the name "mortadella" originates) was a sausage seasoned with myrtle. Currently, production and processing of myrtle berries is common in the islands of Sardinia and Corsica and in Spain, where myrtle berry jams, sweets, liqueurs, and homemade wines are produced.[288]

Myrtle berries have a thick, dark blue skin, containing at least ten anthocyanins.[289-292] Recently, Martin et al.[289] identified delphinidin, petunidin, malvidin, peonidin, and cyanidin 3-glucosides and 3,5-diglucosides in mature berries. This anthocyanin pattern is consistent with the findings of Lowry[293] for other members of myrtle family.

REFERENCES

1. **Jennings, D.L.**, *Raspberries and Blackberries: Their Breeding, Diseases and Growth*, Academic Press, London, 1988.
2. **Torre, L.C. and Barritt, B.H.**, Quantitative evaluation of *Rubus* fruit anthocyanin pigments, *J. Food Sci.*, 42, 488, 1977.
3. **Bate-Smith, E.C.**, Paper chromatography of anthocyanins and related substances in petal extracts, *Nature (London)*, 161, 835, 1984.
4. **Fouassin, A.**, Identification par chromatographie des pigments anthocyaniques des fruits et des legumes, *Rev. Ferment. Ind. Aliment.*, 11, 173, 1956.
5. **Blundstone, H.A.W. and Crean, D.E.C.**, *The Pigments of Red Fruits*, Fruit Vegetable Preservation Research Association, Chipping Campden, England, 1966, 37.
6. **Harborne, J.B. and Hall, E.**, Plant polyphenols. XIII. The systematic distribution and origin of anthocyanins containing branched trisaccharides, *Phytochemistry*, 3, 453, 1964.
7. **Nybom, N.**, Cellulose thin layers for anthocyanin analysis, with special reference to the anthocyanins of black raspberries, *J. Chromatogr.*, 38, 382, 1968.
8. **Jennings, D.L. and Carmichael, E.**, Anthocyanin variation in the genus *Rubus*, *New Phytol.*, 84, 505, 1980.
9. **Casoli, U., Cultrera, R., and Gherardi, S.**, Richerche sugli antociani di ciliegia, di ribes nero e di mora di rovo, *Ind. Conserve*, 42, 255, 1967.
10. **Barritt, B.H. and Torre, L.C.**, Cellulose thin-layer chromatographic separation of *Rubus* fruit anthocyanins, *J. Chromatogr.*, 75, 151, 1973.
11. **Sapers, G.M., Hicks, K.B., Burgher, A.M., Hargrave, D.L., Sondey, S.M., and Bilyk, A.**, Anthocyanin pattern in ripening thornless blackberries, *J. Am. Soc. Hortic. Sci.*, 111, 945, 1986.
12. **Bilyk, A. and Sapers, G.M.**, Varietal differences in the quercetin, kaempferol, and myricetin contents of highbush blueberry, cranberry, and thornless blackberry, *J. Agric. Food Chem.*, 34, 585, 1986.
13. **Naumann, W.D. and Wittemburg, U.**, Anthocyanins, soluble solids, and titratable acidity in blackberries as influenced by preharvest temperature, *Acta Hortic.*, 112, 183, 1980.
14. **Yang, H.Y.**, Pigment precipitation in blackberry wine, *Wines Vines*, 34, 28, 1953.
15. **Yang, H.Y. and Steele, W.F.**, Removal of excessive anthocyanin pigment by enzyme, *Food Technol.*, 12, 517, 1958.
16. **Hong, V. and Wrolstad, R.E.**, Characterization of anthocyanin-containing colorants and fruit juices by HPLC/photodiode array detection, *J. Agric. Food Chem.*, 38, 698, 1990.
17. **Harborne, J.B.**, *Comparative Biochemistry of the Flavonoids*, Academic Press, New York, 1967, 297.
18. **Tanchev, S.S.**, Kinetics of the thermal degradation of anthocyanins during the sterilization and storage of blueberry and blackberry, *Nahrung*, 18, 303, 1974 (CA 81, 134801).
19. **Jennings, D.L. and Carmichael, E.**, Colour changes in frozen blackberries, *Hortic. Res.*, 19, 15, 1979.
20. **Crivelli, G. and Rosati, P.**, Research on suitability of raspberry and blackberry varieties to quick freezing, *Ann. IVTPA*, 6, 61, 1975.
21. **Sapers, G.M., Burgher, A.M., Phillips, J.G., and Galletta, G.J.**, Composition and color of fruit and juice of thorneless blackberry cultivars, *J. Am. Soc. Hortic. Sci.*, 110, 243, 1985.
22. **Shoemaker, J.S.**, *Small Fruit Culture*, 5th ed., AVI Publishing, Westport, CT, 1978, 357.
23. **Hanson, E.J. and Hancock, J.F.**, Highbush blueberry cultivars and production trends, *Fruit Var. J.*, 44, 77, 1990 (HA 1990, 7982).

24. **Woodruff, R.E., Dewey, D.H., and Sell, H.M.**, Chemical changes of Jersey and Rubel blueberry fruit associated with ripening and deterioration, *Proc. Am. Soc. Hortic. Sci.*, 75, 387, 1960.

25. **Suomalainen, H. and Keranen, A.J.A.**, The first anthocyanins appearing during the ripening of blueberries, *Nature (London)*, 191, 498, 1961.

26. **Dekazos, E.D. and Birth, G.S.**, A maturity index for blueberries using light transmittance, *J. Am. Soc. Hortic. Sci.*, 95, 610, 1970.

27. **Ballinger, W.E. and Kushman, L.J.**, Relationship of stage of ripeness to composition and keeping quality of highbush blueberries, *J. Am. Soc. Hortic. Sci.*, 95, 239, 1970.

28. **McClure, W.F., Ballinger, W.E., and Maness, E.P.**, Computerized spectrophotometric analysis of blueberries, *Am. Lab.*, 16, 97, 1984.

29. **Kushman, L.J. and Ballinger, W.E.**, Relation of quality indexes of individual blueberries to photoelectric measurement, *J. Am. Soc. Hortic. Sci.*, 100, 561, 1975.

30. **Kushman, L.J. and Ballinger, W.E.**, Effect of season, location, cultivar, and fruit size upon quality of light-sorted blueberries, *J. Am. Soc. Hortic. Sci.*, 100, 564, 1975.

31. **Ballinger, W.E., Maness, E.P., and Kushman, L.J.**, Anthocyanins in ripe fruit of the highbush blueberry, *Vaccinium corymbosum* L., *J. Am. Soc. Hortic. Sci.*, 95, 283, 1970.

32. **Ballinger, W.E., Maness, E.P., Galletta, G.J., and Kushman, L.J.**, Anthocyanins of ripe fruit of a pink-fruited hybrid of highbush blueberry, *Vaccinium corymbosum* L., *J. Am. Soc. Hortic. Sci.*, 97, 381, 1972.

33. **Makus, D.J. and Ballinger, W.E.**, Characterization of anthocyanins during ripening of fruit of *Vaccinium corymbosum* L. cv. Wolcott, *J. Am. Soc. Hortic. Sci.*, 98, 99, 1973.

34. **Vereshkovski, V.V. and Shapiro, D.K.**, Anthocyanin glycosides in the fruit of highbush blueberry cultivars, *Khim. Prir. Soedin.*, 4, 570, 1985 (CA 104, 165308).

35. **Ballinger, W.E., Maness, E.P., and Galletta, G.J.**, Anthocyanins of albino highbush blueberry fruit, *HortScience*, 6, 131, 1971.

36. **Sapers, G.M., Burgher, A.M., Phillips, J.G., Jones, S.B., and Stone, E.G.**, Colour and composition of highbush blueberry cultivars, *J. Am. Soc. Hortic. Sci.*, 109, 105, 1984.

37. **Schuster, B. and Herrmann, K.**, Hydroxybenzoic and hydroxycinnamic acid derivatives in soft fruits, *Phytochemistry*, 24, 2761, 1985.

38. **Pourrat, H.**, Anthocyanes et flavones des feuilles de *Vaccinium corymbosum* L., *Plant. Med. Phytother.*, 12, 212, 1978.

39. **Blatt, C.R. and Hall, I.V.**, Lowbush blueberry production, Agriculture Canada Pub. No. 1477/E, Ottawa, 1989, 7.

40. **Wood, F.A. and Barker, W.G.**, Stem pigmentation in lowbush blueberry, *Plant Physiol.*, 38, 191, 1963.

41. **Hall, I.V., Forsyth, F.R., and Newbery, R.J.**, Effect of temperature on flower and leaf anthocyanin formation in the lowbush blueberry, *HortScience*, 5, 272, 1970.

42. **Barker, W.G. and Wood, F.A.**, Fruit anthocyanin variation among clones of the highbush blueberry, *Plant Physiol.*, 38, 13, 1963.

43. **Francis, F.J., Harborne, J.B., and Barker, W.G.**, Anthocyanins in the lowbush blueberry, *Vaccinium angustifolium*, *J. Food Sci.*, 31, 583, 1966.

44. **Fuleki, T. and Hope, G.W.**, Effect of various treatments on yield and composition of blueberry juice, *Food Technol.*, 18, 166, 1964.

45. **Hope, G.W.**, A review of the suitability of lowbush blueberry for processing, *Food Technol.*, 19, 115, 1965.

46. **Fuleki, T.**, Fermentation studies on blueberry wine, *Food Technol.*, 19, 105, 1965.

47. **Simard, R.E., Bourzeix, M., and Heredia, N.**, Factors influencing color degradation in blueberry juice, *Lebensm. Wiss. Technol.*, 15, 177, 1982.

48. **Yang, C.S.T. and Yang, P.P.A.**, Effect of pH, certain chemicals and holding time-temperature on the color of lowbush blueberry puree, *J. Food Sci.*, 52, 346, 1987.

49. **Brouillard, R.,** Chemical structure of anthocyanins, in *Anthocyanins as Food Colors,* Markakis, P., Ed., Academic Press, New York, 1982, 1.
50. **Spiers, J.M.,** Rabbiteye blueberry, *Fruit Var. J.,* 44, 68, 1990 (HA 1990, 7980).
51. **Patten, K., Scheurman, M.R., and Burns, E.E.,** Hydro-density sorting of rabbiteye blueberry fruit *(Vaccinium ashei* Reade), *Acta Hortic.,* 241, 354, 1989.
52. **Ballington, J.R., Ballinger, W.E., and Maness, E.P.,** Interspecific differences in the percentage of anthocyanins, aglycones, and aglycone-sugars in the fruit of seven species of blueberries, *J. Am. Soc. Hortic. Sci.,* 112, 859, 1987.
53. **Basiouny, F.M. and Chen, Y.,** Effects of harvest date, maturity and storage intervals on postharvest quality of rabbiteye blueberry *(Vaccinium ashei* Reade), *Proc. Fla. State Hortic. Sci.,* 101, 281, 1988.
54. **Mzhavanadze, V.V.,** Anthocyanins of Caucasian blueberry *(Vaccinium arctostaphylos),* *Biokhim. Rastit.,* 1, 251, 1973 (CA 81, 132840).
55. **Lebedeva, A.M.,** Studies on the chemical composition of some species of *Vacciniaceae* in the USSR, *Rastit. Resur.,* 17, 299,1981 (CA 95, 76821).
56. **Mzhavanadze, V.V.,** Flavonol glycosides from leaves of blueberry *Vaccinium arctostaphylos* L., *Biokhim. Rastit.,* 1, 247, 1973.
57. **Mzhavanadze, V.V., Targamadze, I.L., and Dranik, L.I.,** Phenolic compounds of unripe blueberry, *Vaccinium arctostaphylos* fruits, *Soobshch. Akad. Nauk Gruz. SSR,* 68, 205, 1972 (CA 78, 26516).
58. **Ballinger, W.E., Maness, E.P., and Ballington, J.R.,** Anthocyanin in ripe fruit of the sparkleberry, *Vaccinium arboreum* Marsh., *Can. J. Plant Sci.,* 62, 683, 1982.
59. **Ballinger, W.E., Maness, E.P., and Ballington, J.R.,** Anthocyanin and total flavonol content of *Vaccinium stamineum* L. fruit, *Sci. Hortic.,* 15, 173, 1981.
60. **Ballington, J.R., Kirkman, W.B., Ballinger, W.E., and Maness, E.P.,** Anthocyanin, aglycone, and aglycone-sugar content in the fruits of temperate North American species of four sections in *Vaccinium, J. Am. Soc. Hortic. Sci.,* 113, 746, 1988.
61. **Ishikura, N. and Sugahara, K.,** A survey of anthocyanins in fruits of some Angiosperms, *Bot. Mag. (Tokyo),* 92, 157, 1979.
62. **Martinod, P., Hidalgo, J., Guervara, D., and Pazmino, D.,** Study of pigments present in the mortino, *Vaccinium floribundum* HBK, *Politecnica,* 3, 151, 1975 (CA 88, 3080).
63. **Andersen, O.M.,** Anthocyanins in fruits of *Vaccinium japonicum, Phytochemistry,* 26, 1220, 1987.
64. **Andersen, O.M.,** Anthocyanins in fruits of *Vaccinium oxycoccus* L. (small cranberry), *J. Food Sci.,* 54, 383, 1989.
65. **Zapsalis, C. and Francis, F.J.,** Cranberry anthocyanins, *J. Food Sci.,* 30, 396, 1965.
66. **Hicks, K.B., Sondey, S.M., Hargrave, D., Sapers, G.M., and Bilyk, A.,** Automated preparative HPLC of anthocyanins, *HPLC Mag.,* 3, 981, 1985.
67. **Shawa, A. Y. and Ingalsbe, D. W.,** Anthocyanin enhancement in McFarlin cranberries at optimum maturity, *Proc. Am. Soc. Hortic. Sci.,* 93, 289, 1968.
68. **Eaton, G. W., Zuckerman, B. M., Shawa, A. Y., Eck, P., Dana, M. N., Garren, R., and Lockart, L. L.,** Effect of pre-harvest malathion sprays upon cranberry fruit color, *J. Am. Soc. Hortic. Sci.,* 94, 590, 1969.
69. **Eck, P.,** Colour enhancement of cranberry fruit, *Cranberries,* 4, 6, 1969.
70. **Francis, F. J. and Zuckerman, B. M.,** Effect of fungicide treatment on pigment content and decay in cranberries, *Proc. Am. Soc. Hortic. Sci.,* 81, 288, 1962.
71. **Devlin, R. M. and Demoranville, I. E.,** Influence of Dichlobenil on yield, size, and pigmentation in cranberries, *Weed Sci.,* 16, 38, 1968.
72. **Demoranville, I. E. and Devlin, R. M.,** Some effects of Dichlobenil on the physiology of cranberries in Massachusetts, *Cranberries,* 3, 6, 1969 (JSFA 1971, 1228).
73. **Devlin, R. M. and Demoranville, I. E.,** Influence of 2-chloroethyl phosphonic acid on anthocyanin formation, size, and yield in *Vaccinium macrocarpon* cv. Early Black, *Physiol. Plant,* 23, 1139, 1970.

74. **Eck, P.,** Cranberry yield and anthocyanin content as influenced by ethephon, SADH, and malathion, *J. Am. Soc. Hortic. Sci.,* 97, 213, 1972.
75. **Bramlage, W. J., Devlin, R. M., and Smagula, M. J.,** Effect of preharvest application of ethephon on Early Black cranberries, *J. Am. Soc. Hortic. Sci.,* 97, 625, 1972.
76. **Cracker, L. E.,** Postharvest color promotion in cranberry with ethylene, *HortScience,* 6, 137, 1971.
77. **Devlin, R. M. and Demoranville, I. E.,** Influence of two ethylene releasing compounds on anthocyanin formation, size, and yield in Early Black cranberries, *Proc. Northeast. Weed Sci. Soc.,* 32, 108, 1978.
78. **Francis, F. J. and Atwood, W. M.,** Effect of fertilizer treatments on pigment content of cranberries, *Proc. Am. Soc. Hortic. Sci.,* 77, 351, 1961.
79. **Eaton, G. W.,** Effects of N, P, and K fertilizers on the growth and composition of vines in a young cranberry bog, *J. Am. Soc. Hortic. Sci.,* 96, 426, 1971.
80. **Eck, P.,** Cranberry growth and production in relation to water tables depth, *J. Am. Soc. Hortic. Sci.,* 101, 544, 1976.
81. **Devlin, R. M., Zuckerman, B. M., and Demoranville, I. E.,** Effect of delayed harvest and storage on pigment development in cranberries, *Proc. Am. Soc. Hortic. Sci.,* 92, 793, 1968.
82. **Farag, K. M., Palta, J. P., and Stang, E. J.,** Field application of new Ethrel formulations for early color enhancement in cranberry (*Vaccinium macrocarpon* Ait.) fruit, *Acta Hortic.,* 241, 373, 1989.
83. **Hawker, G. M. and Stang, E. J.,** Characterizing vegetative growth and fruit development in cranberry (*Vaccinium macrocarpon* Ait.) by thermal summation, *Acta Hortic.,* 165, 3111, 1985.
84. **Abdallah, A. Y. and Palta, J. P.,** Changes in biophysical and biochemical properties of cranberry (*Vaccinium macrocarpon* Ait.) fruit during growth and development, *Acta Hortic.,* 241, 360, 1989.
85. **Cansfield, P. E. and Francis, F. J.,** Chemical effects on anthocyanin biosynthesis in stored cranberries, *Can. J. Plant Sci.,* 50, 673, 1970.
86. **Hall, I. V. and Stark, R.,** Anthocyanin production in cranberry leaves and fruit, related to cool temperatures at a low light intensity, *Hort. Res.,* 12, 183, 1972.
87. **Shawa, A.,** The influence of DEPEG on anthocyanin enhancement and keeping quality of McFarlin cranberries, *Acta Hortic.,* 165, 305, 1985.
88. **Lees, D. H. and Francis, F. J.,** Effect of gamma radiation on anthocyanins and flavonol pigments in cranberries *(Vaccinium macrocarpon),* *J. Am. Soc. Hortic. Sci.,* 97, 128, 1972.
89. **Schmid, P.,** Long term investigation with regard to the constituents of various cranberry varieties (*Vaccinium macrocarpon* Ait.), *Acta Hortic.,* 61, 241, 1977.
90. **Sapers, G. M., Graff, G. R., Phillips, J. G., and Deubert, K. H.,** Factors affecting the anthocyanin content of cranberry, *J. Am. Soc. Hortic. Sci.,* 111, 612, 1986.
91. **Vorsa, N. and Welker, W. V.,** Relation between fruit size and extractable anthocyanin content in cranberry, *HortScience,* 20, 402, 1985.
92. **Sapers, G. M., Jones, S. B., Kelley, M. J., Philips, J. G., and Stone, E. G.,** Breeding strategies for increasing the anthocyanin content of cranberries, *J. Am. Soc. Hortic. Sci.,* 111, 618, 1986.
93. **Sapers, G. M., Jones, S. B., and Maher, G. T.,** Factors affecting the recovery of juice and anthocyanin from cranberries, *J. Am. Soc. Hortic. Sci.,* 108, 246, 1983.
94. **Sapers, G. M. and Hargrave, D. L.,** Proportions of individual anthocyanins in fruits of cranberry cultivars, *J. Am. Soc. Hortic. Sci.,* 112, 100, 1987.
95. **Sapers, G. M., Matulaitis, R. M., and Beck, J. A.,** Detection of phenylalanine ammonia-lyase in the skin of blueberry and cranberry fruits, *J. Food Sci.,* 52, 155, 1987.
96. **Grove, K. E. and Robinson, R.,** An anthocyanin of *Oxycoccus macrocarpus* Pers., *Biochem. J.,* 25, 1706, 1931.

97. **Sakamura, S. and Francis, F. J.,** The anthocyanins of the American cranberry, *J. Food Sci.,* 26, 318, 1961.

98. **Fuleki, T. and Francis, F. J.,** The co-occurrence of monoglucosides and monogalactosides of cyanidin and peonidin in the American cranberry *Vaccinium macrocarpon, Phytochemistry,* 6, 1705, 1967.

99. **Camire, A. L. and Clydesdale, F. M.,** High-pressure liquid chromatography of cranberry anthocyanins, *J. Food Sci.,* 44, 926, 1979.

100. **Fuleki, T.,** Individual anthocyanin composition of American cranberry cultivars, *Bull. Liaison Groupe Polyphénols,* 13, 374, 1986.

101. **Puski, G. and Francis, F. J.,** Flavonol glycosides in cranberries, *J. Food Sci.,* 32, 527, 1967.

102. **Cansfield, P. E. and Francis, F. J.,** Quantitative methods for anthocyanins. V. Separation of cranberry phenolics by electrophoresis and chromatography, *J. Food Sci.,* 35, 309, 1970.

103. **Marwan, A. G. and Nagel, C. W.,** Separation and purification of hydroxycinnamic acid derivatives in cranberries, *J. Food Sci.,* 47, 585, 1982.

104. **Wang, P. L., Du, C. T., and Francis, F. J.,** Isolation and characterization of polyphenolic compounds in cranberries, *J. Food Sci.,* 43, 1402, 1978.

105. **Foo, L. Y. and Porter, L. J.,** The structure of tannins of some edible fruits, *J. Sci. Food Agric.,* 32, 711, 1981.

106. **Starr, M. S. and Francis, F. J.,** Oxygen and ascorbic acid effect on the relative stability of four anthocyanin pigments in cranberry juice, *Food Technol.,* 22, 1293, 1968.

107. **Francis, F. J. and Clydesdale, F. M.,** Color measurement of foods. XVIII. Cranberry products, *Food Prod. Dev.,* 4, 54, 1970.

108. **Culpepper, G. W. and Caldwell, J. J.,** The behavior of anthocyanin pigments in canning, *J. Agric. Res.,* 35, 107, 1927.

109. **Kohmann, E. F.,** The color in cranberries, *Food Technol.,* 6, 160, 1952.

110. **Starr, M. S. and Francis, F. J.,** Effect of metallic ions on color and pigment content of cranberry juice cocktail, *J. Food Sci.,* 39, 1043, 1974.

111. **Shrikhande, A. J. and Francis, F. J.,** Effect of flavonol on ascorbic acid and anthocyanin stability in model system, *J. Food Sci.,* 39, 904, 1974.

112. **Kuznetsova, N. A.,** Effect of maturity and conditions of thermal treatment on the rate of phenol compounds disintegration in Byelorussian cranberries, *Konservn. Ovoshchesush. Prom.,* 11, 27, 1975 (CA 84, 29428).

113. **Servadio, G. J. and Francis, F. J.,** Relation between color of cranberries and color and stability of sauce, *Food Technol.,* 17, 632, 1963.

114. **Sapers, G. M., Phillips, J. G., Rudolph, H. M., and Di Vito, A. M.,** Cranberry quality, selection procedures for breeding programs, *J. Am. Soc. Hortic. Sci.,* 108, 241, 1983.

115. **Francis, F. J.,** Detection of enocyanin in cranberry juice cocktail by color and pigment profile, *J. Food Sci.,* 50, 1640, 1985.

116. **Hale, M. L., Francis, F. J., and Fagerson, I. S.,** Detection of enocyanin in cranberry juice cocktail by HPLC anthocyanin profile, *J. Food Sci.,* 51, 1511, 1986.

117. **Hong, V. and Wrolstad, R. E.,** Cranberry juice composition, *J. Assoc. Off. Anal. Chem.,* 69, 199, 1986.

118. **Hong, V. and Wrolstad, R. E.,** Detection of adulteration in commercial cranberry juice drinks and concentrates, *J. Assoc. Off. Anal. Chem.,* 69, 208, 1986.

119. **Francis, F. J.,** Anthocyanins as food colours, *Food Technol.,* 29, 52, 1975.

120. **Chiriboga, C. D. and Francis, F. J.,** An anthocyanin recovery system from cranberry pomace, *J. Am. Soc. Hortic. Sci.,* 95, 233, 1970.

121. **Chiriboga, C. D. and Francis, F. J.,** Ion-exchange purified anthocyanin pigments as a food colorant for cranberry juice cocktail, *J. Food Sci.,* 38, 464, 1973.

122. **Volpe, T.,** Cranberry juice concentrate as a red food coloring, *Food Proc. Dev.,* 10(9), 13, 1976.

123. **Usova, E. M., Redko, E. G., and Dubkova, L. P.,** The use of juice of the berry *Vaccinium macrocarpon* (klikva) as a colouring matter in soft drink production, *Izv. Vyssh. Ucheb. Zaved., Pishch. Tekhnol.,* 5, 105, 1968 (CA 70, 27762).

124. **Stahl, H. D., Bordonaro, M. E., and Nini, D.,** Cranberry color extraction, U.S. Patent 4,775,477, 1988 (FSTA 21, 4V150).

125. **Woo, A. H., von Elbe, J. H., and Amundson, C. H.,** Anthocyanin recovery from cranberry pulp wastes by membrane technology, *J. Food Sci.,* 45, 875, 1980.

126. **Clydesdale, F. M., Main, J. H., and Francis, F. J.,** Cranberry pigments as colorants for beverage and gelatin desserts, *J. Food. Prot.,* 42, 196, 1979.

127. **Robinson, G. M. and Robinson, R.,** A survey of anthocyanins. I, *Biochem. J.,* 25, 1687, 1931.

128. **Mehlitz, A. and Matzik, B.,** Über Untersuchungen an roten und schwarzen Johannis-beersäften. I. Papierchromatographische Untersuchungen der Farbstoffe von roten und schwarzen Johannisbeersäften, *Fluess. Obst.,* 24(3), 7, 1957.

129. **Chandler, B. V. and Harper, K. A.,** Anthocyanins in black currant fruit, *Nature (London),* 181, 131, 1958.

130. **Nybom, N.,** Anthocyansammansaetning hos hodlade nordiska frukter och baer, *Frukt Baer,* 1979, 106.

131. **Øydvin, J.,** Inheritance of four cyanidin 3-glycosides in the red currant, *Hortic. Res.,* 14, 1, 1974.

132. **Keep, E. and Knight, R. L.,** Inheritance of fruit color in blackcurrants and gooseberries, *Rep. East Malling Res. Stn.,* 1969, 139 (PBA 1971, 1562).

133. **Green, A.,** Soft fruits, in *The Biochemistry of Fruits and Their Products,* Vol. 2, Hulme, A. C., Ed., Academic Press, London, 1971, 375.

134. **Dankanits, E.,** Studies on reducing sugars, organic acids and pigments in some currant varieties, *Elelmiszervizsgalati Kozl.,* 22, 48, 1976 (CA 86, 70316).

135. **Averkovic, M. I.,** The effect of the variety on the content of tannins and anthocyanins in black-currant berries, *Nauk. Pr. Lviv. Silskogospod. Inst.,* 53, 77, 1975 (PBA 1977, 5822).

136. **Koeppen, B. H. and Herrmann, K.,** Phenolics of fruits. IX. Flavonoid glycosides and hydroxycinnamic acid esters of blackcurrants *(Ribes nigrum), Z. Lebensm. Unters. Forsch.,* 164, 263, 1977.

137. **Chandler, B. V. and Harper, K. A.,** A procedure for the absolute identification of anthocyanins, the pigments of blackcurrant fruit, *Aust. J. Chem.,* 15, 114, 1962.

138. **Reichel, L. and Reichwald, W.,** Über die Farbstoffe der schwarzen Johannisbeere, *Naturwissenschaften,* 47, 41, 1960 (CZ 1960, 14408).

139. **Demina, T. G.,** Anthocyanins of several varieties of black currants, *Biol. Akt. Soedin. Rastit. Sib. Flory,* 1974, 23 (CA 82, 40688).

140. **LeLous, J., Majoie, B., Moriniére, J. L., and Wulfert, E.,** Étude des flavonoides de *Ribes nigrum, Ann. Pharm. Fr.,* 33, 393, 1975 (CA 84, 176692).

141. **Skrede, G.,** Evaluation of colour quality in blackcurrant fruits grown for industrial juice and syrup production, *Norw. J. Agric. Sci.,* 1, 67, 1987 (HA 1988, 2708).

142. **Simard, R. E., Bourzeix, M., and Heredia, N.,** Factor influencing color degradation in black currant juice *(Ribes nigrum* L.), *Sci. Aliment.,* 1, 389, 1981 (CA 95, 202247).

143. **Taylor, J.,** Color stability of blackcurrant *(Ribes nigrum)* juice, *J. Sci. Food Agric.,* 49, 487, 1989.

144. **Skrede, G.,** Color quality of blackcurrant syrups during storage evaluated by Hunter L', a', b' values, *J. Food Sci.,* 50, 514, 1985.

145. **Skrede, G., Wrolstad, R. E., and Enersen, G.,** Comparing color stability of strawberry and blackcurrant syrup, Paper presented at the 51st Annu. Meet. Inst. Food Technol., Anaheim, CA, June 16–20, 1990.

146. **Rosa, J. and Ichas, H.**, Experiments utilizing natural colour substances from coloured fruit pomaces, *Pr. Inst. Lab. Badaw. Przem. Spozyw.*, 19, 561, 1969 (FSTA 1, 8J832).

147. **Rosa, J.**, Stability of anthocyanin pigment concentrates obtained from blackcurrant press cake. I. Production and storage of concentrates as well as possibilities of application, *Pr. Inst. Lab. Badaw. Przem. Spozyw.*, 23, 269, 1973 (CA 81, 12025).

148. **Rosa, J.**, Stability of anthocyanin pigment concentrates obtained from blackcurrant press cake. II. Studies on the rate of dye destruction during storage of concentrates, *Pr. Inst. Lab. Badaw. Przem. Spozyw.*, 23, 447, 1973 (FSTA 7, 3J454).

149. **Rosa, J.**, Stability of anthocyanin pigment concentrates obtained from blackcurrant press cake. III. Trials of enhancing dye stability in concentrates, *Pr. Inst. Lab. Badaw. Przem. Spozyw.*, 23, 653, 1973 (CA 82, 2902).

150. **Frumkin, M. L., Nakhmedov, F. G., and Borovikova, N. N.**, New natural pigments obtained from black rowanberry and blackcurrant cake, *Konservn. Ovoschchesush. Prom.*, 9, 22, 1974.

151. **Nillson, F. and Traikovski, V.**, Color pigments in the species and hybrids of the genus *Ribes* L., *Landbrukshoegsk. Medd.*, Ser. A, 282, 20, 1977 (CA 88, 186216).

152. **Saburov, N. V. and Ulyanova, D. A.**, Reaction of ascorbic acid and anthocyanin pigments of European blackcurrant and gooseberry juices with the pulp, *Dokl. TSKhA*, 132, 259, 1967 (CA 69, 70707).

153. **Troyan, A. V. and Batunina, A. P.**, Changes in tannin and pigment content of stored gooseberries, *Biol. Akt. Veshchestva Plodov Yagod*, 1976, 136 (HA 1977, 11250).

154. **Stohr, H. and Herrman, K.**, Die phenolischen Inhaltsstoffe des Obstes. VI. Die phenolischen Inhaltssoffe der Johannisbeeren, Stachelbeeren und Kulturheidelbeeren. Veranderungen der Phenolsauren und Catechine wahrend wachstum und Reife von Schwarzen Johannisbeeren, *Z. Lebensm. Unters. Forsch.*, 159, 31, 1975.

155. **Schwab, W. and Herrman, K.**, Hydroxybenzoic and hydroxycinnamic derivatives in soft fruits, *Phytochemistry*, 24, 2761, 1985.

156. **Shoemaker, J. S.**, Bramble fruits, in *Small Fruit Culture*, 5th ed., Shoemaker, J. S., Ed., AVI Publishing, Westport, CT, 1977, 188.

157. **Nybom, N.**, Dunnschicht-chromatographische Anthocyaninanalyse von Fruchtsaften, *Fruchtsaft-Ind.*, 8, 205, 1963.

158. **Daravingas, G. and Cain, R. G.**, The major anthocyanin pigments of black raspberries, *J. Food Sci.*, 31, 927, 1966.

159. **Francis, F. J.**, Anthocyanins of Durham and Heritage raspberry fruits, *HortScience*, 7, 398, 1972.

160. **Misic, P.D.**, Anthocyanin pigments of some red raspberry (*Rubus idaeus*) fruits, *Hort. Res.*, 13, 45, 1973 (CA 80, 57449).

161. **Barritt, B. H. and Torre, L. C.**, Fruit anthocyanin pigments of red raspberry cultivars, *J. Am. Soc. Hortic. Sci.*, 100, 98, 1975.

162. **Hong, V. and Wrolstad, R. E.**, Characterization of anthocyanin-containing colorants and fruit juices by HPLC/photodiode array detection, *J. Agric. Food Chem.*, 38, 698, 1990.

163. **Shamsizade, L. A. and Novruzov, E. N.**, Anthocyanins of fruit of *Rubus caucasicus* Focke., *Rastit. Resur.*, 25, 557, 1989 (CA 112, 73900).

164. **Keep, E.**, Inheritance of fruit colour in a wild Russian red raspberry seedling, *Euphytica*, 33, 507, 1984.

165. **Dimitrov, P., Topchiiski, S., and Komforti, T.**, The results of breeding work with raspberry, *Gradinar. Lozar. Nauka*, 12, 29, 1975 (PBA 1976, 3704).

166. **Blank, F.**, Anthocyanins, flavones, xanthones, in *Encyclopedia of Plant Physiology*, Ruhland, W., Ed., Springer-Verlag, Berlin, 1958, 300.

167. **Sági, F., Kollańyi, L., and Simon, I.**, Changes in the colour and anthocyanin content of raspberry fruit during ripening, *Acta Aliment.*, 3, 397, 1974 (CA 83, 5087).

168. **Robbins, J. A., Sjulin, T. M., and Patterson, M.,** Postharvest storage characteristics and respiration rates in five cultivars of red raspberry, *HortScience*, 24, 980, 1989.

169. **Sjulin, T. M. and Robbins, J.,** Effects of maturity, harvest date, and storage time on postharvest quality of red raspberry fruit, *J. Am. Soc. Hortic. Sci.*, 112, 481, 1987.

170. **Diemair, W. and Schormueller, J.,** The behaviour of coloring matter of the raspberry toward mineral water, *Z. Untersuch. Lebensm.*, 67, 59, 1934 (CA 28, 28054).

171. **Guadagni, D. G., Kelly, S. H., and Ingraham, L. I.,** Kinetics of color, ascorbic acid, and total acid diffusion in frozen syrup-packed raspberries, *Food. Res.*, 25, 464, 1960.

172. **Daravingas, G. and Cain, R. F.,** Thermal degradation of black raspberry anthocyanin pigments in model systems, *J. Food Sci.*, 33, 138, 1968.

173. **Daravingas, G. and Cain, R. F.,** Changes in the anthocyanin pigments of raspberries during processing and storage, *J. Food Sci.*, 30, 400, 1965.

174. **Danchev, M.,** Changes in the color and the amount of anthocyanins during frozen storage of concentrated raspberry juice, *Khranit. Prom. Nauka*, 2, 66, 1986 (CA 107, 6007).

175. **Coffey, D. G., Clydesdale, F. M., Francis, F. J., and Damon, R. A., Jr.,** Stability and complexation of cyanidin 3-glucoside and raspberry juice extract in the presence of selected cations, *J. Food Prot.*, 44, 516, 1981.

176. **Tanchev, S. S., Vladimirov, G., and Ioncheva, N.,** Effect of some pectolytic enzymes on the destruction of anthocyanins, *Nauchni Tr. Vissh. Ist. Khranit. Vkusova Prom. Plovdiv*, 16, 77, 1969 (FSTA 6, 9J1409).

177. **Jiang, J. Patterson, A., and Piggot, J. R.,** Effects of pectolytic enzyme treatments on anthocyanins in raspberry juices, *Int. J. Food Sci. Technol.*, 25, 596, 1990.

178. **Rommel, A., Heatherbell, D. A., and Wrolstad, R. E.,** Red raspberry juice and wine: effect of processing and storage on anthocyanin pigment composition, color and appearance, *J. Food Sci.*, 55, 1011, 1990.

179. **Henning, W.,** Flavonolglykoside der Erdbeeren (*Fragaria* x *ananassa* Duch.), Himbeeren (*Rubus idaeus* L.) and Brombeeren (*Rubus fruticosus* L.), *Z. Lebensm. Unters. Forsch.*, 173, 180, 1981.

180. **Mosel, H. D. and Herrmann, K.,** Die phenolischen Inhaltsstoffe des Obstes. IV. Die phenolischen Inhaltsstoffe der Brombeeren und Himbeeren und deren Veränderungen während Wachstum und Reife der Früchte, *Z. Lebensm. Unters. Forsch.*, 154, 324, 1974.

181. **Mazza, G. and Brouillard, R.,** The mechanism of co-pigmentation of anthocyanins in aqueous solutions, *Phytochemistry*, 29, 1097, 1990.

182. **Woodward, J. R.,** Physical and chemical changes in developing strawberry fruits, *J. Sci. Food Agric.*, 23, 465, 1972.

183. **Haginuma, S., Mizuta, T., and Miura, H.,** Utilization of strawberries. IV. Changes in pectins and anthocyanins of growing and ripening strawberries, *Nippon Shokuhin Kogyo Gakkai-Shi*, 9, 63, 1962 (CA 59, 13274d).

184. **Lundergan, C. A. and Moore, J. N.,** Inheritance of ascorbic acid content and color intensity in fruits of strawberry (*Fragaria* × *ananassa* Duch.), *J. Am. Soc. Hortic. Sci.*, 100, 633, 1975.

185. **Maclachan, J. B.,** The inheritance of colour of fruit and the assessment of plants as source of colour in the cultivated strawberry, *Hortic. Res.*, 14, 29, 1974.

186. **Fowler, C. W., Hughes, H., and Janick, J.,** Anthocyanin markers in strawberry, *HortScience*, 7, 321, 1972.

187. **Hyodo, H.,** Phenylalanine ammonia-lyase in strawberry fruits, *Plant Cell Physiol.*, 12, 989, 1971.

188. **Aoki, S., Araki, C., Kaneko, K., and Katayama, O.,** L-Phenylalanine ammonia-lyase activities of Japanese chestnut, strawberries, apples and brackens, *Nippon Shokuhin Kogyo Gakkai-Shi*, 17, 507, 1971 (CA 76, 1827).

189. **Nelson, J. W., Barritt, B. H., and Wolford, E. R.,** Influence of location and cultivar on color and chemical composition of strawberry fruit, *Tech. Bull. Wash. Agric. Exp. Stn.*, 74, 7, 1972 (HA 1974, 3822).

190. **Ivanov, A. and Stamboliev, M.,** The effect of certain meteorological factors on the length of the picking period and the chemical composition of strawberry fruit, *Gradinar. Lozar. Nauka,* 10, 45, 1973 (HA 1974, 7436).

191. **Guttridge, C. G. and Nunns, A.,** Promotion of reddening in unripe strawberry fruit by fungal extracts, *Nature (London),* 247, 389, 1974.

192. **Cheour, F., Willemot, C., Arul, J., Desjardins, Y., Makhlouf, J., Charest, P. M., and Gosselin, A.,** Foliar application of calcium chloride delays postharvest ripening of strawberry, *J. Am. Soc. Hortic. Sci.,* 115, 789, 1990.

193. **Creasy, L. L., Maxie, E. C., and Chichester, C. O.,** Anthocyanin production in strawberry leaf disk, *Phytochemistry,* 4, 517, 1965.

194. **Robinson, G. M. and Robinson, R.,** A survey of anthocyanins. II, *Biochem. J.,* 26, 1647, 1932.

195. **Robinson, R.,** Über die Synthese von Anthocyanen, *Chem. Ber.,* 67A, 85, 1934.

196. **Akuta, S. and Koda, R.,** Studies on the utilization of strawberries. III. Detection of strawberry components by paper chromatography, *J. Ferment. Technol.,* 32, 257, 1954 (CA 49, 1987d).

197. **Robinson, R. and Smith, H.,** Anthocyanins of the leaf of the copper beech *(Fagus silvatica)* and the fruit of the cultivated strawberry *(Fragaria virginiana), Nature (London),* 175, 634, 1955.

198. **Sondheimer, E. and Kertesz, Z. I.,** The anthocyanin of strawberries, *J. Am. Chem. Soc.,* 70, 3476, 1948.

199. **Lukton, A., Chichester, C. O., and Mackinney, G.,** Characterisation of a second pigment in strawberries *(Fragaria vesca), Nature (London),* 176, 790, 1955.

200. **Sondheimer, E. and Karash, C. B.,** The major anthocyanin pigments of the wild strawberry, *Nature (London),* 178, 648, 1956.

201. **Lamort, C.,** Chromatographie, spectres d'absorption et dégradation d'anthocyanines de fruits, *Rev. Ferment Ind. Aliment. Belg.,* 13, 153, 1958.

202. **Co, H. and Markakis, P.,** Flavonoid compounds in the strawberry fruit, *J. Food Sci.,* 33, 281, 1968.

203. **Poei-Langston, M. S. and Wrolstad, R. E.,** Color degradation in an abscorbic acid-anthocyanin-flavonol model system, *J. Food Sci.,* 46, 1218, 1981.

204. **Wrolstad, R. E., Hildrum, K. I., and Amos, J. F.,** Characterization of an additional anthocyanin pigment in extracts of strawberries, *J. Chromatogr.,* 50, 311, 1970.

205. **Fuleki, T.,** The anthocyanins of strawberry, rhubarb, radish and onion, *J. Food Sci.,* 34, 365, 1969.

206. **Polesello, A. and Rampilli, M.,** Ricerche sul comportamento e la stabilita' dei pigmenti delle fragole. Nota II. Rilievi sulla composizione antocianica prima e dopo conservazione allo stato congelato, *Sci. Tecnol. Alimenti,* 2, 357, 1972 (CA 85, 19291).

207. **Mosorinski, N.,** Changes in anthocyanin content and degradation index in strawberry and blueberry juice during storage at different temperatures, *Hrana Ishrana,* 16, 187, 1975 (CA 83, 191553).

208. **Markakis, P., Livingston, G. E., and Fellers, C. R.,** Quantitative aspects of strawberry pigment degradation, *Food Res.,* 22, 117, 1957.

209. **Pratt, D. E., Balckom, C. M., Powers, J. J., and Mills, L. W.,** Interaction of ascorbic acid, riboflavin, and anthocyanin pigments, *J. Agric. Food Chem.,* 2, 367, 1954.

210. **Haginuma, S.,** Utilization of strawberries. II. Decoloration of strawberry juice, *Nosan Kako Gijutsu Kenkyukai-Shi,* 8, 189, 1961 (CA 59, 13274b).

211. **Chen, W. P. and Wrolstad, R. E.,** A note on the influence of acetaldehyde on colour of strawberry juice, *J. Sci. Food Agric.,* 31, 667, 1980.

212. **Lukton, A., Chichester, C. O., and Mackinney, G.,** The breakdown of strawberry anthocyanin pigment, *Food Technol.,* 10, 427, 1956.

213. **Meschter, E. E.,** Effects of carbohydrates and other factors on strawberry products, *J. Agric. Food Chem.,* 1, 574, 1953.

214. **Tinsley, I. J. and Bockian, A. H.,** Some effects of sugar on the breakdown of pelargonidin 3-glucoside in model system at 90°C, *Food Res.,* 25, 161, 1960.
215. **Sondheimer, E. and Lee, F. A.,** Color changes of strawberry anthocyanin with D-glucose, *Science,* 109, 331, 1949.
216. **Wrolstad, R. E., Skrede, G., Lea, P., and Enersen, G.,** Influence of sugar on anthocyanin pigment stability in frozen strawberries, *J. Food Sci.,* 55, 1064, 1990.
217. **Wesche-Ebeling, P. and Montgomery, N. W.,** Strawberry polyphenoxidase: its role in anthocyanin degradation, *J. Food Sci.,* 55, 731, 1990.
218. **Wrolstad, R. E., Lee, D. D., and Poei, M. S.,** Effect of microwave blanching on the color and composition of strawberry concentrate, *J. Food Sci.,* 45, 1573, 1980.
219. **Flores, J. H. and Heatherbell, D. A.,** Optimizing enzyme and pre-press mash treatment for juice and color extraction from strawberries, *Fluess. Obst.,* 51, 320, 327, 1984 (CA 101, 169350).
220. **Rwbahizi, S. and Wrolstad, R. E.,** Effects of mold contamination and ultrafiltration on the color stability of strawberry juice and concentrate, *J. Food Sci.,* 53, 857, 1988.
221. **Decareau, R. V., Livingston, G. E., and Fellers, C. R.,** Color changes in strawberry jellies, *Food Technol.,* 10, 125, 1956.
222. **Abers, J. E. and Wrolstad, R. E.,** Causative factors of color deterioration in strawberry preserves during processing and storage, *J. Food Sci.,* 44, 75, 1979.
223. **Sistrunk, W. A. and Cash, J. N.,** The effect of certain chemicals on the color and polysaccharides of strawberry puree, *Food Technol.,* 24, 473, 1970.
224. **Wrolstad, R. E. and Erlandson, J. A.,** Effect of metal ions on the color of strawberry puree, *J. Food Sci.,* 38, 460, 1973.
225. **Spayd, S. E. and Morris, J. R.,** Influence of immature fruits on strawberry jam quality and storage stability, *J. Food Sci.,* 46, 414, 1981.
226. **Austin, M. E., Shutak, V. G., and Christopher, E. P.,** Colour changes in harvested strawberry fruits, *Proc. Am. Soc. Hortic. Sci.,* 75, 382, 1960.
227. **Landfald, R.,** Effect of storage temperature and ripeness at harvest on keeping quality of strawberries (*Fragaria × ananassa*), *Meld. Nor. Landbrukshoegsk.,* 62, 10, 1983 (FSTA 16, 8J1280).
228. **Kolesnik, A. A., Shirkanova, V. G., and Efimova, L. E.,** Vitamin C and color change in strawberries during storage, *Tr. Vses. Semin. Biol. Aktiv. (Lech.) Veshchestvam, Plodov Yagod,* 3, 406, 1966 (CA 74, 21962).
229. **Kyzlink, V. and Vit, V.,** Formation and changes of anthocyanin pigments in frozen strawberry pulp, *Sb. Vys. Sk. Chem. Technol. Praze, Potraviny,* E 30, 85, 1971 (CA 76, 152288).
230. **Wrolstad, R. E., Putnam, T. P., and Varseveld, G. W.,** Color quality of frozen strawberries: effects of anthocyanin, pH, total acidity and ascorbic acid variability, *J. Food Sci.,* 35, 448, 1970.
231. **Carballido, A. and Rubio, L. M. J.,** Use of lyophilization for the preservation of strawberries, *An. Bromatol.,* 22, 229, 1970 (CA 75, 74991).
232. **Zhukhova, L. A.,** Changes in anthocyanins of strawberries and raspberries after sublimation drying, *Prikl. Biokhim. Mikrobiol.,* 7, 593, 1971 (CA 75, 139559).
233. **Erlandson, J. A. and Wrolstad, R. E.,** Degradation of anthocyanins at limited water concentration, *J. Food Sci.,* 37, 592, 1972.
234. **Petrova, K. and Peeva, S.,** Farbveraenderung von Erdbeerehalbfabrikaten für die Cocktail Produktion, *Nauch. Tr., Inst. Vinarska Prom., Sofia,* 15, 123, 1977.
235. **Ariuchi, N.,** Studies on the stability of anthocyanins in fruit liqueurs. II. Effect of citric acid and sucrose on stability of anthocyanin pigments in strawberry liqueur, *Tokushima Bunri Daigaku Kankyu Kiyo,* 13, 31, 1975 (CA 89, 4538).
236. **Kosacheva, V. V.,** Change of color of alcoholic strawberry juice, *Spirt. Prom.,* 29, 21, 1963 (CA 60, 4702).

237. **Pilando, L. S., Wrolstad, R. E., and Heatherbell, D. A.,** Influence of fruit compo-
     sition, maturity and mold contamination on the color and appearance of strawberry wine,
     J. Food Sci., 50, 1121, 1985.

238. **Adams, J. B.,** Changes in the polyphenols of red fruits during heat processing, the
     kinetics and mechanism of anthocyanin degradation, Tech. Bull. 22, Campden Food
     Preservation Research Association, Chipping Campden, U.K., 1972.

239. **Adams, J. B. and Ongley, M. H.,** The degradation of anthocyanins in canned straw-
     berries. I. The effect of various processing parameters on retention of pelargonidin 3-
     glucoside, *J. Food Technol.,* 8, 139, 1973.

240. **Adams, J. B. and Ongley, M. H.,** The degradation of anthocyanins in canned straw-
     berries. II. The effect of various additives on the retention of pelargonidin 3-glucoside,
     J. Food Technol., 8, 305, 1973.

241. **Blom, H.,** A method for measuring anthocyanin degradation, *LWT (Progr. Food Eng.),*
     7, 587, 1983.

242. **Blom, H. and Enersen, G.,** What happens to the colour during jam making?, *NINF
     Inf.,* 5, 369, 1983 (FSTA 15, 11J1924).

243. **Blom, H. and Thomassen, M. S.,** Kinetic studies on strawberry anthocyanin hydrolysis
     by a termostable anthocyanin β-glycosidase from *Aspergillus niger, Food Chem.,* 17,
     157, 1985.

244. **Salunkhe, D. K., Gerber, R. K., and Pollard, L. H.,** Physiological and chemical
     effects of gamma radiation on certain fruits, vegetables and their products, *Proc. Am.
     Soc. Hortic. Sci.,* 74, 423, 1959.

245. **Markakis, P., Livingston, G. E., and Fagerson, I. S.,** Effects of cathode ray and
     gamma ray irradiation on the anthocyanin pigments of strawberries, *Food Res.,* 24, 520,
     1959.

246. **Cooper, G. M. and Salunkhe, D. W.,** Effect of gamma-radiation, chemical and pack-
     aging treatments on refrigerated life of strawberries and sweet cherries, *Food Technol.,*
     17, 123, 1963.

247. **Wells, C. E., Tichenor, D. A., and Martin, D. C.,** Ascorbic acid retention and colour
     of strawberries as related to low irradiation and storage time, *Food Technol.,* 17, 77,
     1963.

248. **Johnson, C. F., Maxie, E. C., and Elbert, E. M.,** Physical and sensory tests on fresh
     strawberries subjected to gamma irradiation, *Food Technol.,* 19, 419, 1965.

249. **Herregods, M. and De Proost, M.,** L'effet de l'irradiation gamma sur la conservation
     des fraises, *Food Irradiation,* 4, A38, 1963.

250. **Mahmoud, A. A., Hegazy, R. A., Hussein, M. A., Roushdy, H. M., and Doma,
     M. B.,** Changes in the chemical constituents of ripe strawberry subjected to gamma
     irradiation, *Isot. Radiat. Res.,* 16, 63, 1984 (CA 102, 4660).

251. **Kalinov, V. and van Kooy, J.,** Influence of gamma radiation on the total anthocyanin
     content in strawberries, *Nauch. Tr., Nauchnoissled. Inst. Konserv. Prom., Plovdiv,* 10,
     55, 1973 (CA 81, 24336).

252. **Willstätter, R. and Zollinger, E. H.,** Untersuchung ueber die Anthocyane. XVI. Ueber
     die Farbstoffe der Preiselbeere, *Justus Liebigs Ann. Chem.,* 412, 195, 1915.

253. **Karrer, P. and Widmer, R.,** Untersuchungen über Pflanzenfarbstoffe. I. Über die
     Konstitution einiger Anthocyanidine, *Helv. Chim. Acta,* 10, 5, 1927.

254. **Casoli, U., Cultrera, R., and Dall'Aglio, G.,** Ricerche sugli antociani di fragola e di
     mirtillo, *Ind. Conserve,* 42, 11, 1967.

255. **Bombardelli, E., Bonati, A., Gabetta, B., Martinelli, E. M., Mustich, G., and
     Danieli, B.,** Gas-liquid chromatographic and mass spectrometric identification of antho-
     cyanidins, *J. Chromatogr.,* 120, 115, 1976.

256. **Baj, A., Bombardelli, E., Gabetta, B., and Martinelli, E. M.,** Qualitative and quan-
     titative evaluation of *Vaccinium myrtillus* anthocyanins by high-resolution gas chroma-
     tography and high-performance liquid chromatography, *J. Chromatogr.,* 279, 365, 1983.

257. **Senchuk, G. V. and Borukh, I. F.**, Wild berries of Belorussia, *Rastit. Resur.*, 12, 113, 1976 (CA 84, 132626).
258. **Bounous, G., Albasini, A., Melegari, M., Rinaldi, M., and Pasquero, E.**, Anthocyanin and anthocyanin content in berries of *Vaccinium myrtillus* L. in the Piedmont Alps, *Coll. INRA*, 4, 125, 1981 (CA 98, 50439).
259. **Brenneisen, R. and Steinegger, E.**, Quantitative comparison of polyphenols in *Vaccinium myrtillus* L. fruits at different stages of ripeness, *Pharm. Acta Helv.*, 56, 341, 1981.
260. **Ballington, J. R., Ballinger, W. E., Maness, E. P., and Luby, J. J.**, Anthocyanin, aglycone, and aglycone-sugar content in the fruits of five species of *Vaccinium* section *myrtillus*, *Can. J. Plant Sci.*, 68, 241, 1987.
261. **Melegari, M. and Albasini, A.**, Qualitative characteristics of fruits from *Vaccinium* communities of the Modenese Appennines, with particular reference to black bilberry (*Vaccinium myrtillus* L.), *Atti Soc. Nat. Mat. Modena*, 119, 31, 1988 (CA 113, 229792).
262. **Azar, M., Verette, E., and Brun, S.**, Identification of some phenolic compounds in bilberry juice *Vaccinium myrtillus*, *J. Food Sci.*, 52, 1255, 1987.
263. **Brenneisen, R. and Steinegger, E.**, Zur Analityk der Polyphenole der Fruechte von *Vaccinium myrtillus* L. (Ericaceae), *Pharm. Acta Helv.*, 56, 180, 1981 (HA 1982, 680).
264. **Friedrich, H. and Schonert, J.**, Untersuchungen über Einige Inhaltstoffe der Blatter und Früchte von *Vaccinium myrtillus*, *Planta Med.*, 24, 343, 1979.
265. **Martinelli, E. M. and Bombardelli, A.**, Computer-aided evaluation of liquid-chromatographic profiles for anthocyanins in *Vaccinium myrtillus* fruits, *Anal. Chim. Acta*, 191, 275, 1986.
266. **Albasini, A.**, *Vaccinium myrtillus* L., una ricerca interdisciplinare nell'alto Appennino modenese, in *Lampone, Mirtillo ed Altri Piccoli Frutti*, Atti, Trento, Italy, 4–5 June 1987, 187.
267. **Pourrat, H., Pourrat, A., Lamaison, J. L., and Perrin, B.**, Enzymic extraction of anthocyanin glycosides, *Bull. Liaison Groupe Polyphénols*, 4, 6, 1973.
268. **Azar, M., Verette, E., and Brun, S.**, Comparative study of fresh and fermented bilberry juices — state and modification of the coloring pigments, *J. Food Sci.*, 55, 164, 1990.
269. **Willstatter, R. and Mallison, H.**, Untersuchung über die Anthocyane. III. Über den Farbstoffe der Preiselbeere, *Justus Liebigs Ann. Chem.*, 408, 15, 1915.
270. **Troyan, A. V. and Borukh, I. F.**, Biologically active substances of wild-growing berries of the Carpathians, *Tr. Vses. Semin. Biol. Aktiv. (Lech.) Veshchestvam, Plodov Yagod*, 1966, 244 (CA 73, 119394).
271. **Andersen, O. M.**, Chromatographic separation of anthocyanins in cowberry (lingonberry) *Vaccinium vitis-idaea* L., *J. Food Sci.*, 50, 1230, 1985.
272. **Bandaztiene, Z. and Butkus, V.**, Biological and biochemical characteristics of cowberry. IV. Amount of some organic substances in leaves and berries, *Liet. TSR Mokslu Akad. Darb., Ser. C.*, 3, 13, 1975 (CA 84, 14629).
273. **Grigorova, S., Iocheva, E., and Katsarova, S.**, Amino acid composition and aroma compounds and polyphenols of wild mountain cranberry and whortleberry, *Gradinar. Lozar. Nauka*, 21, 32, 1984 (CA 102, 109894).
274. **Hayashi, K.**, Anthocyanins. XV. Uliginosin, a new dye from the berries of *Vaccinium uliginosum*, *Acta Phytochim.*, 15, 35, 1949 (CA 43, 8450).
275. **Hayashi, K.**, Anthocyanins. XVII. Further studies on uliginosin, *Acta Phytochim.*, 15, 45, 1949 (CA 43, 8450).
276. **Andersen, O. M.**, Anthocyanins in fruits of *Vaccinium uliginosum* L. (bog whortleberry), *J. Food Sci.*, 52, 665, 1987.
277. **Butkus, V. F., Butkene, Z. P., and Tamulis, T. P.**, Biological and biochemical characteristics of bog bilberry. VIII. Contents and dynamics of anthocyanins, leucoanthocyanins and amino acids in berries, *Liet. TSR Mokslu Akad. Darb., C, Ser. B*, 321, 1989 (CA 112, 73895).

278. **Larsson, E. G. K.**, *Rubus arcticus* L. subsp. *stellarcticus*, a new raspberry, *Acta Hortic.*, 112, 143, 1980.

279. **Willstaedt, H.**, The pigment of the berry *Rubus chamaemorus* L., *Scand. Arch. Physiol.*, 75, 155, 1936.

280. **Keränen, A. J. A. and Suomalainen, H.**, Anthocyanins of arctic bramble, *Rubus arcticus* L., *Suom. Kemistil. B*, 33, 155, 1960 (CA 55, 10590e).

281. **Luh, B. S., Kamber, P. J., and Hasan, A. U.**, Stabilitaet von Saccharose und Anthocyanin in Boysenbeeresyrup, *Fruchtsaft-Ind.*, 6, 292, 1961 (CA 56, 15900a).

282. **Dirdjokusumo, S. and Luh, B. S.**, Packaging of foods in laminate and aluminum foil combination pouches. II. Boysenberry puree, *Food Technol.*, 19, 120, 1965.

283. **Luh, B. S., Stachowicz, K., and Hsia, C. L.**, The anthocyanin pigments of boysenberries, *J. Food Sci.*, 30, 300, 1965.

284. **Monro, J. A. and Sedcole, J. R.**, Boysenberries, relationship between properties of developing fruit, *N.Z. J. Exp. Agric.*, 14, 327, 1986 (CA 106, 135398).

285. **Monro, J. A. and Lee, J.**, Changes in elements, pectic substances and organic acids during development of boysenberry fruit, *J. Sci. Food Agric.*, 38, 195, 1987.

286. **Given, N. K., Given, H. M., and Pringle, R. M.**, Boysenberries, preliminary studies on effect of ripening on composition, *N.Z. J. Exp. Agric.*, 14, 319, 1986 (FSTA 19, 7J78).

287. **Given, N. K. and Pringle, R. M.**, Preliminary studies on mechanical harvesting of boysenberries, *N.Z. J. Exp. Agric.*, 13, 169, 1985 (FSTA 18, 3J83).

288. **Scortichini, M.**, *Frutti da Riscoprire*, Edizioni Agricole, Bologna, Italy, 1990, 20.

289. **Martin, T., Vilaescusa, L., De Sotto, M., Lucia, A., and Diaz, A. L.**, Determination of anthocyanin pigments in *Myrtus communis* berries, *Fitoterapia*, 61, 85, 1990.

290. **Diaz, A. M. and Martin, T.**, Malvidin 3,5 diglucoside from the berries of *Myrtus communis*, *Fitoterapia*, 60, 282, 1989.

291. **Diaz, A. M. and Martin, T.**, Estudio de los pigmentos antocianicos de las bayas de *Myrtus communis* L., *Int. Congr. Cien. Farm.*, 1987, 2348.

292. **Diaz, A. M. and Abeger, A.**, *Myrtus communis*, composicion quimica y actividad biologica de sus extractos, una revision, *Fitoterapia*, 58, 167, 1987.

293. **Lowry, J. B.**, Anthocyanins of the Melastomataceae, Myrtaceae and some allied families, *Phytochemistry*, 15, 511, 1976.

Chapter 5

TROPICAL FRUITS

I. ACEROLA

Acerola, also known as West Indian cherry and Barbados cherry, is the fruit of *Malpighia marginata* (previously *M. punicifolia* L.). It weighs about 4 g, is green when immature, and changes to yellow-orange, and dark red when fully ripe.[1,2] It is a rich source of vitamin C, providing about 1.3 mg of ascorbic acid per kilogram of fresh fruit.[2-4]

Jackson[5] reported a range of 0.29 to 5.07 g of ascorbic acid per kilogram of fruit of different species of *Malpighia*. The ascorbic acid content also depends on the stage of fruit maturity. The partially mature fruit has more ascorbic acid than when fully ripe.[6] Acerola fruit is produced commercially, primarily in Puerto Rico, and processed into juice. The juice may be used as an ascorbic acid supplement and to modify the color of passion fruit, guava, and pineapple juices.

The anthocyanin responsible for the red color of the fruit and juice of acerola is malvidin 3,5-diglucoside.[1-7] The presence of anthocyanin and ascorbic acid in the juice results in a decrease in the color and nutritional quality of the juice during processing and storage.[8,9] The exact nature of the reaction(s) responsible for the loss of color in the presence of ascorbic acid is still not well understood. It is known, however, that autooxidation of ascorbic acid yields H_2O_2 which mediates the degradation of the flavylium cation to colorless species.[10]

II. BANANA

Almost all the edible bananas originate from one or both of two species of the *Eumusa* section of the genus *Musa*: *M. acuminata* and *M. balbisiana*.[11-13] The different cultivars of banana are classified according to the genomic constitution and ploidy level: diploid *acuminata* cultivars AAcv (cv. to differentiate from AAw, wild forms of *M. acuminata*); triploid and tetraploid *acuminata* cultivars AAA, AAAA; and triploid and tetraploid hybrid cultivars AAB, ABB, AAAB, AABB, ABBB.

In all bananas, the anthocyanins are always accompanied by leucoanthocyanins.[14] The distribution of the anthocyanins in the bracts, which protect the sterile male flowers, is characteristic of *Musa* and has taxonomic value. Genetic factors regulate the hydroxylation and methylation of the anthocyanins. The lack of anthocyanin pigmentation, corresponding to a yellow coloration of the bracts, appears to be due to the presence of a recessive gene. A low level of methylation of the anthocyanins is characteristic of the edible varieties.[15] Simmonds[16] reported glucosides of pelargonidin and cyanidin in *M. coccinea*, and glucosides of cyanidin and delphinidin in *M. laterita, M. balbisiana*, and *M. velutina*. In addition, methylated anthocyanin glucosides of peonidin, petunidin, and malvidin were also found in *M. acuminata*. This anthocyanic arrangement was also correlated to the geographic distribution of the species.[15,16] From the presence/absence of methylated anthocyanins and/or ratio between delphinidin and cyanidin, Horry and Jay[12,13] classified 59 varieties of *M. acuminata* and *M. balbisiana* in five main chemotypes. Their findings also suggest the occurrence of two centers of domestication of *M. acuminata* cultivars in southeast Asia. One of these groups, containing primarily simple cyanidin derivatives, is essentially restricted to *M. acuminata* ssp. *zebrina*. The other four chemotypes contain rutinosides of cyanidin and delphinidin in different ratios, groups 3 and 4; and rutinosides of cyanidin, malvidin, and peonidin, groups 1 and 2.

Harborne[17] identified 3-glucosides and 3-rutinosides of cyanidin and delphinidin in *M. velutina*. The best-known edible banana, the yellow *M. sapientum*, lacks anthocyanins. However, anthocyanins formed in these bananas after exposure to UV rays have been identified as 3-deoxyanthocyanins.[18] Those in other parts of the banana have received limited attention. It is, however, known that the red color in the seeds of *M. acuminata* is due to pelargonidin,[19] while that in the skin of certain banana fruit is due to peonidin and malvidin.[17]

III. CACAO

The fruit of the cacao tree (*Theobroma cacao* L., family Sterculiaceae) is a pod which contains 30 to 60 seeds immersed in a mucilaginous pulp.[20,21] Chocolate is obtained from the seeds of cacao. There are two types of cacao: 'Criollo' and 'Forastero' or 'Trinatario'. 'Forastero' has purple beans in smooth, rounded pods, while 'Criollo' has unpigmented beans in a rough, yellow pod with pointed tips. The Forastero variety is, by far, the most economic type, accounting for over 90% of the world production.[20]

The nature of the purple substance in cacao was attributed to cyanidin by Fincke[21] and Knapp[22] but it was Lawrence et al.[23] who found that it had reactions characteristic of cyanidin 3-glucoside. Birch,[24] however, was not able to isolate the cyanidin because of its instability. Fincke[25,26] attempted to define the role of anthocyanins in cacao, and described a procedure for their determination.[27] Forsyth[28-30] fractionated the cacao polyphenolics on a cellulose column and followed the changes in these compounds during the fermentation process. Five groups were characterized: catechins (5%), epicatechins (5%), anthocyanins (about 3%), procyanidins (32%), and leucoanthocyanins (15%). The anthocyanins were identified as cyanidin 3-galactoside and cyanidin 3-arabinoside.[31] These results were later confirmed by Schubiger et al.[32] Kharlamova[33] assigned the role of precursor of the aromas to catechin, epicatechin, and gallocatechin. The phenolic composition of hybrid cacao beans from the Ivory Coast is so different from those reported in the literature that Cros[34,35] hypothesized a biochemical differentiation of the cacao beans.

The anthocyanin composition in other parts of the cacao plant (leaves, flowers, pods) is similar to that of the fruit.[36,37] Jacquemin[38] reported that the anthocyanins in the cacao leaves are suitable for the botanical classification of the species, and Pence et al.[39] studied their formation *in vitro* in cotyledons grown in a sugar-rich medium. Kononovicz and Janick[40] observed an increased production of anthocyanins, alkaloids, and lipids in *T. cacao* embryos *in vitro* when placed in a culture medium with a high sugar content.

The anthocyanins of 'Forastero' cacao beans are located in special pigmented cells.[41] Three types of cells were identified: long, brown epidermal cells (2 to 4 μm); pigmented cells containing anthocyanins, polyhydroxyphenols, and purines; and reserve cells containing cocoa butter, starch, and aleurone granules. The polyphenols containing cells of purple beans are structurally different from those of brown beans.[42] According to Brown,[43] reserve cells isolated from dried cotyledons contain theobromine and caffeine; and their content, on a dry weight basis, accounts for 10 to 13% of the total cotyledon tissue. There is no correlation between seed color and levels of caffeine and theobromine in 'Criollo' beans. This indicates that the production of anthocyanins in this variety of cacao is not related to the biosynthesis of alkaloids.[44] Several authors[45-50] have studied the effect of anthocyanins and other phenolic compounds on the flavor of processed cocoa. Kharlamova[49] pointed out that a high content of anthocyanin is generally responsible for the acidic and bitter tastes in chocolate. Lange[50] examined the nutritional aspect of polyphenols in cacao. Others have used anthocyanin content of the beans as an index of quality, and to follow and optimize processing of cacao beans.[51-55] Rohan[51,52] classified cacao bean quality on the basis of anthocyanin content. This system has since become the official method of analysis of the International Office of Cacao and Chocolate (OICC).[53] The external color of the cotyledon is related to fermentation. Fermented beans with less than 10% of the original anthocyanin content possess optimal quality.[54,55] The analytical and sensory procedures currently available for color assessment include the colorimetric method of OICC[53] (with its variations and simplifications[56,57]), the recently published HPLC method of Pettipher,[58] and the classical visual evaluation with its well-known limitations of low precision and low accuracy.

In addition to anthocyanins, cacao beans contain a polyflavone glucoside pigment that appears to be a promising natural red food colorant.[59-63] Enzymatic,[59,60] physicochemical,[61] and chemical[62] methods have been used to obtain this colorant from the by-products of cocoa processing. The pigment has a very low toxicity level[63] and outstanding antioxidant activity.[64] An in-depth review of the biochemistry and technology of cacao has been published by Minifie.[20]

IV. COFFEE

The coffee plant, *Coffea* spp. (Rubiaceae), originated in Ethiopia and spread through Arabia into many tropical regions of the world, where it is grown on a large scale. The plant has white flowers that produce a red or crimson fruit about the size of a small cherry. Normally, each fruit contains two slightly convex seeds covered by a membrane. Like a cherry, the fruit has a pulp that surrounds the two seeds and a skin covering.[65,66] After roasting, the seeds are used in the preparation of the well-known drink, coffee. Only 3 of the 70 species of coffee are cultivated: *Coffea arabica*, which provides 75% of the world's production; *C. canephora*, about 25%; and *C. liberica* and others, less than 1%.[66]

Tschirsch[67] was the first to investigate the nature of the pigments of coffee beans. Kroemer[68] noted that these pigments were anthocyanins and were located in the epidermal layer of the fruit. Sotero[69] observed that unroasted seeds also contained anthocyanins. Sondahl et al.[70] reported that Ethrel improves the general quality and favors fruit development with a uniform anthocyanin pigmentation.

Coffea arabica has cultivars with red fruit, Caturra Vermelho and Bourbon Vermelho; with yellow fruits, Caturra Amarelo and Bourbon Amarelo; and cultivars with pigmented leaves, *purpurascens* Cramer. Cultivar effects on anthocyanins and other phenolic compounds of *Coffea arabica* are summarized in Table 1. The beans from the red-fruited cultivars contain anthocyanins. However, while Bourbon beans contain two cyanidins, the Caturra beans contain two pelargonidins and one peonidin. The yellow fruit cultivars are characterized by the presence of leucoanthocyanins and flavonols and the lack of anthocyanins. The Caturra cultivars differ from the Bourbon cultivars by showing different anthocyanin, flavone, and phenolic acid profiles. According to Lopes et al.,[71] the yellow-fruited cultivars also contain a higher concentration of flavonols than the red-fruited cultivars. Cyanidin 3-glucoside and 3-rutinoside were also found in the pulp of an unspecified cultivar of *Coffea arabica* by Zualaga and Tabacchi,[72,73] and a phenolic acid profile similar to that shown of Table 1 has recently been reported.[74] Until 1870, only the variety of *C. arabica* with its mature red fruit was known. A mutant with yellow flowers was then discovered which is now widely cultivated in the Sao Paulo region of Brazil. This form, known as *C. arabica* var. *typica* Cramer *forma xanthocarpa*, contains no anthocyanins in the fruit.[77] Lopes and Monaco[78] found 75 different flavonoids in the genus *Coffea* of which only 5 were common to all species.

V. DATE

The date (*Phoenix dactylifera* L.) is one of the oldest cultivated plants, probably originating in western Asia and North Africa. It is commercially

TABLE 1
Anthocyanins and Other Phenolics of Four Cultivars of *Coffea arabica*

Component	Cultivar			
	Bourbon Vermelho	Bourbon Amarelo	Caturra Vermelho	Caturra Amarelo
Cyanidin 3-glycoside	+	−	−	−
Cyanidin 3,5-diglycoside	+	−	−	−
Pelargonidin 3,7-diglycoside	−	−	+	−
Pelargonidin 3-trioside	−	−	+	−
Peonidin 3-5-diglycoside	−	−	+	−
Leucocyanidin	−	+	−	+
Quercetin 3-glycoside	+	+	+	+
Quercetin 3-diglycoside	−	−	+	+
Quercetin 3-trioside	+	+	−	−
Kaempferol 3-triosiode	+	+	+	+
Myricetin 3-glycoside	+	+	+	+
Luteolin ?-glycoside	−	+	−	−
Apigenin 7-glycoside	−	−	+	+
Chlorogenic acid(s)	+	+	+	+
Caffeic acid	+	+	+	+
Ferulic acid	+	+	+	+
Vanillic acid	−	−	+	+
Protocatechuic acid	−	−	+	+

+, Present; −, absent.

Adapted from Lopes, C. R., Musche, R., and Hec, M., *Rev. Bras. Genet.*, 7, 657, 1984.

grown for its fruit in Iraq, Saudi Arabia, North Africa, and in the southwest United States.[79] Saudi Arabia is the leading producer with over 400 different cultivars and a yearly production of over 350,000 ton.[80] The fruit is an oblong 3 to 8 cm long berry with a thick, sweet yellow or red flesh, and a slender, often pointed seed. The Deglet Noor variety of California is red.

Ragab et al.[81] reported that the color intensity and the ratio of reducing/nonreducing sugars increase with ripening of fruit, while the tannins continuously decrease. Color is a valid index of fruit ripeness. The red color of the fruit is due to polyphenolic constituents (Table 2). Tannins are the most complex polyphenols of dates; their presence has been recognized for many years.[81] Maier and Metzler[82-83] characterized the tannins of Deglet Noor dates as both water-soluble and water-insoluble condensed tannins of the leucoanthocyanidin type. The water-soluble tannins gave color tests and absorption maxima typical of 5-, 7-, 3'-, 4'-tetrahydroxy flavans. When treated with hot acid in butanol, the major anthocyanidin produced was cyanidin chloride. The water-insoluble tannins also produced cyanidin chloride under these con-

TABLE 2
Polyphenolic Constituents of Date Fruit and Leaves

Constituent	Tissue
Soluble proanthocyanidin tannin	Fruit, leaves
Insoluble proanthocyanidin tannin	Fruit, leaves
(+)-Catechin	Fruit, leaves
(−)-Epicatechin	Fruit, leaves
3-*O*-Caffeoylshikimic acid	Fruit, leaves
4-*O*-Caffeoylshikimic acid	Fruit, leaves
5-*O*-Caffeoylshikimic acid	Fruit, leaves
Quercetin-3—glucoside	Fruit, leaves
Isorhamnetin-3-glucoside	Fruit, leaves
Luteolin 7-glucoside	Leaves
Luteolin 7-rutinoside	Leaves
Glycosylapigenin	Leaves
Luteolin	Leaves
Tricin	Leaves
Isorhamnetin-3-rutinoside	Leaves
Quercetin 3-rutinoside	Leaves

Adapted from Vandercook, C. G., Hosegana, H., and Maier, V. P., *Tropical and Subtropical Fruits*, Nagy, S. and Shaw, P. E., Eds., AVI Publishing, Westport, CT, 1980, 506.

ditions. These are characteristic leucoanthocyanidin reactions which indicate the presence of flavan 3-ol oligomers. The major ethyl acetate-soluble polyphenols in green dates are the dactylifric acids, the 3-position isomers of monocaffeoylshikimic acid. The presence of these compounds in Deglet Noor dates was reported by Maier.[84] Maier et al.[85] established the structures of dactylifric acid, isodactylifric acid, and neodactylifric acid to be 3-*O*-, 4-*O*-, and 5-*O*-caffeoylshikimic acid, respectively.

Harborne et al.[86] found 3-*O*-caffeoylshikimic to be a major phenol in many palm flowers, often occurring in greater amounts than other flavonoids. Maier and Metzler[87] isolated from immature date fruits a quercetin 3-glucoside and isorhamnetin 3-glucoside; and Williams et al.[88] found luteolin 7-glucoside, luteolin 7-rutinoside, and glycosylapigenin in date leaves. Subsequently, Williams et al.[89] reported the presence of negatively charged flavonoids, flavone C-glycosides, and leucoanthocyanidins. The flavone aglycones luteolin and tricin were identified after acid hydrolysis of tissue extracts.[89]

VI. LYCHEE

The lychee or litchi, *Litchi chinensis* Sonn., syn. *Nephelium litchi*, is a fruit of Chinese origin. It is 4 to 5 cm long and contains a large seed within a whitish, gelatinous, edible pulp. The fruit is covered by a peeling ranging in color from pink to wine-red. At present, lychee is grown commercially in many countries including South Africa, Australia, Japan and the U.S. (Florida and Hawaii).[90] Species with fruit characteristics similar to the lychee include longan (*Euphaia langoma* Lam.), rambutan (*Nephelium lappaceum* L.), mangosteen (*Garcinia mangostana* L.), and durian (*Durio zibethinus* L.).

Prasad and Jha,[91] using TLC, reported that the red pigmentation of the lychee was due to cyanidin 3-glucoside, cyanidin 3-galactoside, pelargonidin 3-rhamnoside, and pelargonidin 3,5-diglucoside. According to these authors, the cyanidin glycosides appear first and the glycosides of pelargonidin, at a later date.[91]

Recently, however, Lee and Wicker,[92] using HPLC and spectral analysis, characterized the anthocyanins of 'Brewster' lychee as cyanidin 3-rutinoside, cyanidin 3-glucoside, and malvidin 3-acetylglucoside. These investigators also found that cyanidin 3-rutinoside was the major pigment and comprised over 67% of the total anthocyanin in lychee skin extracts. Malvidin 3-acetylglucoside was second in quantity at 14.7%, and cyanidin 3-glucoside was less than 10%. During refrigerated storage in the dark, the total and individual anthocyanins decrease considerably (Figure 1).[92] The concentration of the major pigment, cyanidin 3-rutinoside, in 'Brewster' lychee is approximately 1.2 mg/g fresh weight (FW) on the second day after harvest and decreases to 0.9 mg/g after 15 d of storage (Figure 1). After 48 d, the content of this pigment decreases to 0.27 mg/g, a loss of more than 77%.[92] The level of total anthocyanin in 1.68 mg/g FW on the second day after harvest increases to 1.79 mg/g after 8 d and to 2.06 mg/g after 15 d of storage, and then decreases gradually thereafter. There are also flavonols, glucosides of kaempferol and quercetin, chalcones, a flavanone, an aurone, and condensed tannins in lychee fruit.[91,93,94] During cooking and canning of the fruit, the condensed tannins slowly break down, yielding catechin and leucoanthocyanin. During refrigerated storage, the olymeric pigments of lychee fruit increase, and after 48 d at 4°C they comprise over 50% of the lychee pigments.[92]

Anthocyanins are an excellent index of ripeness. They progressively increase during fruit development, reach a maximum concentration at the same time as the fruit reaches the maximum size, decrease with overripening, and disappear with senescence.[94,95] The relationship between skin color and maturity is, however, cultivar dependent. A close relationship between the degree of red skin and optimum eating quality has been reported for the red-skinned cultivars — Hei Ye, Zhuang Yuan Hong, and Yu Ke Bao fruit — with 50 to 70% red skin being optimum.[95]

Fruit of the lychee is prized in fresh, canned, and dried forms.[90] Fruit freshness, especially color, can be maintained if treated with 2-aminobutane or washed with 3% NaCl before drying.[96] Fruit quality can also be improved somewhat by spraying the fruit with putrescine during ripening.[97]

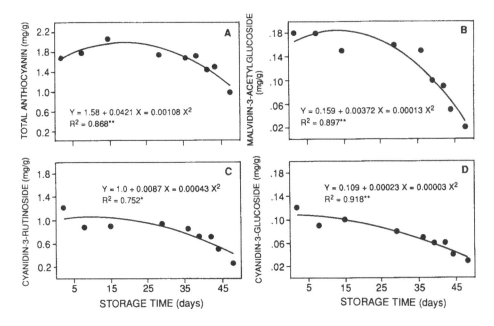

FIGURE 1. Changes in total anthocyanin (A), malvidin 3-acetylglucoside (B), cyanidin 3-rutinoside (C), and cyanidin 3-glucoside (D) contents of 'Brewster' lychee during storage. Regressions are significant at $p = 0.01$ (**) or at $p = 0.05$ (*). (From Lee, H. S. and Wicker, L., *Food Chem.*, 40, 263, 1991. With permission.)

VII. JAVA PLUM

Many tropical fruits of the Myrtaceae or Java plum family are used to make juices, preserves, and gelatins. Trevisan et al.[98] identified the sugars of *Myrciaria jaboticaba* as glucose, fructose, and sucrose; the organic acids as citric and oxalic; and the anthocyanin of the fruit skin as peonidin 3-glucoside. Bobbio and Scamparini[99] conducted an analytical study to identify the most important components of the fruit of *Eugenia jambolana* Lamarck. The sugars are glucose, fructose, and mannose; malic and citric acids are also present. The anthocyanin pigment in the fruit is cyanidin 3-glucoside. Sharma and Seshadri[100] separated two anthocyanin pigments from *E. jambolana* fruit and identified them as the glycosides of petunidin and malvidin. Jain and Seshadri[101] identified the heterosidic structure of these pigments as delphinidin and petunidin 3-gentiobioside and malvidin 3-laminaribioside, as well as petunidin 3-gentiobioside. Guimares et al.[102] studied the physicochemical characteristics of pitanga fruit or the Surinam cherry or *E. uniflora*. Benk[103] reported the characteristics and uses of jabotica *(Myrciaria caulifora)*. Lowry[104] identified cyanidin 3-glucoside in the fruit of *E. aquea*, while the flower of *E. malaccensis* is pigmented by malvin.

The fruit of jamun or duhat (*Sysyum cumini* Linn.) is extensively used in part of India and the Philippines to make juices and nectars. Khurdiya and Roy[105] studied its stability over time as well as its anthocyanin content. Martinez and Del Valle[106] extracted anthocyanins from the fruit and studied the possibility of its use as a natural food colorant.

In Myrtaceae there are two other important food species: the feijoa (*Feijoa sellowiana* Berg.), which has cyanidin 3-glucoside in its flower pigments; and guava (*Psidium guajavica* L.), whose fruit contains cyanidin 3-glucoside.

VIII. MANGO

The mango, *Mangifera indica* L., is considered to be the king of tropical fruits. It has been known as a cultivated species for 4 to 6 thousand years,[107] and is one of the most important commercial crops of the world. At present, world production of the mango fruit is about 14 million metric tons, with India producing about 70%. Other major producing countries include Pakistan, Brazil, the Philippines, and Mexico.

Fruit of the mango is a laterally compressed, fleshy drupe produced on long stalks in small clusters. The shape varies from round to ovate-oblong, with the length varying from 2 to 30 cm in different varieties and the weight from several grams to more than a kilogram. It contains three major types of pigments: chlorophyll, carotenoids, and anthocyanins.[108] Fruits of some cultivars are externally green throughout development (e.g., Bombay Green); others turn greenish-yellow (Mulgoa, Totapuri) or yellow (Alphonso) during development. Cultivars such as Haden, Irwin, and Tommy Atkins have a pronounced mixture of yellow or violet, with or without green color and with red flesh from fruit set to maturity.[108] The external color of the fruit is an important quality factor determining its maximum level of desirability between 13 and 15 weeks after fruit set. The most significant external color changes during ripening are the disappearance of chlorophyll, the increase of total and β-carotenoids, and the apparent increase of anthocyanin pigments.[108,109]

Anthocyanins may be found in the fruit and/or in the leaves of mangos.[110,111] The anthocyanin composition of the fruit is rather simple because a single anthocyanin, peonidin 3-galactoside, so far has been identified in the peel of the Haden mango.[110] This anthocyanin is, however, rather rare because it is found elsewhere only as a constituent of the red cranberry[110] and sugarcane.[112] Colored fruit surface area and intensity of red color were increased by spraying the fruit before harvest with 3 to 6% of the commercial antitranspirant preparation, Vapor Gard.[109] The anthocyanin composition of the leaves of mangos is much more complex than that of the fruit because there are glycosides of pelargonidin, delphinidin, peonidin, and/or cyanidin in varying concentrations depending upon the variety, age, and health of the plant.[111]

Other phenolic compounds reported in fruit of the mango include gallic acid, *m*-digallic acid, β-glucogallin, gallotannin, quercetin, isoquercitrin, mangiferin, and ellagic acid.[113,114] These compounds contribute to the flavor of the mango, as they do to most other tropical fruits.

IX. MANGOSTEEN AND KOKUM

The mangosteen (*Garcina mangostana* L.) is a southeast Asian fruit of considerable appeal. It is an attractively colored fruit with translucent pulp that is sugary, sweet, and aromatic.[115] The fruit is covered with a thick cortex rich in bitter substances, mostly tannins and xanthones.[115-117] Mangosteens are grown in Vietnam, southern Thailand, southeastern Burma, and Sri Lanka.[118] Du and Francis[119] identified the anthocyanins of the fruit as cyanidin 3-sophoroside (the most abundant) and as cyanidin 3-glucoside (the minor component). Total anthocyanin content in the skin, as measured in hydrochloric acid-methanol, is 92 mg/100 g.[120]

Red pigments are present in the skin of kokum fruit, *G. indica* Chois. (*G. purpurea* Roxb.). Cyanidin 3-glucoside and cyanidin 3-sambubioside, in a 1:4 ratio, make up 2.4% (d.w.b.) of the fruit rind or cortex.[121] In view of the high pigment content, the kokum could be a good source of water-soluble natural red color as well as a source of a yellow fat-soluble color, garcinol, for use in various foods.[121,122] Based on the stability in model systems of anthocyanin pigment from the kokum, Padival and Sastry[123] found promising uses for it in products such as preserves and beverages. Kokum rind is also a rich source of tannins (2.85% d.w.b.), pectin (5.71% d.w.b.), and hydroxycitric acid (22.8% d.w.b.).[121]

X. PASSION FRUIT

The edible fruited species of passion fruit, the purple passion fruit *(Passiflora edulis)* and the yellow passion fruit (*P. edulis* f. *flavicarpa*) are the two major varieties that are processed. *P. edulis,* a native of southern Brazil, has round or egg-shaped fruits that are 4 to 9 cm long, 3.5 to 7 cm in diameter, and deep purple when ripe. The shell is 3 to 6 mm thick and moderately hardy, with yellow to orange pulp that has a pleasant aroma.[124] *P. edulis* f. *flavicarpa* has a somewhat larger fruit (6 to 12 cm long and 4 to 7 cm in diameter). The shell of the fruit is hard and from 3 to 10 mm thick. The pulp is yellow to orange, highly aromatic, and juicy.[124] Fruit of both species is processed into a uniquely flavored juice. Passion fruit juice, however, has no anthocyanins.[125] The anthocyanin pigmentation in purple passion fruit is located primarily in the outer cortex which contains pelargonidin 3-diglucoside at a concentration of about 1.4 mg of pigment per 100 g of fresh material.[126]

Harborne[127] reported finding delphinidin 3-glucoside in *P. edulis* fruit. This report, however, was not confirmed in subsequent studies by Ishikura and Sugahara[128] who reported finding cyanidin 3-glucoside. The anthocyanin pigment can be removed from the passion fruit skin that remains after juice extraction and can be used as a natural colorant.[126]

The purple pigmentation in passion fruit flowers is controlled by a single gene which is dominant over green. In the fruit, two dominant genes govern color expression.[129]

Halim and Collins[130] identified four pigments in another edible variety, *P. quadrangularis*, known as giant granadilla because of the fruit size (200 to 400 g). They reported a complex and rich anthocyanin composition made up of cyanidin 3-glucoside, malvidin 3,5-diglucoside, petunidin 3,5-diglucoside, and delphinidin 3,5-diglucoside. The flowers contain 3-glucoside and 3,5-diglucosides of delphinidin, petunidin, and malvidin; cyanidin derivatives are absent.[131]

XI. OTHERS

A. AVOCADO

Fruit of the avocado, *Persea americana* Mill., is consumed primarily as a fresh fruit product. There are three recognized races of avocados based on their origin: Mexican, Guatemalan, and West Indian. Many cultivars of commercial importance are hybrids of these three races.[132] The reported anthocyanin pigments of avocado are cyanidin 3-galactoside and cyanidin 3,5-diglucoside acylated with *p*-coumaric acid.[133] The avocado also contains a large number of phenolic acids including ferulic; caffeic; sinapic; *o-*, *m-*, and *p*-coumaric; syringic; vanillic; isovanillic; gallic; protocatechuic; hydroxybenzoic; and resorcilic acid.[134] Total phenolic content can vary widely between fruit tissues and in the same tissue at different stages of growth (Table 3).[134,135]

B. PHALSA

Phalsa, *Grewia* spp. L., is native to southwestern Africa and is widespread throughout India. The fruit possesses a very attractive color, ranging from crimson red to dark purple. The fruit deteriorates easily and is, therefore, used principally for juice which also possesses a deep red color and has a pleasant flavor. It is considered to have a cooling effect during the hot weather. Gangwar and Tripathi[136] observed that ripe *G. asiatica* fruit could be kept for only 24 h. The greatest amount of juice was obtained from *G. subinequalis* L. fruit by thermal treatment at 50°C; this allowed the optimal composition of acids, anthocyanins, sugar, and mineral elements. The juice remained stable for 100 d if stored at low temperatures (3°C) and for only 10 d if stored at room temperature.[137]

Two anthocyanin pigments have been identified in *G. subinequalis* fruit as delphinidin 3-glucoside and cyanidin 3-glucoside.[138]

C. PINEAPPLE

The pineapple (*Ananas comosus* [L.] Merrill) is a tropical fruit native of Central and South America, but is also cultivated in tropical Asia and Africa. The most important producers in recent years have been Thailand, the U.S.

TABLE 3
Total Phenolics (mg/g FW ± SE) of Avocado Tissues as
Gallic Acid Equivalents

Cultivar	Mesocarp	Cotyledon	Young leaf	Mature leaf
Hass	1.8 ± 0.0[a]	24.6 ± 0.0[b]	19.3 ± 0.2[c]	17.5 ± 0.1[d]
Gwen	1.8 ± 0.0[a]	26.5 ± 0.1[d]	17.0 ± 0.1[d]	16.5 ± 0.1[d]
Fuerte	1.1 ± 0.0[a]	29.8 ± 0.1	24.2 ± 0.1[b]	15.0 ± 0.1

Note: a-d, Values with different letters are different from each other at the
0.05 significance level. Those with the same letter are not significantly
different from each other. Unmarked values are significantly different
from each other and from all others.

Reprinted with permission from Torres, A. M., Mau-Lastovicka, T., and Rezaaiyan, R., *J. Agric. Food Chem.*, 35, 921, 1987. Copyright 1987, American
Chemical Society.

(Hawaii and Puerto Rico), the Philippines, Mexico, Brazil, the Ivory Coast,
and South Africa.[118] Some cultivars of pineapple are red pigmented by anthocyanins (i.e., variety Kew), and this color disappears at full maturity.
Gortner,[139] following the evolution of pigments in the pineapple, observed a
continuous drop of chalcones and anthocyanins in the fruit shell during ripening. Carotenoids, in contrast, decreased during the first stage of development of the fruit and increased rapidly 2 weeks before full maturity. The
chlorophylls also showed fluctuations, indicating the existence of definite
stages of ripening. A close correlation between fruit ripening and content of
chlorophyll, anthocyanin, and chalcone in the skin of the fruit was also
reported by Lodh et al.[140] No variations have been observed in pulp anthocyanins in the Kew variety of *Ananas comosus* during ripening.[140] Fresh
'Mauritius' pineapple pulp contains 19.8 μg/g dry weight of anthocyanin
expressed as cyanidin 3-galactoside.[141] Both blanching and drying of pineapple
cubes results in a significant decrease in anthocyanin content attributed to
leaching and heat damage of the pigment.[141] At this time the identity of these
anthocyanins is not known.

REFERENCES

1. **Asenjo, C. F.,** Acerola in *Tropical and Subtropical Fruits,* Nagy, S. and Shaw, P. E.,
   Eds., AVI Publishing, Westport, CT, 1980, 341.
2. **Herrman, K.,** Uebersicht ueber die chemische Zusammensetzung und die Inhalstoffe
   einer Reihe wichtiger exotischer Obstarten, *Z. Lebensm. Unters. Forsch.,* 173, 47, 1981.
3. **Floch, H. and Gelard, A.,** Estimation of ascorbic acid in the fruits of Guiana, *Arch.
   Inst. Pasteur Guyane Fr.,* 337, 1954.

4. **Floch, G. and Gelard, A.,** *Malpighia punicifolia:* its exceptional richness in vitamin C, *Arch. Inst. Pasteur Guyane Fr.,* 368, 1955.

5. **Jackson, G. C.,** Ascorbic acid content of five *Malpighia* spp., *J. Agric. Univ. Puerto Rico,* 47, 193, 1963.

6. **Singh-Dhaliwal, T. and Torres-Sepulveda, A.,** Performance of azerole, *Malpighia punicifolia,* in the coffee region of Puerto Rico, *J. Agric. Univ. Puerto Rico,* 46, 195, 1962.

7. **Santini, R. and Huyke, A. J.,** Identification of the anthocyanin present in the acerola which produces color changes in the juice on pasteurization and canning, *J. Agric. Univ. Puerto Rico,* 40, 171, 1956.

8. **Nieva, F. S.,** Extraction, processing, canning and keeping quality of acerola (West Indian cherry, *Malpighia punicifolia* L.) juice, *J. Agric. Univ. Puerto Rico,* 39, 175, 1955.

9. **Fitting, K. O. and Miller, C. D.,** Stability of ascorbic acid in frozen bottled acerola juice alone or combined with other fruit juices, *Food Res.,* 25, 303, 1960.

10. **Iacobucci, G. A. and Sweeny, J. G.,** The chemistry of anthocyanins, anthocyanidins and related flavylium salts, *Tetrahedron,* 39, 305, 1983.

11. **Forsyth, W. G. C. and Hamilton, K. S.,** Bract pigmentation in bananas, *Nature (London),* 211, 325, 1966.

12. **Horry, J. P. and Jay, M.,** Distribution of anthocyanins in wild and cultivated banana varieties, *Phytochemistry,* 27, 2667, 1988.

13. **Horry, J. P.,** Chimiotaxonomie et organization genetique dans le genre *Musa, Fruits,* 44(10), 509, 1989.

14. **Seshadri, T. R.,** Some results of the study of Indian fruits, *Sci. Cult.,* 31, 444, 1965.

15. **Simmonds, N. W.,** Anthocyanins in bananas, *Nature (London),* 173, 402, 1954.

16. **Simmonds, N. W.,** Anthocyanins in bananas, *Ann. Bot. (London),* 28, 471, 1954.

17. **Harborne, J. B.,** Flavonoids of the Monocotyledonae, in *Comparative Biochemistry of the Flavonoids,* Academic Press, London, 233.

18. **Singh, N. S.,** Influence of UV light on the biosynthesis of anthocyanin-like pigments in ripening bananas, *Phytochemistry,* 11, 163, 1972.

19. **Mahey, S., Mukerjee, S. K., Saroja, T., and Seshadri, T. R.,** Proanthocyanidin glycosides in *Musa acuminata, Indian J. Chem.,* 9(4), 381, 1971.

20. **Minifie, B. W.,** *Chocolate, Cocoa and Confectionery: Science and Technology,* AVI Publishing, Westport, CT, 1980, 736.

21. **Fincke, H.,** Kleine beitrage zur Untersuchung von Kakaobohnen und Kakaoerzeugnisse, V. Mitt. *Z. Unters. Lebensm.,* 55, 559, 1928.

22. **Knapp, A. W.,** Wissenschaftliche Gesichpunkte der Kakaofermentierung, V. Enzyme, Gerbstoffe und Theobromine, *Bull. Imp. Inst. London,* 34, 154, 1936.

23. **Lawrence, W. J. C., Price, J. R., Robinson, G. M., and Robinson, R.,** A survey of anthocyanins, V. *Biochem. J.,* 32, 1661, 1938.

24. **Birch, H. F.,** Investigation of the purple coloring matter of cacao bean, *Annu. Rep. Cacao Res.,* 9, 51, 1939.

25. **Fincke, H.,** The origin of the palatable qualities of cacao and the relation of external properties, chemical composition and the palatibility of cacao bean. III. The pigments and tannins of cacao, *Kazett,* 27, 392, 1938.

26. **Fincke, H.,** The origin of the palatable qualities of cacao and the relation of external properties, chemical composition and the palatability of cacao bean. III. The influence of tannin-forming substances, tannin and tannin-like substances on the palatable qualities of the cacao, *Kazett,* 27, 452, 1938.

27. **Finke, H.,** Detection and determination of anthocyanins. Suitable group reactions for catechin, tannins and analogous plant constituents, *Z. Unters. Lebensm.,* 82, 209, 1941.

28. **Forsyth, W. G. C.,** Cacao polyphenolic substances. I. Fractionation of fresh bean, *Biochem. J.,* 51, 511, 1952.

29. **Forsyth, W. G. G.,** Cacao polyphenolic substances. II. Changes during fermentation, *Biochem. J.,* 51, 516, 1952.

30. **Forsyth, W. G. C.,** Purple beans, *Cocoa Conf.,* London, U.K., 1953, 32.
31. **Forsyth, W. G. C. and Quesnel, V. C.,** Cacao polyphenolic substances. IV. Anthocyanin pigments, *Biochem. J.,* 65, 177, 1957a.
32. **Schubiger, G. F., Roesch, E., and Egli, H. H.,** Beitrag zur Kenntnis der Kakao-polyphenole und verwandten Substanzen unter Anwendung der Spektrophotometrie, Chromatographie und Elecktrophorese, *Fette Seifen Anstrichm.,* 59, 631, 1957.
33. **Kharlamova, O. A.,** Etude de la composition des tanins de la feve de cacao Accra et Bahia, *Khlebopek. Konditer. Prom.,* 3, 15, 1964.
34. **Cros, E., Rouly, M., Villeneuve, F., and Vincent, J. C.,** Search for a cocoa fermentation index. I. Evolution of tannins and total phenolics in the cocoa bean, *Cafe, Cacao, The,* 26, 109, 1982.
35. **Cros, E., Rouly, M., Villeneuve, F., and Vincent, J. C.,** Search for a cocoa fermentation index. II. Estimation of the red colouring matter of cocoa beans, *Cafe, Cacao, The,* 26, 115, 1982.
36. **Griffiths, L. A.,** Phenolic acids and flavonoids of *Theobroma cacao* L. Separation and identification by paper chromatography, *Biochem. J.,* 70, 120, 1958.
37. **Scogin, R.,** Anthocyanins of the *Sterculiaceae, Biochem. Syst. Ecol.,* 7, 35, 1979.
38. **Jacquemin, H.,** Foliar anthocyanins of three tropical trees *(Mangifera indica, Theobroma cacao, Lophira alata). Trav. Lab. Matiere Med. Pharm. Galenique Fac. Pharm. Paris,* 56(1), 118, 1971.
39. **Pence, V. C., Hasegawa, P. M., and Janick, J.,** *In vitro* cotyledonary development and anthocyanin synthesis in zygotic and asexual embryos of *Theobroma cacao, J. Am. Soc. Hortic. Sci.,* 106, 381, 1981.
40. **Kononowicz, A. K. and Janick, J.,** In vitro development of zygotic embryos of *theobroma cacao, J. Am. Soc. Hortic. Sci.,* 109, 266, 1984.
41. **Roesch, E., Schubiger, G. F., and Egli, R. H.,** Etude histochimique des fèves de cacao, *Rev. Int. Choc.,* 1, 21, 1959.
42. **Biehl, B.,** Changes in the subcellular structure of the cotyledons of cocoa beans *(Theobroma cacao* L.) during fermentation and drying, *Z. Lebensm. Unters. Forsch.,* 153, 137, 1973.
43. **Brown, H. B.,** Separation of pigment cells of cacao, *Nature (London),* 173, 492, 1954.
44. **Paiva, M., Wright, D. C., and Janick, J.,** Alkaloid variation in *Theobroma cacao, Proc. 8th Int. Cocoa Res. Conf.,* Cocoa Producers Alliance 1982, 229.
45. **Liebig, A. W.,** Kakaorot und Kakaobraun in ihren Bedeutung für Farbe und aussehen bei Tafelschokoladen, *Rev. Int. Choc.,* 8, 320, 1953.
46. **Kleinert, J.,** Cocoa beans and chocolate, *Rev. Int. Choc.,* 20, 418, 1965.
47. **Rohan, T. A.,** The flavour of chocolate, its precursors and study of their reactions, *Gordian,* 69(6), 433, 1969.
48. **Karlamova, O. A.,** Effect of the degree of fermentation of cocoa beans on the quality of chocolate and on phenolic compounds in chocolate, *Khlebopek. Konditer. Prom.,* (6), 17, 1972.
49. **Kharlamova, O. A.,** *Chemical Composition, Properties and Flavour of Cocoa Beans,* Pishchevaya Promyshlennost, Moscow, 1974, 118.
50. **Lange, H.,** Die Inhaltbestandteile der Kakaobohne und ihre Ernaehrungsphysiologischen Eigenschaften, *Suesswaren,* 13, 8, 1969.
51. **Rohan, T. A.,** Processing of raw cacao. I. Small scale fermentation, *J. Sci. Food Agric.,* 9, 104, 1958.
52. **Rohan, T. A.,** Processing of raw cocoa. II. Uniformity in heap fermentation and development of methods for rapid fermentation of West African Amelonado cocoa, *J. Sci. Food Agric.,* 9, 542, 1958.
53. **O.I.C.C.,** Methodes d'Analyse de l' O.I.C.C., Dosage colorimetrique des pigments de cacao, *Rev. Int. Choc.,* 18, 11F, 1963.
54. **Rohan, T. A.,** Processing of raw cocoa for the market, Food and Agricultural Organization Study No. 60, Rome, 1963.

55. **Roelofsen, P. A.**, Fermentation, drying and storage of cacao beans, *Adv. Food Res.*, 8, 225, 1958.

56. **Chassevenet, F. and D'Ornano, M.**, Photometric determination of pigments in cocoa, Method published by the International Office of Cocoa and Chocolate and several variations, *Cafe, Cacao, The*, 10, 243, 1966.

57. **Gureva, K. B. and Tserevitinov, O. B.**, Evaluating the Degree of Fermentation of Cocoa Beans, *U.S.S.R. Patent* 646, 254, 1979.

58. **Pettipher, G. L.**, An improved method for the extraction and quantitation of anthocyanins in cocoa beans and its use as an index of the degree of fermentation, *J. Sci. Food Agric.*, 37, 289, 1986.

59. Morinaga Co. Ltd., Red Pigment, Japanese Patent 5,248,190, 1977.

60. **Motoda, S.**, Production of flavonoid pigments by polyphenol-oxidase, *Nippon Shokuhin Kogyo Gakkai-Shi*, 29, 25, 1982 (FSTA 15, 3T140).

61. **Eggen, I. B.**, Berry-Like Flavoring, U.S. Patent 4, 156, 030, 1979.

62. **Hua, Z., Yu, Q., and Zhou, H.**, Preliminary study on extraction of an edible coloring material from cocoa bean shells, *Linchan Huaxue Yu Gongye*, 2(3), 32, 1982.

63. **Ma, F., Zie, H., Wang, J., Zhao, Y., and Quin, H.**, Toxicological study of food coloring materials from cocoa bean shells, *Shipin Xexue (Beijing)*, 44, 24, 1983.

64. **Yamaguchi, N., Naito, S., and Yokoo, Y.**, Studies on natural antioxidants. V. Antioxidative activity of pigments extracted from cocoa bean husk, *Nippon Shokuhin Kogyo Gakkai-Shi*, 29, 534, 1982.

65. **Clifford, M. N. and Willson, K. C.**, *Coffee: Botany, Biochemistry and Production of Beans and Beverage*, AVI Publishing, Westport, CT, 1985.

66. **Belitz, H.-D. and Girosch, W.**, *Food Chemistry*, Springer-Verlag, Berlin, 1987.

67. **Tschirsch, A.**, Violette Chromatophoren in der Fruechtschalen der Kaffee, *Schweiz. Wochenschr. Chem. Phys.*, 36, 452, 1888.

68. **Kroemer, K.**, Angebl. Vorkommen von violetten Chromatophoren, *Bot. Centralbl.*, 84, 33, 1900.

69. **Sotero, A.**, Photosynthesis of pigments of the coffee seed. Chemistry of the seed anthocyanins, *Rev. Bras. Quim.*, 4, 368, 1937.

70. **Sondahl, M. R., Teixiera, A. A., Fazuoli, L. C., and Monaco, L. C.**, Effect of ethylene on the type and quality of coffee, *Turrialba*, 24, 17, 1974.

71. **Lopes, C. R., Musche, R., and Hec, M.**, Identification of flavonoid pigments and phenolic acids in the cultivars Bourbon and Caturra of *Coffea arabica* L., *Rev. Bras. Genet.*, 7, 657, 1984.

72. **Zualaga, V. J. and Tabacchi, R.**, Contribution à l'étude des composés phénoliques de la pulpe de Café, *Séminaire C.P.C.I.A.*, Narbonne, France, 17 to 19, Juin, 1981.

73. **Zualaga, V. J. and Tabacchi, R.**, Contribution à l'étude de la composition chimique de la pulpe de Café, in *9e Coll. Sci. Int. Café*, Londres, ASIC, Paris, 1980, 33.

74. **Ramirez-Martinez, J. R.**, Phenolic compounds in coffee pulp: quantitative determination by HPLC, *J. Sci. Food Agric.*, 43, 135, 1988.

75. **Krug, C. A. and Carvalho, A.**, Genetica de "Coffea", *Bol. Tec. Inst. Agric. Campinas*, 82, 1, 1940.

76. **Krug, C. A., Mendez, J. E. T., and Carvalho, A.**, Taxonomia de *C. arabica* L., *Bragantia*, 9, 157, 1948.

77. **Krug, C. A. and Carvalho, A.**, The genetics of *Coffea*, *Adv. Genet.*, 4, 127, 1951.

78. **Lopes, C. R. and Monaco, L. C.**, Chemotaxonomic studies of some species of the genus *Coffea*, *J. Plant Crops*, 7, 6, 1979.

79. **Vandercook, C. E., Hosegana, H., and Maier, V. P.**, Dates, in *Tropical and Subtropical Fruits*, Nagy, S. and Shaw, P. E., Eds., AVI Publishing, Westport, CT, 1980, 506.

80. **Sawaya, W. N., Khalil, J. K., Safi, W. N., and Al-Shalhat, A.**, Physical and chemical characterization of three Saudi date cultivars at various stages of development, *Can. Inst. Food Sci. Technol. J.*, 16, 87, 1983.

81. **Ragab, M. H. H., El-Tabey Shehata, A. M., and Sedky, A.,** Egyptian dates. II. Chemical changes during development and ripening of six varieties, *Food Technol.,* 10, 405, 1956.

82. **Maier, V. P. and Metzler, D. M.,** Quantitative changes in date polyphenols and their relation to browning, *J. Food Sci.,* 30, 80, 1965.

83. **Maier, V. P. and Metzler, D. M.,** Changes in individual date polyphenols and their relation to browning, *J. Food Sci.,* 30, 747, 1965.

84. **Maier, V. P.,** Hydroxycinnamoyl esters of quinic and shikimic acids, in *Aspects of Plant Phenolic Chemistry,* Proc. 3rd Annu. Symp. Plant Phenolics Group N.A., Toronto, 1963.

85. **Maier, V. P., Metzler, D. M., and Huber, A. H.,** 3-*O*-caffeoylshikimic acid (dactylliferic acid) and its isomers, a new class of enzymic browning substrates, *Biochem. Biophys. Res. Commun.,* 14, 124, 1964.

86. **Harborne, J. B., Williams, C. A., and Greenham, J.,** Distribution of charged flavones and caffeylshikimic acid in *Palmae, Phytochemistry,* 13, 1557, 1974.

87. **Maier, V. P. and Metzler, D. M.,** Isolation of catechins from Deglet Noor dates, unpublished data, 1965.

88. **Williams, C. A., Harborne, J. B., and Clifford, H. T.,** Flavonoid patterns in the monocotyledons. Flavonols and flavones in some families associated with the *Poaceae, Phytochemistry,* 10, 1059, 1971.

89. **Williams, C. A., Harborne, J. B., and Clifford, H. T.,** Negatively charged flavones and tricin as chemosystematic markers in the *Palmae, Phytochemistry,* 12, 2417, 1973.

90. **Cavaletto, C. G.,** Lychee, in *Tropical and Subtropical Fruits,* Nagy, S. and Shaw, P. E., Eds., AVI Publishing, Westport, CT, 1980, 469.

91. **Prasad, U. S. and Jha, O. P.,** Changes in pigmentation patterns during litchi ripening, *Plant Biochem. J.,* 5, 44, 1978.

92. **Lee, H. S. and Wicker, L.,** Anthocyanin pigments in the skin of lychee fruit, *J. Food Sci.,* 56, 466, 1991.

93. **Lee, H. S. and Wicker, L.,** Quantitative changes in anthocyanin pigment of lychee fruit during refrigerated storage, *Food Chem.,* 40, 263, 1991.

94. **Paull, R. E., Chen, N. J., Deputy, J., Huang, H., Cheng, G., and Gao, F.,** Litchi growth and compositional changes during development, *J. Am. Soc. Hortic. Sci.,* 109, 817, 1984.

95. **Fang, T.-T. and Huang, Y.-C.,** A study on the picking maturity of litchi (*Litchi chinensis* Sonn.), *Hortic. Sci. China,* 16, 27, 1970.

96. **Lu, R., Cai, L., Wu, J., Zhan, X., and Huang, W.,** Study of the preservation of freshness and color of litchi fruit, *Shipin Kexue (Beijing),* 62, 13, 1985 (CA 103, 21427).

97. **Mitra, S. K. and Sanyal, D.,** Effect of putrescine on fruit set and fruit quality of litchi, *Gartenbauwissenschaft,* 55, 83, 1990.

98. **Trevisan, L. M., Bobbio, F. O., and Bobbio, P. A.,** Carbohydrates, organic acids and anthocyanins of *Myrciaria jaboticaba* Berg., *J. Food Sci.,* 37, 818, 1972.

99. **Bobbio, F. O. and Scamparini, A. R. P.,** Carbohydrates, organic acids and anthocyanin of *Eugenia jambolana* Lamarck, *Ind. Aliment.,* 21, 296, 1982.

100. **Sharma, J. N. and Seshadri, T. R.,** Survey of anthocyanins from Indian sources. II. *J. Ind. Chem. Res. (India),* 14B, 211, 1955.

101. **Jain, M. C. and Seshadri, T. R.,** Anthocyanins of *Eugenia jambolana* fruits, *Indian J. Chem.,* 13, 20, 1975.

102. **Guimares, F. A., de Holanda, L. F. F., Maia, G. A., and Moura Fe, J. de A.,** Physical and chemical studies on Pitanga (*Eugenia uniflora*), *Cien. Technol. Aliment.,* 2, 208, 1982.

103. **Benk, E.,** Weiterer Beitrag zur Kenntnis auslaendischer Obstfruechte, *Reichstoffe, Aromen, Kosmet.,* 31, 234,103, 1981.

104. **Lowry, J. B.,** Anthocyanins of the Melastomataceae, Myrtaceae and some allied families, *Phytochemistry,* 15, 513, 1976.

105. **Khurdiya, D. S. and Roy, S. K.,** Storage studies on jamun *(Syzigium cumini)* juice and nectar, *J. Food Sci. Technol. (India),* 22, 217, 1985.

106. **Martinez, S. B. and Del Valle, M. J.,** Storage stability and sensory quality of duhat *(Sysyum cumini* Linn.) anthocyanins as food colorant, *UP Home Econ. J.,* 9(1), 7, 1981.

107. **Singh, L. B.,** *The Mango,* Leonard Hill, London, 1968.

108. **Lakshminarayana, S.,** Mango, in *Tropical and Subtropical Fruits,* Nagy, S. and Shaw, P. E., Eds., AVI Publishing, Westport, CT, 1980, 184.

109. **Barmore, C. R., Spalding, D. H., and McMillan, R. T.,** Anthocyanin development in mango fruit in response to a preharvest antitranspirant spray, *HortScience,* 9, 114, 1974.

110. **Proctor, T. A. and Creasy, L. L.,** The anthocyanins of mango fruit, *Phytochemistry,* 8, 2108, 1969.

111. **Paris, R. and Jacquemin, H.,** Polyphenolic pigments (anthocyanins and flavonoids) of young *Mangifera indica* (mango) leaves, *C.R. Acad. Sci., Ser. D,* 270, 3232, 1970.

112. **Timberlake, C. F. and Bridle, P.,** Distribution of anthocyanins in food plants, in *Anthocyanins as Food Colors,* Markakis, P., Ed., Academic Press, New York, 1982, 125.

113. **El Ansari, M. A., Reddy, K. K., Sastry, K. N. S., and Nayudamma, Y.,** Polyphenolic constituents of mango *(Mangifera indica)* fruit, *Leather Sci. (Madras),* 16, 13, 1969.

114. **El Ansari, M. A., Reddy, K. K., Sastry, K. N. S., and Nayudamma, Y.,** Polyphenols of *Mangifera indica, Phytochemistry,* 10, 2239, 1971.

115. **Martin, F. W.,** Durian and mangosteen, in *Tropical and Subtropical Fruits,* Nagy, S. and Shaw, P. E., Eds., AVI Publishing, Westport, CT, 1980, 407.

116. **Sen, A. K., Sarkar, K. K., Mazumber, P. C., and Banerji, N.,** Minor xanthones of *Garcinia mangostana, Phytochemistry,* 20, 183, 1981.

117. **Sen, A. K., Sarkar, K. K., Mazumder, P. C., Banerji, N., Uusvuori, R., and Hase, T. A.,** The structure of garcinones A, B and C: three new xanthones from *Garcinia mangostana, Phytochemistry,* 21, 1747, 1982.

118. **Knight, R.,** Origin and world importance of tropical and subtropical fruit crops, in *Tropical and Subtropical Fruits,* Nagy, S. and Shaw, P. E., Eds., AVI Publishing, Westport, CT, 1980, 2.

119. **Du, C. T. and Francis, F. J.,** Anthocyanins of mangosteen, *Garcinia mangostana, J. Food Sci.,* 42, 1667, 1977.

120. **Tulyatan, V., Subhimaros, S., Sukcharoen, O., and Dulyapirunhasilp, S.,** Extraction of anthocyanin from mangosteen rind, *Food,* 19, 25, 1989.

121. **Krishnamurthy, N., Lewis, Y. S., and Ravindranath, B.,** Chemical constituents of kokam fruit rind, *J. Food Sci. Technol. (India),* 19, 97, 1982.

122. **Krishnamurthy, N. and Sampathu, S. R.,** Application of garcinol (pigment from the rind of kokum, *Garcinia indica* fruit) as a colorant for butter and ghee and spectrophotometric method of its estimation, *J. Food Sci. Technol. (India),* 24(6), 312, 1987.

123. **Padival, R. A. and Sastry, L. V. L.,** Stability of safflower, kokum and beetroot colours in model systems and some fruit products, *J. Food Sci. Technol. (India),* 22, 346, 1985.

124. **Chan, H. T.,** Passion fruit, in *Tropical and Subtropical Fruits,* Nagy, S. and Shaw, P. E., Eds., AVI Publishing, Westport, CT, 1980, 300.

125. **Pruthi, J. S.,** Physiology, chemistry and technology of passion fruit, *Adv. Food Res.,* 12, 203, 1963.

126. **Pruthi, J. S., Susheela, R., and Lal, G.,** Anthocyanin pigments in passion fruit skin, *J. Food Sci.,* 26, 385, 1961.

127. **Harborne, J. B.,** *Comparative Biochemistry of Flavonoids,* Academic Press, London, 1967, 132.

128. **Ishikura, N. and Sugahara, K.,** A survey of anthocyanins in fruits of some Angiosperms, *Bot. Mag. (Tokyo),* 92, 157, 1979.

129. **Nakasone, H. Y., Hirano, R., and Ito, P.,** Preliminary observations on the inheritance of several factors in the passion fruit (*Passiflora edulis* L. and *forma flavicarpa*), Tech. Prog. Rep., Hawaii Agric. Exp. Stn., No. 161, 1967, 11.

130. **Halim, M. M. and Collins,** Anthocyanins of *Passiflora quadrangularis*, *Bull. Torrey Bot. Club*, 97, 247, 1970.

131. **Billot, J.,** Pigments anthocyaniques des fleur de *Passiflora quadrangularis*, *Phytochemistry*, 13, 2886, 1974.

132. **Ahmed, E. M. and Barmore, C. R.,** Avocado, in *Tropical and Subtropical Fruits,* Nagy, S. and Shaw, P. E., Eds., AVI Publishing, Westport, CT, 1980, 127.

133. **Prabha, T. N., Ravindranath, B., and Pathwardhan, M. V.,** Anthocyanins of avocado *(Persea americana)*, *J. Food Sci. Technol.*, 17, 241, 1980.

134. **Torres, A. M., Mau-Lastovicka, T., and Rezaaiyan, R.,** Total phenolics and high performance chromatography of phenolic acids of avocado, *J. Agric. Food Chem.*, 35, 921, 1987.

135. **Biale, J. B. and Young, R. E.,** The avocado pear, in *The Biochemistry of Fruits and Their Products*, Vol. 2, Hulme, A. C., Ed., Academic Press, London, 1971, 2.

136. **Gangwar, B. M. and Tripathi, R. S.,** A study on chemical changes during ripening and storage of phalsa, *Indian Food Packer*, 26, 5, 1972.

137. **Khurdiya, D. S. and Anand, J. C.,** Effect of extraction method, container and storage temperature on phalsa fruit juice (*Grewia subinequalis* L.), *Indian Food Packer*, 35, 68, 1981.

138. **Khurdiya, D. S. and Anand, J. C.,** Anthocyanins in phalsa (*Grewia subinequalis* L.), *J. Food Sci. Technol. (India)*, 18, 112, 1981.

139. **Gortner, W. A.,** Chemical and physical development of the pineapple fruit. IV. Plant pigments constituents, *J. Food Sci.*, 30, 31, 1965.

140. **Lodh, S. B., Divakar, N. G., Chadha, L. K., Melanta, K. R., and Selvaraj, Y.,** Biochemical changes associated with growth and development of the pineapple fruit variety Kew. II. Changes in plant pigments and enzyme activity, *Indian J. Hortic.*, 30, 381, 1973.

141. **Sian, N. K. and Ishak, S.,** Carotenoid and anthocyanin contents of papaya and pineapple: influence of blanching and predrying treatments, *Food Chem.*, 39, 175, 1991.

Chapter 6

GRAPES

Grapes are the world's largest fruit crop with approximately 65 million metric tons produced annually.[1] About 80% of the total crop is used in wine making; 13% is sold as table grapes; and the balance is grown largely for raisins, juice, and other products. Most grapes are grown in Europe, with Italy, France, and Spain accounting for approximately 40% of the world's 9,000,000 ha of grapevines.[1,3] California produces about 20% of the world's raisins and 10% of the world's table grapes.[2]

There are two major types of grapes: European and North American. European grapes belong to the species *Vitis vinifera* L., and over 95% of all the grapes produced are of this species. These grapes are characterized by a relatively thick skin that adheres to a firm pulp which is sweet throughout. Varieties of *V. vinifera* grow best in a Mediterranean type of climate with long, relatively dry summers and mild winters.[2,3] Certain varieties are used principally for wine and others, for raisins or table use. Leading wine varieties include Cabernet Sauvignon, Chardonnay, Pinot Noir, Zinfandel, and Carignane. Table grapes are large and tasty and have an attractive color. The chief varieties include Emperor and Tokay, which grow in large clusters of red berries. Other table grapes include the greenish-white, seedless 'Perlette' and the black 'Ribier'. Most raisin grapes are seedless, and have a soft texture when dried. Leading raisin varieties include Thompson Seedless, Black Corinth, and Muscat of Alexandria.[2,3] North American grapes belong to two main species: *V. labrusca* and *V. rotundifolia*. Both species can be consumed fresh or processed into juice, wine, or jelly. The *V. labrusca* grapes are grown principally in the lower Great Lakes region of the U.S. and Canada. The most important variety is the Concord, which has large purple berries. *V. rotundifolia* or 'Muscadina' grapes grow throughout the southeastern United States, from North Carolina to eastern Texas. They are exemplified by the Scuppernong variety which is long lived, disease resistant, and vigorous. The bunches are relatively small; and the berries have a strong musky odor and large seeds, ripen unevenly, and fall off singly as they ripen.[3]

The principal components of ripe grapes are the carbohydrates glucose and fructose. These sugars, which are converted into alcohol during fermentation, are major factors in the taste of table grapes, raisins, juices, and other products. The organic acids, although present in small amounts, contribute significantly to the taste of grapes and grape products; and their concentration has marked effect on the color intensity of anthocyanins.

The anthocyanins are fundamentally responsible for all the color differences between grapes and the resultant wines.[4] The amount and composition of the anthocyanins present in red grapes vary greatly with the species, variety, maturity, seasonal conditions, production area, and yield of fruit.[4-15] Also, the conditions of fermentation[16-23] and aging of wine such as temperature, duration, sulfur dioxide, and alcohol concentrations affect the concentration of anthocyanins in the wines.[14-30] Furthermore, wine color (hue, brightness, and saturation) varies with the total concentration and composition of anthocyanin in the wine.[3,4] Anthocyanins also contribute to the organoleptic and chemical qualities of wine because of their interaction with other phenolic compounds as well as with proteins and polysaccharides.[4,31]

I. ANTHOCYANINS IN *VITIS VINIFERA*

Many works have been published on the anthocyanins of grapes. Ribéreau-Gayon[4] reviewed data on the pigment composition of *V. vinifera* as determined largely by the use of paper chromatography. It is now well known that the distribution of anthocyanins in the fruit is complex and varies according to variety. Since the mid-1970s, investigations on anthocyanins have been facilitated by the use of high-performance liquid chromatography (HPLC). Manley and Shubiak[32] separated 3-glucosides of malvidin, petunidin, and peonidin in concentrates of commercial grape skin extracts (Enocianina) on Pellidon (polyamide bonded on glass beads) by elution with chloroform-methanol in 1976. Later reports of anthocyanin separations have favored the use of reverse phase chromatography. In 1978, Wulf and Nagel[33] separated and quantified 20 discrete anthocyanin components in skins of *V. vinifera* grown in Washington State on a Lichrosorb Si-60 reversed-phase column with a formic acid-water-methanol elution gradient. The HPLC chromatogram of 'Cabernet Sauvignon' pigments obtained by these authors is shown in Figure 1. Peaks 1 to 5 represent 3-glucosides of the five anthocyanidins found in grapes. Delphinidin, cyanidin, petunidin, peonidin, and malvidin 3-monoglucosides were the only anthocyanins found in 'Pinot Noir' grapes, which is consistent with the conclusion of Rankine et al.[5] and Fong et al.[34] that 'Pinot Noir' does not contain acylated anthocyanins. The 3-acetylglucosides and 3-*p*-coumarylglucosides of 'Cabernet Sauvignon' are based on malvidin; with lesser amounts of delphinidin, petunidin, and peonidin; and only traces of cyanidin. In addition, malvidin 3-caffeylglucoside and three unknown pigments were separated and quantified by Wulf and Nagel.[33] Using a similar HPLC method, Piergiovanni and Volonterio[35] separated and quantified the 3-glucosides and their acetyl esters of delphinidin, cyanidin, petunidin, peonidin, and malvidin; as well as petunidin, peonidin, and malvidin 3-*p*-coumarylglucosides and peonidin and malvidin 3-caffeylglucosides from the skin of 'Merlot' grapes from the Pavese region of Italy. These cultivars also provide evidence for the presence of two additional anthocyanidin cinnamates, namely, delphinidin and cyanidin, although the structure of the acylating acid was not determined.

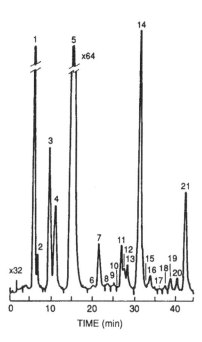

FIGURE 1. Anthocyanins of *V. vinifera*, cv. 'Cabernet Sauvignon' grapes, Dp = delphinidin, Cy = cyanidin, Pt = petunidin, Pn = peonidin, Mv = malvidin, Gl = monoglucoside, Ac = acetate, Coum = *p*-coumarate, Caf = caffeoate. 1 = Dp-3-Gl; 2 = Cy-3-Gl; 3 = Pt-3-Gl; 4 = Pn-3-Gl; 5 = Mv-3-Gl; 6 = unknown; 7 = Dp-3-Gl-Ac; 8 = Cy-3-Gl-Ac; 9 = unknown; 10 = unknown; 11 = Pt-3-Gl-Ac; 12 = Mv; 13 = Pn-3-Gl-Ac; 14 = My-3-Gl-Ac; 15 = unknown; 16 = Dp-3-Gl-Coum; 17 = Cy-3-Gl-Coum; 18 = Mv-3-Gl-Caf; 19 = Pt-3-Gl-Coum; 20 = Pn-3-Gl-Coum; 21 = Mv-3-Gl-Coum. (From Wulf, L. W. and Nagel, C. W., *Am. J. Enol. Vitic.*, 29, 42, 1978. With permission.)

Roggero et al.[9] also used HPLC to determine the anthocyanin pattern in Cabernet Sauvignon as well as in five other *V. vinifera* cultivars (Grenache, Carignan, Syrah, Mourvedre, and Cinsault) from southern France. Qualitatively, the anthocyanin patterns were found to be very similar to those reported by Wulf and Nagel[33] and Piergiovanni and Volonterio;[35] although, quantitatively, the relative amounts of the individual anthocyanins in each cultivar were considerably different. For instance, the content of 3-glucosides in the skin ranged from 57.0% of the total anthocyanins for 'Syrah' grapes, to 65.3% in 'Cabernet Sauvignon' grapes, and to 84.2% in 'Grenache' grapes. Likewise, the 3-acetylglucosides ranged from 1.9% in 'Grenache', to 14.3% in 'Syrah', and to 26.2% in 'Cabernet Sauvignon' grape skins. The malvidin derivatives represented 54.5 to 74.2% of the nonacylated monoglucosides, 65.7% to 78.9% of the monoglucoside acetates, and 75.6 to 80.7% of the monoglucoside caffeoates and *p*-coumarates in the varieties under investigation. In 1988, Hebrero et al.[36] found a similar qualitative anthocyanin pattern in 'Tempranillo' grapes from the Toro region of Spain. Table 1 provides the distribution of anthocyanins in 'Cabernet Sauvignon', 'Merlot', 'Syrah', and

TABLE 1
Anthocyanin Distribution in *V. vinifera* Grapes

Peak No.[a]	Anthocyanin	'Cabernet Sauvignon'[b]	'Merlot'[c]	'Syrah'[d]	'Tempranillo'[e]
1	Delphinidin 3-monoglucoside	9.99	9.38	5.2	16.64
2	Cyanidin 3-monoglucoside	1.33	2.29	1.0	6.53
3	Petunidin 3-monoglucoside	6.07	7.94	6.1	12.33
4	Peonidin 3-monoglucoside	5.27	6.98	8.3	12.53
5	Malvidin 3-monoglucoside	42.57	40.69	36.4	36.68
6	Unknown	0.29	—	—	—
7	Delphinidin 3-monoglucoside-acetate	2.51	1.62	0.7	0.42
8	Cyanidin 3-monoglucoside-acetate	0.13	0.42	0.2	0.11
9 & 10	Unknowns	0.08	—	—	—
11	Petunidin 3-monoglucoside-acetate	2.20	1.84	1.3	0.30
12	Malvidin	—	—	—	—
13	Peonidin 3-monoglucoside-acetate	0.93	1.62	2.7	0.15
14	Malvidin 3-monoglucoside-acetate	20.46	11.53	9.4	1.21
15 & 16	Delphinidin 3-monoglucoside-*p*-coumarate	0.49	—	1.0	2.21
17	Cyanidin 3-monoglucoside-*p*-coumarate	0.08	—	2.4	0.97
18	Malvidin 3-monoglucoside-caffeoate	0.14	t	2.4	0.15
19	Petunidin 3-monoglucoside-*p*-coumarate	0.44	t	0.6	1.70
20	Peonidin 3-monoglucoside-*p*-coumarate	0.59	1.96	2.0	1.47
21	Malvidin 3-monoglucoside-*p*-coumarate	6.38	8.90	21.9	6.47

Note: t = Trace.

[a] Refers to peak numbers in Figure 1.
[b] From Wulf, L. W. and Nagel, C. W., *Am. J. Enol. Vitic.*, 29, 42, 1978.
[c] From Piergiovanni, L. and Volonterio, G., *Riv. Vitic. Enol.*, 33, 289, 1980.
[d] From Roggero, J. P., Ragonnet, B., and Coen, S., *Vignes & Vins*, 327, 38, 1984.
[e] From Hebrero, E., Santos-Buelga, C., and Rivas-Gonzalo, J. C., *Am. J. Enol. Vitic.*, 39, 227, 1988.

'Tempranillo' grapes from Washington State (U.S.), Italy, France, and Spain. Compared with the other varieties, Tempranillo shows a higher percentage of 3-monoglucosides of delphinidin, cyanidin, and petunidin and a lower percentage of malvidin 3-monoglucoside and malvidin 3-acetylglucoside.

Bakker and Timberlake[10] used HPLC to assess the distribution of anthocyanins in methanolic skin extracts of 26 grape cultivars at five sites in the Douro Valley in northern Portugal for port wine production. Their results (Table 2) show that malvidin-based anthocyanins predominate. Of these, malvidin 3-glucoside was the major pigment (33 to 60%), being exceeded by malvidin 3-*p*-coumarylglucoside (2 to 51%) in only two cultivars; malvidin 3-acetylglucoside (1 to 15%) was consistently the lowest. Peonidin 3-glucoside was prominent in four cultivars; and the 3-glucosides of delphinidin, petunidin, and cyanidin were at low proportions throughout. The ratio of malvidin 3-acetylglucoside to total malvidin glucosides is a useful aid in identification of the grape cultivar because it seems to be characteristic of cultivar and independent of the site of production. Piergiovanni and Volonterio[37] examined 20 samples of 15 cultivars of *V. vinifera* grapes from Italy and found that the proportion of acetate forms for each anthocyanidin offered the best characterization of the cultivars. According to Roggero et al.,[38] characterization of the cultivar can best be achieved by factorial analysis of the data on contents of malvidin monoglucoside, anthocyanin acetic ester, anthocyanin cinnamic ester, and ratio of malvidin and peonidin glucoside.

II. ANTHOCYANINS IN OTHER *VITIS* SPECIES

In addition to *V. vinifera*, there are probably 50 other species of *Vitis*.[39] Ribéreau-Gayon[4] described the distribition of anthocyanins in 15 species. Table 3 gives the relationships between *V. rotundifolia, V. amurensis,* and 11 American species. The important differences between species may be summarized as follows: in *V. lincecumii, V. aestivalis,* and *V. coriacea* cyanidin and peonidin predominate; in *V. riparia* and *V. rupestris* diglucosides are the main pigments; and in other species, including *V. vinifera,* monoglucosides are present in the largest amounts (Tables 1 to 3).

In *V. rotundifolia* (muscadine grape) Ribéreau-Gayon[4,40,41] reported contents of nonacylated 3,5-diglucosides of malvidin, peonidin, petunidin, cyanidin, and delphinidin. Of these pigments delphinidin and petunidin 3,5-diglucosides accounted for 67% of the total anthocyanin content (Table 3). More recently, Goldy et al.[13] surveyed 84 wild selections of *V. rotundifolia* from Alabama, Arkansas, North Carolina, and Virginia. They found that all plants contained the 3,5-diglucosides of delphinidin, cyanidin, petunidin, peonidin, and malvidin; and percent ranged from 13.5 to 68.9%, 2.6 to 43.5%, 8.6 to 34.7%, 1.6 to 25.1%, and 0.8 to 30% total pigment, respectively. Nine selections produced the 3-monoglucosides of delphinidin, cyanidin, petunidin, peonidin, and malvidin pigments not found in studies of *V. rotundifolia* by Ribéreau-Gayon[4] and others.[42-44] To date, however, no acylated

TABLE 2
Area Percent of Known Anthocyanins (A) in Grape Skin Extracts of 1982 Port Wine Grape Cultivars from Northern Portugal

Cultivar	Site[a]	Mv[b] (%)	MV-ac (%)	MV-cou (%)	Pn (%)	Dp (%)	Pt (%)	Cy (%)	Tot Mv/ known A (%)	Mv-ac/ tot Mv (%)	Mv-ac/ Mv+Mv-ac (%)	Known A/ tot A (%)
Tinta Cao	Tua	33	11	51	1	2	3	tr	94	11	24	88
Tinta Barroca	BC	35	5	51	4	2	3	tr	91	6	13	92
Tinta Barroca	Tua	37	5	47	6	2	3	tr	89	6	12	92
Touriga Nacional	P	36	14	38	6	2	3	tr	89	16	28	95
Touriga Nacional	V	38	15	36	6	2	3	tr	89	17	28	93
Touriga Nacional	Tua	41	15	31	6	3	4	tr	87	17	26	93
Tinta da Barca	Tua	37	13	39	4	3	4	tr	88	15	26	93
Touriga Francesa	BC	41	11	36	3	4	5	tr	87	15	23	93
Touriga Francesa	SB	43	13	30	4	4	6	tr	86	12	21	93
Tinta Francisca	Tua	46	5	35	4	4	5	1	86	5	9	94
Tinta Francisca	SB	48	1	32	5	6	7	1	82	2	3	92
Rufete	Tua	45	5	30	5	7	7	1	80	6	10	91
Malvesia Preta	SB	53	4	20	6	7	9	1	77	5	7	95
Souzao	Tua	57	4	15	6	8	8	2	76	5	7	96
Tinta Roriz	Tua	41	4	29	3	11	11	1	74	6	10	88
Tinta Santarem	Tua	49	3	19	23	2	3	1	72	5	7	96
Viera da Natividade	V	52	4	14	27	1	2	1	69	5	7	98
Tinta Amarela	Tua	49	4	18	10	8	9	2	71	6	8	94
Tinta Amarela	BC	47	4	18	7	11	11	2	69	5	7	93
Mourisco	Tua	48	2	16	16	7	9	3	66	3	4	94
Cornifesto	P	60	1	2	22	6	7	2	63	2	2	99
Donzelinho	Tua	44	1	6	25	9	9	6	50	1	1	98

[a] BC = Baixo Corgo, P = Pinhao, V = Vilariça, SB = Santa Barbara.

[b] Mv = malvidin 3-glucoside; Mv-ac = malvidin 3-acetylglucoside; Mv-cou = malvidin 3-p-coumarylglucoside; Pn = peonidin 3-glucoside; Dp = delphinidin 3-glucoside; Pt = petunidin 3-glucoside; Cy = cyanidin 3-glucoside; Tot = total.

From Bakker, J. and Timberlake, C. F., *J. Sci. Food Agric.*, 36, 1315, 1985. With permission.

TABLE 3
Major Anthocyanins of Selected Species of the Genus *Vitis*[a]

Anthocyanin	V. rotundifolia	V. riparia	V. rupestris	V. labrusca	V. arizonica	V. berlandieri	V. monticola	V. cordifolia	V. rubra	V. lincecumii	V. aestivalis	V. coriacea	V. amurensis
Total number of anthocyanins	5[a]	15	12	10[a]	11	9	6	9	12	17	9	7	5
Percent of Each													
Cyanidin													
Monoglucoside		2		5	8	8	3	10	20	29	31	58	
Acylated monoglucoside										7			
Diglucoside	9	5	2		1				1	2	3	4	
Acylated diglucoside										5			
Peonidin													
Monoglucoside				10	14	16	5	11	5	7	11	6	13
Acylated monoglucoside				3	1	1			1				
Diglucoside	6	2	8	1	10	2		2	1	3	4	4	15
Acylated diglucoside										2		4	
Delphinidin													
Monoglucoside		14	9	21	13	23	36	15	30	17	31	20	
Acylated monoglucoside		1	3							6			
Diglucoside	38	12	34						1	1			
Acylated diglucoside		2	6							3			
Petunidin													
Monoglucoside		10	3	15	10	20	26	18	20	8	10	4	5
Acylated monoglucoside		1											
Diglucoside	29	17	22	1	1	1		2	2	1			
Acylated diglucoside		3	2							3			
Malvidin													
Monoglucoside		6	2	34	29	25	27	30	16	4	6		27
Acylated monoglucoside		2		8	3	2	3	5	1		2		
Diglucoside	18	21	8	2	10	2		5	2	1	2		40
Acylated diglucoside		2	1					2		1			

[a]   More recent works[11,13,43] put the total number of anthocyanins of *V. rotundifolia* at 10 and that of *V. labrusca* var. Concord at 31, respectively.

Adapted from Ribéreau-Gayon, P., in *Anthocyanins as Food Colors*, Markakis, P., Ed., Academic Press, New York, 1982, 209.

anthocyanins have been found in *V. rotundifolia*. By contrast, *V. vinifera* is rich in acylated anthocyanins (Tables 1 and 2), which appear to possess more color stability with increasing temperature and light.[45,46]

In an effort to combine the disease resistance of *V. rotundifolia* with the quality of *V. vinifera*, grape breeders have been producing hybrids from these two species for more than a century.[44] By using HPLC techniques Goldy et al.[11] analyzed the anthocyanin content of 12 *Euvitis* × *V. rotundifolia* hybrids; three *V. vinifera* × *V. rotundifolia*; and one combination of *V. labrusca*, *V. rotundifolia* and *V. vinifera*. All intersubgeneric hybrids contained mono- and diglucosides, and all but one produced acylated anthocyanins. A total of 40 pigments were separated, with the Concord variety containing the highest number of pigments at 31. The relative content of acylated anthocyanins in the hybrids ranged from 0 to 51.1%. Similar results reported by Lamikanra[12] from a study of 14 black *V. rotundifolia* hybrids grown in central Florida are shown in Table 4. Only one anthocyanin, cyanidin 3(6-*O*-*p*-coumarylglucoside)-5-glucoside was common to 13 of the 14 hybrids at relative concentrations ranging from 0.4 to 30.6%. Monoglucosides and diglucosides were evenly distributed among the fruits, occurring at relative concentrations between 14 to 89% and 11 to 68%, respectively. Most of the cultivars had no delphinidin 3,5-diglucoside, and the relative amounts of the other nonacylated anthocyanins were very low. This appears to be indicative of recessive contributions of anthocyanins by *V. rotundifolia* to the hybrids or the extent of their involvement in the development of these genetic materials.

V. labrusca grapes are known to contain a mixture of acylated and nonacylated mono- and diglucosides of the six common anthocyanidins (Table 3). The anthocyanins of these grapes, especially those of the Concord variety, have been investigated by a number of workers.[12,47-52] Early investigations described the presence of 14 anthocyanins in 'Concord' grapes, some which were acylated with *p*-coumaric acid. In more recent work, Hrazdina[51,52] characterized 20 anthocyanin pigments in 'Concord' grapes, which include 3-glucoside, 3-(6-*O*-*p*-coumarylglucoside), 3,5-diglucoside, and 3(6-*O*-*p*-coumarylglucoside) of delphinidin and cyanidin. As previously mentioned, Goldy et al.[11] separated 31 anthocyanins from 'Concord' grapes; however, they fully characterized only 12. These were the 3-glucosides and 3,5-diglucosides of cyanidin, peonidin, delphinidin, petunidin, and malvidin; and the 3-acetylglucoside and 3-*p*-coumarylglucoside of malvidin.

V. riparia and *V. rupestris* are useful as breeding material of rootstocks because they confer vine resistance, particularly against the root louse *Phylloxera*. The anthocyanins of these species are primarily glucosides of delphinidin, petunidin, and malvidin (Table 3). According to Ribéreau-Gayon,[4,40] the diglucoside characteristic of *V. riparia* and *V. rupestris* is dominant over the monoglucoside character of *V. vinifera*. Thus, a cross between *V. riparia* and *V. vinifera* yields an F_1 population of hybrids, all with a diglucoside character; however, if one of the F_1 hybrid is crossed with *V. vinifera*, one half of the subsequent F_2 hybrids will not possess 3,5-diglucosides.[4] This genetic feature of these species has resulted in some difficulties with the differentiation of red wines from *V. vinifera* and red wines from hybrids.[4]

The anthocyanin skin pigments of the American grape (*V. cinerea* Engelman) not given in Table 3 are qualitatively similar to the pigment profiles of most *V. vinifera* varieties.[52] Anthocyanins acylated with acetic acid were

TABLE 4

Anthocyanins of *Vitis rotundifolia* Hybrid Grapes

Anthocyanin	LE	BL	MS	MB	M4	M6	HE	BS	CO	C8	H1	D2	C2	AD
Total number of anthocyanins	13	9	12	13	9	16	6	13	14	11	17	15	12	6
							Percent of Each							
Delphinidin 3,5-diglucoside				1.9		0.2				3.1				
Cyanidin 3,5-diglucoside				17.9		0.4				3.6	0.5		0.2	
Petunidin 3,5-diglucoside			16.5	10.1	7.9					15.0	1.2		0.2	
Delphinidin 3-glucoside	3.6	4.3			13.9	1.1			1.9		1.1			0.9
Peonidin 3,5-diglucoside		9.3	26.4	6.9		2.3		3.1	2.0	5.4	0.7	3.0	18.6	
Malvidin 3,5-diglucoside			4.7			16.2		2.1	2.8	53.0	0.2	6.8	41.2	3.0
Cyanidin 3-glucoside	2.6	26.9	11.9		50.4	7.6	29.3	6.2	15.6			3.1		8.4
Petunidin 3-glucoside	6.1	44.7	18.1	21.7	7.5	3.0		1.3	24.9	7.7	1.9	0.5		
Peonidin 3-glucoside									2.0	0.1	0.4	1.8		
Malvidin 3-glucoside								2.5	15.5	5.6	5.4			
Delphinidin 3(6-O-p-coumarylglucoside)-5 glucoside	2.1		7.0	0.7	15.0	28.5	30.3		21.6	6.0	16.6	4.6		73.5
Cyanidin 3(6-O-p-coumarylglucoside)-5-glucoside	30.6	1.4	11.9	0.6	3.9	11.9	19.2	22.7	0.8		0.5	5.4	0.4	12.5
Petunidin 3(6-O-p-coumarylglucoside)-5-glucoside	1.6	3.1	0.2	0.9	0.8			3.5			5.5	0.2	3.9	
Peonidin 3(6-O-p-coumarylglucoside)-5-glucoside	5.8		0.4	4.7		1.6		8.3	1.8		1.9	1.9	34.7	
Malvidin 3(6-O-p-coumarylglucoside)-5-glucoside	0.1		0.4	9.8		1.4		3.8			22.9	1.6	0.2	1.5
Delphinidin 3(6-O-p-coumarylglucoside)	2.56					10.2	1.6	27.2				6.0		

TABLE 4 (continued)
Anthocyanins of *Vitis rotundifolia* Hybrid Grapes

Anthocyanin	LE	BL	MS	MB	M4	M6	HE	BS	CO	C8	H1	D2	C2	AD
Unknown				0.3					1.5	0.4	3.6	4.0	0.2	
Cyanidin 3(6-*O*-*p*-coumarylglucoside)	2.7	5.8	1.8	7.9	0.2	1.6		3.7	0.5		4.1	13.4	0.3	
Unknown	1.2		0.6			3.8					7.3			
Petunidin 3(6-*O*-*p*-coumarylglucoside)				16.4			2.2	3.1	2.4			12.9	0.1	
Unknown	1.4										27.7	35.1		
Unknown					0.4									
Peonidin 3(6-*O*-*p*-coumarylglucoside)		3.5				2.3		14.0		0.3				
Malvidin 3(6-*O*-*p*-coumarylglucoside)	15.7	1.1				8.1	9.2		6.1					
Unknown													0.1	

Note: LE, BL, MS, MB, M4, M6, HE, BS, CO, C8, H1, D2, C2, and AD are Lenoir, Blue Lake, Midsouth, Miss Blue M4-83, M6-7E, Herbermont, Black Spanish, Conquistador, CD8-23, H17-22, DC2-23, CD12-72, and AD2-75 cultivars, respectively.

Adapted from Lamikanra, O., *Food Chem.*, 33, 225, 1989.

TABLE 5
Total Anthocyanin Content of Selected Grape Cultivars

Cultivar	Total anthocyanin (mg/100 g)[a]		Cultivar	Total anthocyanin (mg/100 g)[b]
	Range	Mean		
Cabernet Sauvignon	86–98	92	Lenoir	51.5
Gamay	27–59	39	Blue Lake	12.4
Pinot Noir	33	33	Midsouth	5.5
Baco Noir	148–186	163	Miss Blue	18.9
Chelois	76–165	130	M4-83	37.4
De Chaunac	144–257	198	M6-7E	50.4
Marechal Foch	115–204	154	Herbermont	37.0
Festivee	230–350	300	Black Spanish	40.0
Vincent	396–474	439	Conquistador	105.5
Lomanto × Colobel	480–751	603	CD8-23	54.0
Colobel	429–626	528	H17-22	35.9
Lomanto	381–450	421	DC2-23	85.9
Ives	159–297	245	CD12-72	29.7
Concord	61–112	76	AD2-75	59.0

[a] From Fuleki, T. and Babjak, L. J., *Highlights Agric. Res. Ont.*, 9, 6, 1986.
[b] From Lamikanra, O., *Food Chem.*, 33, 225, 1989.

first discovered in skins of *V. cinerea*.[53] They are difficult to characterize because their absorption spectra are identical to the corresponding 3-gluco-sides. The anthocyanins of two "native" Japanese grape varieties, 'Koshu' and 'Ryugan', have recently been characterized in order to determine the origin of these varieties.[55,56] Yokotsuka et al.[55,56] found that both varieties have a very similar phenolic composition, although the relative amounts of individual phenolic compounds are somewhat different.[55] In the 'Koshu' grape the 3-glucosides of malvidin (20 to 37%), peonidin (39 to 53%), cyanidin (<20%), and delphinidin (<1%) represented the major anthocyanins, and two unusual formic acid acylated peonidin 3-glucosides were present in trace amounts.[56] This pigment pattern is very different from that of grapes of *V. vinifera* (Tables 1 and 2) but more similar to that of 'Muscat Hamburg' which is believed to be an Oriental-European variety.[56] The 'Koshu' grape skin has no 3,5-diglucoside and a very small amount of delphinidin 3-glucoside, in-dicating that it is not from an American species (Table 3), although its viti-cultural characteristics are somewhat similar to American varieties with the exception that it is susceptible to *Phylloxera*. There is no clear evidence supporting 'Koshu' being a *V. vinifera*. As indicated by Akuta et al.,[57] the overall data suggest that it is a unique pink skin variety intermediate between those of European and American varieties.

Total anthocyanin content of red grapes ranges from about 30 to 750 mg/100 g of ripe berries (Table 5), and total phenolics range from 260 to 900 mg/100 g.[14] Cultivar, maturity, year of production, and other environmental factors affect the total anthocyanin and total phenolic content of the fruit.[4,11,12,14]

TABLE 6
**Effect of Cinnamic Acid and Its Derivatives on Anthocyanin
Production by Cultured Callus Tissue of *Vitis* sp. on B5
Medium**

Added substrate	Conc (mg/l)	Cell growth (grams per test tube)	Ratio of Cy3G:Pn3G (μg/g FCW)	
Cinnamic acid	10	0.86	155:12	(93:7)[a]
	50	0.80	115:16	(88:12)
	100	0.81	62:9	(87:13)
p-Coumaric acid	10	0.98	111:5	(96:4)
	50	0.95	84:3	(97:3)
	100	0.89	100:2	(98:2)
Caffeic acid	10	0.92	95:5	(95:5)
	50	0.97	112:5	(96:4)
	100	1.04	91:4	(96:4)
Control[b]		0.81	264:17	(94:6)

[a]   Ratio of cyanidin 3-glucoside (Cy3G):peonidin 3-glucoside (Pn3G) in %.
[b]   Periodically subcultured callus (sixth subculture) on B5 medium at 40 d.

From Tamura, H., Kumaoko, Y., and Sugisawa, H., *Agric. Biol. Chem.*, 53, 1969, 1989. With permission.

III. ANTHOCYANINS IN CULTURED CELLS OF *VITIS* SPECIES

Formation and identification of anthocyanins in cultured cells of *Vitis* species have been investigated by several groups. Yamakawa et al.[58] separated six anthocyanins from the callus and suspension-cultured cells of 'Bailey Alicant A' grapes ([*V. lincecumii* × *V. labrusca* × *V. vinifera*] × [*V. vinifera* × *V. vinifera*]). Using thin-layer chromatography (TLC) and spectral analysis, the three major pigments were characterized as cyanidin 3-glucoside, peonidin 3-glucoside, and peonidin 3,5-diglucoside. The content of anthocyanins in the cultured cells was 0.25 mg/g fresh weight (FW), and the yield of pigments was 156 mg/l of cultured broth incubated in the dark for 16 d.[59]

Tamura et al.[60] used HPLC, ^{1}H-NMR, and fast atom bombardment-mass spectrophotometry (FAB-MS) to characterize the anthocyanins produced by a callus culture of 'Muscat Bailey A' incubated on B5 medium (Gamborg et al.[61]) as cyanidin 3-β-glucopyranoside and peonidin 3-β-glucopyranoside. The content of total anthocyanin 40 d after subculture was 0.5 mg/g of fresh callus weight (FCW), and the ratio of cyanidin- to peonidin 3-glucoside decreased significantly with the addition of cinnamic acid to the B5 medium (Table 6). For instance, addition of 10, 50, and 100 mg of acid per liter of medium lowered the total production of anthocyanins by 41, 53, and 75%, respectively; and the content of cyanidin 3-glucoside from 94 to 87% of the total anthocyanin content. Addition of p-coumaric or caffeic acid to the medium, also lowered production of total anthocyanin content, but had little

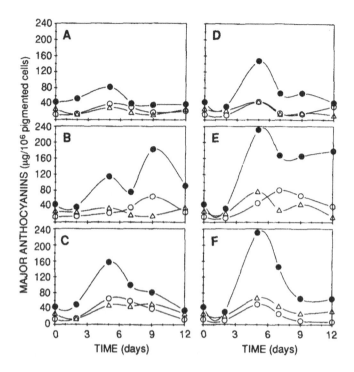

FIGURE 2.   Concentration of cyanidin 3-glucoside (○), peonidin 3-glucoside (●), and peonidin 3-p-coumarylglucoside (△) in pigmented cells during the growth cycle of *Vitis vinifera* cell suspension inoculated in (A) medium containing 88 mM sucrose and 25 mM nitrate, (B) 88 mM sucrose and 6.25 mM nitrate, (C) 88 mM sucrose and 0 mM nitrate, (D) 132 mM sucrose and 25 mM nitrate, (E) 132 mM sucrose and 6.25 mM nitrate, (F) 132 mM sucrose and 0 mM nitrate. (From Do, C. B. and Cormier, F., *Plant Cell Rep.*, 9, 500, 1991. With permission.)

effect on the ratio of cyanidin- to peonidin 3-glucoside. Conversely, these same authors found that the production of anthocyanins by callus culture could be increased 1.4 times by incubating the tissue in the dark for about 30% of the incubation period.

In pigmented cells of *V. vinifera* cv. Gamay Fréaux var. teinturier grown on Gamborg B5 medium, Do and Cormier[62] identified cyanidin 3-glucoside, peonidin 3-glucoside, and peonidin 3-p-coumarylglucoside. The levels of these anthocyanins in pigmented cell suspensions incubated in media containing 88 and 132 mM sucrose; and 0, 6.25, and 25 mM nitrates are shown in Figure 2. In all media, intracellular anthocyanin concentration peaked on day five of culture. Cells grown in medium containing 33 mM sucrose and 6.25 mM nitrates showed a second maximum in anthocyanin concentration at day nine. At both levels of sucrose, decreasing the concentration of nitrate in the culture medium significantly enhanced the intracellular accumulation of total anthocyanins, particularly of peonidin 3-glucoside. Maximum accumulation of anthocyanins was obtained when nitrate concentration was reduced from 25 to 6.25 mM and when sucrose concentration was increased

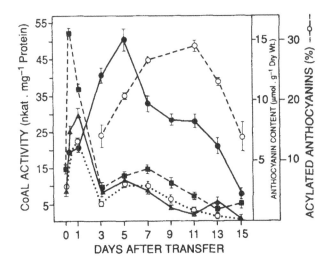

FIGURE 3.   Changes in anthocyanin level (●), acylated anthocyanin (○), and coenzyme A ligase activity with *p*-coumaric (■), ferulic (▲), and caffeic (○) acids during growth of grape cell culture. (From Lofty, S., Fleuriet, A., Ramos, T., and Macheix, J. J., *Plant Cell Rep.*, 8, 93, 1989. With permission.)

from 88 to 132 m*M*. Similar findings on the effects of low inorganic nitrogen and high sucrose concentrations on anthocyanin accumulation were reported by Yamakawa et al.[63]

In cell suspension cultures of *V. vinifera* cv. Gamay Fréaux, Lofty et al.[64] also found that anthocyanin content peaked at day 5 of culture and the maximum acylated anthocyanin concentration was recorded 11 d after transfer (Figure 3). In these cell suspensions, coenzyme A (CoA) ligase activity increased considerably in the hours following seed culturing, and this is believed to provide the *p*-coumaryl CoA required for the synthesis of the anthocyanins.[64]

The presence of peonidin 3,5-diglucoside in the callus and cultured cells of the *Vitis* hybrid 'Bailey Alicant A'[58] and its absence in cell suspensions of *V. vinifera* cv. Gamay Fréaux[62] are consistent with the generally accepted view that *V. vinifera* lacks 3,5-diglucosides.[4] On the other hand, the occurrence of cyanidin 3-glucoside as the major anthocyanin of 'Bailey Alicant' grape cells and peonidin 3-glucoside as the major anthocyanin of 'Gamay Fréaux' grape cells is not entirely consistent with the fact that peonidin derivatives account for less than 30% of the anthocyanin content of *V. vinifera* (Tables 1 and 2); and cyanidin derivatives are generally not major anthocyanins of *V. lincecumii, V. labrusca,* and *V. vinifera* (Tables 1 to 3). Thus it may be suggested that in cultured cells of *Vitis,* the composition of anthocyanins is very likely determined by the source of the cell culture and by the composition of the medium in which the cells are incubated.[60,62,63]

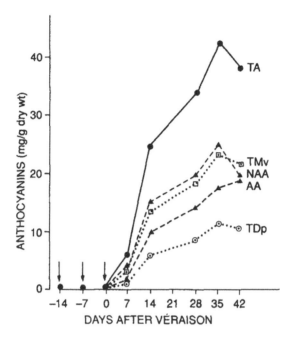

FIGURE 4. Changes in concentration of anthocyanins in the skin of ripening 'Cabernet Sauvignon' grapes. TA = total anthocyanins; TMv = total malvidin derivatives; NAA = nonacylated anthocyanins; AA = acylated anthocyanins; TDP = total delphinidin derivatives; (↓) = traces of anthocyanins detected before véraison. (From Darné, G., *Connaiss. Vigne Vin*, 22, 225, 1988. With permission.)

IV. CHANGES IN ANTHOCYANINS DURING RIPENING OF GRAPES

Anthocyanin content of grapes increases during ripening, and accumulation begins at véraison, an event signaling the end of the development period of the berries and the onset of the ripening process.[65] Development of anthocyanins from véraison to complete maturity is generally characterized by three phases. The first phase exhibits a slow increase which is followed by a rapid increase ending in a stabilization phase before a decrease at the end of ripening.[4,65-72] Figure 4 shows the changes in concentration of total anthocyanins, total malvidin derivatives, total nonacylated anthocyanins, total acylated anthocyanins, and total delphinidin derivatives in the skin of ripening 'Cabernet Sauvignon' grapes from the Bordeaux area.[71] The changes in the major monoglucosides and acylated anthocyanin pigments are shown in Figures 5 and 6. Nonacylated anthocyanins, representing about 60% of the total anthocyanins, showed a clearly defined maximum level about 35 d after véraison. Somers[67] found that, in grapes grown in South Australia, maximum anthocyanin concentration was recorded 20 to 30 d after véraison, when the

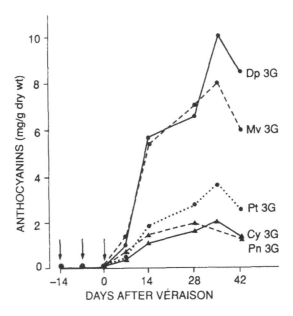

FIGURE 5. Changes in concentration of anthocyanins in the skin of ripening 'Cabernet Sauvignon' grapes. Dp 3G = delphinidin 3-glucoside; Mv 3G = malvidin 3-glucoside; Pt 3G = petunidin 3-glucoside; Cy 3G = cyanidin 3-glucoside; Pn 3G = peonidin 3-glucoside. (From Darné, G., *Connaiss. Vigne Vin*, 22, 225, 1988. With permission.)

sugar content was 20 to 24%. The acylated pigments increased steadily during ripening and neither the total (Figure 4) nor the individual acylated anthocyanins (Figure 6) displayed the characteristic decrease at the end of ripening. The most probable reason for the lack of a decrease in acylated anthocyanins at this physiological stage is the larger number of steps required for the synthesis of these pigments.[73] It is expected that the maximum content and a decrease in the acylated pigments may have been detected had sampling and analysis of the grapes continued for 5 to 10 additional days. In 'Gamay Fréaux' grape cell suspensions cultured *in vitro*,[65] acylated anthocyanins peaked 6 d after peaking of the nonacylated pigments (Figure 3).

Roggero et al.[70] related the variation of the anthocyanins in 'Syrah' grapes to their position in the biosynthetic pathway. These authors found that cyanidin 3-glucoside peaks 19 d after véraison (Figure 7), followed by delphinidin and petunidin glucosides, and later by peonidin and malvidin derivatives. Thus, there is successive accumulation of anthocyanins with increasing degree of substitution in the B-ring. The important acylated anthocyanins in 'Syrah' grape skins, malvidin 3-glucoside acetate and *p*-coumarate, parallel the changes in concentration of malvidin 3-glucoside.[70]

In 'Barbera' and 'Bonarda' grapes from the Pavese and Monferrato regions of Italy, Piergiovanni and Volonterio[69] found only moderate changes in the relative proportions of anthocyanins during ripening. It was pointed out, however, that ripe grapes contain considerable quantities of acylated antho-

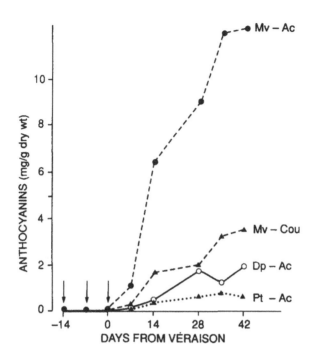

FIGURE 6. Changes in concentration of the major acylated anthocyanins in the skin of ripening 'Cabernet Sauvignon' grapes. Mv-Ac = malvidin 3-acetylglucoside; Mv-Cou = malvidin 3-*p*-coumarylglucoside; Dp-Ac = delphinidin 3-acetylglucoside; Pt-Ac = petunidin 3-acetylglucoside. (From Darné, G., *Connaiss. Vigne Vin*, 22, 225, 1988. With permission.)

cyanins and that the most abundant anthocyanin of the grape is malvidin 3-glucoside. Significant positive correlations between content of anthocyanins (total, malvidin 3-glucoside, malvidin 3-acetylglucoside, peonidin 3-glucoside, total acylated anthocyanins) and ripening index (% sugar per parts per million tartaric acid) were also reported by these authors. These correlations, however, indicate no interdependence between accumulation of anthocyanins in the skin and sugars in the pulp of grapes, because the correlation coefficients vary between cultivars and from year to year.

González-San José et al.[72] reported that in 'Tempranillo' grapes from Spain, the anthocyanin content increases during ripening, but not in a steady manner because of the occurrence of a drop in concentration a few days prior to physiological maturity (Figure 8). The accumulation pattern of petunidin 3-glucoside fits a third degree polynomial regression model, while the other monoglucosides could be described by a second degree polynomial model. Development of acyl glucosides in 'Tempranillo' grapes is characterized by a phase of steady accumulation, followed by a drop in concentration until a discernible minimum is reached and a phase of rapid accumulation until physiological maturity is reached (Figure 8 [b–d]). Noteworthy is the lack of a decrease in the concentration of these pigments at the end of ripening. In addition, the acyl glucosides in 'Tempranillo' grapes do not evolve in the

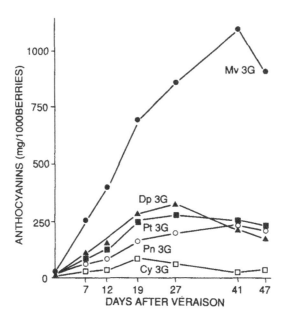

FIGURE 7. Changes in anthocyanin content in the skins of ripening 'Syrah' (clone ANTV 73) grapes. Mv 3G = malvidin 3-glucoside; Pn 3G = peonidin 3-glucoside; Pt 3G = petunidin 3-glucoside; Dp 3G = delphinidin 3-glucoside; Cy 3G = cyanidin 3-glucoside. (Adapted from Roggero, J. P., Coen, S., and Ragonnet, B., *Am. J. Enol. Vitic.*, 37, 77, 1986.)

same way as the monoglucosides or as the acyl glucosides of 'Cabernet Sauvignon' grapes (Figure 6).

The pattern of development of anthocyanins is thus strongly influenced by cultivar, season, and site of production, as well as cultural practices and difficulty in establishing the exact length of the ripening period.[4,72,76] The level of solar radiation has long been known to be an important parameter in the red coloration of grapes.[74-76] Reduction of radiation to 15% of normally full sunlight results in a 58 to 59% decrease in the total anthocyanin content (Table 7), whereas total soluble solids are very little affected. Pirie[76] attributes the loss in anthocyanin content with reduced level of solar radiation to the influence of light on the sugar level of grape skin, which is considerably reduced. Conversely, the light has little effect on sugar level in the pulp of the berry. Significant differences in both anthocyanin concentration and total content per berry between sun-exposed and shaded fruit prior to ripening were found by Crippen and Morrison[77] in clusters of 'Cabernet Sauvignon'. At the final harvest date, however, there were no significant differences between sun-exposed and shaded berries. As in other fruits, it would appear that the major regulation features in grapes is phenylalanine ammonia-lyase (PAL) activity, which occurs mainly in berry skins.[65,78] When whole isolated berries are exposed to light for 72 h, PAL activity and anthocyanins increase, whereas in the dark PAL activity decreases and no accumulation of anthocyanin is observed.[78]

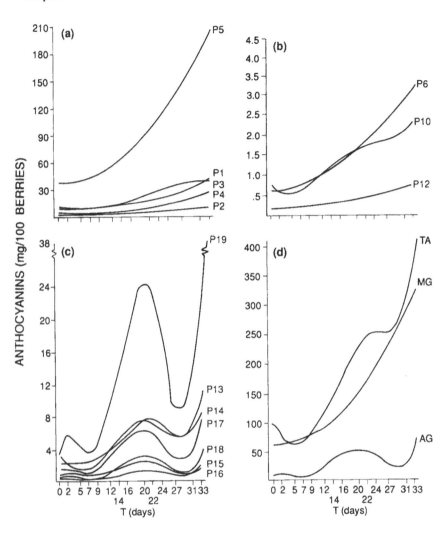

FIGURE 8. Evolution of anthocyanins during ripening of 'Tempranillo' grapes. The time values (T) correspond to days after first harvest (T = 0). (a) Monoglucosides of delphinidin (P_1), cyanidin (P_2), petunidin (P_3), peonidin (P_4), and malvidin (P_5). (b) Acetylglucosides of delphinidin (P_6), petunidin (P_{10}), and peonidin (P_{12}). (c) Coumaryl glucosides of delphinidin (P_{14}), cyanidin (P_{16}), petunidin (P_{17}), peonidin (P_{18}), and malvidin (P_{19}); caffeoyl glucoside of malvidin (P_{15}) and acetyl glucoside of malvidin (P_{13}). (d) Total monoglucosides (MG), total major acyl glucosides (AG), and total anthocyanins (TA). (From González-San José, M. L., Barron, L. J. R., and Diez, C., *J. Sci. Food Agric.*, 51, 337, 1990. With permission.)

Grapes tend to develop the highest fruit coloration at 15 to 25°C day and 10 to 20°C night temperature.[77] At high temperature (35 to 37°C day and 30 to 32°C night) the anthocyanin biosynthesis is completely inhibited in sensitive cultivars such as 'Tokay' or 'Emperor' grapes,[75,77] and greatly reduced in the

TABLE 7
Influence of Level of Solar Radiation on Coloration of
'Emperor' Grapes

Level of solar radiation (% of full sunlight)	Total soluble solids (°Brix)		Fruit coloration (anthocyanin in mg/cm² berry surface)	
	Sept. 2	Sept. 14	Sept. 2	Sept. 14
100	17.42 a	19.85 a	0.32 a	0.31 a
54	17.37 a	19.55 a	0.34 a	0.33 a
15	16.50 ab	17.90 b	0.17 b	0.17 b
3	16.62 ab	17.67 b	0.13 c	0.13 c
0.8	15.75 b	17.15 b	0.09 c	0.11 c
0	15.60 b	16.95 b	0 d	0 d

Note: Within a column, means followed by the same letter are not significantly different at the
5% level by Duncan's multiple-range test.

From Kliewer, W. M., *Am. J. Enol. Vitic.*, 28, 96, 1977. With permission.

more deeply colored 'Cardinal' and 'Pinot Noir' cultivars.[74] Buttrose et al.[80]
found that the skins of 'Cabernet Sauvignon' fruits from vines grown at 20°C
day temperature had two to three times the anthocyanin concentration as fruits
ripened at 30°C. In warm and dry regions, such as California, sprinkling
grapevines from bloom to fruit maturity greatly increases fruit coloration,
although the response depends to a considerable extent on cultivar.[81] The
limiting factor in anthocyanin synthesis at high temperature appears to be a
reduced accumulation of soluble sugars in the fruits.[82]

Other factors affecting anthocyanin accumulation during grape maturation
are soil fertility, water supply, disease, and level of plant growth regulators.
Experiments have shown that using excess nitrogen fertilizer in soil can reduce
anthocyanin content of berries by 20 to 30%, delay maturity, and reduce color
of the wine produced from these grapes.[4,75] Limited water supply results in
berries being smaller but richer in anthocyanins and tannins.[83,84] The presence
of the mold *Botrytis cinerea* on the berries induces production of an oxidase
with anthocyanase activity,[85,86] which partially destroys the anthocyanins in
the berries and in wine produced from infected grapes.[85] Protection can be
achieved by application of sodium bisulfite.[4] Ethephon ([2-chloroethyl] phosphonic acid or Ethrel) application to grapevines has been shown to accelerate
ripening, increase color, and reduce vegetative growth; however, responses
have been variable depending on the time and rate of application, cultivar,
and location.[87-91] For instance, ethephon applied to 'Pinot Noir' grapes in
1977 and 1978 in central Washington, at 500 ppm at véraison produced fruit
5 and 18% greater in color than in the control.[91] However, the pigment makeup
of the fruit at harvest was altered considerably by ethephon (Table 8). The
3-glucosides of cyanidin, delphinidin, and petunidin accounted for a significantly larger proportion of anthocyanin pigments in the control than in the

TABLE 8
Effect of Ethephon on Anthocyanins of 'Pinor Noir' Grapes

Anthocyanin	Control (% of total)	Treated (% of total)
Delphinidin 3-glucoside	5.35[a]	2.88
Cyanidin 3-glucoside	3.22[a]	1.84
Petunidin 3-glucoside	5.99[b]	4.06
Peonidin 3-glucoside	21.1	24.8[a]
Malvidin 3-glucoside	64.3	66.3

[a],[b] Significantly greater at the 1 and 5% level, respectively, by *t*-test.

From Powers, J. R., Shively, A. E., and Nagel, C. W., *Am. J. Enol. Vitic.*, 31, 203, 1980. With permission.

ethephon-treated fruit. Conversely, in fruit from treated vines, peonidin 3-glucoside accounted for a significantly larger proportion of the total pigments. Malvidin 3-glucoside, the major anthocyanin of 'Pinot Noir', also occurred in a greater proportion in treated fruit, but the differences were not significant.

Treatment of grapes with 1000 ppm abscisic acid 3 to 4 d after véraison resulted in a significant increase in anthocyanin content of the skin of 'Olympia' grapes 2 to 3 weeks after application.[92] The increase in anthocyanin content was accompanied by a comparable increase in glucose, fructose, rhamnose, and total sugar content of the skin.[92] Abscisic acid also stimulated anthocyanin accumulation in cell cultures of 'Olympia' grapes, although this stimulation required relatively high temperatures and light. The mechanism(s) by which abscisic acid stimulates anthocyanin production in grapes remain, however, essentially unknown.[82]

V. CHANGES IN ANTHOCYANINS DURING PROCESSING AND STORAGE OF GRAPES

A. WINE MAKING
The conditions of wine making and of maturation processes have a very large impact on the color of rosé and red wines. Many reports have been published on the effect of methods of processing and storage on color characteristics of wines.[4,93-98]

1. Extraction Process
The extraction of anthocyanins and related phenolics from the grape solids begins with the crushing of grapes and continues through the maceration, pressing, and fermentation operations. Raising the temperature to 50 to 60°C prior to fermentation of the mash or to 30 to 40°C after the main fermentation

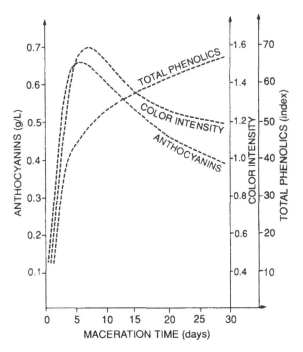

FIGURE 9. Idealized changes in anthocyanins, color intensity, and total phenolics during maceration of crushed grapes. (From Ribéreau-Gayon, P. and Glories, Y., *Enotecnico*, 22, 545, 1980. With permission.)

may be carried out to increase extraction of red pigments, as well as flavor components which contribute to the character of red wines.

Vigorous crushing enhances the extraction and diffusion of the antho-cyanins and other phenolic compounds;[97] however, the resultant wines are frequently too astringent and bitter. Furthermore, enzymatic oxidation of phenolic compounds is enhanced, which may lead to browning of the must or juice and development of undesirable sensory characteristics of the wine. The use of sulfur dioxide (SO_2) as an antioxidant and antiseptic is widely accepted as an indispensable aid in wine making.[98] However, if too much SO_2 is added to musts, fermentation may be incomplete and the color will be bleached.[99]

Maceration, characterized by the diffusion of anthocyanins and other phenolics from the grape solids into the must-wine, may occur either prior to fermentation as in thermovinification or during the alcoholic fermentation, either with crushed grapes (traditional vinification) or with whole grapes (carbonic maceration). In all cases, maceration time is most important. In wines produced by the traditional method, 5 to 6 d of maceration is sufficient for a colored, fruity, low tannin wine for early drinking. If the maceration period is extended, anthocyanin content and color intensity decrease, while total phenolics continue to increase[4,82,100] (Figure 9).

Extended pomace contact or maceration time is practiced in France and northern Italy, where in order to secure wines of very high tannin content, the wine may be allowed to remain with the skins and seeds for a week or more after fermentation is completed. In California wineries, the crushed grapes are in the vats usually only about 3 to 4 d and rarely more than 5 d.[101]

Timberlake and Bridle,[93] compared wines from 'Cascade' grapes made by carbonic maceration, thermovinification, and traditional vinification methods. They found that wine made by thermovinification (60°C for 30 min) was much more colored than that fermented on its skins (10 to 15°C for 2 d); however, it contained less anthocyanin and more polymeric pigments. The wine made by carbonic maceration (30 to 35°C for 7 d) was the least colored, despite containing less of anthocyanins similar to those of wine produced by thermovinification. The differences in color were attributed to variations of the physicochemical state of the anthocyanins.

Somers and Evans[94] followed the changes in anthocyanins, total phenolics, and color density of heat treated 'Syrah' (syn. Shiraz) grape juice during thermovinification at pH 3.40 and 3.83 and during traditional fermentation on the skins. Their results reveal that by the end of fermentation at pH 3.83, color density ($E_{420} + E_{520}$) had decreased fivefold, whereas the anthocyanin content and total phenolics declined only by approximately 30%. At pH 3.4, the color density decreased threefold and the total anthocyanins and total phenolics declined by 20%. The authors attributed the majority of the loss in color density to the destructive effect of ethanol upon structures of deeply colored pigment aggregates present in the juice prior to fermentation.

The changes in 15 anthocyanins in musts and wines from three cultivars of grapes from southern Italy were reported by Leone et al.[95] Table 9 gives the evolution of the anthocyanins in 'Troia' must and wine produced by the conventional fermentation method. As in other *Viniferas* (Table 1), malvidin 3-glucoside and its acetate and coumarate derivatives were the major anthocyanins of 'Troia' grapes. Within 24 h from crushing of the grapes, the percentage of malvidin 3-glucoside had changed from 28.6 to 44.4%, malvidin 3-acetylglucoside remained near 20%, and malvidin 3-*p*-coumarylglucoside decreased from 27.6 to 12.3% of the total anthocyanin content. Seven days after the beginning of wine making, at the time of drawing off of the must and pressing of the pomace, the total anthocyanin content of the must had increased from 446 mg/l at the beginning of maceration to 1079 mg/l; and the percent distribution of the pigments had undergone only limited variations. In wine at the third racking, 7 months after vinification, the malvidin derived anthocyanins accounted for 85% of the total anthocyanins in the wine. At this stage, the delphinidin, cyanidin, petunidin, and peonidin derivatives were only minor components or had completely disappeared from the wine. The amount of the major anthocyanins transferred from the grapes to the 7-month-old wine were 41% malvidin 3-glucoside, 39% malvidin 3-acetylglucoside, and only 9% malvidin 3-*p*-coumarylglucoside. The corresponding values for a 20-d-old 'Syrah' wines are 31, 32, and 8.5%, respectively.[9]

TABLE 9
Anthocyanin Composition of 'Troia' Grapes during Wine Making

Anthocyanin	Grapes at harvest		24 h		48 h		72 h		Drawing off and pressing		First racking		Second racking		Third racking	
	ppm	%	mg/l	%	mg/l	%	mg/l	%	mg/l	%	mg/l	%	mg/l	%	mg/l	%
Dp	85	4.9	26	5.8	48	3.9	40	3.2	32	3.0	22	2.0	21	2.7	12	2.7
Cn	8	0.5														
Pt	83	4.8	24	5.4	61	5.0	68	5.4	60	5.6	53	4.9	41	5.2	24	5.4
Pn	55	3.1	16	3.6	22	1.8	13	1.0	21	1.9	15	1.4	6	0.8	6	1.3
Mv	500	28.6	198	44.4	560	45.4	551	43.7	464	43.0	523	48.1	362	45.8	206	46.3
Dp acetate	18	1.0					14	1.1	29	2.7	23	2.1	6	0.8	6	1.3
Cy acetate	tr.								13	1.2	23	2.1	5	0.6	6	1.3
Pt acetate	19	1.1	8	1.8	15	1.2	33	2.6	20	1.9	14	1.3	17	2.1	6	1.3
Pn acetate	13	0.7	6	1.3	10	0.8	32	2.5	12	1.1	261	24.0	6	0.8	6	1.3
Mv acetate	335	19.2	93	20.9	286	23.2	300	23.8	256	23.7	13	1.2	204	25.8	129	29.0
Dp coumarate	67	3.8			19	1.5	11	0.9	6	0.5	19	1.8	6	0.8		
Mv caffeoate	15	0.9									9	0.8				
Pt coumarate	42	2.4	13	2.9	23	1.9	30	2.4	20	1.9	112	10.3	14	1.8		
Pn coumarate	25	1.4	7	1.6	19	1.5	16	1.3	16	1.5			9	1.1		
Mv coumarate	482	27.6	55	12.3	170	13.8	153	12.1	130	12.0			94	11.9	44	9.9
Total	1747		446		1233		1261		1079		1087		791		445	

Note: Crushing of grapes: 10-19-82; pressing: 10-23-82; first racking: 10-27-82; second racking: 11-30-82; third racking: 5-5-83.

Dp = delphinidin 3-glucoside; Cy = cyanidin 3-glucoside; Pt = petunidin 3-glucoside; Mv = malvidin 3-glucoside.

From Leone, A. M., La Notte, E., and Gambacorta, G., *Vignevini*, 4, 17, 1984. With permission.

The changes in anthocyanin 3-monoglucosides during traditional and carbonic maceration of 'Terrano' grapes from the Veneto region of Italy are given in Table 10. In must-wine processed by both methods, the content of total and individual anthocyanins increased with time of extraction. The quantity of monoglucosides extracted was, however, considerably larger in products processed by the traditional method of maceration. The percent distribution of the pigments extracted by two methods also varied; and malvidin 3-glucoside — the major anthocyanin pigment — accounted for 83.3% of the total anthocyanins in must-wine processed by CO_2 extraction, but only for 68.7% of the anthocyanins in products processed by the traditional method. Wines produced by carbonic maceration also contain about 60% of the total phenolics and 50% of the nontannin phenolics of wines processed by the traditional method.[96] This indicates that increasing contact time may lead to a desirable increase in anthocyanin content, but to an undesirable accumulation of tannins. The polyphenol composition of wines produced by carbonic maceration is closely linked to the combination of temperature and the duration of the fermentation. It has been reported that it requires 10 d at 30 to 32°C to obtain wines rich in tannins and 6 d at 32°C or 12 d at 25°C to obtain a wine more colored than tannic.[102]

Changes in anthocyanins and other phenolics as a result of their adsorption/desorption by grape solids and yeast have been reported;[103] however, such effects are minimal. Nonetheless, seed and stalk phenols are qualitatively and quantitatively different from those of the berry; and under conditions of vigorous crushing and extended pomace contact they will contribute to the quality of the finished product.[104,105]

2. Storage and Aging

Wines are stored in oak, concrete, redwood, lined iron or steel, and stainless steel containers.[101] For red table wines and for quality dessert wines, oak is the preferred material for storage. The average oak cask or redwood wine tank is not airtight, but is rather porous. Air can, and does, enter slowly; the oxygen contacting the wine plays an important role in the aging process of most wines.[100,101] Red wines of moderate to high phenolic content 1100 to 1800 mg gallic acid equivalent (GAE) per liter, are commonly stored in oak barrels or casks for up to 2 years. This storage period, called the maturation phase, is followed by the aging phase in which the wine is stored in bottles.[24,100,106-109]

During maturation in barrels or tanks, where oxygen is present, the redness ($A_{520 nm}$) of red young wine decreases and the absorbance in the yellow/brown region, near 420 nm, rises.[110] These changes in color characteristics reflect the progressive displacement of grape anthocyanins by more stable polymeric pigments, which account for up to 50% of the color density of 1-year-old wine.[24,111] These polymers are also responsible for alterations in taste and flavor, especially reduced astringency.[112] The reactions considered responsible

TABLE 10
Changes in Selected Anthocyanins during Traditional and Carbonic Maceration of 'Terrano' Grapes

Anthocyanin	Traditional								Carbonic							
	0 d		4 d		7 d		14 d		0 d		4 d		7 d		14 d	
	mg/l	%	mg/l	%	mg/l	%	mg/l	%	mg/l	%	mg/l	%	mg/l	%	mg/l	%
Delphinidin 3-glucoside	—	—	0.3	0.6	1.9	1.6	12.0	6.8	—	—	—	—	—	—	0.4	0.6
Cyanidin 3-glucoside	—	—	0.8	1.6	3.5	3.0	3.6	2.1	—	—	0.3	8.5	0.3	1.82	0.9	1.4
Petunidin 3-glucoside	—	—	0.3	0.6	6.9	5.9	17.8	10.2	—	—	—	—	—	—	1.6	2.5
Peonidin 3-glucoside	0.7	29.2	6.8	14.1	16.5	14.3	21.0	12.1	—	—	0.8	22.8	1.8	10.9	7.8	12.2
Malvidin 3-glucoside	1.7	70.8	40.0	82.9	86.5	75.0	119.7	68.7	t	—	2.4	68.6	14.3	87.2	53.4	83.3
Total monoglucosides	2.4		48.2		115.3		174.1		t	—	3.5		16.4		64.1	

Note: t, Trace.

Adapted from Da Porto, C., Bortolomeazzi, R., and Sensidoni, A., *Vignevini*, 5, 51, 1989.

for the formation of these polymeric pigments include: acetaldehyde-mediated condensation, copigmentation, and self-association. As discussed in Chapter 1, these reactions lead to color enhancement and pigment complexes less sensitive to pH change than the anthocyanins. Their stability and reactivity in wine is, however, affected by many factors including: oxygen, SO_2, acetaldehyde, temperature, pH, and concentration of molecules with the ability to act as copigments.

The role of oxygen, believed to be essential in wine maturation by some authors,[100,106,113] has recently been seriously questioned.[107] Results of experiments by Somers and Evans[107] on the effects of the presence or absence of oxygen on the color composition of two red wines having low and high phenolic contents revealed that the reactions responsible for formation of polymeric pigments during maturation of red wines can proceed under anaerobic conditions. Although under commercial wine storage conditions, the normal presence of dissolved oxygen influences the color and phenolic composition of the wine by affecting the accumulation of acetaldehyde from autooxidation of ethanol.[106] Self-oxidation of the o-diphenol group in wine leads to intermediate formation of H_2O_2 and a coupled oxidation of ethanol to acetaldehyde.[114] In wines, acetaldehyde can also be formed by microbial action during fermentation and by photochemical decomposition of tartaric acid catalyzed by traces of iron.[93]

The role of acetaldehyde in maturation and aging of red wine has been recognized since the early 1900s, when Trillat[115] identified acetaldehyde as the factor responsible for wine color instability and formation of precipitates causing both a decrease in astringency and an evolution in color. These findings were confirmed by several other authors including, Joslyn and Comar,[116] Cantarelli,[117] and Mareca-Cortés and del Amo Gili.[118] Singleton et al.[119] reported that red color (absorbance at 520 nm and sensory color rating) was considerably greater for a lot of port fortified with aldehydic brandy than for another subsample of the same wine treated with brandy without added aldehyde or heads. Singleton et al.[119] reported that wines which exhibited more color because of the addition of aldehydic spirits were characterized by a resistance to color increase upon lowering the pH and the polyphenol content of the wine. They advanced the hypothesis that acetaldehyde reacts with anthocyanins via an acid-catalyzed Baeyer (phenol-formaldehyde-Novolak) reaction to form polymers whose resistance to pH effect would seem reasonable.

Timberlake and Bridle[121] proposed a reaction mechanism initiated by the generation of an acetaldehyde carbonium ion, which is capable of reacting with the active positions (6 and 8) of the catechin-type compound, whereby the carbonyl carbon readily undergoes nucleophilic attack by position 8 of catechin.[122] The proposed reaction[121] of acetaldehyde with catechin and malvidin 3,5-diglucoside, including modification of the initiating step, is shown in Figure 10. The acetaldehyde-catechin complex forms a reactive carbonium

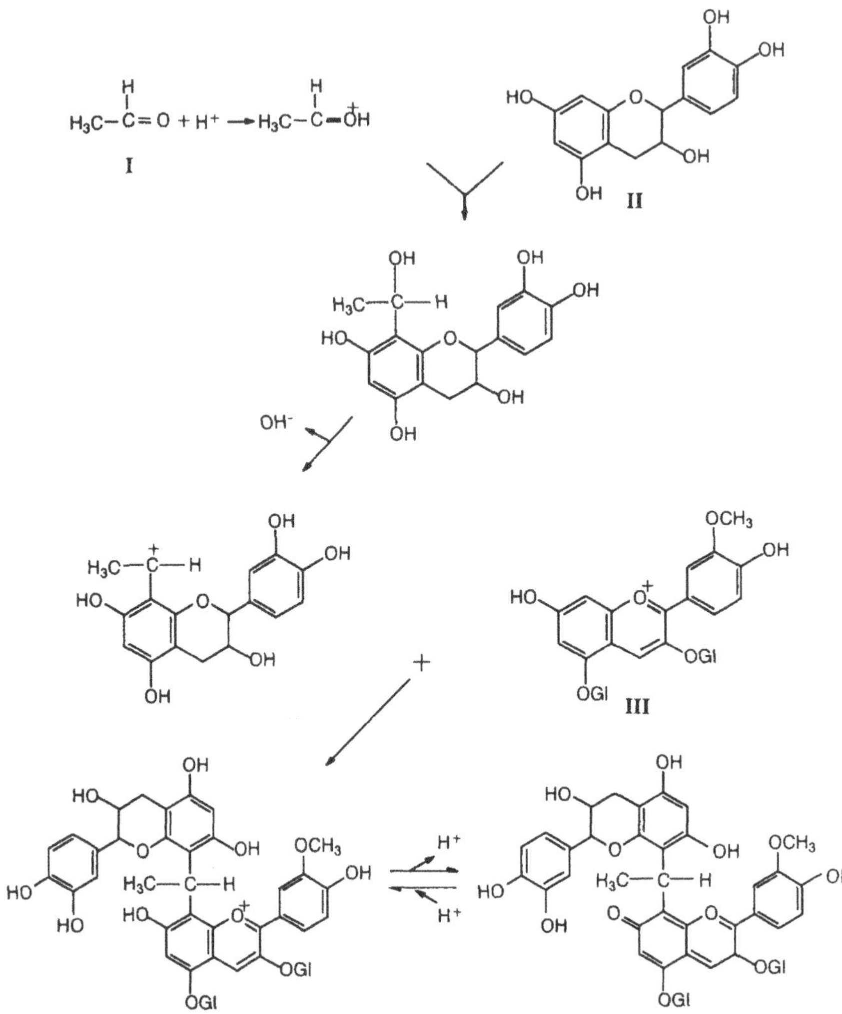

FIGURE 10. Reaction of acetaldehyde (I) with catechin (II) and malvidin 3,5-diglucoside (III). (Adapted from Timberlake, C. F. and Bridle, P., *Am. J. Enol. Vitic.*, 27, 97, 1976.)

ion which combines with the anthocyanin flavylium ion, most probably at position 8. Formation of the quinonoidal form occurs more readily as a result of the substitution pattern which accounts for the subsequent increase in color and the violet shift. Further condensation onto the polymer could occur via the reactive 6 position. By analogy with the behavior of simple flavylium salts, the 6 and 8 positions of the anthocyanin A-ring have been proposed as the reactive positions, with position 8 being the most reactive.[121,125] However, Dournel[125] reported that CH_3CH bridges occur with the same probability between the C_8/C_8, C_8/C_6, and C_6/C_6 positions of anthocyanin and catechin.

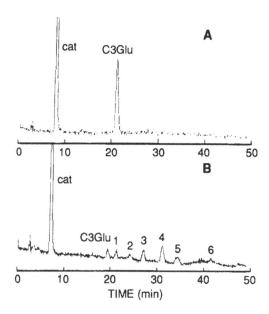

FIGURE 11. HPLC profile at 280 nm of cyanidin 3-glucoside model system (1×10^{-4} M) containing acetaldehyde (2.5×10^{-2} M) and catechin (9.5×10^{-4} M) after initial preparation (A) and 20 d at 23 ± 1°C in the dark (B). Peaks: catechin (cat), cyanidin 3-glucoside (C3G), and unidentified condensation products (1-6). Solvent: aqueous H3PO$_4$-NaOAc; T = 20°C; pH = 3.5. (From Green, R. C. and Mazza, G., *Can. Inst. Food Sci. Technol. J.*, 21, 537, 1988.)

Baranowski and Nagel[126] used wine model systems containing equimolar quantities of malvidin 3-glucoside and *d*-catechin (5×10^{-4} M), with (2.5×10^{-3} M) and without acetaldehyde to investigate the kinetics of the CH$_3$CH bridged polymerization reaction. Their results point to the occurrence of two types of polymerization: acetaldehyde bridged condensation, and copigmentation between the anthocyanins and *d*-catechin. At low concentrations of acetaldehyde, both seem to occur simultaneously. When excess acetaldehyde is present, the bridged polymerization reaction predominates. When no acetaldehyde is present, copigmentation of anthocyanin with *d*-catechin would occur.[127-130]

Green and Mazza[131] reported on the influence of acetaldehyde and catechin on aqueous solutions of cyanidin 3-glucoside (1×10^{-4} M) and aqueous and alcoholic extracts of saskatoon berries. Their results confirm previous findings[121,126] and reveal the presence of six condensation products which were separated by HPLC (Figure 11). Four components were detected by Roggero et al.[132] during condensation of catechin or epicatechin with malvidin 3-glucoside in the presence of acetaldehyde. However, the structural characteristics of these products have not yet been established. Acetaldehyde was assigned a primary role in the phenolic condensation reactions of red wines by Ribéreau-Gayon and co-workers,[100,106,113] who also recommended in-

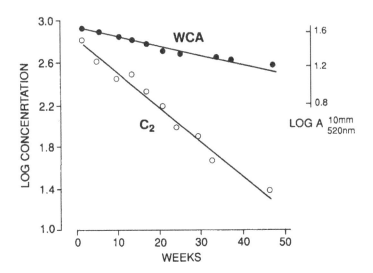

FIGURE 12.   Logarithmic changes of total pigment color (WCA as $A_{520\ nm}^{10\ mm}$ and total antho-
cyanins determined by HPLC (C_2) as mg/l) of port wine. The scales are such that log concentration
$= \log 18.9\ A_{520\ nm}^{10\ mm}$. (From Bakker, J., Preston, N. W., and Timberlake, C. F., *Am. J. Enol.
Vitic.*, 37, 121, 1986. With permission.)

creased aeration and monthly racking of the wine to reduce astringency and
improve color. However, Somers and Wescombe[133] have recently concluded
that any increase in acetaldehyde during processing and storage of red wine
may be detrimental to sensory properties and stability. Also, the wide variation
in acetaldehyde formation and consumption, resulting from variations in wine
processing and storage practices, makes quality control of wine extremely
difficult and contributes largely to the range in red wine styles and quality.[133]
    The role of acetaldehyde is most significant in the maturation and aging
of port wines, where acetaldehyde is derived primarily from the brandy spirit
used for fortification.[134-136] Because of the high alcohol content (19 to 20%),
the acetaldehyde-mediated condensation reaction is likely the single most
important reaction in terms of color variation and astringency of port wines,
as increasing concentrations of alcohol is damaging to the copigmentation
reaction.[127-130] In the aging of port wines, made by fermentation at 28°C and
fortified to 19.5% v/v ethanol with brandy, the losses in total pigment color
— measured as absorbance at 520 nm in 1 *M* HCl — and total anthocyanins
— determined by HPLC — are logarithmic with time (Figure 12).[29,135] The
rate of loss of individual anthocyanins in port wine depends on their structure,
with malvidin 3-glucoside disappearing more slowly than malvidin 3-acetyl-
glucoside or malvidin 3-*p*-coumarylglucoside.[135] The absorbance due to polymeric pigments, calculated by subtracting the color of the monomeric anthocyanins in 1 *M* HCL from total pigment color, is shown in Figure 13. In
port with a normal content of aldehyde, the absorbance due to polymeric
compounds increased during the first 4 weeks, reached a plateau, and de-

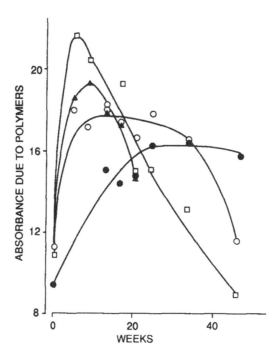

FIGURE 13. Absorbance due to polymers during aging of port wines. (○) Control with a normal content of aldehyde; (▲) very low aldehyde level; (□) with spirit and 200 mg/l acetyl-glucoside; (●) with spirit and 100 mg/l SO_2. (From Bakker, J., *Vitis*, 25, 203, 1986. With permission.)

creased slowly after 20 weeks. In wine fortified with spirit and 200 mg/l aldehyde, the absorbance due polymers went up rapidly, but once the highest value was reached it decreased rapidly. Port with spirit and 100 mg/l SO_2 showed a slow increase in polymer absorbance over 24 weeks, which decreased slightly after 33 weeks. The absorbance of polymers in port with a very low aldehyde level displayed a pattern similar to that of port fortified with spirit and 200 mg/l aldehyde but with a significantly lower maximum level. The rapid decrease in polymeric pigment color in port high in acetaldehyde is attributed to precipitation of large polymeric molecules and/or to reduced color intensity of the larger polymers.

A significant role of the copigmentation reaction of anthocyanins with other polyphenols in the maturation and aging of wine was recognized by Somers and Evans.[94] However, to date no quantitative data on the contribution of this unique molecular interaction to the color characteristics of red wines have been reported. Although known for a long time,[137] the copigmentation reaction has only recently been investigated at the molecular level.[127-129,137,138] It is now known that noncovalent complex formation between the anthocyanin-colored structures and the copigment is responsible for intensification and change in the initial solution color.[127,138] Molecules capable of acting as copigments include a large variety of structurally unrelated compounds, such as flavonoids, other polyphenols, alkaloids, amino acids, and anthocyanins themselves.[127-130,137-139] Red wines contain many phenolic compounds besides the anthocyanins including: *p*-hydroxybenzoic, *o*-hydroxybenzoic, salicylic, gallic, cinnamic, *p*-coumaroyl tartaric (coutaric), and caffeoyl tartaric (caftaric) acids; feruloyl tartrate; *p*-coumaroyl-glucoside; feruloylglucoside; kaempferol 3-glucoside, 3-glucuronide, galactoside, and 3-glucosylarabinoside; quercetin 3-glucoside, 3-glucuronide, and 3-rutinoside, 3-glucosylgalactoside, and 3-glucosylxyloside; myricetin and isorhamnetin 3-glucosides; myricetin 3-glucuronide; (+)-catechin; (−)-epicatechin; (+)-gallocatechin; (−)-epigallocatechin; epicatechin-gallate; catechin-gallate; catechin-catechingallate; dihydroquercetin 3-rhamnoside; dihydrokaempferol 3-rhamnoside; and a variety of tannins.[25,140-160] Therefore, the copigmentation reaction is probably a major molecular interaction mechanism involved in variations of color and astringency during production and aging of wine. Pigment-copigment complexes in wine remain, however, to be isolated; this may not, in fact, be possible as the hydrophobic stacking interactions are not normally able to resist chromatographic separation.

In aqueous solutions, the magnitude of the copigment effect has been demonstrated to be dependent on the type and concentration of both the anthocyanin and the copigment, the temperature, and the composition and pH of the aqueous medium.[127-130,137-139] In particular, the copigmentation reaction takes place in the entire acidic pH range and even extends into the alkaline range.[127-129,137] In acidic media (pH ≤ 1.5), the copigment effect always gives rise to an increase in the absorbance at the visible wavelength (hypochromic shift) and a shift to longer wavelength of the maximum of absorption of the visible band (bathochromic shift). As the pH rises, the bathochromic shift remains essentially unchanged, whereas the hypochromic shift is replaced by the reverse effect; this is, a strong increase in the absorption at all wavelengths in the visible range (hyperchromism). Under alkaline conditions, the copigment effect is generally reduced.[129] Depending on the copigment, the maximum copigment effect occurs within the 3 to 6 pH range.[127,129,130] The pH of musts and wines varies from approximately 2.8 to 4.0 depending on the variety, region, and season.[101]

The addition of alcohol to an aqueous copigmented anthocyanin solution reduces the copigment effect.[127,129,130] However, copigmentation does take place in water-alcohol solutions provided that water is the main component of the mixture. Figure 14 illustrates the reduction in copigmentation equilibrium between malvin and chlorogenic acid caused by the addition of methanol and ethanol. As can be noted, for 6×10^{-4} dm^3 malvin solutions copigmented by chlorogenic acid at the copigment-pigment molar ratio of 12.8 at 20°C and pH 5, the magnitude of the copigment effect of 10 and 20% ethanolic solutions were approximately 85 and 70% of the value of pure water, respectively. Somers and Evans[94] also found a negative linear relationship be-

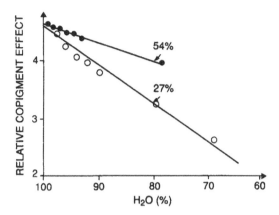

FIGURE 14. Plot of the relative magnitude of the copigment effect for 6×10^{-4}/mol dm^3 malvin solutions copigmented by chlorogenic acid vs. the mass percent of water. Copigment to pigment molar ratio: 12.8; T = 20°C; color = 520 nm; pH = 5. Cosolvent added: methanol (●) and ethanol (○). The relative copigment effect is expressed as ln {{[(A − A$_o$)/A$_o$]mix/[(A − A$_o$)/A$_o$]water} × 100%}, where A and A$_o$ are absorbance of anthocyanin solutions at 520 nm in the presence and absence of chlorogenic acid. The curves drawn on the figure theoretically meet at ln 100 = 4.60, which is the pure water reference value. (Adapted from Brouillard, R., Wigand, M. C., Dangle, O., and Cheminat, A., *J. Chem. Soc., Perkin Trans.* II, 8, 1235, 1991.)

tween color and ethanol content of wine, and attributed the phenomenon of major color loss during the making of red wine to the destructive effect of ethanol on the structures of deeply colored pigment aggregates. These authors also noted that the color loss caused by addition of 10 to 15% v/v ethanol to juices was restored by the removal of ethanol. Somers and Evans[94] proposed that the anthocyanin-copigment complexes are formed by electrostatic association of the red and blue forms of the anthocyanin which are stabilized by charge interactions and by hydrogen bonding to related phenolics. More recent evidence, however, indicates that copigmentation is a molecular interaction between the anthocyanin-colored structures and the copigment. The main role of the copigment is to control the extent of hydration to the flavylium cation to the carbinol pseudobase.[127,138] Also, available evidence further indicates that the phenomenon of copigmentation is due to hydrophobic stacking of the aromatic nuclei of the anthocyanin and the copigment. There is little evidence to support the view that copigmentation is due to hydrogen bonding.

According to Somers and co-workers,[24,107,109] storage temperature is the primary environmental factor influencing changes in the color characteristics of red wine during maturation. In 'Shiraz' wine stored under nitrogen, anthocyanins decreased and polymeric pigments increased much faster at 25°C than at 3°C.[107] These preliminary findings from ampoule experiments were confirmed with batch lots of young 'Shiraz', 'Cabernet Sauvignon', and 'Malbec' wines stored near 45°C in 250 to 5000-l insulated stainless steel tanks for 25 to 40 d. At these storage conditions, progressive change in the color

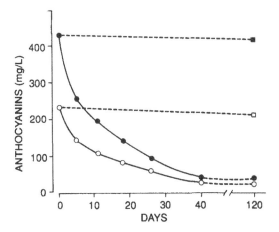

FIGURE 15. Decrease in total anthocyanins (●) and in malvidin 3-glucoside (○) during anaerobic storage of 'Shiraz' wine (1986) at 42 to 45°C for 40 d. Residual concentrations after 120 d are shown for the control wine (■,□) and for the treated wine. (From Somers, T. C. and Pocock, K. F., *Vitis*, 29, 109, 1990. With permission.)

of wine was found to correspond to that normally seen over a much longer period of storage. Spectral measurements revealed a shift in color from the purple hues of young red wines (characterized by low A_{420}/A_{520}) to the ruby red appearance (high A_{420}/A_{520}) typical of more mature wines. These effects were accompanied by a rapid decrease in anthocyanin concentration during heat treatment compared with marginal alteration in control wine, which had been stored at 10 to 20°C[109] (Figure 15).

The effect of storage temperature on wine color can be attributed to two related physicochemical processes. The first is the endothermic nature of the structural transformations of anthocyanins and the shift in the anthocyanin equilibria at higher temperatures toward the colorless chalcone from the colored flavylium and quinonoidal base.[162,163] The second physicochemical process is related to the increased dissociation of the flavylium cation-copigment complexes, and the acetaldehyde-mediated condensation products at higher temperature.[127,137] Therefore, due to the endothermic nature of the structural transformation, as the storage temperature of wine is increased the degree of redness of red young wine decreases due to the decreased concentration of flavylium cations, and the yellowness increases due to the higher concentration of chalcones. Concurrently, the concentration of the relatively more stable anthocyanin-copigment complexes decreases with increasing temperature because complex formation is favored by low temperature.[127,129] Upon dissociation of the anthocyanin-copigment complex, the anthocyanins undergo hydration and tautomeric reactions; and these lead to a decrease in redness and a subsequent increase in yellow/rubyness of the wine.

In attempts to accelerate aging of wine by increasing the storage temperature, researchers have generally observed large decreases in anthocyanin

TABLE 11
Changes in Total Anthocyanin, Catechin and Epicatechin Content (mg/l) during Fermentation and Aging of 'Merlot' and 'Cabernet Sauvignon' Wine as Measured by HPLC

	Cabernet Sauvignon			'Merlot'		
Days	Total anthocyanin	*d*-Catechin	Epicatechin	Total anthocyanin	*d*-Catechin	Epicatechin
0	—	12.8	9.2	306.6	21.7	15.3
0.92	399.7	49.3	22.4	365.1	53.3	54.3
2.13	438.3	60.1	34.1	341.4	81.1	70.4
3.04	474.4	67.9	—	329.9	77.4	61.3
4.25	382.1	103.4	38.5	298.0	110.8	84.7
8[a]	—	104.4	34.4	237.7	108.7	73.4
26	294.0	—	—	221.0	109.6	74.8
48	255.1	100.1	33.0	200.7	—	—
95	156.7	83.1	24.9	103.5	87.4	56.0
165	75.1	73.6	19.5	47.7	78.4	51.3
200	50.7	65.7	17.5	36.9	71.9	43.8
240	22.3	—	—	25.0	—	—

[a] End of fermentation and first racking; wine stored at 2 to 3°C from end of fermentation (day 8) to day 48. At the end of 48 d, the wine was racked again and placed at 20 to 22°C.

Adapted from Nagel, C. W. and Wulf, L. W., *Am. J. Enol. Vitic.*, 30, 111, 1979.

content of wine with storage time.[25] In Cabernet Sauvignon and Merlot wines stored at 2 to 3°C for 48 d and then at 20 to 22°C, Nagel and Wulf[25] observed large decreases in anthocyanins and other phenolic compounds (Table 11). From the changes in anthocyanins and polyphenols, these authors concluded that at 20 to 22°C, wine aging proceeds two to three times faster than at approximately 13°C; consequently, 240 d of storage at room temperature corresponds to 1.3 to 2 years of normal aging. Accelerated aging of dessert wines by the use of heat treatment has been advocated for some time.[101,163] However, results have often been unsatisfactory because the effect of temperature has been masked and confused with the influence of other environmental factors, particularly oxygen.[109] A satisfactory explanation for the accelerated rate of aging and for the disappearance of anthocyanins and other phenolics was not provided by Nagel and Wulf.[25] However, in light of recent advances in the understanding of the intermolecular copigmentation of anthocyanins,[127-130,138,139] it is now possible to explain the observed changes more adequately. For instance, in model solutions at pH 3 at 20°C, malvin concentration 6×10^{-4} M, and copigment to pigment molar ratio of 10, the copigment effect of epicatechin is 71 and that of (+)-catechin is 50.[129] The disappearance values for these two flavans in Cabernet Sauvignon wine stored at room temperature (day 48 to day 200) (Table 11), are 47 and 34%, respectively. The disappearance of these and other colorless polyphenols during fermentation, maturation, and aging of wine parallels the loss of anthocyanins. In Tables 9 and 11, the major losses of anthocyanins and flavans are due to the formation and removal (during racking of wine) of colored and colorless polymers. As already mentioned, formation of anthocyanin-polyphenol complexes is greatly enhanced at low temperatures. Thus, when young red wines are stored at 2 to 3°C, polymers too large to remain in solution are formed. With the racking of wine, these polymers are removed and a relatively large drop in anthocyanin content may be observed (Table 11). Also, at high storage temperature, the equilibrium between the colored and colorless forms of the monomeric and complexed anthocyanins is strongly shifted toward the colorless forms, with the reverse occurring at low temperature.[127,138,163] Thus, the more rapid aging of wine at a higher storage temperature is primarily due to the temperature-dependency of the copigmentation and hydration reactions of anthocyanins.

Another phenomenon associated with maturation and aging of the wine is self-association of anthocyanins and of other polyphenols. The self-association of anthocyanins, first suggested by Asen et al.,[137] has been extensively studied by Hoshino and associates.[164-170] However, to date, its role in maturation and aging of wine has received little attention. Recently, however, we have demonstrated self-association of pelargonin, malvin, and cyanin (albeit to a lesser extent) in model food systems containing ethanol.[130] From ^{1}H-NMR and CD data of anthocyanins in aqueous solutions, Hoshino[170] has proposed a helically stacked conformation for the homo-association of the neutral and ionized quinonoidal bases of anthocyanins. According to this model, the stacked anthocyanin chromophores form a hydrophobic core surrounded by hydrophilic glucosyl residues. In the resultant hydrophobic area, the benzopyrylium ring would be protected from water attack and as such retain its color.

B. JUICE

Grape juice, the liquid expressed from suitably ripened grapes, differs little in composition from the grapes except for the content of crude fiber and the oils which are present primarily in the seeds. In North America, the majority of red grape juice is prepared from 'Concord' grapes (*V. labrusca*).[171,172] *V. vinifera* grapes are much higher in sugar, lower in acidity, and not as flavorful as *V. labrusca* grapes. This lack of tartness and flavor is believed to be the reason for the limited quantity of *V. vinifera* grapes processed into juice.[171,172] Juice produced from the Muscat varieties has a strong flavor, but is colorless and insufficiently tart and must be blended with suitable red wine grapes (Barbera, Valdepenas . . .) to obtain an acceptable product.

A typical purple-red color is associated with high quality in grape juice, and changes in color during processing and storage from purple-red to brown result in a drastic decline in quality.[173] The color of 'Concord' juice is largely due to over 30 anthocyanin pigments located in and adjacent to the skin (see previous section on anthocyanins of other *Vitis* species). Maturity influences the color of the grapes and, consequently, the juice made from them. That is, mature grapes yield juice with superior sensory quality and more desirable color than less-ripe grapes. The development of the typical purple-red color in 'Concord' grapes begins at véraison and continues through maturation. However, since the pH of 'Concord' grapes increases from about 3.1 at véraison to 3.8 or higher at maturity, a bathochromic shift in the maximum visible absorption occurs as the pigment color changes from purple-red to blue.[174-177]

'Concord' juice is acceptable only when hot pressed. In this process, the stemmed/crushed grapes are heated to 60°C by passing through a heat exchanger. The heating step is followed by addition of 50 to 200 ppm of pectinase enzyme and a press aid to reduce the slipperiness of the pulp, thus permitting the effective use of a screw press. The mash is generally hot pressed to maximize yield and color extraction. The pressed juice is cleared with the aid of a pectolytic enzyme at around 0°C. Potassium metabisulfite may be used to prevent browning. The juice is flash pasteurized at 85°C and stored in bulk at approximately 0°C for a few weeks to precipitate the excess potassium bitartrate. The tartrate precipitates are known as argols and constitute approximately 4 kg/tonne of original grape juice. Unlike the tartrates from wine, which tend to appear in the form of large pure crystals (wine stones), tartrates from grape juice are much smaller and contain substantial amounts of occluded pigments.[172] The detartrate juice which is to be preserved by pasteurization must be fixed with bentonite in order to remove proteins which form an undesirable haze upon heating. The juice is then filtered and concentrated or bottled utilizing a pasteurization procedure to prevent microbial spoilage. Grape juice is most often concentrated by evaporation or freezing of water; however, concentration is also possible using a combination of reverse osmosis and evaporation.[172]

During the period between harvesting and heating of the stemmed/crushed grapes, enzyme treatment, pressing, and pasteurization, a significant loss of color occurs in 'Concord' grape juice.[178] During low-temperature bulk tank storage, further losses occur as a result of tartration, precipitation of complexed anthocyanins, and changes in pH. The mechanisms involved in the variations of color and flavor in the juice during processing and storage are similar to those discussed for wine except that neither acetaldehyde nor alcohol are involved. Certain chemicals such as reducing and chelating agents and antioxidants have been shown to stabilize the color of grape juice.[179,180] However, the color-stabilizing effect of these chemicals can be attributed to shifts in the equilibrium of the anthocyanin colored and colorless forms induced previously by changes in pH, temperature, and concentration of pigments and copigments.[127-130,138,162,163]

TABLE 12
Composition of 'Concord' and 'De Chaunac' Pomace Samples
Obtained from Commercial Pressing Operations

	'Concord'			'De Chaunac'		
Component	Cold pressed	Hot pressed	Fermented skins	Cold pressed	Hot pressed	Fermented skins
Seed, %	15.1	21.9	17.7	29.8	35.9	37.0
Total soluble solids, %	12.7	12.1	5.3	10.2	9.3	4.0
Total acidity, g/l	12.8	9.4	10.9	7.4	7.4	7.6
Tannin, mg/100 g[a]	391	254	178	444	120	120
Alcohol, %	0.04	0.04	5.88	0.02	0.03	2.38
Methyl anthranilate, ppm	8.04	5.20	5.40	—	Nil	—
Total volatile esters, ppm	57	52	75	3.0	3.0	37.7
Volatile acidity, g/l	0.15	0.10	0.27	0.70	0.50	1.10
Total anthocyanin, mg/100 g	460	200	360	425	79	211

[a] Does not include tannin from the seed.

From Fuleki, T. and Babjak, L. J., *Highlights Agric. Res. Ont.*, 9, 6, 1986. With permission.

VI. RECOVERY OF ANTHOCYANINS AND OTHER PRODUCTS FROM GRAPE POMACE

Grape pomace or marc consists of processed skins, seeds, and stems, and represents as much as 20% of the weight of grapes processed into wine and juice.[101,181] Table 12 gives the composition of 'Concord' and 'De Chaunac' pomace samples from commercial wineries and juice-processing operations in the Niagara peninsula.[182]

In the past, grape pomace has been disposed of in a number of means. Some wineries have dehydrated the pomace to low moisture content and ground it for use in feeding livestock, particularly dairy cows.[101] Growers have also spread it on vineyards as a soil conditioner, but this practice is declining in spite of the fact that growers may be paid to truck the pomace away from wineries. Consequently, considerable quantities of pomace are taken to disposal sites, and these sites are becoming more difficult to locate and expensive to operate. A variety of products, including ethanol, tartrates, citric acid, grape seed oil, hydrocolloids, and anthocyanins, can be produced from grape pomace.[101,181,182] Figure 16 shows an outline of the process for the production of several products from grape pomace. The first step is the separation of the seeds from grape pomace for oil recovery by mechanical pressing or by solvent extraction of the ground seeds. Fermentation of pomace yields ethanol, and this can be used for fortification of dessert wines and for fuel alcohol and other purposes. Tartrates can be extracted by a continuous method or batch process.[101] In the continuous method, the pomace is conveyed through a tank of hot water into which water continuously flows and from

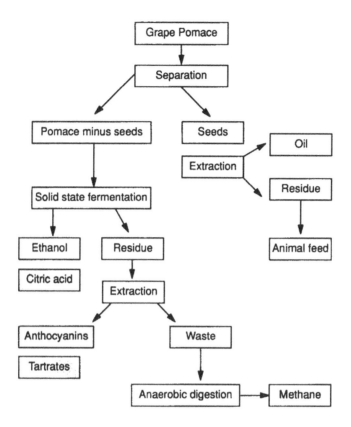

FIGURE 16. Flow diagram for resource recovery from grape pomace. (From Hang, Y. D., *Process Biochem.*, 23, 2, 1988. With permission.)

which extract also continuously flows. In the batch process, the pomace is first mixed with three parts of water, heated to 60 to 100°C, drained, and heated again with two parts of water to one part of pomace; then the drained pomace is pressed.[101] Tartrates can also be extracted by the beet diffusion battery principle, in which pomace is placed in several tanks in series, using hot water extraction.[101]

Anthocyanins are probably the most valuable components of grape pomace, and a large number of methods for their extraction have been reported. Enocyanin or "enocianina", a deeply colored extract of red grape pomace, has been produced commercially in Italy since 1879.[183-186] Initially, the extraction of enocianina was achieved with the use of ethanol. It was soon realized, however, that the recovery of pigment could be improved considerably with the use of a mild solution of sulfur dioxide in water.[184] Unfermented or fermented pomace from intensely colored grapes ('Raboso del Piave', 'Friularo', 'Lancellotta', 'Teinturier') is used for the preparation of enocyanin.[183-186] The commercial method of producing enocyanin may vary slightly from one manufacturer to another, depending on the characteristics of the pomace available and the market requirements. At present, the majority of enocyanin is produced in the Reggio Emilia region of Italy, where grape residues with very high anthocyanin content are available in large quantities.[186,187] The general methods for the commercial production of enocyanin involves three major steps: (1) maceration of the pomace in a mild solution of sulfur dioxide for 48 to 72 h; (2) separation of the water phase from the spent pomace; and (3) vacuum concentration of the water extract, with or without recovery of ethanol. The maceration or steeping process is normally carried out by mixing one part of pomace with one part of 0.1 to 0.2% aqueous sulfur dioxide solution, or its equivalent in bisulfite or metabisulfite. To maximize the extraction of anthocyanins, the pomace-SO_2 solution is thoroughly mixed several times over the 48 to 72 h maceration period. Aqueous sulfur dioxide is used to improve the extraction of the anthocyanins, and to protect the pigments from oxidation and microbial spoilage. The separation of the aqueous pigment extract from the pomace is normally achieved by allowing the free-run extract to flow into a tub and pumping it to a storage tank. The pomace is then pressed using a vertical or a horizontal basket or a Willmes press, and both the press and free-run extracts are combined to give a practically colorless liquid which remains as such for as long as sufficient SO_2 is present. This diluted, aqueous, pigment extract is then filtered, desulfurized, and concentrated by vacuum evaporation at 40 to 45°C to produce an intensely colored concentrate; or spray-dried to produce a dark red water-soluble powder. The quality and anthocyanin content of commercial enocyanin products varies widely. Cappelleri[188] cites the following values for liquid Italian enocyanin products: density at 15°C, 1.0183 to 1.1426; total acidity (as tartaric acid), 6.40 to 33.75 g/l; total solids, 48.1 to 288.5 g/l; ash content, 5.64 to 61.53 g/l; alcohol, absent or in trace amounts; sulfur dioxide, 200 to 600 mg/l; anthocyanin content, 0.3 to 4 g/100 g; and tannins, 4 to 24 g/100 g. Originally, enocyanin was used to intensify the color of red wines; however, in later years it has been found to have applications as a colorant in cosmetics, pharmaceuticals, fruit juices, and many other food products.[183-191]

Many attempts have been made to improve or replace the over 100-year-old sulfur dioxide-aided extraction method used in the production of Italian enocyanin. Peterson and Jaffe[191] patented a process by which anthocyanins were extracted from grape pomace with an alcoholic solvent containing 200 to 2000 mg/l sulfur dioxide. These researchers found that the addition of SO_2 to the extraction solvent resulted in an increased pigment yield. This observation was similar to that made by Carpentieri (and reported by Garoglio[186]) in the late 1800s, when he found that anthocyanins were five times more soluble in must containing 1000 to 1200 mg/l SO_2 than in the same must free of SO_2. Philip[192] reported a process for extracting anthocyanins from grape pomace with 0.1 to 1.0% tartaric acid in methanol or ethanol. In this process, the dried pomace was packed in a column and extracted with acidified alcohol at a flow rate of 25 ml/min. The alcohol extract was neutralized with an appropriate amount of 40% KOH solution to prevent the degradation of anthocyanins. The extract was then cooled to 15°C and the precipitate (cream of tartar) was filtered off, followed by vacuum concentration of the filtrate at 40°C. The concentration of anthocyanins in the concentrate was 0.65 g/100 ml. Nedelchev et al.[193] reported a process using 10 to 12 in-line diffusers for recovery of anthocyanins and alcohol from wine pomace. Inagami and Koya[194] used an aqueous solution containing 60% ethanol and 630 mg/l SO_2 to extract the pigments. Addition of glycerol, citric acid, and HCl facilitated the removal of excess SO_2 with nitrogen stripping. Calvi and Francis[195] recovered anthocyanins from 'Concord' grape juice pomace with 0.01% citric acid in ethanol for pigment stability studies in beverages. Main et al.[196] used the same process to produce a spray-dried anthocyanin-containing product from 'Concord' grape juice sludge. Mitin and Paraska[197] used 0.05 to 0.07% Pectavamorin PIOX enzyme to improve the yield of extracted anthocyanins with aqueous HCl. Yokoyama and Ono[198] patented a process for pigment extraction using an aqueous solution containing 10,000 mg/l SO_2.

In 1984 a U.S. patent was granted to Shrikhande[199] for the extraction and intensification of anthocyanins from grape pomace. In this process, the pomace (or marc) is ground and extracted with 3 parts of water containing 1200 ppm SO_2 at 71°C. After 15 min mixing, the mixture is pressed to recover the liquid. The solids are reduced by treatment with 100- to 150-ppm pectinase or 250-ppm fungal amylase for 4 to 14 d. The mixture is then filtered and stripped with nitrogen, and peroxide is added to reduce the SO_2 concentration from 1200 to 100 to 350 ppm. Sulfuric acid (about 4.7 l to 3786 l of extract) is then added. After mixing, 500 ppm of acetaldehyde is added at 37.7°C and mixed for 5 min. The extract is then placed on a Duolite 5861 affinity ion exchange column to which the pigment adheres, thus allowing many impurities to be washed through the column. The pigments can then be eluted with an ethanol/water mixture, concentrated, filtered, and marketed as a concentrated liquid.

Langston[200] has patented a process for the production of a pigment extract free of sugars, organic acids, polymerized anthocyanins, and other water-soluble material. This product is obtained using an aqueous extraction solvent containing bisulfite, followed by chromatographic separation on a nonionic adsorbent. Fuleki and Babjak[182] developed a process for the commercial production of a natural food colorant, Enovit, from grape pomace (Figure 17). This product can be manufactured in both powder and liquid forms, with an anthocyanin concentration of 40 g/kg in the powder and 10 g/kg in the liquid form. Enovit is completely soluble in water and alcohol. The colors produced depend on the pH and the concentration of the colorant, with color intensification from light red to deep red occurring with decreasing pH. Excessive concentrations of sulfur dioxide in the product will reduce the intensity of color while metals such as iron, aluminum, and tin will form complexes with the anthocyanins yielding blue hues. Canada, the United States, Japan, and countries in the European Economic Community (EEC) permit the use of this

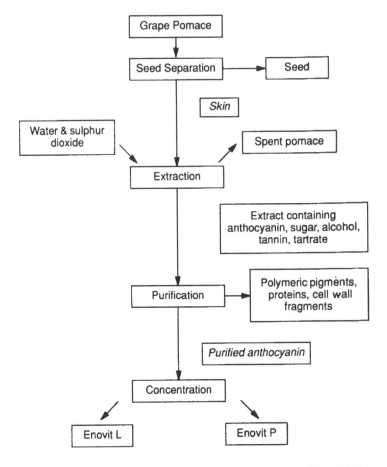

FIGURE 17. Scheme for the production of Enovit from grape pomace. (From Fuleki, T. and Babjak, L. J., *Highlights Agric. Res. Ont.*, 9, 6, 1986. With permission.)

colorant in food. The manufacturer recommends it for use in beverages, jams, jellies, syrups, drink-mix crystals, ice creams, pastries, confectioneries, and also some cosmetics and pharmaceuticals.[182]

REFERENCES

1. FAO Production Yearbook, FAO Statistics Series No. 43, Food and Agriculture Organization of the United Nations, Rome, 1989.
2. **Nelson, K. E.**, The grape, in *Quality and Preservation of Fruits*, Eskin, N. A. M., Ed., CRC Press, Boca Raton, FL, 1991, 125.
3. **Childers, N. F.**, *Modern Fruit Science*, Horticultural Publications, Rutgers University, New Brunswick, NJ, 1976, 666.
4. **Ribéreau-Gayon, P.**, The anthocyanins of grapes and wines, in *Anthocyanins as Food Colors*, Markakis, P., Ed., Academic Press, New York, 1982, 209.

5. **Rankine, B. C., Kepner, R. E., and Webb, A. D.**, Comparison of anthocyanin pigments of vinifera grapes, *Am. J. Enol. Vitic.*, 9, 105, 1958.

6. **Albach, R. F., Kepner, R. E., and Webb, A. D.**, Comparison of anthocyanin pigments of red vinifera grapes. II. *Am. J. Enol. Vitic.*, 10, 164, 1959.

7. **Akiyoshi, M., Webb, A. D., and Kepner, R. E.**, The major anthocyanin pigments of *Vitis vinifera* varieties, Flame Tokay, Emperor, and Red Malaga, *J. Food Sci.*, 28, 177, 1963.

8. **Van Buren, J. P., Bertino, J. J., Einset, J., Remaily, G. W., and Robinson, W. B.**, A comparative study of the anthocyanin pigment composition in wines derived from hybrid grapes, *Am. J. Enol. Vitic.*, 21, 117, 1970.

9. **Roggero, J. P., Ragonnet, B., and Coen, S.**, Analyse fine des anthocyanines des vins et des pellicules de raisin par la technique HPLC, *Vignes & Vins*, 327, 38, 1984.

10. **Bakker, J. and Timberlake, C. F.**, The distribution of anthocyanins in grape skin extracts of Port wine cultivars as determined by high performance liquid chromatography, *J. Sci. Food Agric.*, 36, 1315, 1985.

11. **Goldy, R. G., Ballinger, W. E., and Maness, E. P.**, Fruit anthocyanin content of some *Euvitis × Vitis rotundifolia* hybrids, *J. Am. Soc. Hortic. Sci.*, 111, 955, 1986.

12. **Lamikanra, O.**, Anthocyanins of *Vitis rotundifolia* hybrid grapes, *Food Chem.*, 33, 225, 1989.

13. **Goldy, R. G., Maness, E. P., Stiles, H. D., Clark, J. R., and Wilson, M. A.**, Pigment quantity and quality characteristics of some native *Vitis rotundifolia* Michx., *Am. J. Enol. Vitic.*, 40, 253, 1989.

14. **Fuleki, T.**, Anthocyanin composition of some highly pigmented grape cultivars, Paper 4th Int. Conf. Groupe Polyphenols, Strasbourge, France, 1990.

15. **Yokotsuka, K. and Nishino, N.**, Extraction of anthocyanins from Muscat Bailey A grape skins, *J. Ferment. Bioeng.*, 69, 328, 1990.

16. **Berg, M. W. and Akiyoshi, M.**, The effect of various must treatments on the color and tannin content of red grape juices, *Food Res.*, 22, 373, 1957.

17. **Ough, C. S. and Amerine, M. A.**, Studies with controlled fermentation. VII. Effect of ante-fermentation blending of red must and white juice on color, tannins, and quality of Cabernet Sauvignon wine, *Am. J. Enol. Vitic.*, 13, 181, 1962.

18. **Berg, H. W. and Akiyoshi, M.**, Color behavior during fermentation and aging of wines, *Am. J. Enol. Vitic.*, 13, 126, 1962.

19. **Rankine, B. C.**, Heat extraction of color from red grapes for wine making, *Aust. Wine, Brewing Spirit Rev.*, 82, 40, 1964.

20. **Pifferi, P. G. and Zamorani, A.**, Contributo alla conoscenza della sostanza colorante dei vini. II. Azione dell' anidride solforosa e della rifermentazione sugli antociani del vino, *Riv. Vitic. Enol.*, 17, 115, 1964.

21. **Harvalia, A.**, La couleur des vins rouges, *Chim. Chronika (Athens)*, 20(9), 155, 1965.

22. **Flora, L. F.**, Time-temperature influence on Muscadine grape juice quality, *J. Food Sci.*, 41, 1312, 1976.

23. **Flora, L. F.**, Influence of heat, cultivar and maturity on the anthocyanin 3,5-diglucosides of Muscadine grapes, *J. Food Sci.*, 43, 1819, 1978.

24. **Somers, T. C. and Vérette, E.**, Phenolic composition of natural wine types, in *Modern Methods of Plant Analysis, Wine Analysis*, Linskens, H. F. and Jackson, J. F., Eds., Springer-Verlag, Berlin, 1988, 219.

25. **Nagel, C. W. and Wulf, L. W.**, Changes in the anthocyanins, flavonoids and hydroxycinnamic acid esters during fermentation and aging of Merlot and Cabernet Sauvignon, *Am. J. Enol. Vitic.*, 30, 111, 1979.

26. **McCloskey, L. P. and Yengoyan, L. S.**, Analysis of anthocyanins in *Vitis vinifera* wines and red color versus aging by HPLC and spectrophotometry, *Am. J. Enol. Vitic.*, 32, 257, 1981.

27. **Sims, C. A. and Morris, J. R.,** Effects of pH, sulfur dioxide, storage time, and temperature on the color and stability of red Muscadine grape wine, *Am. J. Enol. Vitic.,* 35, 35, 1984.

28. **Sims, C. A. and Morris, J. R.,** A comparison of the color components and color stability of red wine from Noble and Cabernet Sauvignon at various pH levels, *Am. J. Enol. Vitic.,* 36, 181, 1985.

29. **Bakker, J., Preston, N. W., and Timberlake, C. F.,** The determination of anthocyanins in aging red wines: comparison of HPLC and spectral methods, *Am. J. Enol. Vitic.,* 31, 121, 1986.

30. **Ellis, L. P. and Kok, C.,** Color changes in Blanc de Noir wines during aging at different temperatures and its color preference limits, *S. Afr. J. Enol. Vitic.,* 8, 16, 1987.

31. **Haslam, E. and Lilley, T. H.,** Natural astringency in food stuffs. A molecular interpretation, *CRC Crit. Rev. Food Sci. Nutr.,* 27, 1, 1988.

32. **Manley, C. H. and Shubiak, P.,** High pressure liquid chromatography of anthocyanins, *Can. Inst. Food Sci. Technol. J.,* 8, 35, 1976.

33. **Wulf, L. W. and Nagel, C. W.,** High-pressure liquid chromatographic separation of anthocyanins in *Vitis vinifera, Am. J. Enol. Vitic.,* 29, 42, 1978.

34. **Fong, R. A., Kepner, R. E., and Webb, A. D.,** Acetic acid acylated anthocyanin pigments in the grape skins of a number of varieties of *Vitis vinifera, Am. J. Enol. Vitic.,* 22, 150, 1971.

35. **Piergiovanni, L. and Volonterio, G.,** Tecniche cromatografiche nello studio della frazione antocianica delle uve, *Riv. Vitic. Enol.,* 33, 289, 1980.

36. **Hebrero, E., Santos-Buelga, C., and Rivas-Gonzalo, J. C.,** High performance liquid chromatography-diode array spectroscopy identification of anthocyanins of *Vitis vinifera* variety Tempranillo, *Am. J. Enol. Vitic.,* 39, 227, 1988.

37. **Piergiovanni, L. and Volonterio, G.,** Studio della frazione antocianica delle uve, *Vignevini,* 8, 49, 1981.

38. **Roggero, J. P., Coen, S., and Larice, J. L.,** Etude comparative de la composition anthocyanique des cépages. Essai de classification, *Bull. Liaison Groupe Polyphénols,* 13, 380, 1986.

39. **Bailey, L. H.,** *Manual of Cultivated Plants,* Macmillan, New York, 1977, 648.

40. **Ribéreau-Gayon, P.,** Recherches sur les anthocyannes des vegetaux (application au genre *Vitis*), Doctoral thesis, Paris, Librarie Generale de L'Enseignment, 4, Rue Dante, 4, 1959.

41. **Ribéreau-Gayon, P.,** Les composés phénoliques du raisin et du vin, *Ann. Physiol. Veg.,* 6, 211, 1964.

42. **Ballinger, W. E., Maness, E. P., Nesbitt, W. B., and Carroll, D. E., Jr.,** Anthocyanins of black grapes of 10 clones of *Vitis rotundifolia* Michx., *J. Food Sci.,* 38, 909, 1973.

43. **Ballinger, W. E., Maness, E. P., Nesbitt, W. B., Makus, D. J., and Carroll, D. E., Jr.,** A comparison of anthocyanins and wine color quality in black grapes of 39 clones of *Vitis rotundifolia* Michx., *J. Am. Soc. Hortic. Sci.,* 99, 338, 1974.

44. **Nesbitt, W. B., Maness, E. P., Ballinger, W. E., and Carrol, D. E., Jr.,** Relationship of anthocyanins of black muscadine grapes (*Vitis rotundifolia* Michx.) to wine color, *Am. J. Enol. Vitic.,* 25, 30, 1974.

45. **Robinson, W. B., Weirs, L. D., Bertino, J. J., and Mattick, L. R.,** The relation of anthocyanin composition to color stability of New York state wines, *Am. J. Enol. Vitic.,* 17, 178, 1966.

46. **Van Buren, J. P., Bertino, J. J., and Robinson, W. B.,** The stability of wine anthocyanin on exposure to heat and light, *Am. J. Enol. Vitic.,* 19, 147, 1968.

47. **Anderson, R. J.,** Concerning the anthocyanins of Norton and Concord grapes. A contribution to the chemistry of grape pigments, *J. Biol. Chem.,* 57, 795, 1923.

48. **Sastry, L. V. L. and Tischer, R. G.,** Behaviour of anthocyanin pigments in Concord grapes during heat processing and storage, *Food Technol.,* 6, 82, 1952.

49. **Ingalsbe, D. W., Neubert, A. M., and Carter, G. H.,** Concord grape pigments, *J. Agric. Food Chem.,* 11, 263, 1963.

50. **Shewfelt, R. L. and Ahmed, E. M.,** The nature of the anthocyanin pigments in South Carolina Concord grapes, *S.C., Agric. Exp. Stn. Tech. Bull.,* No. 1025, 1966.

51. **Hrazdina, G.,** Anthocyanin composition of Concord grapes, *Lebensm. Wiss. Technol.,* 8, 111, 1975.

52. **Williams, M., Hrazdina, G., Wilkinson, N. M., Sweeny, J. G., and Iacobucci, G. A.,** High-pressure liquid chromatographic separation of 3-glycosides, 3,5-diglucosides, 3-(6-*O-p*-coumaryl)glucosides and 3-(6-*O-p*-coumarylglucoside)-5-glucosides of anthocyanidins, *J. Chromatogr.,* 155, 389, 1978.

53. **Anderson, D. W., Julian, E. A., Kepner, R. E., and Webb, A. D.,** Chromatographic investigation of anthocyanin pigments in *Vitis cinerea, Phytochemistry,* 9, 1569, 1970.

54. **Anderson, D. W., Gueffroy, D. E., Webb, A. D., and Kepner, R. E.,** Identification of acetic acid as an acylating agent of anthocyanin pigments in grapes, *Phytochemistry,* 9, 1579, 1970.

55. **Yokotsuka, K., Nozaki, K., and Kushida, T.,** Comparison of phenolic compounds including anthocyanin pigments between Koshu and Ryugan grapes, *J. Ferment. Technol.,* 62, 477, 1984.

56. **Yokotsuka, K., Nishino, N., and Singleton, V. L.,** Unusual Koshu grape skin anthocyanins, *Am. J. Enol. Vitic.,* 39, 288, 1988.

57. **Akuta, S., Ohta, H., Osajima, Y., Matsudomi, M., and Kobayashi, K.,** Anthocyanin pigments in *Vitis vinifera* varieties, Koshu and Gros Colman, *Nippon Shokuhin Kogyo Gakkai-Shi,* 24, 521, 1977.

58. **Yamakawa, T., Ishida, K., Koto, S., Kodama, T., and Minoda, Y.,** Formation and identification of anthocyanins in cultured cells of *Vitis* sp., *Agric. Biol. Chem.,* 47, 997, 1983.

59. **Yamakawa, T., Ohtsuka, H., Onomichi, K., Kodama, T., and Minoda, Y.,** Production of anthocyanin pigments by grape cell culture, Proc. 5th Int. Congr. Plant Tissue Cell Cult., *Plant Tissue Cult.,* p. 273, 1982.

60. **Tamura, H., Kumaoka, Y., and Sugisawa, H.,** Identification and quantitative variation of anthocyanins produced by cultured callus tissue of *Vitis* sp., *Agric. Biol. Chem.,* 53, 1969, 1989.

61. **Gamborg, O. L., Miller, R. A., and Ojma, K.,** Nutrient requirements of suspension cultures of soybean root cells, *Exp. Cell Res.,* 50, 151, 1968.

62. **Do, C. B. and Cormier, F.,** Effects of low nitrate and high sugar concentrations on anthocyanin content and composition of grape (*Vitis vinifera* L.) cell suspension, *Plant Cell. Rep.,* 9, 500, 1991.

63. **Yamakawa, T., Kato, S., Ishida, K., Kodama, T., and Minoda, Y.,** Production of anthocyanins by *Vitis* cells in suspension culture, *Agric. Biol. Chem.,* 47, 2185, 1983.

64. **Lofty, S., Fleuriet, A., Ramos, T., and Macheix, J. J.,** Biosynthesis of phenolic compounds in *Vitis vinifera* cv. Gamay Fréaux cell suspension cultures: study on hydroxycinnamoyl CoA:ligase, *Plant Cell Rep.,* 8, 93, 1989.

65. **Hrazdina, G., Parsons, G. F., and Mattick, L. R.,** Physiological and biochemical events during development and maturation of grape berries, *Am. J. Enol. Vitic.,* 35, 220, 1984.

66. **Pirie, K. and Mullins, M. G.,** Interrelationships of sugars, anthocyanins, total phenols, and dry weight in the skins of grape berries during ripening, *Am. J. Enol. Vitic.,* 28, 204, 1977.

67. **Somers, T. C.,** Pigment development during ripening of the grape, *Vitis,* 14, 269, 1976.

68. **Flora, L. F. and Lane, R. P.,** Effect of ripeness and harvest date on several physical and composition factors of Cowart Muscadine grapes, *Am. J. Enol. Vitic.,* 30, 241, 1979.

69. **Piergiovanni, L. and Volonterio, G.,** Studio della frazione antocianica delle uve. Nota II. Variazione di composizione durante la maturazione, *Tecnol. Aliment. Imbottigliamento,* 6, 22, 1983.

70. **Roggero, J. P., Coen, S., and Ragonnet, B.**, High performance liquid chromatography survey on changes in pigment content in ripening grapes of Syrah. An approach to anthocyanin metabolism, *Am. J. Enol. Vitic.*, 37, 77, 1986.

71. **Darné, G.**, Évolution des différentes anthocyanes des pellicules de Cabernet-Sauvignon au cours du développement des baies, *Connaiss. Vigne Vin*, 22, 225, 1988.

72. **González-San José, M. L., Barron, L. J. R., and Diez, C.**, Evolution of anthocyanins during maturation of Tempranillo grape variety (*Vitis vinifera*) using polynomial regression models, *J. Sci. Food Agric.*, 51, 337, 1990.

73. **Greisebach, H.**, Biosynthesis of anthocyanins, in *Anthocyanin as Food Colors*, Markakis, P., Ed., Academic Press, London, 1982, 69.

74. **Kliewer, W. M.**, Effect of day temperature and light intensity on coloration of *Vitis vinifera* L. grapes, *J. Am. Soc. Hortic. Sci.*, 95, 693, 1970.

75. **Kliewer, W. M.**, Influence of temperature, solar radiation and nitrogen on coloration and composition of Emperor grapes, *Am. J. Enol. Vitic.*, 28, 96, 1977.

76. **Pirie, A. J. G.**, Phenolics Accumulation in Red Wine Grapes (*Vitis vinifera* L.), thesis, University of Sydney, Sydney, Australia, 1977.

77. **Crippen, D. D. and Morrison, J. C.**, The effects of sun exposure on the phenolic content of Cabernet Sauvignon berries during development, *Am. J. Enol. Vitic.*, 37, 4, 1986.

78. **Roubelakis-Angelakis, K. A. and Kliewer, W. M.**, Effects of exogenous factors on phenylalanine ammonialyase activity and accumulation of anthocyanins and total phenolics in grape berries, *Am. J. Enol. Vitic.*, 37, 275, 1986.

79. **Kliewer, M. W. and Torres, R. E.**, Effect of controlled day and night temperatures on grape coloration, *Am. J. Enol. Vitic.*, 23, 71, 1972.

80. **Buttrose, M. S., Hale, C. R., and Kliewer, W. M.**, Effect of temperature on the composition of Cabernet Sauvignon berries, *Am. J. Enol. Vitic.*, 22, 71, 1971.

81. **Kliewer, W. M. and Schultz, H. B.**, Effect of sprinkler cooling of grapevines on fruit growth and composition, *Am. J. Enol. Vitic.*, 24, 77, 1973.

82. **Macheix, J. J., Fleuriet, A., and Billot, J.**, *Fruit Phenolics*, CRC Press, Boca Raton, FL, 1991, 199.

83. **Bourzeix, M., Heredia, N., Meriaux, S., Rollin, H., and Rutten, P.**, De l'influence de l'alimentation hydrique de la Vigne sur les caractéristiques anatomiques des baies de raisins et leur richesse en couleur, tannins et autres constituants phénoliques, *C.R. Acad. Sci., Ser. D.*, 284, 365, 1977.

84. **Duteau, J., Guilloux, M., Glories, Y., and Seguin, G.**, Influence de l'alimentation en eau de la vigne sur la teneur en sucre réducteurs, acides organiques et composés phénoliques des raisins, *C.R. Acad. Sci., Ser. C*, 292, 965, 1981.

85. **Dubernet, M.**, Recherches sur la tyrosinase de *Vitis vinifera* et la laccase de *Botrytis cinerea*, thése, Université de Bordeaux, Bordeaux, France, 1974.

86. **Dubernet, M., Ribéreau-Gayon, P., Lerner, H. R., Harel, E., and Mayer, A. M.**, Purification and properties of laccase from *Botrytis cinerea*, *Phytochemistry*, 16, 191, 1977.

87. **Jensen, F. L., Kissler, J. J., Peacock, W. L., and Leavitt, G. M.**, Effect of ethephon on color and fruit characteristics of 'Tokay' and 'Emperor' table grapes, *Am. J. Enol. Vitic.*, 26, 79, 1975.

88. **Weaver, R. J. and Montgomery, R.**, Effect of ethephon on coloration and maturation of wine grapes, *Am. J. Enol. Vitic.*, 25, 39, 1974.

89. **Weaver, R. J. and Pool, R. M.**, Effect of (2-chloroethyl) phosphoric acid (ethephon) on maturation of *Vitis vinifera* L., *J. Am. Soc. Hortic. Sci.*, 96, 725, 1971.

90. **Eynard, I., Gray, G., and Rissone, M.**, Effects of ethephon on vine grapes, *Plant Growth Regul. Proc. Int. Symp.*, 2, 635, 1975.

91. **Powers, J. R., Shively, A. E., and Nagel, C. W.**, Effect of ethephon on color of Pinot noir fruit and wine, *Am. J. Enol. Vitic.*, 31, 203, 1980.

92. **Matsushima, J., Hiratsuka, S., Taniguchi, N., Uada, R., and Suzaki, N.,** Anthocyanin accumulation and sugar content in the skin of grape cultivar 'Olympia' treated with abscisic acid, *J. Jpn. Soc. Hortic. Sci.,* 58, 551, 1989.

93. **Timberlake, C. F. and Bridle, P.,** The effect of processing and other factors on the color characteristics of some red wines, *Vitis,* 15, 37, 1976.

94. **Somers, T. S. and Evans, M. E.,** Grape pigment phenomena: interpretation of major color losses during vinification, *J. Sci. Food Agric.,* 30, 623, 1979.

95. **Leone, A. M., La Notte, E., and Gambacorta, G.,** Gli antociani nelle fasi di macerazione e di elaborazione del vino. L'influenza della tecnica diffusiva sulla loro estrazione, *Vignevini,* 4, 17, 1984.

96. **Da Porto, C., Bortolomeazzi, R., and Sensidoni, A.,** Le antocianine monoglucosidi nelle uve Terrano e nei vini ottenuti per vinificazione tradizionale e per precondizionamento con CO_2, *Vignevini,* 5, 51, 1989.

97. **Riva, M., Cantarelli, C., and Cassani, L.,** Estrazione dei polifenoli dalla bacca dell'uva e riduzione delle dimensioni, *Ind. Bevande,* 10, 33, 1980.

98. **Somers, T. C. and Evans, M. E.,** Color composition and red wine quality — the importance of low pH and low SO_2, *Aust. Grape Grower and Wine Maker,* 136, 1, 1975.

99. **Markakis, P.,** Stability of anthocyanins in foods, in *Anthocyanins as Food Colors,* Markakis, P., Ed., Academic Press, New York, 1982, 245.

100. **Ribéreau-Gayon, P. and Glories, Y.,** La tipicizzazione dei vini rossi: caratteri chimici ed organolettici, *Enotecnico,* 22, 545, 1986.

101. **Amerine, M. A., Berg, H. W., and Cruess, W. V.,** *The Technology of Wine Making,* 2nd ed., AVI Publishing, Westport, CT, 288, 367, 659, 1967.

102. **Flanzy, C., Flanzy, M., and Benard, P.,** *La Vinification par Macération Carbonique,* INRA, Paris, 1987.

103. **Ribéreau-Gayon, P.,** The chemistry of red wine color, in *The Chemistry of Wine Making,* Webb, A. D., Ed., American Chemical Society, Washington, D. C., 1974, 50.

104. **Bourzeix, M., Weyland, D., and Heredia, N.,** Étude des catechines et des procyanidols de grappe de raisin, du vin et d'autres dérivés de la vigne, *Bull. OIV,* 669–670, 1171, 1986.

105. **Haslam, E.,** In vino veritas: oligomeric procyanidins and the aging of red wines, *Phytochemistry,* 19, 2577, 1980.

106. **Ribéreau-Gayon, P., Pontallier, P., and Glories, Y.,** Some interpretations of color changes in young red wines during their conservation, *J. Sci. Food Agric.,* 34, 505, 1983.

107. **Somers, T. C. and Evans, M. E.,** Evolution of red wines. I. Ambient influences on color composition during early maturation, *Vitis,* 25, 31, 1986.

108. **Macheix, J. J., Fleuriet, A., and Billot, J.,** *Fruit Phenolics,* CRC Press, Boca Raton, FL, 1990, 327.

109. **Somers, T. C. and Pocock, K. F.,** Evolution of red wines. III. Promotion of the maturation phase, *Vitis,* 29, 109, 1990.

110. **Bakker, J., Bridle, P., and Timberlake, C. F.,** Tristimulus measurements (CIELAB 76) of port wine color, *Vitis,* 25, 67, 1986.

111. **Somers, T. C.,** The polymeric nature of wine pigments, *Phytochemistry,* 10, 2175, 1971.

112. **Singleton, V. L. and Noble, A. C.,** Wine flavor and phenolic substances, *ACS Symp. Ser.,* 23, 47, 1976.

113. **Pontallier, P. and Ribéreau-Gayon, P.,** Influence de l'aération et du sulfitage sur l'évolution de la matiére colorante des vins rouge au cours de la phase d'élevage, *Connaiss. Vigne Vin,* 17, 105, 1983.

114. **Wildenradt, H. L. and Singleton, V. L.,** Production of acetaldehydes as a result of oxidation of polyphenolic compounds and its relation to wine aging, *Am. J. Enol. Vitic.,* 25, 119, 1974.

115. **Trillat, A.,** Sur le diverses destinations de l'aldéhyde acetique dans le vin rouge, *Bull. Soc. Chim.,* 4, 546, 1909.

116. **Joslyn, M. A. and Comar, C. L.,** Role of acetaldehyde in red wines, *Ind. Eng. Chem.,* 33, 919, 1941.

117. **Cantarelli, C.,** Invecchiamento naturale lento ed invecchiamento accelerato del vino in rapporto di caratteri organolettici ed al valore biologico del prodotto, *Accad. Ital. Vite Vino, Siena, Atti,* 10, 197, 1958.

118. **Mareca Cortés, I. and del Amo Gili, E.,** Evolution de la matiére colorante des vins de Rioja au cours du vieillissement, *Ind. Agric. Aliment.,* 76, 601, 1959.

119. **Singleton, V. L., Berg, H. W., and Guymon, J. F.,** Anthocyanin color level in port-type wines as affected by the use of wine spirits containing aldehydes, *Am. J. Enol. Vitic.,* 15, 75, 1964.

120. **Singleton, V. L. and Guymon, F. F.,** A test of fractional addition of wine spirits to red and white port wines, *Am. J. Enol. Vitic.,* 14, 129, 1963.

121. **Timberlake, C. F. and Bridle, P.,** Interactions between anthocyanins, phenolic compounds, and acetaldehyde and their significance in red wines, *Am. J. Enol. Vitic.,* 27, 97, 1976.

122. **Timberlake, C. F. and Bridle, P.,** Anthocyanins: color augmentation with catechin and acetaldehyde, *J. Sci. Food Agric.,* 28, 539, 1977.

123. **Hillis, W. E. and Urbach, G.,** The reaction of (+)-catechin with formaldehyde, *J. Appl. Chem.,* 9, 474, 1959.

124. **Bendz, G., Martensson, O., and Nilsson, E.,** Studies of flavylium compounds. I. Some flavylium compounds and their properties, *Ark. Kemi,* 27, 65, 1967.

125. **Dournel, J.,** Recherches sur le combinaisons anthocyanes — flavonols. Influence de ces reactions sur la couleur du vin rouge, Ph.D. thesis, Université de Bordeaux II, Bordeaux, France, 1985.

126. **Baranowski, E. S. and Nagel, C. W.,** Kinetics of malvidin 3-glucoside condensation in wine model systems, *J. Food Sci.,* 48, 419, 1983.

127. **Mazza, G. and Brouillard, R.,** The mechanism of co-pigmentation of anthocyanins in aqueous solutions, *Phytochemistry,* 29, 1097, 1990.

128. **Cai, Y., Lilley, T. H., and Haslam, E.,** Polyphenol-anthocyanin copigmentation, *J. Chem. Soc., Chem. Commun.,* 5, 380, 1990.

129. **Brouillard, R., Wigand, M. C., Dangle, O., and Cheminat, A.,** pH and solvent effects on the copigmentation reaction of malvidin by polyphenols, purine and pyrimidine derivatives, *J. Chem. Soc., Perkin Trans.,* II, 8, 1235, 1991.

130. **Miniati, E., Damiani, P., and Mazza, G.,** Copigmentation and self-association of anthocyanins in food model systems, *Ital. J. Food Sci.,* 4, 109, 1992.

131. **Green, R. C. and Mazza, G.,** Effect of catechin and acetaldehyde on color of saskatoon berry pigments in aqueous and alcoholic solutions, *Can. Inst. Food Sci. Technol.,* 21, 537, 1988.

132. **Roggero, J. P., Coen, S., Archier, P., and Rocheville-Divorne, C.,** Étude par C.L.H.P. de la réction glucoside de malvidin-acetaldéhyde-composé phenolique, *Conn. Vigne Vin,* 21, 163, 1987.

133. **Somers, T. C. and Wescombe, L. G.,** Evolution of red wines. II. An assessment of the role of acetaldehyde, *Vitis,* 26, 27, 1987.

134. **Bakker, J. and Timberlake, C. F.,** The mechanism of color changes in aging port wine, *Am. J. Enol. Vitic.,* 37, 288, 1986.

135. **Bakker, J.,** HPLC of anthocyanins in port wines: determination of aging rates, *Vitis,* 25, 203, 1986.

136. **Bakker J. and Timberlake, C. F.,** The distribution and content of anthocyanins in young port wines as determined by high performance liquid chromatography, *J. Sci. Food Agric.,* 36, 1325, 1985.

137. **Asen, S., Stewart, R. N., and Norris, N. K.,** Co-pigmentation of anthocyanins in plant tissues and its effect on color, *Phytochemistry,* 11, 1139, 1972.

138. **Brouillard, R., Mazza, G., Saad, Z., Albrecht-Gary, A. M., and Cheminat, A.,** The copigmentation reaction of anthocyanins: a microprobe for the structural study of aqueous solutions, *J. Am. Chem. Soc.,* 111, 1604, 1989.

139. **Rüedi, P. and Hutter-Beda, B.,** An additional aspect of intermolecular anthocyanin-co-pigmentation, *Bull. Liaison Groue Polyphénols,* 15, 332, 1990.

140. **Ribéreau-Gayon, P.,** Les composés phénoliques du raisin et du vin. I. Les acides phénols, *Ann. Physiol. Veg.,* 6, 119, 1964.

141. **Ribéreau-Gayon, P.,** Les composés phénoliques du raisin et du vin. II. Les flavonosides et les anthocyanosides, *Ann. Physiol. Veg.,* 6, 210, 1964.

142. **Ribéreau-Gayon, P.,** Les composés phénoliques du raisin et du vin. III. Les tannins, *Ann. Physiol. Veg.,* 6, 259, 1964.

143. **Ribéreau-Gayon, P.,** Les flavonosides de la baie dans le genre *Vitis, C.R. Acad. Sci., Paris,* 258, 1335, 1964.

144. **Ribéreau-Gayon, P.,** Identification d'esters des acides cinnamiques et de l'acide tartrique dans les limbes et les baies de *V. vinifera, C.R. Acad. Sci., Paris, Ser. D,* 260, 341, 1965.

145. **Ribéreau-Gayon, P.,** *Plant Phenolics,* Oliver & Boyd, Edinburgh, 1972.

146. **Singleton, V. L.,** Grape and wine phenolics, background and prospects, *C.R. Coll. Centenaire Paris,* p. 215, 1980.

147. **Singleton, V. L. and Esau, P.,** *Phenolic Substances in Grapes and Wine, and Their Significance,* Academic Press, New York, 1969.

148. **Singleton, V. L., Zaya, J., and Trousdale, E.,** Caftaric and coutaric acids in fruits of *Vitis, Phytochemistry,* 25, 2127, 1986.

149. **Singleton, V. L., Salgues, M., Zaya, J., and Trousdale, E.,** Caftaric acid disappearance and conversion to products of enzymic oxidation in grape must and wine, *Am. J. Enol. Vitic.,* 36, 50, 1985.

150. **Singleton, V. L., Zaya, J., Trousdale, E., and Salgues, M.,** Caftaric acid in grapes and conversion to a reaction production during processing, *Vitis,* 23, 113, 1984.

151. **Su, C. T. and Singleton, V. L.,** Identification of three flavan-3-ols from grapes, *Phytochemistry,* 8, 1553, 1969.

152. **Trousdale, E. K. and Singleton, V. L.,** Astilbin and engeletin in grapes and wine, *Phytochemistry,* 22, 619, 1983.

153. **Romeyer, F. M., Macheix, J. J., Goiffon, J. P., Reminiac, C. C., and Sapis, J. C.,** The browning capacity of grapes. III. Changes and importance of hydroxycinnamic-acid-tartaric esters during development and maturation of the fruit, *J. Agric. Food Chem.,* 31, 346, 1983.

154. **Romeyer, F. M., Sapis, J. C., and Macheix, J. J.,** Hydroxycinnamic esters and browning potential in mature berries of some grape varieties, *J. Sci. Food Agric.,* 36, 728, 1985.

155. **Romeyer, F., Macheix, J. J., and Sapis, J. C.,** Changes and importance of oligomeric procyanidins during maturation of grape seeds, *Phytochemistry,* 25, 219, 1986.

156. **Food, L. Y. and Porter, L. J.,** The structure of tannins of some edible fruits, *J. Sci. Food Agric.,* 32, 711, 1981.

157. **Ong, B. Y. and Nagel, C. W.,** High-pressure liquid chromatographic analysis of hydroxycinnamic acid-tartaric acid esters and their glucose esters in *Vitis vinifera, J. Chromatogr.,* 157, 345, 1978.

158. **Ong, B. Y. and Nagel, C. W.,** Hydroxycinnamic acid-tartaric acid ester content in mature grapes and during maturation of White Riesling grapes, *Am. J. Enol. Vitic.,* 29, 277, 1978.

159. **Cheynier, V. and Rigaud, J.,** HPLC separation and characterization of flavonols in the skins of *Vitis vinifera* var. Cinsault, *Am. J. Enol. Vitic.,* 37, 248, 1986.

160. **Cheynier, V. and Rigaud, J.,** Identification et dosage de flavonols du Raisin, *Bull. Liaison Groupe Polyphénols,* 13, 442, 1986.

161. **Asen, S., Stewart, R. N., Norris, K. H., and Massie, D. R.,** A stable blue non-metallic copigment complex of delphanin and C-glycosylflavones in Prof. Blaauw iris, *Phytochemistry,* 9, 619, 1970.

162. **Brouillard, R. and Delaporte, B.,** Chemistry of anthocyanin pigments. II. Kinetic and thermodynamic study of proton transfer, hydration and tautomeric reactions of malvidin 3-glucoside, *J. Am. Chem. Soc.,* 99, 8461, 1977.

163. **Brouillard, R. and Delaporte, B.,** Degradation thermique des anthocyanes, *Bull. Liaison Groupe Polyphénols,* 8, 305, 1978.

164. **Hoshino, T., Matsumoto, U., and Goto, T.,** Evidences of the self-association of anthocyanins. I. Circular dichroism of cyanin anhydrobase, *Tetrahedron Lett.,* 21, 1751, 1980.

165. **Hoshino, T., Matsumoto, U., and Goto, T.,** The stabilizing effect of the acyl group on the co-pigmentation of acylated anthocyanins with C-glucosylflavones, *Phytochemistry,* 19, 663, 1980.

166. **Hoshino, T., Matsumoto, U., and Goto, T.,** Self-association of some anthocyanins in neutral aqueous solution, *Phytochemistry,* 20, 1971, 1981.

167. **Hoshino, T., Matsumoto, U., Goto, T., and Harada, N.,** Evidence for the self-association of anthocyanins. IV. PMR spectroscopic evidence for the vertical stacking of anthocyanin molecules, *Tetrahedron Lett.,* 23, 433, 1982.

168. **Hoshino, T., Matsumoto, U., Harada, N., and Goto, T.,** Chiral exciton couples stacking of anthocyanins: intepretation of the origin of anomalous CD induced by anthocyanin association, *Tetrahedron Lett.,* 22, 3621, 1981.

169. **Goto, T., Tamura, H., Kawai, T., Hoshino, T., Harada, N., and Kondo, T.,** Chemistry of metalloanthocyanins, *Ann. N.Y. Acad. Sci.,* 471, 155, 1986.

170. **Hoshino, T.,** An approximate estimate of self-association and the self-stacking conformation of malvin quinonoidal bases studied by ^{1}H-NMR, *Phytochemistry,* 30, 2049, 1991.

171. **Pederson, C. S.,** Grape juice in *Fruit and Vegetable Juice Processing Technology,* Tressler, D. K. and Joslyn, M. A., Eds., Van Nostrand Reinhold/AVI, New York, 1971, 234.

172. **McLellan, M. R. and Rau, E. J.,** Grape juice processing, in *Production and Packaging of Non-Carbonated Fruit Juices and Beverages,* Hicks, D., Ed., Blackie & Sons, Glasgow, U.K., 1990, 226.

173. **Morris, J. R., Sistrunk, W. A., Junek, J., and Sims, C. A.,** Effects of fruit maturity, juice storage, and juice extraction temperature on quality of 'Concord' grape juice, *J. Am. Soc. Hortic. Sci.,* 111, 742, 1986.

174. **Sistrunk, W. A.,** Enzymatic and nonenzymatic reactions affecting the color of Concord grape juice, *Arkansas Farm Res.,* 21(5), 8, 1972.

175. **Sistrunk, W. A. and Cash, J. N.,** Processing factors affecting quality and storage stability of Concord grape juice, *J. Food Sci.,* 39, 1120, 1974.

176. **Spayd, S. E. and Morris, J. R.,** Maturation and quality of 'Concord' grapes as influenced by the pre-harvest complex, *Arkansas Farm Res.,* 27, 5, 1978.

177. **Spayd, S. E. and Morris, J. R.,** Influence of irrigation, pruning severity, and nitrogen on yield and quality of 'Concord' grapes in Arkansas, *J. Am. Soc. Hortic. Sci.,* 103, 211, 1978.

178. **Sistrunk, W. A. and Gascoigne, H. L.,** Stability of color in Concord grape juice and expression of color, *J. Food Sci.,* 48, 430, 1983.

179. **Skalski, C. and Sistrunk, W. A.,** Factors influencing color degradation in Concord grape juice, *J. Food Sci.,* 38, 1060, 1973.

180. **Morris, J. R., Sims, C. A., and Cawthon, D. L.,** Effects of excessive potassium levels on pH, activity and color of fresh and stored grape juice, *Am. J. Enol. Vitic.,* 34, 35, 1983.

181. **Hang, Y. D.,** Recovery of food ingredients from grape pomace, *Process Biochem.,* 23, 2, 1988.

182. **Fuleki, T. and Babjak, L. J.,** Natural food colourants from Ontario grapes, *Highlights Agric. Res. Ont.,* 9, 6, 1986.
183. **Dieci, E.,** Sull'enocianina tecnica, *Riv. Vitic. Enol.,* 12, 567, 1967.
184. **Garoglio, P. G.,** *Nuovo Trattato di Enologia,* Vol. 3, Sansoni, Florence, Italy, 1953, 64.
185. **Garoglio, P. G.,** *La Nuova Enologia,* Instituto di Industrie Agrarie, Florence, Italy, 1965, 502.
186. **Garoglio, P. G.,** L'Enocianina, in *Enciclopedia Vitivinicola Mondiale,* AEB, Brescia, 1980, 130–131, 427–429.
187. **Lancrenon, X.,** Recent trends in the manufacturing of natural red colours, *Process Biochem.,* 13, 16, 1978.
188. **Cappelleri, G.,** Utilizzazione dei sottoprodotti della vinificazione, *Vini d'Italia,* 23, 213, 1981.
189. **Markakis, P.,** Anthocyanins as food additives, in *Anthocyanins as Food Colors,* Markakis, P., Ed., Academic Press, New York, 1982, 245.
190. **Francis, F. J.,** Food colorants: anthocyanins, *CRC Crit. Rev. Food Sci. Nutri.,* 28, 273, 1989.
191. **Peterson, R. G. and Jaffe, E. B.,** Berry and Fruit Treating Process, U.S. Patent 3,484,259, 1969.
192. **Philip, T.,** An anthocyanin recovery system from grape wastes, *J. Food Sci.,* 39, 859, 1974.
193. **Nedelchev, N., Getov, G., Nikova, Z., and Miskov, P.,** Preparation of a colorant concentrate (oenocyanin) for food industry purposes, *Khranit. Prom.,* 24, 10, 1975.
194. **Inagami, K. and Koya, T.,** Sulphite Removal from Anthocyanin Pigments, Japanese Patent No. 75,018,524, 1975 (Karupisu Shokuhin Kiogyo K.K.).
195. **Calvi, J. P. and Francis, F. J.,** Stability of Concord grape *(V. labrusca)* anthocyanins in model systems, *J. Food Sci.,* 43, 1448, 1978.
196. **Main, J. H., Clydesdale, F. M., and Francis, F. J.,** Spray drying anthocyanin concentrates for use as food colorants, *J. Food Sci.,* 43, 1693, 1978.
197. **Mitin, O. N. and Paraska, P. I.,** Food Dye from Grape Residues, U.S.S.R. patent, 660,993, 1979.
198. **Yokoyama, I. and Ono, T.,** Process for Extracting Anthocyanin-Type Colors from Natural Products, U.S. Patent 4,302,200, 1981.
199. **Shrikhande, A. J.,** Extraction and Intensification of Anthocyanins from Grape Pomace and Other Material, U.S. Patent 4,452,822, 1984.
200. **Langston, M. S. K.,** Anthocyanin Colorant from Grape Pomace, U.S. Patent 4,500,556, 1985.

Chapter 7

OTHER FRUITS

I. BARBERRY

Barberry, *Berberis* spp., is a spiny-leafed shrub cultivated for its ornamental foliage which assumes brilliant colors in autumn. There are nearly
175 species native to North and South America, Asia, Europe, and North
Africa. Kuilman[1] determined the role of anthocyanins on the physiology of
B. neuberti plants. Vakula[2] reported that the biosynthesis of anthocyanins in
the leaves of *Berberis* spp. is inhibited by shading. Plants grown at higher
elevation have higher leaf pigmentation,[3] and the highest anthocyanin content
occurs in spring.[4] The rate of respiration of anthocyanin-containing leaves is
1.5 to 2 times higher than that of green leaves,[5] and red-pigmented leaves
display a lower photosynthetic activity.[6] Semkina[7,8] identified five anthocyanins in the leaves of common barberry *(B. vulgaris)*. The major pigments
were characterized as monoglucosides of peonidin, cyanidin, and delphinidin.
Red-leafed varieties contain primarily derivatives of peonidin, while the green-
leafed types contain principally cyanidin derivatives. The anthocyanin content
in red-pigmented leaves can be as high as 156 mg/g of fresh weight (FW),
while green leaves contain less than 3.0 mg/g of fresh tissue. When exposed
to reduced light intensity, plants with the highest content of anthocyanins may
have greater resistance to gas and salts damage.[8] Cyanidin 3-glucoside is the
predominant anthocyanin in the autumn leaves of *B. thunbergii*, also known
as Japanese barberry.[9] The total anthocyanin content in the leaves of *B.
thunbergii* is 24.1 mg/g of FW, while that of *B. vulgaris* is 18.7 mg/g of
fresh tissue.[10]

Barberry shrubs produce edible, red, oval berries rich in anthocyanins,
tannins, ascorbic acid, organic acids, and pectin.[11,12] The berries have been
suggested for use as a source of natural food colorants[13] and for use in liqueurs,
spirits,[14] and pharmaceutical products.[12] Chandra and Todaria[15] followed the
changes in the pigments during fruit growth and ripening. They found that
anthocyanins increase while chlorophyll and berberin (an alkaloid) decrease.
The major pigment in *B. vulgaris* fruit was identified by Lawrence et al.[16]
as a monoside of cyanidin; however, according to Fouassin,[17] this is a monoside of pelargonidin. Boboreko et al.[18] reported a rare form of *B. vulgaris*
which they called *alba*. It has white fruit lacking anthocyanins. The berries
of *B. thunbergii* contained pelargonidin 3-glucoside and cyanidin 3-gluco-
side.[19]

Ten anthocyanins make up the more complex pigmentation of the fruit
of *B. buxifolia*, or Magellian barberry. They are the 3-glucosides; 3-rutinoside-
5-glucosides of peonidin, malvidin, delphinidin, and petunidin; and petunidin
3-rutinoside and petunidin 3-gentiobioside.[20-22] Ishikura and Sugahara[23] iden-

TABLE 1
Anthocyanins of Barberries (*Berberis* spp.)

Species	Anthocyanin[a]	Ref.
Berberis buxifolia	Pn, Mv, Dp, Pt 3-glucosides	20
	Pn, Mv, Dp, Pt 3-rutinoside-5-glucosides	
	Pt 3-rutinoside	
	Pt 3-gentiobioside	
B. koreana	Cy, Pn, Pt 3-glucoside	25
B. cretica	Dp, Pt, Mv, Pn 3-glucosides	24
B. crataegina	Dp, Pt, Mv, Pn 3-glucosides	24
B. darwinii	Pt, Dp, Mv, Pn 3-glucosides	26, 27
	Pt, Dp 3-rutinosides	
	Pn 3-gentiobioside	
B. farreri	Cy 3-glucoside	28
B. fortunei	Cy, Dp, Pt, Mv 3-glucosides	33
	Cy, Dp, Pt, Mv 3-rutinosides	
	Pn glucoside	
B. ruscifolia	Dp, Pt, Cy, Pn 3-glucosides	29
	Dp 3-rutinoside	
B. sieboldii	Cy, Pn, Pt 3-glucosides	25
B. thunbergii	Pg 3-glucoside, Cy 3-glucoside	19
B. vulgaris	Cy, Pn, Pt 3-glucosides	25
	Pg, Cy 3-glucosides	28

[a]   Pn = peonidin, Mv = malvidin, Dp = delphinidin, Pt = petunidin,
Cn = cyanidin, and Pg = pelargonidin.

tified 3-glucoside of cyanidin, delphinidin, petunidin, and malvidin; in addition to the respective 3-rutinosides; as well as a glucoside of peonidin in *B. fortunei*. Four anthocyanins were identified in fruits of *B. crataegina* and *B. cretica*, which are widely distributed through the Middle East.[24] They are the 3-glucosides of delphinidin, petunidin, malvidin, and peonidin. Vereskovskii and Shapiro[25] reported the anthocyanin composition of five *Berberis* species. According to these authors, *B. vulgaris* or common barberry contains cyanidin 3-glucoside, peonidin 3-glucoside, and petunidin 3-glucoside; *B. sieboldii* and *B. koreana* contain the 3-glucosides of cyanidin, petunidin, and peonidin; and the other two species also contain delphinidin 3-glucoside and cyanidin 3-rutinoside. In the fruit of *B. darwinii*, Medrano et al.[26] found peonidin 3-gentiobioside, petunidin 3-glucoside, petunidin 3-rutinoside, delphinidin 3-glucoside, delphinidin 3-rutinoside, malvidin 3-glucoside, and peonidin 3-glucoside. Glucosides of petunidin had been reported earlier by Robinson and Robinson[27] in the bluish berries of *Berberis* × *stenophylla*, *B. stenophylla*, and *B. darwinii*, along with glucosides of malvidin and cyanidin. According to Suomalainen and Eriksson,[28] the orange berries of *B. farreri* contain cyanidin 3-glucosides while the berries of *B. vulgaris* contain the 3-glucosides of pelargonidin and cyanidin. In the fresh fruit of the medicinal plant *B. ruscifolia*, Frontera et al.[29] found delphinidin 3-glucoside, delphinidin 3-rutinoside, petunidin 3-glucoside, cyanidin 3-glucoside, and peonidin 3-

glucoside. Considering the complexity and heterogeneity of anthocyanins of the few species studied (Table 1) and that the genus *Berberis* is made up of a very large number of species, it would appear that much more work to fully characterize the anthocyanins of *Berberis* spp. is required.

II. BUCKTHORN

Buckthorn (*Hippophae rhamnoides* L.) is a willowlike deciduous shrub of the continental regions of Europe and Asia, which produces red berries used to make juices, jams, jellies, and preserves rich in acid.[30-33] Buckthorn berry pulp or juice may also be added to apple and plum products to increase their vitamin C content,[34] which accounts for 350 to 475 mg/100 g of fresh fruit,[35] and 500 mg/100 ml of juice.[36] Stocker[37] reported that small buckthorn berries (0.10 to 0.12 g) contain 1300 mg/g vitamin C compared to an average content of 835 mg. This author also observed a positive relationship between color and ascorbic acid content of berries.

The pigments in the skin of buckthorn berry are chlorophyll, carotenoids, and anthocyanins. The fruit skin also contains catechin, flavonols, and chlorogenic acid. Shapiro et al.[38] reported an anthocyanin content of 140 to 380 mg/100 g of fresh berries, with peonidin 3,5-diglucoside being the only identified anthocyanin.[39] Another major flavonoid of buckthorn berries is isorhamnetin 3-glucoside.[39] Harborne,[40] however, noted that the presence of isorhamnetin does not normally occur with the corresponding anthocyanin peonidin, and thus the identification of this anthocyanin may be incorrect. The presence of isorhamnetin 3-glucoside and isorhamnetin 3-rutinoside was confirmed by Friedrich,[41] Grigorescu and Contz,[42] and Hoerhammer et al.[43] The anthocyanins need to be identified with certainty. Other phenolics of buckthorn berry include rutin, quercetin 7-rhamnoside, narcissin, and myricetin.[44-49]

III. CORNELIAN CHERRY

Cornelian cherry (*Cornus mas* L.) is one of approximately 40 species of the genus *Cornus*, which produces an oblong, 1 to 2 cm long, scarlet, edible fruit used in preserves and sweets.[50] The fruit grows rapidly after fruitset; and ripening coincides with maximum chlorophyll content, with anthocyanin formation following later (Table 2).[51] Lawrence et al.[52] reported that the anthocyanin pigments of *C. mas* are made up of equal quantities of a monoside of cyanidin and a monoside of pelargonidin. A comparative study of the anthocyanins of yellow and red fruit of *C. florida* revealed the presence of glucosides of peonidin, petunidin, and malvidin in the red fruit. In the yellow fruit, only traces of malvidin were detected.[53] In the floral bracts of *C. florida*, Santamour and Lucente[54] identified cyanidin 3-glucoside and a small quantity of petunidin 3-glucoside. Only cyanidin 3-glucoside was found in the fruit. In the fruit of *C. mas*, these authors characterized pelargonidin 3-glucoside in the pulp and cyanidin 3-glucoside and pelargonidin 3-glucoside in the skin

TABLE 2
Anthocyanins of Cornelian Cherries and Other
Cornaceae

Species	Anthocyanin[a]	Refs.
Cornus alba	Cy, Dp 3-galactosides	58
	Cy 3-arabinoside	
C. alternifolia	Dp 3-glucoside	57
	Dp, Pt 3-rutinoside	
C. canadensis	Pg, Cy 3-glucosides, Pg, Cy 3-galactosides	61
	Pg, Cy 3-rutinosides	
	Pg 3-sophoroside	
C. florida	Cy, Dp 3-glucosides	54, 60
	Cy, Dp, Pg 3-galactosides	
	Cy 3-arabinoside	
C. kousa	Cy, Dp, Pg 3-glucosides	60
C. mas	Cy, Dp, Pg 3-galactosides	56, 57
	Cy, Pg 3-rhamnogalactosides	

[a]   Cy = cyanidin, Dp = delphinidin, Pt = petunidin, Pg = pelargonidin.

of the fruit. According to Lazar et al.,[55] the flavonoid composition of fruits and flowers of *C. sanguinea* is the same. Du and Francis[56] found cyanidin 3-galactoside in fruits of *C. mas*. In another study, the same fruit was shown to also contain pelargonidin 3-galactoside and pelargonidin 3-rhamnogalac-toside.[57] Anthocyanins (cyanidin 3-galactoside and delphinidin monogluco-side) are also found in the petioles of *C. alternifolia* whose fruit contains delphinidin 3-glucoside and 3-rutinoside with lesser quantities of petunidin 3-rutinoside. The bark of *C. alba* contains mainly cyanidin 3-galactoside with smaller quantities of cyanidin 3-arabinoside and monoglucoside of delphin-idin.[58] Tamas and Stoleriu[59] reported four anthocyanins in *C. mas*. Du et al.[60] found cyanidin 3-glucoside, delphinidin 3-glucoside, and pelargonidin 3-glu-coside in *C. kousa*. In *C. florida*, Du et al.[60] found cyanidin 3-galactoside and cyanidin 3-glucoside to be the major pigments; delphinidin 3-glucoside and 3-galactoside were present in minor amounts and pelargonidin 3-galac-toside and cyanidin 3-arabinoside were identified as trace pigments. Other studies by Du et al.[61] on the *Cornus* genus revealed the presence of four principal anthocyanins. In bunchberry *(C. canadensis)* they found pelargon-idin 3-sophoroside; the 3-galactoside, 3-glucoside, and 3-rutinoside of pelar-gonidin; and traces of the analogous cyanidins.

IV. CROWBERRY

Crowberry, *Empetrum nigrum* Coll., is a northern plant producing edible berries with a high anthocyanin content. There are three well-known subspe-cies of this fruit, *E. nigrum* L. or southern crowberry, *E. nigrum* ssp. *her-maphroditum* or northern crowberry, and *E. nigrum* ssp. *japonicum* or Japanese crowberry. *E. nigrum* ssp. *hermaphroditum* (Hagerup) Bocker is a species typical of the circumpolar regions, while *E. nigrum* L. and *E. nigrum* ssp. *japonicum* K. Koch are found between 40 and 70° N latitude. All species have been used for food by the people native to the subarctic region for centuries, and in recent years considerable interest in the composition and commercial utilization of this fruit has been expressed.

Hayashi et al.[62,63] identified delphinidin 3-galactoside as the principal pigment in *E. nigrum* spp. *japonicum* which they called empetrin. Chromatographic analysis by Fouassin[17] showed six anthocyanin pigments with traces of a seventh. Suomalainen and Eriksson[28] reported two malvidin derivatives, malvidin 3-glucoside and malvidin 3-galactoside; a monoside of delphinidin; two monosides of cyanidin; and a monoside of petunidin. Moore et al.[64] identified two anthocyanins already noted by Fouassin[17] as well as petunidin, peonidin, and cyanidin 3-galactosides with traces of the corresponding 3-glucosides. More recently, Kärppä et al.[65] separated and characterized 12 nonacylated anthocyanins. They are glucosides, galactosides, and arabinosides of delphinidin and petunidin; and galactosides and arabinosides of cyanidin, malvidin, and petunidin. The presence of malvidin 3-glucoside was not confirmed. The three principal pigments which represent 60% of the total are galactosides of cyanidin, delphinidin, and malvidin. The anthocyanin content changes with the maturity of the fruit.[66] In immature northern crowberries, Kärppä[66] found only 3 mg of anthocyanin per 100 g of fruit and the only anthocyanins present were glucosides of cyanidin. As ripening proceeds, the glucosides of cyanidin decrease and those of malvidin increase. The other glucosides appear later in the maturation process reaching their maximum when the fruit is ripe[66] (Table 3). These changes in anthocyanin composition of the crowberry reveal a clear evolution in the biosynthetic capacity of this fruit in relation to its physiological age, leading to maximum methoxylation of the aglycone as maturation proceeds.

The dark-colored crowberry juice contains about 45 to 50% of the total anthocyanins of the fruit; and because of its relatively bland taste, it may have application as natural food colorant. The anthocyanin content, calculated on a fresh weight basis, varies from 300 to 420 and from 350 to 500 mg/100 g in the northern crowberry and in the southern crowberry, respectively.[67] Since the fruit contains little pectin, the use of pectinase increases the amount of juice extracted by pressing by only 3 to 17%.[67] The stability of the anthocyanin pigments was studied by Kallio et al.,[68,69] who reported that in untreated juice the half-life of the individual anthocyanins is 4 to 6 weeks. When 90% of the oxygen is removed, the stability improves by a factor of 3 to 4. The glucosides and galactosides have longer half-lives than the arabinosides. The presence of Fe^{3+} and Al^{3+} improves the stability but also changes the color shade of crowberry anthocyanins.[67,70] In Eastern Europe where *E. nigrum* is readily available, the fruit anthocyanins are apparently extracted and used as natural food colorants.[71]

TABLE 3
Anthocyanin Contents and Relative Amounts of Different Anthocyanins of Crowberries at Different Stages of Ripeness

Anthocyanin contents[b]	Stage of ripeness[a]			
	UR (3)	HR (180)	R (460)	OR (450)
Anthocyanin[c]	t[e]	23.4	23.2	24.4
Dp-glu + Dp-gal[d]	t	3.6	3.5	3.4
Dp-ara	t	27.0	26.7	27.8
Cn-gal	76.6	31.5	27.8	25.0
Cn-ara	23.4	7.9	6.8	6.0
	100.0	39.4	34.6	31.0
Pt-glu + Pt-gal[d]	—	8.4	9.3	10.1
	—	1.7	1.5	1.5
	—	10.1	10.8	11.6
Pn-gal	t	6.4	7.1	6.4
Pn-ara	—	2.7	2.4	2.4
	t	9.1	9.5	8.8
Mv-gal	t	12.4	16.2	18.3
Mv-ara	—	1.8	2.2	2.6
	t	14.2	18.4	20.9

[a] UR = unripe, HR = half ripe, R = ripe, OR = overripe.
[b] In mg/100 g FW.
[c] Dp = delphinidin, Cn = cyanidin, Pt = petunidin, Pn = peonidin, Mv = malvidin; glu = glucose, ara = arabinose, gal = galactose.
[d] Unseparable in the HPLC analysis.
[e] t = Trace <0.1%.

From Kärppä, J., *Lebensm. Wiss. Technol.*, 17, 175, 1984. With permission.

V. DANEWORT

The anthocyanins of danewort (*Sambucus ebulus* L.) were first studied by Fouassin,[17] who found the pigment to be exclusively cyanidin 3-xylosyl-glucoside. Chiarlo et al.[72] identified cyanidin 3-sambubioside and cyanidin 3-sambubioside-5-glucoside as well as caffeic and p-coumaric acids. According to Ogynanov et al.,[73] this *Sambucus* species has an anthocyanin content one and a half times that of *S. nigra*; and is therefore used as a source of food colorant for juices, syrups, and gelatins. Novruzov and Aslanov[74] reported a pigment content of 3.9 to 4.5 g/100 g of dry fruit and the presence of six other anthocyanins. Two anthocyanins were identified as cyanidin 3,5 diglucoside and cyanidin 3-glucoside. Lachman et al.[75] isolated and identified three pigments: cyanidin 3,5-diglucoside (80%), cyanidin 3-arabinoglucoside

TABLE 4
Anthocyanins of Elderberries
(*Sambucus nigra*)

	Conc (% of total)	
Anthocyanin	**BHH[a]**	**DD[b]**
Cy 3-glucoside	65.7	32.6
Cy 3-sambubioside	32.4	55.2
Cy 3-sambubioside-5-glucoside	1.1	10.2
Cy 3,5-diglucoside	0.8	2.0

[a] From Brønnum-Hansen, K. and Hansen, S. H., *J. Chromatogr.*, 262, 385, 1983.
[b] From Drdak, M. and Daucik, P., *Acta Aliment.*, 19, 3, 1990.

(15%), and cyanidin 3-glucoside (5%). In addition, they reported the presence of glucosides of quercetin and hydroxycinnamic acids.

VI. ELDERBERRY

Elderberry, *Sambucus nigra* L., is a large and rather coarse shrub or small tree widely distributed throughout the temperate and continental zones of Europe. Although they can be eaten raw, elderberries are used most extensively for juices, wines, and jams.[76] The black, soft-skinned berries are one of the richest sources of anthocyanins, of a relatively simple composition, and with few other phenolic compounds.

Nolan and Casey[77] found a 3-monoglycoside and a pentosylglycoside of cyanidin in berries of *S. nigra.* Fouassin,[17] using paper chromatography, separated four pigments and characterized the two principal ones as monosides of cyanidin. Harborne,[78] and Reichel and Reichwald,[79] and Reichel et al.[80] later reported that one of the pigments of *S. nigra* was cyanidin 3-xylosylglucoside. The structure of the principal pigment was characterized as cyanidin 3-(2 G-xylosylglucoside) and called sambycyanin by Reichel and Reichwald.[81]

Using a variety of chromatographic techniques including HPLC, LC, and column chromatography, Brønnum-Hansen and Hansen[82] isolated and characterized cyanidin 3-sambubioside-5-glucoside, cyanidin 3,5-diglucoside, cyanidin 3-sambubioside, and cyanidin 3-glucoside in elderberries from Denmark. Table 4 lists the percentage distribution of these anthocyanins as determined by HPLC. As can be noted, the two 3,5-diglycosides amount to less than 2% of the total and the amount of cyanidin 3-sambubioside is roughly half that of cyanidin 3-glucoside. Recently, Drdak and Daucik[83] confirmed the occurrence of these four pigments in *S. nigra*; however, the relative concentration of the anthocyanins reported by these authors is significantly different from that reported by Brønnum-Hansen and Hansen[82] (Table 4), and suggests that different varieties were probably used in the two studies. In their investigation on changes of individual anthocyanins during production of elderberry pigment concentrates, Drdak and Daucik[83] found that in the course of pasteurization, fermentation, and concentration cyanidin 3-sambubioside was the most stable pigment. Brønnum-Hansen and associates[84,85] conducted studies on recovery of elderberry anthocyanins by an extraction and freeze-drying process, as well as storage stability of the product obtained.[86]

Another *Sambucus* species related to *S. nigra* is *S. racemosa* L., or red elderberry. The fruit of this small tree-shrub contains cyanidin 3(p-coumaryl)-sambubioside-5-glucoside and cyanidin 3-sambubioside-5-glucoside.[87] Traces of two other anthocyanins have been detected, but their identity has not been established, although their chromatographic characteristics resemble those of raphanusin C and perillanin.[87] In an unidentified *Sambucus* species containing 3.3 mg of anthocyanin per 100 g of fresh fruit, Shin and Ahn[88] characterized cyanidin 3-sambubioside-5-glucoside, cyanidin 3,5 diglucoside, cyanidin 3-xylosylglucoside-5-glucoside acylated with p-coumaric acid, cyanidin 3,5 diglucoside acylated with p-coumaric acid, and cyanidin 3-glucoside. From the fruits of *Sambucus canadensis*, Johansen et al.[89] isolated four anthocyanin glycosides. The major pigment, representing 69.8% of total anthocyanin content, was characterized as cyanidin 3-[6-(p-coumaryl)-2-(xylosyl)-glucoside]-5-glucoside. The other anthocyanins were identified as cyanidin 3-sambubioside-5-glucoside (22.7%), cyanidin 3-sambubioside (2.3%), and cyanidin 3-glucoside (2.1%).[89]

As previously mentioned, elderberries are used to manufacture a variety of products including preserves, jams, juices, and food colors. Pigment extracts are generally prepared from *S. nigra* and *S. ebulus* and may be used for coloring apple and blackcurrant juices and wines, particularly port wines.[90-93] The addition of elderberries accelerates grape wine fermentation, and in some cases, it has been done to make red wines from white wines.[91] Addition of elderberry juice to grape wine is, however, illegal in most countries, and this fraud can easily be detected by chromatographic analysis of the adulterated wine.[92] Elderberry juice may also be added to apple and black currant juices to improve their color as well as their taste.[93] Pfannhauser and Riedl[94] used HPLC analysis of the anthocyanins to detect adulteration of black currant juice with elderberry juice. Residues remaining after extraction of the juice from elderberries may be used to produce natural food colorants.

A known manufacturing facility of these colorants is the Michuring Canning Plant in Kishinev.[95] According to Chaplin et al.,[95] the process used at this facility involves: grinding of the fruit residues remaining after juice extraction, extraction of the pigments with hot water, pressing of the mixture, and concentration of the water extract to 40 to 50% dry weight. With this process, a tonne of raw material yields about 70 to 75 kg of colorant suitable for coloring a wide variety of goods. To remove pectins, proteins, carbohydrates, and other undesirable water soluble substances from the crude pigment, the extract may be treated with acidified alcohol and partially purified by filtration.[97] Absorption of the pigments on talc and starch after the impurities have been removed from the pigments with alcohol has also been suggested.[98] At the experimental level, a number of other techniques have been used for extraction/purification of elderberry pigments. They include ultrafiltration,[99] fermentation coupled with concentration,[100-102] aqueous extraction coupled with centrifugation,[103] and chromatography.[104] All of these techniques can be used to produce functional products, but their commercialization is often hindered by the high cost of the finished product.

VII. FIG

Some varieties of *Ficus carica* (Moraceae) fruits have a violet skin due to the presence of anthocyanins.[105] The first studies on pigments of the fig were carried out by Robinson and Robinson,[27] who found cyanidin 3-glucoside. More recently, Puech et al.[106] separated and characterized by chromatography and spectrophotometry four anthocyanins in the intense black-blue fig. The main pigment is cyanidin 3-rhamnoglucoside, accounting for about 75% of the total. The other pigments present are cyanidin 3,5-diglucoside (11%), cyanidin 3-glucoside (11%), and pelargonidin 3-rhamnoglucoside (3%). Cyanidin 3-rhamnoglucoside was found in the skin of the fruit as well as in the pericarp and pedicels of the fruit seeds. In *F. carica* (Nigra variety) Duro and Condorelli[107] found only one pigment, cyanidin 3-glucoside, which is the same pigment found in *F. pumila*.[23,108] Duro and Condorelli[107] also reported the occurrence of cyanidin 3-glucoside, cyanidin 3-rutinoside, and cyanidin 3,5-diglucoside in *F. nipponica*. Fruits of *F. wightiana* and *F. erecta* have been reported as containing cyanidin 3-rutinoside and cyanidin 3-glucoside[23,108] (Table 5). Application of ethephon to 'Mission' fig promotes accumulation of anthocyanins and fruit growth, and both of these features are strongly affected by light.[109] Treatment with ethephon in light increases the rate of anthocyanin accumulation, but a smaller amount is found in the fruit. Growth of ethephon treated fruit is strongly stimulated in the dark, but the anthocyanin level remains near zero.[109] Sharma and Seshadri[110] found cyanidin rutinoside in the fruit stem of *F. bengalensis* and cyanidin diglucoside in the leaf. Other phenolics of figs include: the 3-glucosides and 3-rutinosides of kaempferol and quercetin, apigenin-6-*C*-glucosyl-8-*C*-arabinoside, and apigenin-6-*C*-arabinosyl-8-*C*-glucoside.[111]

VIII. HONEYSUCKLE

In the amethyst fruit of honeysuckle, *Lonicera nitida* L., Robinson and Robinson[27] identified the anthocyanin pigment as a 3-bioside of cyanidin. Recently, anthocyanins have also been found in leaves of *L. japonica* var.

TABLE 5
Anthocyanins of Figs (*Ficus* spp.)

Species	Anthocyanin[a]	Refs.
Ficus carica cv. Mission	Cy, Pg 3-rutinoside	106
	Cy 3-glucoside	
	Cy 3,5-diglucoside	
F. carica cv. Nigra	Cy 3-glucoside	107
F. erecta	Cy 3-rutinoside	23, 108
	Cy 3-glucoside	
F. nipponica	Cy 3-rutinoside	107, 108
	Cy3-glucoside	
	Cy3,5-diglucoside	
F. pumila	Cy 3-glucoside	107, 108
F. wightiana	Cy 3-rutinoside	23,108
	Cy 3-glucoside	

[a]   Cy = cyanidin, Pg = pelargonidin.

aureo-reticulata.[112] *L. edulis* and *L. kamtschatica* contain cyanidin 3-rutinoside and cyanidin 3-glucoside in a 3:1 ratio.[113] A variety of *L. altaica* from the Irkutsk region of Russia contains only cyanidin 3-glucoside in a concentration of 4% dry weight basis.[114] Other varieties of *Lonicera* contain cyanidin 3-rutinoside in a concentration of about 2 g/100 of dry fruit.[115] Cyanidin 3-glucoside has been identified in *L. coerulea.*[116] A phytochemical study of the flavonoids from Caprifoliaceae was carried out by Glennie.[117]

In addition to anthocyanins, fruits of honeysuckle also contain catechin, flavonols, procyanidins, and tannins. Kieselev et al.[118] have recommended their use as a raw material for the pharmaceutical industry. A Japanese patent[119] proposes the use of anthocyanins extracted from the seeds of *L. coerulea* for food colors.

IX. MULBERRY

The common or black mulberry (*Morus nigra* L.), originally from China, is widespread throughout countries with a moderate climate. The fruit is used to produce juice, jam, and jelly. The earliest report on the pigments of this fruit is by Yamamoto[120] who identified cyanidin 3-glucoside as the fruit pigment. Fouassin[17] found two monosides of cyanidin and a monoside of pelargonidin. Subsequently, Toscano and Lamonica[121] reported the presence of another pigment besides cyanidin 3-glucoside. Maki and Inamoto[122] identified cyanidin 3-glucoside as well as two other minor pigments, pelargonidin 3-glucoside and petunidin 3-rutinoside. This apparent confusion on the identity of pigments of the mulberry is probably due to a mistaken identification of the species analyzed by some of the investigators. According to Timberlake,[123] *M. nigra* contains cyanidin 3-glucoside, and the purple fruits of the Asian white mulberry *(M. alba)* have this and other pigments. Ishikura[108] found cyanidin 3-rutinoside as the main component of *M. alba* fruit. Miyakawa and Takehana[124] extracted cyanidin 3-glucoside from *M. bombycis* and used it as a model pigment in a study on quality of canned peaches. According to Maki et al.,[125] the mulberry anthocyanins possess good stability (80 d at 30°C) and can be used to color carbonated drinks.

X. ORANGE

In addition to the carotenoids and liposoluble chlorophylls which are normally present in oranges, some varieties of orange (*Citrus sinensis* L., Rutaceae) also contain water-soluble pigments that give a characteristic red color, from which the name "blood oranges" is derived. Matlack[126] showed that these pigments are anthocyanins, and Yuasa et al.[127] described the cytological aspects of these pigments. Carrante[128] investigated factors that influence the production of anthocyanins in orange varieties from Sicily, while Ajon[129-131] published results on the concentration of pigments in juice from different varieties of blood oranges.

Different factors are known to influence the development of color in blood oranges. The most important are varietal characteristics[132] and climatic conditions during fruit maturation, especially temperature.[133,134] Chandler[135] used chromatography to examine blood oranges from California ('Moro' variety) and found that 95% of the anthocyanins were made up of cyanidin 3-glucoside. There were traces of a second pigment tentatively identified as delphinidin 3-glucoside. Koch and Haase-Sajak[136] confirmed the presence of cyanidin 3-glucoside and delphinidin 3-glucoside in the Moro variety, and reported that other varieties of colored orange (such as 'Sanguinello', 'Paternó', and 'Spanish Mezzosangue') all contain cyanidin 3-glucoside; however, only the 'Sanguinello' oranges contain traces of delphinidin 3-glucoside. They also found that all the varieties tested contain, in small quantities, two other unidentified pigments. In the 'Moro' variety, Porretta et al.[137] found four other pigments in addition to cyanidin 3-glucoside, but these minor pigments were not identified because they were present in trace amounts. In the juice of oranges from Calabria, Kunkar[138] identified cyanidin 3-glucoside and delphinidin 3-glucoside, and reported the presence of four additional pigments which were not identified. The author hypothesized that one of these could be cyanidin 3,5-diglucoside because glucose was the only sugar detected in all pigments after hydrolysis. Licastro and Bellomo[139] separated seven pigments from the juice of 'Moro' oranges, five of which were identified as peonidin 3-glucoside, cyanidin 3-glucoside, cyanidin 3,5-diglucoside, petunidin 3-glucoside, and delphinidin 3-glucoside.

Maccarone et al.[140,141] used high-performance liquid chromatography to isolate 14 pigments from the Moro variety, of which 10 were identified with spectrophotometric techniques. Of these, cyanidin 3-glucoside makes up about 50% of the total. The 3-glucosides of delphinidin, petunidin, and pelargonidin are present in lesser quantities than the 3,5-diglucosides of cyanidin, delphinidin, peonidin, and petunidin. An acylated anthocyanin, cyanidin 3-(4''-acetyl)-glucoside, was identified for the first time in the orange. This pigment is the second most abundant (about 18%) after cyanidin 3-glucoside.

Benk[142] found that in addition to differences in pigmentation, the juice of blood oranges has higher sugar and protein contents but lower citric acid and ash contents compared to those of normal varieties. 'Moro' oranges from Italy yield a very dark juice which may contain over 200 mg anthocyanins per 100 ml of juice.[143] Significant differences in anthocyanin content of blood oranges from Calabria and Sicily were reported by Di Giacomo et al.,[144] who found that oranges from Calabria had a much lower content of anthocyanins (199.7 ± 106.6 mg/l) than oranges from Sicily (721.0 ± 153.3 mg/l). Recently, Lee et al.[145] found that while California blood orange juice contains 96 to 166 mg/l total anthocyanin pigments, orange juice made from Florida-grown 'Moro' oranges contains only 3.7 mg/l total anthocyanin pigments. The anthocyanin profiles of Florida and California blood orange juices are, however, similar, containing 83 to 91% cyanidin, 6 to 10% delphinidin, 1 to 6% peonidin, and below 1% petunidin and pelargonidin. During storage of blood oranges, the development of the red color is favored by low temperature.[146] Anthocyanin content of the juice also increases when the fruit is stored in an atmosphere rich in ethylene and/or even more so in an atmosphere rich in oxygen.[147]

The problem of color stability in the juice of blood oranges has been studied by a number of authors with the aim of removing or stabilizing the anthocyanins present.[148-155] Casoli et al.[148] investigated the action of enzymatic extracts of fungal origin on the removal of color from the juice. Similarly, Dall'Aglio et al.[149] tested the effect of commercial preparations of glucose oxidase on color removal and found that although this enzyme was effective in decolorizing the juice, it adversely affected its flavor. The purpose of removing the anthocyanin/red color from the juice of blood oranges is to produce stable juice from this type of orange, so as to be able to market it through the same channels as regular orange juice. A recently patented process[150] uses ultrafiltration to remove the anthocyanins as well as the bitter-tasting limonin. The anthocyanins which are recovered are highly prized colors for drinks, foods, and pharmaceuticals.

Approaches aimed at stabilizing the anthocyanins of blood orange fruit juice have included acidification and antioxidant additives, pasteurization with microwave, and complexation of anthocyanins with rutin and phenolic acids.[137,151-155] Some of these methods have shown considerable promise, but to our knowledge none has been commercialized yet.

XI. POMEGRANATE

Pomegranate (*Punica granatum* L.) contains anthocyanins in the flowers, the pulp surrounding the seed, and the fruit peel. The fleshy seed coat, which

TABLE 6
Anthocyanins of Pomegranates
(*Punica granatum* L.)

	Relative conc	
Anthocyanin[a]	Seedcoats	Peel
Cy 3-glucoside	+ + + +	+ + + +
Cy 3,5-diglucoside	+ +	+ +
Dp 3-glucoside	+ +	−
Dp 3,5-diglucoside	+	−
Pg 3-glucoside	+ +	+
Pg 3,5-diglucoside	+	+

[a] Cy = cyanidin, Dp = delphinidin, Pg = pelargonidin.

Adapted from Du, C. T., Wang, P. L., and Francis, F. J., *J. Food Sci.*, 40, 417, 1975.

is the edible part of the fruit, is used to make juices and fermented fruit drinks.[156] Saxena et al.[157] have recently reviewed the chemistry and technology of the pomegranate.

The first report on the anthocyanins of *P. granatum* was by Karrer and Widmer,[158] who identified pelargonidin 3,5-diglucoside in the flowers. Sharma and Seshadri[110] reported the presence of a pentoside of malvidin in the seeds and two pentosides of malvidin and petunidin in the skin. Harborne[159] found delphinidin 3,5-diglucoside in pomegranate juice and pelargonidin 3,5-diglucoside in the flowers. Lowry[160] confirmed the presence of pelargonidin 3,5-diglucoside in the flowers. Pifferi et al.[161] found six pigments, of which four were identified as delphinidin 3-glucoside, delphinidin 3,5-glucoside, cyanidin 3-glucoside, and cyanidin 3,5-diglucoside. The other two were not identified because they were present only in trace amounts. Du et al.[162] reported cyanidin 3-glucoside as the major anthocyanin of the seed coat and fruit peel, accompanied by pelargonidin 3-glucoside, pelargonidin 3,5-diglucoside, cyanidin 3,5-diglucoside, delphinidin 3-glucoside, and delphinidin 3,5-diglucoside. In the fruit peel, the derivatives of pelargonidin are more predominant than those of cyanidin, and there are no delphinidin glycosides. Santagati et al.[163] also identified delphinidin and cyanidin 3-glucosides and 3,5-diglucosides in the seed coat, and observed that the ratio between the delphinidin and cyanidin derivatives favored the cyanidins in fruit from the hilly regions but not in fruit from trees grown in the plains. To date, there have been no reports of acylated anthocyanins in the pomegranate (Table 6).

Anthocyanins appear in the fruit about 90 d after flowering, and according to Lee et al.[164] the first anthocyanins to appear are the 3-glucosides of cyanidin and pelargonidin. Markh and Lysogor[165] examined 16 varieties grown in Crimea and Central Asia; they reported that the fruit contained from 0.22 to

1.05% polyphenols, including cyanidin 3,5-diglucoside, cyanidin 3-gluco-side, delphinidin 3-glucoside, and various phenolic acids. Some highly colored varieties may contain as much as 600 to 765 mg of anthocyanins per 100 g of juice.[166] Botrus et al.[167] found that the most representative polyphenol in pomegranate juice is catecol, with lesser quantities of anthocyanins and leucoanthocyanins.

XII. ROSEHIP

Rosehip is the fruit of *Rosa canina* L. and other *Rosa* species. It is used primarily for making jam, jelly, preserve, and juice. In the Republic of Georgia, rosehips are abundant and apparently over 3 million jars of juice are produced annually.[168,169] Hips vary in weight between 1.25 and 2.93 g each, with the seeds accounting for 20 to 64% of the weight depending on the variety.[168] The pulp of rosehips is a rich source of ascorbic acid and can also be used to supplement and improve peach and apricot nectars.[170,171]

The red color of rosehips is due primarily to carotenoids (*trans-* and *cis-*lycopene, rubixanthin, zeaxanthin, and carotene), although anthocyanins may also contribute to rosehip color.

Koncalova[172] reported the presence of cyanidin and peonidin in the hips of *R. pimpinellifolia* and the presence of cyanidin in hips of the hybrid *R.* × *reversa*. Sobolevskaya and Demina[173] found cyanidin 3-rutinoside in ripe fruit of dog rose (*R. canina* L.), and Plekhanova et al.[174] reported the presence of delphinidin glycosides and procyanidins in fresh fruit of *R. spinosissima* or Scotch rose. Delphinidin derivatives are, however, highly uncommon in Rosaceae, and their identification in the fruit of *R. spinosissima* may be incorrect. Demina,[175] however, identified cyanidin 3-glucoside and cyanidin 3,5-diglucoside as well as a glycoside of delphinidin; leucoanthocyanins; and ascorbic, citric, malic, and oxalic acids in the pulp of *R. spinosissima*.

R. damascena (Damask or Kazanlik) rose is used to make rose petal jam and attar of rose. Velioglu and Mazza[176] using HPLC identified cyanidin 3,5-diglucoside as the main pigment, representing 95% of the 285 mg total anthocyanin per kilogram of fresh petals. Also identified were six flavan 3-ols, six kaempferols, and three quercetin glycosides (Table 7). In petals of 670 cultivars and 8 species of ornamental roses, Yokoi[177] identified cyanidin 3-glucoside and 3,5-diglucoside, pelargonidin 3-glucoside and 3,5-diglucoside, and peonidin 3,5-diglucoside. Of these, cyanidin 3,5-diglucoside was present in all of the cultivars analyzed. The inheritance of anthocyanins in rose petals has been studied by Marshall et al.[178]

Recently, Kuliev et al.[179] have patented a process for the extraction of anthocyanin-based food colorants from the fruits of *R. spinosissima*, and a brown pigment with potential as a food colorant has been produced from ripe fruits of *R. laevigata*.[180]

TABLE 7
Flavonoids of Rose Petals (*Rosa damascena* Mill.)

Peak	Compound	Relative conc
1, 3–7	flavan 3-ols	+
8	Cyanidin 3,5-diglucoside	+ +
9–13, 15	Flavan 3-ols	+
16	Quercetin 3-glucoside	+ + +
17	Quercetin 3-glucosylgalactoside	+ +
18	Quercetin 3-arabinoside	+
20	Kaempferol 3-galactosylarabinoside	+
21	Kaempferol 3-glucosylarabinoside	+ + + +
23	Kaempferol 3-arabinoside	+ + +
24	Kaempferol 3-rhamnosylglucoside	+ + +
25	Kaempferol 3-rhamnosylgalactoside	+ +
26	Kaempferol 3-rutinosyl-7-arabinoside	+ +

Adapted from Velioglu, Y. S. and Mazza, G., *J. Agric. Food Chem.*, 39, 463, 1991.

XIII. OTHERS

A. STRAWBERRY TREE FRUIT

Strawberry tree (*Arbutus unedo* L.), distributed throughout the Mediterranean area, is an evergreen tree that produces round, red fruits (2 to 4 cm in diameter) which resemble strawberries. A delicious alcoholic cider and jams can be made with the fruit.[181] The leaves, skins, fruits, and flowers of the strawberry tree contain the flavonoids afzelin, juglalin, avicularin, quercetin, and hyperin.[182] The red pigment of the fruit was identified as cyanidin 3-glucoside by Proliac and Raynaud.[183] However, Maccarone et al.[184] found — in addition to cyanidin 3-glucoside — two other pigments, delphinidin 3-galactoside and cyanidin 3-galactoglucoside.

B. MAQUI

Aristotelia chilensis Mol. & Stuntz. is native to South America, where it is known as maqui. Its fruit is a berry rich in anthocyanins, and in Chile it may be used to enrich the color of the locally produced red wine. Schmidt-Hebbel et al.[185] identified the anthocyanin of the maqui fruit as the diglucosides of cyanidin, malvidin, delphinidin, and petunidin in a concentration similar to that of wine made from grape hybrids. More recently, Diaz et al.[186] reported the presence of nine anthocyanins which are cyanidin 3,5-diglucoside, delphinidin 3,7-diglucoside, delphinidin 3-glucoside acylated with *p*-coumaric acid, delphinidin 3-glucoside, malvidin 3,5-diglucoside, malvidin 3,7-diglucoside, malvidin 3,5-diglucoside acylated with *p*-coumaric acid, petunidin 3,5-diglucoside, and petunidin 3-glucoside.

C. SHEEPBERRY

Viburnum (Caprifoliaceae) is represented by numerous species widespread throughout North America, Asia, and Europe. They are mostly rather large shrubs or rarely small trees often used as ornamental plants. Their common names are sheepberry, wayfarer trees, or cranberry bush or nanny berry.

In the fruit of *V. trilobum*, Wang and Francis[187] found a 3-arabinoglucoside of cyanidin. This anthocyanin was later characterized as cyanidin 3-arabinosylsambubioside (arabinose + xylose + glucose).[188] Appreciable quantities of the same pigment are also found in *V. dentatum*.[189] In nine *Viburnum* spp., Ishikura[108,190] identified a single anthocyanin, cyanidin 3-xylosylglucoside (sambubioside). *V. lantana* has ovoid berries, 4 to 6 mm in diameter, which contain kaempferol 3-diglucoside and a single anthocyanin (cyanidin 3-glucoside).[191] The fruit of *V. tinus* has two anthocyanins, cyanidin 3-glucoside and cyanidin 3-sambubioside, and several other flavonols including quercetin, isoquercetin, and rutin.[192]

REFERENCES

1. **Kuilman, L. W.,** Physiologische Untersuchung über die Anthocyane, *Rec. Trav. Bot. Neerl.,* 17, 953, 1930.
2. **Vakula, V. S.,** The effect of light on leaf colour in ornamental plants, *Bjull. Glav. Bot. Sada,* 78, 37, 1968 (HA 1969, 5074).
3. **Chandra, P. and Todaria, N. P.,** Leaf pigmentation in *Berberis* species from different altitudes, *Photosynthetica,* 18, 414, 1984 (CA 102, 42907).
4. **Semkina, L. A.,** Pigment content of leaves in purple-leaved trees, *Bjull. Glav. Bot. Sada,* 72, 78, 1969 (HA 1970, 4471).
5. **Ketsoskhoveli, E. N. and Dzaparidze, I. G.,** Respiration of anthocyanin-containing and green forms of tree species, *Soobshch. Akad. Nauk Gruz. SSR,* 62, 661, 1971 (CA 75, 85248).
6. **Lebedev, S. I. and Litvinenko, L. G.,** Comparative study of photochemical activity of chromoplasts of anthocyanin-containing and green forms of plants, *Puti Povysh. Intensivn. Prod. Fotosint. Akad. Nauk, Uk., Respub. Mezhvedom. Sb.,* 1966, 85 (CA 67, 51082).
7. **Semkina, L. A.,** Intraspecific chemical variation in content of anthocyanin pigments in *Berberis vulgaris* (common barberry), *Ekologiya,* 2, 45, 1971 (CA 75, 85143).
8. **Semkina, L. A.,** On the origin and evolutionary role of purple-leafed varieties of arboreal plants, *Tr. Inst. Ekol. Rast. Zhivotn., Ural Nauchn. Tsentr. Akad. Nauk SSSR,* 91, 104, 1975 (CA 84, 147646).
9. **Murrell, J. R., Jr. and Wolf, F. T.,** The anthocyanin in the autumn leaves of Japanese barberry, *Bull. Torrey Bot. Club,* 96, 594, 1969 (HA 1970, 4473).
10. **Mamaev, S. A. and Semkina, L. A.,** Anthocyanin pigments in purple-leaved forms of some trees and shrubs, *Rastit. Resur.,* 7, 280, 1971 (CA 75, 95377).
11. **Novruzov, E. N.,** Chemical composition of fruits and berries of plants growing wild in Azerbaijian, *Rastit. Resur.,* 24, 48, 1988 (HA 1988, 4028).
12. **Palamarchuk, A. S. and Bondarenko, V. E.,** Chemical composition of the fruit of common barberry *(Berberis vulgaris), Rastit. Resur.,* 10, 237, 1974 (CA 81, 60875).
13. **Kasumov, M. A. and Kuliev, V. B.,** Natural dyes used to color food products, *Izv. Akad. Nauk Az. SSR, Ser. Biol. Nauk,* 4, 126, 1981 (CA 96, 102655).
14. **Benk, E.,** Weniger bekannte Wildfruechte, *Fluess. Obst.,* 53, 58, 1986 (FSTA 18, 12J3).

15. **Chandra, P. and Todaria, N. P.,** Maturation and ripening of three *Berberis* species from different altitudes, *Sci. Hortic.,* 19, 91, 1983 (CA 98, 212980).
16. **Lawrence, W. J. C., Price, J. R., Robinson, G. M., and Robinson, R.,** Distribution of anthocyanins in flowers, fruits and leaves, *Philos. Trans. R. Soc. London, Ser. B,* 230, 149, 1939.
17. **Fouassin, A.,** Identification par chromatographie des pigments anthocyaniques des fruits et des legumes, *Rev. Ferment. Ind. Aliment.,* 11, 173, 1956.
18. **Boboreko, E. Z., Shapiro, D. K., Narizhnaya, T. I., and Starkova, N. Y.,** A rare form of *Berberis vulgaris, Vestsi Akad. Navuk, SSS Ser. Biyal. Navuk,* 5, 7, 1984 (CA 102, 21248).
19. **Du, C. T. and Francis, F. J.,** Anthocyanins of *Cotoneaster* and barberry, *HortScience,* 9, 40, 1974.
20. **Pomilio, A. B.,** Anthocyanins in fruits of *Berberis buxifolia, Phytochemistry,* 12, 218, 1973.
21. **Diaz, L. S. N. and Olave, R. N.,** Estudio de la posicion del grupo azucar en los pigmientos antocianicos mediante relaciones espectrofotometricas, *Alimentos,* 4, 15, 1979 (FSTA 14, 7A556).
22. **Diaz, N. L. S. and Olave, N. R.,** New methods for acid and alkaline hydrolysis in the spectrophotometric identification of anthocyanins, *Rev. Agroquim. Tecnol. Aliment.,* 21, 419, 1981 (CA 96, 160899).
23. **Ishikura, N. and Sugahara, K.,** A survey of anthocyanidins in fruits of some Angiosperms. II. *Bot. Mag (Tokyo),* 92, 157, 1979 (CA 91, 87324).
24. **Cubucku, B. and Dortunc, T.,** Study of polyphenolic compounds of *Berberis crataegina* DC and *B. cretica* L., *Doga, Seri C* 6, 11, 1982 (CA 97, 3629).
25. **Vereskovskii, V. V. and Shapiro, D. K.,** Chromatographic study of anthocyanin pigments in the fruit of some barberry species, *Khim. Prir. Soedin.,* 4, 569, 1985 (CA 104, 165307).
26. **Medrano, M. A., Tomas, M. A., and Frontera, M. A.,** Aislamiento e identificacion de antocianinas en frutos de la provincia de Chubut (Argentina). Frutos de *Ribes aureum* Pursh y *R. magellanicum* Poir y frutos de *Berberis darwini,* Hook., *Rev. Latinoam. Quim.,* 16, 84, 1985 (CA 104, 85426).
27. **Robinson, G. M. and Robinson, R.,** A survey of anthocyanins. II. *Biochem. J.,* 26, 1647, 1932.
28. **Suomalainen, H. and Eriksson, C.,** Anthocyanine in Nordischen und in Einiger anderen Beerenfruechten, *Z. Lebensm. Unters. Forsch.,* 112, 197, 1961.
29. **Frontera, M. A., Tomas, M. A., and Tombesi, O. L.,** Chromatographic and chemical study of *Berberis ruscifolia* Lam. of regional origin, *J. High Resolut. Chromatogr.,* 12(10), 691, 1989.
30. **Griebel, C. and Hess, G.,** Die Sanddornbeere, ein C Vitamin reiche zur Herstellung von Marmalade geeingnete Fruecht, *Z. Unters. Lebensm.,* 79, 469, 1940 (CZ 1940, II, 418).
31. **Gontea, I. and Barduta, Z.,** The nutrient value of fruits from *Hippophae rhamnoides, Igiena,* 23, 13, 1974 (FSTA 6, 12J1692).
32. **Mrozewski, S. and Bakowska, E.,** Utilisation des fruits de *Hippophae rhamnoides* dans l'industrie alimentaire, *Przem. Spozyw.,* 14, 23, 1960 (IAA 953, 22).
33. **Gontea, I., Barduta, Z., and Sepeteanu, A.,** Nutritive value of jam made from pulped buckthorn *(Hippophae rhamnoides), Igiena,* 22, 463, 1973.
34. **Bat, S. and Zamyansan, J. A.,** Verarbeitung von Sanddorn in der Mongolischen Volksrepublik, *Lebensmittelindustrie,* 35, 113, 1988 (FSTA 22, 3J26).
35. **Quiao, T., Geng, J., Feng, J., and Xu, W.,** Nutritional composition of the fruit of *Hippophae rhamnoides, Yingyang Xuebao,* 10, 146, 1988 (CA 110, 74110).
36. **Lipparini, L.,** Il contenuto in acido ascorbico dei frutti dell'olivello spinoso, *Quad. Merceol.,* 3, 147, 1964.

37. **Stocker, O.,** Tiroler Sanddorn *(Hippophae rhamnoides* L.) als Vitamin C-Hochleis-trungpflanze, *Zuechter,* 19, 9, 1949 (HA 1949, 2549).
38. **Shapiro, D. K., Anikhimovskaya, L. V., and Narizhnaya, T. I.,** Chemical composition of fruits of sea buckthorn *(Hippophae rhamnoides)* grown in Belorussia, *Konserv. Ovoshchesush. Prom.,* 10, 23, 1979 (FSTA 13, 1J68).
39. **Bielig, H. J.,** Die Farbstoffe der Sanddornbeere *(Hippophae rhamnoides), Chem. Ber.,* 77, 748, 1944 (CA 44, 517h).
40. **Harborne, J. B.,** Chemicogenetical studies of flavonoid pigments, in *The Chemistry of Flavonoid Compounds,* Geissman, T. A., Ed., Pergamon Press, Oxford, 1962, 608.
41. **Friedrich, H.,** The flavones of the fruit of *Hippophae rhamnoides, Kongr. Pharm. Wiss., Vortr. Originalmitt.,* 23, 175, 1963 (CA 62, 16630f).
42. **Grigorescu, E. and Contz, O.,** Flavones of *Hippophae rhamnoides, Pharmazie,* 21, 116, 1966 (CA 64, 18031h).
43. **Hoerhammer, L., Wagner, H., and Khalil, E.,** Flavonol glycosides of the fruit of the seabuckthorn *(Hippophae rhamnoides), Lloydia,* 29, 225, 1966 (CA 65, 17361c).
44. **Krolikowska, M.,** Investigation of the flavonoid fraction of the fruit *Hippophae rhamnoides* L. (Eleagnaceae), *Rocz. Chem.,* 45, 115, 1971 (CA 75, 31210).
45. **Krolikowska, M.,** Ueber die Flavonolglykoside von Sanddornbeeren, *Hippophae rhamnoides, Planta Med.,* 22, 418, 1972.
46. **Mukhamedyarova, M. M. and Chumbalov, T. K.,** Polyphenols of *Hippophae rhamnoides, Khim. Prir. Soedin.,* 6, 854, 1979 (CA 93, 41502).
47. **Potapova, I. M., Gachechiladze, N. D., Glazunova, E. M., Yusufbekov, K. Y., and Isobae, M. D.,** Flavonoids in the berries of *Hippophae rhamnoides* growing in the Pamirs, *Khim. Prir. Soedin.,* 6, 837, 1980 (CA 94, 136167).
48. **Xiao, Z. Y. and Ziao, R.,** Studies on the flavonoids of *Hippophae rhamnoides* L. I, *Ssu-chuan I Hsueh Yuan Hsueh Pao,* 11, 174, 1980 (CA 94, 171083).
49. **Purve, O., Zhamyansan, Y., Malikov, V. M., and Baldan, T.,** Flavonoids from *Hippophae rhamnoides* growing in Mongolia, *Khim. Prir. Soedin.,* 3, 403, 1978 (CA 89, 143362).
50. **Bailey, L. H.,** *Manual of Cultivated Plants,* Macmillan, New York, 1977, 755.
51. **Sivtsev, M. V. and Abramovich, I. V.,** Pigment changes during fruit development in *Rosa canina* and *Cornus mas, Rastit. Resur.,* 15, 230, 1979 (HA 1978, 9598).
52. **Lawrence, W. J. C., Price, J. R., Robinson, G. M., and Robinson, R.,** A survey of anthocyanins, *Biochem. J.,* 32, 1661, 1939.
53. **Chester, W. and Stone, C.,** Comparative study of the anthocyanins in the red-fruited and yellow-fruited flowering dogwood, *Bull. Torrey Bot. Club.,* 91, 406, 1964.
54. **Santamour, F. S. and Lucente, R. A.,** Anthocyanins in *Cornus, Bull. Torrey Bot. Club.,* 94, 108, 1967.
55. **Lazar, M., Pita, M., and Grigorescu, E.,** Flavonoid fraction in *Cornus sanguinea* flowers and fruits, *Farmacia* (Bucharest), 19, 39, 1971 (CA 74, 136370).
56. **Du, C. T. and Francis, F. J.,** Anthocyanins from *Cornus mas, Phytochemistry,* 12, 2487, 1973.
57. **Du, C. T. and Francis, F. J.,** A new anthocyanin from *Cornus mas, HortScience,* 8, 29, 1973.
58. **Du, C. T., Wang, P. L., and Francis, F. J.,** Anthocyanins from Cornaceae, *Cornus kousa* and *Cornus florida, HortScience,* 9, 243, 1974.
59. **Tamas, M. and Stoleriu, S.,** Chromatographic identification of anthocyanidins in some flowers and fruits in indigenous plants, *Stud. Cercet. Biochim.,* 19, 113, 1976.
60. **Du, C. T., Wang, P. L., and Francis, F. J.,** Anthocyanins of *Cornus alternifolia* and *Cornus alba, HortScience,* 10, 36, 1975.
61. **Du, C. T., Wang, P. L., and Francis, F. J.,** Anthocyanins of Cornaceae, *Cornus canadensis, Phytochemistry,* 13, 2002, 1974.
62. **Hayashi, K., Suzushino, G., and Ouchi, K.,** Anthocyanins. XX. Empetrin, a new pigment from the Japanese crowberry, *Proc. Jpn. Acad.,* 27, 430, 1951 (CA 46, 7567g).

63. **Hayashi, K., Suzushino, G., and Ouchi, K.,** Anthocyanins. XX. Empetrin, a new pigment from the Japanese crowberry, *Misc. Rep. Res. Inst. Nat. Resour.,* 23, 8, 1951 (CA 47, 478b).

64. **Moore, D. M., Harborne, J. B., and Williams, C. A.,** Chemotaxonomy, variation and geographical distribution of the Empetraceae, *Bot. J. Linn. Soc.,* 63, 277, 1970.

65. **Kärppä, J., Kallio, H., Peltonen, I., and Linko, R.,** Anthocyanins of crowberry, *Empetrum nigrum* Coll., *J. Food Sci.,* 49, 634, 1984.

66. **Kärppä, J.,** Analysis of anthocyanins during ripening of crowberry, *Empetrum nigrum* Coll., *Lebensm. Wiss. Technol.,* 17, 175, 1984.

67. **Linko, R., Kärppä, J., Kallio, H., and Ahtonen, S.,** Anthocyanin content of crowberry and crowberry juice, *Lebensm. Wiss. Technol.,* 16, 343, 1983.

68. **Kallio, H., Pallasaho, S., Kärppä, J., and Linko, R. R.,** Stability of anthocyanins in crowberry juice, in *The Shelf Life of Foods and Beverages,* Charalambous, G. E., Ed., Elsevier, Amsterdam, 1986, 285.

69. **Kallio, H., Pallasaho, S., Kärppä, J., and Linko, R. R.,** Comparison of the half-lives of the anthocyanins in the juice of crowberry, *Empetrum nigrum, J. Food Sci.,* 51, 408, 1986.

70. **Kallio, H., Pallasaho, S., Kärppä, J., Linko, R., and Shibamoto, R.,** Stability of anthocyanins of crowberry, *Empetrum nigrum* , ACS Meet., Chicago, IL, Sept. 8–13, 1985, AGFD 155.

71. **Kharlamova, O. A. and Kafka, B. U.,** *Natural Food Colorants,* Pishchevaya Promishlennost, Moskwa, U.S.S.R., 1979.

72. **Chiarlo, B., Cajelli, E., and Piazzai, G.,** Sui costituenti delle drupe di *Sambucus ebulus.* I. Pigmenti antocianici e acidi fenolici, *Fitoterapia,* 49, 99, 1978 (CA 90, 51398).

73. **Ogynanov, I., Popov, A., Ivanova, B., Dinkov, D., Petkov, V., and Manolov, P.,** *Sambucus ebulus* Linnaeus, phytochemical and pharmacological screening, *Riv. Ital. EPPOS,* 61, 114, 1979 (FSTA 14, 8J1150).

74. **Novruzov, E. N. and Aslanov, S. M.,** Some data on the chemical composition of ripe danewort fruits, *Dokl. Akad. Nauk Az. SSR,* 40, 61, 1984 (CA 102, 128871).

75. **Lachman, J., Pivec, V., Rehakova, V., and Hubacek, J.,** Flavonoids, anthocyanins, phenolic and carboxylic acids and saccharides in fruits of elderberry (*Sambucas racemosa* L.), *Sb. Uvtiz, Potravin. Vedy,* 7, 29, 1989 (CA 111, 171121).

76. **Benk, E.,** Composition of the edible parts and juice of some wild fruits, *Reichstoffe Aromen,* 8, 386, 1958 (JSFA 1959, 332i).

77. **Nolan, T. J. and Casey, H. M. T.,** The pigment of the berry of the elder *(Sambucus nigra), Proc. R. Irish Acad.,* 40, 56, 1931.

78. **Harborne, J. B.,** Plant polyphenols. IX. The glycosidic pattern of anthocyanin pigments, *Phytochemistry,* 2, 85, 1963.

79. **Reichel, L. and Reichwald, W.,** Ueber die Farbstoffe der schwarzen Holunderbeere, *Naturwissenschaften,* 47, 40, 1960 (CA 54, 16557b).

80. **Reichel, L., Stroh, H., and Reichwald, W.,** Ueber die Farbstoffe der schwarzen Holunderbeere, *Naturwissenschaften,* 44, 468, 1957 (CA 52, 2191b).

81. **Reichel, L. and Reichwald, W.,** Ueber die Struktur des Sambycyanins, *Pharmazie,* 32, 40, 1977 (CA 87, 53517).

82. **Brønnum-Hansen, K. and Hansen, S. H.,** High-performance liquid chromatographic separation of anthocyanin of *Sambucus nigra, J. Chromatogr.,* 262, 385, 1983.

83. **Drdak, M. and Daucik, P.,** Changes of elderberry *(Sambucus nigra)* pigments during the production of pigment concentrates, *Acta Aliment.,* 19, 3, 1990 (IAA 698, 21).

84. **Brønnum-Hansen, K., Jacobsen, F., and Flink, J. M.,** Anthocyanin colorants from elderberry (*Sambucus nigra* L.) I. Process considerations for production of liquid extract, *J. Food Technol.,* 20, 703, 1985.

85. **Brønnum-Hansen, K. and Flink, J. M.,** Anthocyanin colorants from elderberry (*Sambucus nigra* L.) II. Process considerations for production of a freeze dried product, *J. Food Technol.,* 20, 713, 1985.

86. **Brønnum-Hansen, K. and Flink, J. M.,** Anthocyanin colorants from elderberry *(Sambucus nigra* L.) III. Storage stability of the freeze dried product, *J. Food Technol.,* 20, 725, 1985.

87. **Lamaison, J. L., Guichard, J. P., and Pourrat, H.,** Anthocyanins from the fruits of *Sambucus racemosa* L. (Caprifoliaceae), *Plant. Med. Phytother.,* 13, 188, 1979 (CA 92, 124888).

88. **Shin, M. S. and Ahn, S. Y.,** Studies on identification of the anthocyanins in elderberry *(Sambucus), Hanguk Sikp. Kwahakhoe Chi,* 12, 305, 1980 (CA 95, 93789).

89. **Johansen, O. P., Andersen, Ø. M., Nerdal, W., and Aksnes, D. W.,** Cyanidin 3-[6-(*p*-coumaroyl)-2-(xylosyl)-glucoside]-5-glucoside and other anthocyanins from fruits of *Sambucus canadensis, Phytochemistry,* 30, 4137, 1991.

90. **Amerine, M. A.,** Composition of wines. II. Organic constituents, *Adv. Food Res.,* 5, 20, 1954.

91. **Saller, W. and De Stefani, C.,** Holunderzusatz zu Traubenmost, *Weinberg Keller,* 7, 45, 1960.

92. **Schneider, J. and Epp, F.,** Paper chromatographic separation of colored constituents of European hybrid and berry-fruit red wines, *Mitt. Rebe Wein,* 9A, 111, 1959 (JSFA 1961, 189i).

93. **Duggen, H.,** Versuche mit Holunder *(Sambucus nigra), Erwebstobstbau,* 19, 99, 1977 (FSTA 11, 1J11).

94. **Pfannhauser, W. and Riedl, O.,** Anthocyane: Bildung, Gewinnung und Analyse Natürlicher Lebensmittelfarbstoffe, *Ernaehrung,* 7, 560, 1983.

95. **Chaplin, M. A., Frumkin, M. L., Nakhmedov, F. G., and Golubeva, Z. F.,** Natural food dues from blackcurrant and elderberry wastes, and their utilization for coloured starters made from berry and fruits, *Tr. Vses. Nauchno-Issled. Inst. Konservn. Ovoshchesush. Prom.,* 17, 134, 1974 (FSTA 6, 6J749).

96. **Chaplin, M. A., Nakhmedov, F. G., Zaglodin, G. F., and Frumkin, M. L.,** Economic aspects for obtaining food dyes from the waste of coloured berries used as starters, *Tr. Vses. Nauchno-Issled. Inst. Konservn. Ovoshchesush. Prom.,* 17, 141, 1983 (FSTA 6, 6J750).

97. **Chkeidze, Z. K., Kobaladze, O. R., and Burachevskii, I. I.,** Method of Obtaining a Red Food Dye from Vegetable Matter, U.S.S.R. Patent 460,286, 1975 (FSTA 7, 9H1307).

98. **Chkeidze, Z. K.,** Method of Obtaining a Red Food Dye from Vegetable Matter, U.S.S.R. Patent 460,287, 1975 (FSTA 7, 9H1307).

99. **Chkeidze, Z. K., Kadzhaya, M. P., Medzmariashvili, F. V., Macharashvili, T. G., and Kobaladze, O. R.,** Refinement of a method for determining pigments in a dye produced from elder, *Khlebopek. Konditer. Prom.,* 6, 15, 1973 (CA 78, 40019).

100. **Drdak, M., Malik, F., Greif, G., and Silhakova, A.,** Coloring Agent from the Common Elder, Czechoslovakian Patent 225,530, 1982 (CA 103, 36424).

101. **Drdak, M., Malik, F., Silharova, A., and Greif, G.,** Pigments of the common elderberry *(Sambucus nigra).* II. Methods for preparing pigment concentrates, *Bull. Potravin Vysk,* 22, 29, 1983 (FSTA 18, 2T3).

102. **Mikova, K., Havlikova, L., and Klezlova, J.,** Production of colour concentrates from elderberries, *Prum. Potravin,* 37, 295, 1986 (FSTA 20, 9J56).

103. **Malgarini, G. and Peri, C.,** Estrazione di antociani dalla bacche di *Sambucus nigra, Tecnol. Aliment.,* 1, 69, 1978.

104. **Marcheva, D., Kovachev, I., and Dilova, N.,** Anthocyanin separation and quantification by chromatography on gels, *IUPAC Int. Symp. Chem. Nat. Prod.,* 2, 436, 1978 (CA 91, 209567).

105. **Karagozova, M. D.,** Histological and histochemical studies of several medicinal plants, *Farmatsiya (Sofia),* 24, 5, 26, 1974.

106. **Peuch, A. A., Rebeiz, C. A., Catlin, P. B., and Crane, J. C.,** Characterization of anthocyanins in the fig *(Ficus carica* L.) fruits, *J. Food Sci.,* 40, 775, 1975.

107. **Duro, F. and Condorelli, P.**, Research on the natural pigments in the fruits of *Ficus carica* var. nigra, *Quad. Merceol.*, 16, 37, 1977 (CA 87, 164201).

108. **Ishikura, N.**, A survey of anthocyanins in fruits of some angiosperms. I, *Bot. Mag. (Tokyo)*, 88, 41, 1975 (CA 83, 25080).

109. **Puech, A. A., Rebeiz, C. A., and Crane, J. C.**, Pigment changes associated with application of ethephon ((2-chloroethyl-) phosphonic acid) to fig (*Ficus carica* L.) fruits, *Plant Physiol.*, 57, 503, 1976.

110. **Sharma, J. N. and Seshadri, T. R.**, Survey of anthocyanins from Indian sources, *J. Ind. Chem. Res.*, 14B, 211, 1955 (CA 50, 5846g).

111. **Sieuek, F., Herrmann, K., Grotjahn, L., and Wray, V.**, Isomeric di-C-glucosylflavones in fig (*Ficus carica* L.), *Z. Naturforsch., Teil C*, 40, 8, 1985.

112. **Sul, J. H., Lee, K. M., and Kwack, B. H.**, Leaf colour change in variegated *Lonicera japonica* var. *aureo-reticulata* at various nitrogen and light intensity levels, Abstr. Comm. Papers, *Korean Soc. Hortic. Sci.*, 7, 168, 1989 (HA 1990, 502).

113. **Sobolevskaya, K. A. and Demina, T. G.**, Anthocyanins of the ripe fruits of several wild bush plants, *Tr. Vses. Semin. Biol. Aktiv. (Lech.) Veshchestvam Plodov Yagod*, 4, 129, 1970, (CA 81, 74951).

114. **Fedoseeva, G. M.**, Anthocyanins of fruits of *Lonicera altaica* grown in the Irkutsk region, *Usp. Izuch. Lek. Rast. Sib., Mater. Mezhvuz. Nauch. Konf.*, 1973, 89 (CA 81, 117087).

115. **Torchinskaya, V. M., Bochkarnikova, N. M., and Merkuleva, T. D.**, Ascorbic acid and substances with vitamin P activity in honeysuckle fruit, *Tr. Biol.-Pochv. Inst., Dal'nevost. Nauchn. Tsentr. Akad. Nauk SSSR*, 13, 97, 1973 (HA 1975, 8302).

116. **Ma, Z. C. and Foo, L. Y.**, Isolation and identification of red pigment in edible deep-blue honeysuckle, *J. Nanjing Forestry Univ.*, 4, 67, 1987 (HA 1990, 5989).

117. **Glennie, C. W.**, Comparative phytochemical study of the Caprifoliaceae, *Diss. Abstr. Int. B*, 31, 6454, 1971 (CA 75, 126636).

118. **Kieselev, V. E., Demina, T. G., Azovtsev, G. R., Polyakova, L. V., and Vysochina, G. I.**, Flavonoids of plants of Siberian flora, Rastit. Bogatstva Sib., p. 93, 1971 (CA 77, 85643).

119. **S.S. Pharmaceutical Co. Ltd.**, Natural Dye for Foods, Japanese Patent 21,019, 1967 (CA 68, 67919).

120. **Yamamoto, K.**, Coloring matter and organic acids in the fruits of the mulberry, *J. Agric. Chem. Soc. Jpn.*, 10, 1046, 1934 (CA 29, 831).

121. **Toscano, M. A. and Lamonica, G.**, Sui pigmenti di frutti di *Morus nigra* L. Nota preliminare, *Atti Soc. Peloritana Sci. Fis. Mat. Nat.*, 18, 155, 1972 (CA 83, 55669).

122. **Maki, Z. and Inamoto, H.**, Anthocyanins of mulberry, *Kyoto Furitsu Daigaku Gakujutsu Hokoku: Rigaku, Seikatsu Kagaku*, 1972, 17 (CA 78, 82073).

123. **Timberlake, C. F.**, Anthocyanins in fruits and vegetables, in *Recent Advances in the Biochemistry of Fruits and Vegetables*, Friend, J. and Rhodes, M. J. C., Eds., Academic Press, London, 1981, 221.

124. **Miyakawa, H. and Takehana, H.**, Quality of canned peaches. I. Preventing the change of the anthocyanin color to purple, *Chiba Daigaku Engeigakubu Gakujutsu Hokoku*, 14, 43, 1966 (CA 68, 21029).

125. **Maki, Z., Tashiro, M., and Inamoto, H.**, Stability of anthocyanin pigments in mulberry, *Kyoto Furitsu Daigaku Gakujutsu Hokoku: Rigaku, Seikatsu Kagaku*, 1981, 23 (CA 96, 102760).

126. **Matlack, M. B.**, Observations on the red color of blood orange, *Plant Physiol.*, 6, 729, 1931 (CA 26, 1009).

127. **Yuasa, A., Osumi, M., and Nakamura, H.**, Cytological studies on plastid. I. Chromoplasts in oranges, *Sci. Pap. Coll. Genet. Educ., Univ. Tokyo*, 7, 202, 1957.

128. **Carrante, V.**, I pigmenti delle arance italiane, *Ann. Stn. Sper. Fruttic. Agrumicult., Acireale*, 16, 193, 1941 (CA 42, 695).

129. **Ajon, G.,** The determination of carotenoids and anthocyanins in orange juice, *Riv. Ital. EPPOS,* 23, 355, 1941 (CA 37, 35197).

130. **Ajon, G.,** The determination of carotenoids and anthocyanins in some fruits, *Chim. Ind. Agric. Biol.,* 18, 205, 1942 (CA 40, 32011).

131. **Ajon, G.,** The pigments of citrus juice during ripening, *Riv. Ital. EPPOS,* 25, 150, 1943 (CA 40, 35001).

132. **Chapot, H.,** Quelques oranges sanguines, *Cahiers Rech. Agron.,* 18, 61, 1964 (PBA 1966, 4899).

133. **Meredith, F. J. and Young, R. H.,** Effect of temperature on pigment development in Red Blush grapefruit and Ruby blood oranges, *Proc. 1st Int. Citrus Symp.,* 1, 271, 1969 (HA 1970, 7127).

134. **Reuther, W. and Rios-Castano, D.,** Comparison of growth, maturation, and composition of Citrus fruits in subtropical California and tropical Colombia, *Proc. 1st Int. Citrus Symp.,* 1, 227, 1969 (HA 1975, 7123).

135. **Chandler, B. V.,** Anthocyanins of blood orange, *Nature (London),* 182, 933, 1958 (CA 53, 6350g).

136. **Koch, J. and Haase-Sajak, E.,** Natural coloring matters in Citrus fruits. II. Anthocyanins in blood-oranges, *Z. Lebensm. Unters. Forsch.,* 127, 1, 1965 (JSFA 1965, 248ii).

137. **Porretta, A., Casoli, U., and Dall'Aglio, G.,** Richerche sugli antociani del succo di arancia, *Ind. Conserve,* 41, 175, 1966.

138. **Kunkar, A.,** Le antocianine del succo d'arancio Sanguinello calabrese, *Riv. Ital. EPPOS,* 50, 180, 1968 (CA 69, 75738).

139. **Licastro, F. and Bellomo, A.,** Identificazione degli antociani nelle arance Moro, *Rass. Chim.,* 25, 306, 1973 (CA 80, 58557).

140. **Maccarone, E., Maccarrone, A., Perrini, G., and Rapisarda, P.,** Anthocyanins of the Moro orange juice, *Ann. Chim.,* 73, 533, 1983 (CA 100, 50204).

141. **Maccarone, E., Maccarrone, A., and Rapisarda, P.,** Acylated anthocyanins from oranges, *Ann. Chim.,* 75, 79, 1985 (CA 103, 21426).

142. **Benk, E.,** Ueber Herstellung, Zusammensetzung und Beurteilung von Blutorangelikoer, *Alkohol.-Ind.,* 72, 264, 1959 (IAA 20, 2473).

143. **Russo, C. and Galoppini, C.,** La formazione degli antociani nelle arance Moro, Tarocco e Sanguinello, *Essenze Deriv. Agrum.,* 39, 67, 1969 (CA 72, 89012).

144. **Di Giacomo, A., Calvarano, M., Calvarano, I., Di Giacomo, G., and Belmusto, D.,** Il succo delle arance sanguigne italiane, *Essenze Deriv. Agrum.,* 59, 273, 1989 (FSTA 23, 3H4).

145. **Lee, H. S., Carter, R. D., Barrows, S. M., Dezman, D. J., and Castle, W. S.,** Chemical characterization by liquid chromatography of Moro blood orange juices, *J. Food Comp. Anal.,* 3, 19, 1990.

146. **Ildis, P.,** Cambiamenti della costituzione fisica di alcuni agrumi durante il magazzinaggio, *Rev. Gen. Froid,* 33, 1139, 1955.

147. **Aharoni, Y. and Houck, L. G.,** Changes in rind, flesh, and juice color of blood oranges stores in air supplemented with ethylene or in oxygen-enriched atmospheres, *J. Food Sci.,* 47, 2091, 1982.

148. **Casoli, U., Dall'Aglio, G., and Leoni, C.,** Azione di estratti enzimatici vegetali sui pigmenti antocianici delle arance, delle more di rovo e dei mirtilli, *Ind. Conserve,* 44, 102, 1969 (FSTA 2, 7J685).

149. **Dall'Aglio, G., Versitano, A., Gherardi, S., and Porretta, A.,** Decolorazione del succo d'arancia Sanguinello, *Ind. Conserve,* 49, 13, 1974 (CA 81, 62340).

150. **AID S.p.A.,** Process for the Selective Extraction of Anthocyanins from Orange Juice, *Italian Patent,* 1,169,170, 1985.

151. **Galoppini, C. and Russo, C.,** La conservazione del colore nei succhi di arancia solfitati, *Essenze Deriv. Agrum.,* 39, 69, 1969 (CA 72, 89012).

152. **Trifiro, E. and Landi, S.**, Gli antociani nei succhi di frutta e l'impiego dell'acido ascorbico, *Ind. Conserve*, 40, 3, 1965.

153. **Maccarone, E., Maccarrone, A., and Rapisarda, P.**, Stabilization of anthocyanin of blood orange fruit juice, *J. Food Sci.*, 50, 901, 1985.

154. **Maccarone, E., Longo, M. L., Leuzzi, U., Maccarrone, A., and Passerini, A.**, Stabilizzazione del succo d'arancia pigmentata con trattamenti fisici e additivi fenolici, *Chim. Ind.*, 70, 95, 1985 (CA 110, 93741).

155. **Maccarrone, E., Maccarrone, A., and Rapisarda, P.**, Technical note: colour stabilization of anthocyanins in blood orange fruit juice by tannic acid, *Int. J. Food Sci. Technol.*, 22, 159, 1987.

156. **Tressler, R. K. and Joslyn, M. A.**, *Fruit Juice Technology*, AVI Publishing, Westport, CT, 1971, 284.

157. **Saxena, A. K., Manan, J. K., and Berry, S. K.**, Pomegranates: post-harvest technology, chemistry, and processing, *Indian Food Packer*, 8, 43, 1987 (FSTA 21, 2J39).

158. **Karrer, P. and Widmer, R.**, Ueber die Pflanzenfarbstoffen, *Helv. Chim. Acta*, 10, 67, 1927.

159. **Harborne, J. B.**, Anthocyanins and their sugar components, *Fortschr. Chem. Org. Naturst.*, 17, 175, 1962.

160. **Lowry, J. B.**, Anthocyanins of the Melastomataceae, Myrtaceae and some allied families, *Phytochemistry*, 15, 513, 1976.

161. **Pifferi, P. G., Cultrera, R., and Baldassarri, L.**, Studi sui pigmenti naturali Nota VII. Gli antociani del melograno (*Punica granatum*), *Sci. Tecnol. Aliment.*, 2, 307, 1972.

162. **Du, C. T., Wang, P. L., and Francis, F. J.**, Anthocyanins of pomegranate, *Punica granatum*, *J. Food Sci.*, 40, 417, 1975.

163. **Santagati, N. A., Duro, R., and Duro, F.**, Studio dei pigmenti presenti nei semi di melograno, *Riv. Merceol.*, 23, 247, 1984 (CA 102, 75698).

164. **Lee, J. W., Kim, K. S., and Kim, S. D.**, Studies on changes in the composition of pomegranate fruit during maturation. II. Changes in polyphenol compounds and anthocyanin pigments, *J. Korean Soc. Hortic. Sci.*, 15, 64, 1974 (HA 1975, 4469).

165. **Markh, A. T. and Lysogor, T. A.**, Pomegranate polyphenols, *Izv. Vyssh. Uchebn. Zaved. Pishch. Tekhnol.*, 2, 36, 1973 (CA 79, 41100).

166. **Kriventsov, V. I. and Arendt, N. K.**, Anthocyanins of pomegranate juice, *Tr. Gos. Nikitsk. Bot. Sad.*, 83, 110, 1981 (HA 1986, 1441).

167. **Botrus, D., Zykina, T. V., Kostinskaya, L. I., and Golovchenko, G. A.**, Polyphenol compounds in pomegranate, *Izv. Vyssh. Uchebn. Zaved. Pishch. Tekhnol.*, 3, 117, 1984 (CA 101, 109241).

168. **Nizharadze, A. N., Kupatadze, I. V., and Gelashvili, E.**, The dog rose, a valuable raw material for the preserves industry, *Konservn. Ovoshchesush. Prom.*, 4, 36, 1977 (FSTA 10, 4J386).

169. **Galeb, S. P.**, Processing of rose hips (*Rosa Aff. rubiginosa* L.), *Invest. Agric.*, 2, 39, 1976 (FSTA 10, 10J1413).

170. **Kurt, A. and Yamankaradeniz, R.**, Composition of wild rose hips in Erzurum Province and their potential for processing, *Doga Bilim Dergisi*, 7, 243, 1983 (FSTA 17, 5J141).

171. **Mapson, L. W. and Tomalin, A. W.**, Preservation of ascorbic acid in rose hips during storage, *J. Sci. Food Agric.*, 7, 424, 1958.

172. **Koncalova, M. N.**, Anthocyanidins from hips of *Rosa pimpinellifolia* L., *Preslia Praha*, 43, 198, 1971 (HA 1972, 1909).

173. **Sobolevskaya, K. A. and Demina, T. G.**, Anthocyanins of the ripe fruits of several wild bush plants, *Tr. Vses. Semin. Biol. Aktiv. (Lech.), Veshchestvam Plodog Yagod*, 129, 1970 (CA 81, 74951).

174. **Plekhanova, T. I., Bandyukova, V. A., and Bairamkulova, F. K.**, Chemical components of *Rosa spinosissima* fruits, *Khim. Prir. Soedin.*, 3, 398, 1978 (CA 89, 126160).

175. **Demina, T. G.,** Anthocyanins in ripe fruits from thorny dog rose (*Rosa spinosissima* var. *vulgaris*), *Izv. Sib. Otd. Akad. Nauk SSSR, Ser. Biol. Med. Nauk,* 133, 1966 (CA 66, 26581).

176. **Velioglu, Y. S. and Mazza, G.,** Characterization of flavonoids in petals of *Rosa damascena* by HPLC and spectral analysis, *J. Agric. Food Chem.,* 39, 463, 1991.

177. **Yokoi, M.,** Colour and pigment distribution in ornamental plants. V. Anthocyanin distribution in rose cultivars, *Tech. Bull. Fac. Hortic. Chiba Univ.,* 22, 13, 1974 (HA 1976, 1470).

178. **Marshall, H. H., Campbell, C. G., and Collicutt, L.,** Breeding for anthocyanin colors in *Rosa, Euphytica,* 32, 205, 1983.

179. **Kuliev, V. B., Mamedov, F. A., Filonova, G. L., Ermakova, R. A., Komrakova, N. A., and Treis, V. V.,** Production of a Red Food Dye, *U.S.S.R. Patent,* 1,592,318, 1990 (FSTA 23, 5V128).

180. **Li, H. and Feng, X.,** Extraction and toxicological estimation test of Jinyingzi brown pigment, *Linchan Hauxue Yu Gongye,* 10, 195, 1990 (CA 114, 182110).

181. **Benk, E.,** Ein Beitrag zur Kenntnis Auslaendischer Obstfruechte, *Fluess. Obst.,* 48, 184, 1981 (FSTA 14, 9J1398).

182. **Daguet, J. C. and Foucher, J. P.,** Les flavonoides de *Arbutus unedo* L. (Ericaceae), *Plant. Med. Phytother.,* 16, 185, 1982 (HA 1983, 6675).

183. **Proliac, A. and Raynaud, J.,** Anthocyanin pigments of fruits from *Arbutus unedo* L., *Plant. Med. Phytother.,* 15, 109, 1981.

184. **Maccarone, E., Cuffari, G., Passerini, A., and Rapisarda, P.,** The anthocyanins of *Arbuts unedo* fruits (Ericaceae), *Ann. Chim.,* 80, 171, 1990 (CA 114, 78596).

185. **Schmidt-Hebbel, H., Michelson, W., Masson, L., and Steltzer, H.,** Nachweis der Anthocyanfarbstoffe in Chilenischer Trauben, rotweinen und deren Hybriden sowie in Maqui und Mora durch dunnschicht Chromatographie, *Z. Lebensm. Unters. Forsch.,* 137, 169, 1968.

186. **Diaz, L. S., Rosende, C. G., and Antunez, M. I.,** Spectrophotometric identification of anthocyanin pigments from maqui fruits (*Aristotelia chilensis* Mol. & Stuntz.), *Rev. Agroquim. Tecnol. Aliment.,* 24, 538, 1984 (CA 102, 201170).

187. **Wang, P. L. and Francis, F. J.,** New anthocyanin from *Viburnum trilobum, HortScience,* 7, 87, 1972.

188. **Du, C. T., Wang, P. L., and Francis, F. J.,** Cyanidin 3-arabinosyl sambubioside in *Viburnum trilobum, Phytochemistry,* 13, 1998, 1974.

189. **Francis, F. J.,** Anthocyanins as food colors, *Food Technol.,* 29(5), 52, 1975.

190. **Ishikura, N.,** A further survey of anthocyanins and other phenolics in *Ilex* and *Euonymus, Phytochemistry,* 14, 1439, 1975.

191. **Guichard, J. P., Regerat, F., and Pourrat, H.,** Flavones et anthocyanes des fruits de *Viburnum lantana* L., *Plant. Med. Phytother.,* 10, 105, 1976 (CA 86, 2385).

192. **Godeau, R. P., Pellissier, Y., and Fouraste, I.,** Constituents of *Viburnum tinus* L. III. Anthocyanic and flavonoid compounds of fruits, *Plant. Med. Phytother.,* 13, 37, 1979 (CA 91, 16731).

Chapter 8

CEREALS

I. BARLEY

Anthocyanins occur in pigmented varieties of barley, *Hordeum vulgare* L., and in plants subjected to nutritional or pathological stresses. In the purple-seeded varieties, the spermoderm and aleurone may be strongly pigmented and the diffusion of the anthocyanins may obscure the cell walls; the pigmentation of the aleurone is genetically independent while that of the lemma and palea is linked to that of the caryopsis.[1,2] Accumulation of anthocyanins following nitrogen deficiency was reported by Gassner and Straib,[3] and production of anthocyanins due to phosphorous deficiency was observed by Hamy.[4] Takahashi and Yamamoto[5] reported on the distribution of anthocyanin and melanin pigmentation in the beard, glume, seed, auricle, and ligule of barley grown in different areas. According to Narula and Mathur,[6] the purple pigmentation of the auricle is controlled by a single dominant gene and is associated with the genes that control pigment accumulation in the straw, lemma, and pericarp.

The pigmentation of various parts of the barley plant can be black (generally due to the presence of melanins) or blue, red, or purple (due to anthocyanins).[7] Mullick et al.[8] found glycosides of cyanidin and delphinidin in three blue and two purple varieties of barley and in the lemma and caryopsis of one black variety. The purple varieties also contained pelargonidin derivatives. In all, Mullick et al.[8] provided evidence for the presence of six anthocyanins in barley: two cyanidin, two delphinidin, and two pelargonidin derivatives. In barley malt, McFarlane et al.[9] found cyanidin, delphinidin, and several unidentified anthocyanidins. More recently, Jende-Strid[10] found that the yellow, blue, or black varieties of barley seeds contained delphinidin and cyanidin, while the purple varieties also contained pelargonidin. Metche and Urion,[11] however, found cyanidin 3-arabinoside in the seed coat of barley.

According to Woodward and Thieret,[12] the presence of anthocyanin pigmentation in seeds and ears of barley is controlled by two complementary genes, *P* and *C*. The origin of these genes is believed to be the purple forms of a wild species, *H. agriocrithon* var. *dawoense*, native to Tibet. The blue-colored aleurone is governed by the complementary genes, *B1* and *B12*.[13] The pigmentation is present in almost all wild species and in many cultivated ones. It is a dominant character over white and is inherited monofactorily.[14] Other plant parts containing anthocyanins are the auricle, node, and stem.[15,16]

The presence of anthocyanins may be an important characteristic in barley used for beer production. In fact, when pigmented barley is fermented into beer, the resulting product is pinkish and has to be removed by a clarification treatment.[17] In addition to anthocyanins, barley contains several other fla-

TABLE 1
Anthocyanins and Other Phenolics of Barley
(*Hordeum vulgare* L.)

Compound	Plant part	Ref.
Anthocyanins		
Pelargonidin	Pericarp and aleurone	8
Pelargonidin glycosides	Pericarp and aleurone	8
Cyanidin	Pericarp and aleurone	8
Cyanidin glycosides	Pericarp and aleurone	8
Delphinidin	Pericarp and aleurone	8
Delphinidin glycosides	Pericarp and aleurone	8
Proanthocyanins		
Procyanidin B-3 dimer	Aleurone cells	21
(+)-Catechin	Aleurone cells	21
Leucocyanidin	Whole grains	11
Leucodelphinidin	Aleurone cells	21
Flavonols		
Chrysoeriol	Whole grains	22
Phenolic acids		
p-Hydroxybenzoic acid	Whole grains	23
Salicylic acid	Whole grains	24
Protocatechuic acid	Whole grains	24
Vanillic acid	Whole grains	24
Syringic acid	Flour	25
p-Coumaric acid	Whole grains	24
m-Coumaric acid	Whole grains	24
o-Coumaric acid	Whole grains	23
Ferulic acid	Whole grains	23
Sinapic acid	Flour and whole grain	24, 25
Phenol glucosides		
p-Hydroxybenzoic acid 4-*O*-β-glucoside	Whole grain	23
Vanillic acid 4-*O*-β-glucoside	Whole grain	23
o-Coumaric acid 2-*O*-β-glucoside	Whole grain	23
Ferulic acid 4-*O*-β-glucoside	Whole grain	23

vonoids, phenolic acids, and phenol glucosides[21-25] (Table 1). The occurrence of proanthocyanins can cause major technological difficulties during production of beer because these phenolics are responsible for turbidity in beer.[18] Jende-Strid[19] reported on the relationship between the genetic arrangement and the composition of the phenolic fractions in various cultivars and mutants of malt barley, and Nilan et al.[20] developed barley mutants lacking anthocyanins and proanthocyanins.

II. BUCKWHEAT

Buckwheat is commonly grown for its black or gray triangular seeds. It can also be grown as a green manure crop, a companion crop, a cover crop, a source of buckwheat honey, and a pharmaceutical plant yielding rutin.[26]

There are three known species of buckwheat: common buckwheat, *Fagopyrum esculentum* Moench; tartary buckwheat, *F. tataricum* Gaertn.; and perennial buckwheat, *F. cymosum* L. Common buckwheat is also known as *F. sagittatum* Gilib. and is by far the most important species, accounting for over 90% of the world buckwheat production.[27]

In eastern Europe, buckwheat is a basic food item in porridges and soups. In North America, it is marketed primarily in pancake mixes, which may contain buckwheat mixed with wheat, maize, rice, or oat flours, plus a leavening agent. In Japan, buckwheat is marketed primarily as flour for manufacturing a variety of noodles (soba) and as groats. Groats (the part of the grain left after the hulls are removed from the seeds) and farina (made from groats) are used for breakfast food; porridge; and thickening materials in soups, gravies, and dressings. Buckwheat is also used with vegetables and spices in kasha and with wheat, maize, or rice in bread and pasta products.[26,27]

One of the earliest reports on the anthocyanins of buckwheat is that of Kaerstens[28] who found the red pigments in the epidermis, subepidermal layer, and in cells of the cortex of the hypocotyl of common buckwheat. More recent studies by Inoue et al.[29] led to the identification of cyanidin 3-glucoside, cyanidin 3-galactoside, and at least two other biosides (one of which is believed to be cyanidin 3-rhamnosylgalactoside) in the hypocotyls of common buckwheat. Troyer[30] characterized the anthocyanins in buckwheat hypocotyls as derivatives of cyanidin and found that biosynthesis of anthocyanins in *F. sagittatum* depends on light intensity.[31] According to Harraschain and Mohr,[32] the light-induced synthesis of anthocyanins in buckwheat depends on the phytochrome and the high-energy photomorphogenic system and not on photosynthesis. The biosynthesis of flavonoid precursors is inhibited by aminooxyacetate[33] and stimulated by nitrogen deficiency.[34] Extensive studies have dealt with flavonoid metabolism in buckwheat, utilizing anthocyanins as markers;[35-38] and it has been established that dihydrokaempferol (aromadendrin) is a very efficient precursor for cyanidin and quercetin in buckwheat seedlings.

Other reported phenolic compounds of buckwheat include: rutin, quercetin, hyperin, vitexin, isovitexin, orientin, and isoorientin. The phenolic acids of buckwheat seeds are the hydroxybenzoic acids: syringic, *p*-hydroxybenzoic, vanillic, and *p*-coumaric acids. Also, in buckwheat seeds are found soluble oligomeric condensed tannins which, along with the phenolic acids, provide astringency and affect the color and nutritive value of buckwheat products.[39-42]

III. MAIZE OR CORN

The inheritance of anthocyanin pigmentation in maize or corn (*Zea mays* L.) is well documented. In 1921, Emerson[43] showed that the *R* locus is responsible for anthocyanin pigmentation in both aleurone (outer layer of the endosperm) and plant tissues. He designated four classes of pigmentation

TABLE 2
Anthocyanins of Maize (*Zea mays* L.)

Anthocyanin	Plant part	Ref.
Cyanidin 3-glucoside	Seed coat	58,59
	Throughout plant	64
	Seed coat and cobs	62
	Leaves	66
Cyanidin 3-galactoside	Shoot tissues and seed	60,61
Cyanidin 3-(6″-malonylglucoside)	Leaves	66
Cyanidin 3-dimalonylglucoside	Leaves	66
Pelargonidin 3-glucoside	Seed coat	58,59
Pelargonidin glycoside	Throughout plant	69

patterns: *R-r* (pigment in both aleurone and plant), *R-q* (pigmented aleurone, green plant), *r-r* (nonpigmented aleurone, pigment in plant), and *r-q* (non-pigmented aleurone, green plant). Stadler[44,45] confirmed that the locus consists of two elements that can independently change from pigment-producing (dominant) to nonpigmented (recessive). *S* (*R*, dominant; *r*, recessive) is the seed component that functions in the aleurone and embryo; and *P* (*-r*, dominant; *-g*, recessive) is the plant component that determines color in seedlings and certain mature tissues, notably anthers. The genetic model of complexity was further studied by Rhoades[46] and McClintock[47] and recently was confirmed at the molecular level by correlating the presence of distinctive cross-hybridizing DNA restriction fragments with the genetic determination of plant (P) and seed (S) function in the *R-r* standard allele.[48] Two other regulatory genes (*B* and *Pl*) function in both the seed and plant, but there is no evidence for complexity at either locus.[49] Some *B* and *R* alleles can substitute for each other in regulating kernel pigmentation. However, this relationship is not true for anthocyanin accumulation in plant tissues. Only plants with particular dominant alleles at *B* are intensely pigmented at maturity in tissues such as leaves, husks, and tassel glumes. When the *Pl* gene is also present, the color is most intense and extends to inner husk and sheath regions not normally exposed to light. In contrast, in the presence of the recessive allele *pl*, plants containing the *B* gene develop pigmentation only in those tissues directly exposed to light.[49]

Recently, several genes controlling the synthesis of anthocyanins in maize have been identified and cloned. These include the structural genes for chalcone synthase (*C2*);[50] dihydroquercetin reductase (*A1*);[51] uridine 5′-diphosphate (UDP)-glucose-flavonol glucosyltransferase (bronze, *Bz1*)[52] and (bronze-2, *Bz2*);[53] and the regulatory genes, *C1, R, Pl*, and *B*, which control the expression of the unlinked structural genes.[48]

Despite the significant advances in the genetic regulation of anthocyanin biosynthesis, surprisingly the exact structures of the maize anthocyanins have not been fully clarified (Table 2). Straus[58] identified cyanidin and pelargonidin

3-glucosides in endosperm tissue of 'Black Mexican Sweet Corn'. Harborne and Gavazzi[59] found the same pigments in the seed coat or aleurone layer of the endosperm of *PrPr prpr* corn, but they were accompanied by two unstable aliphatic acylated derivatives of each. In the shoots of maize seedlings, La-wanson and Osude[60] found cyanidin 3-galactoside, rather than the 3-glucoside, accompanying pelargonidin 3-glucoside. Cyanidin 3-galactoside was also found in Peruvian dark-seeded corn, where it may be acylated with *p*-coumaric acid.[61] Nakatani et al.[62] identified cyanidin 3-glucoside as the major antho-cyanin in cobs and kernels of Bolivian purple corn variety Maize Morado. Pelargonidin and peonidin 3-glucosides were minor components, and more than seven other anthocyanins were detected.

Seah and Styles[63] isolated seven pigments by chromatography of *PrPr* strains of the W22 selection. Two of these pigments were cyanidin and pe-largonidin derivatives; the third was a peonidin glycoside. Later, Styles and Ceska[64] identified the 3-deoxyanthocyanidin luteolinidin as well as the gly-cosides of cyanidin, pelargonidin, and peonidin and five additional pigments. According to Reddy and Reddy,[65] the 3-deoxyanthocyanidin is present in the form of luteoforol which is the precursor of the "bronze" pigment. Recently, using a mild extraction procedure and TLC, HPLC, and fast atomic bom-bardment-mass spectrometry (FAB-MS) Harborne and Self[66] found cyanidin 3-glucoside together with two acylated derivatives in reddened maize leaves. The two acylated pigments were identified as cyanidin 3-(6″-malonylgluco-side) and cyanidin 3-dimalonylglucoside.

Environmental factors, particularly light, influence the concentration and perhaps the composition of anthocyanin in maize.[67-71] In seedlings, antho-cyanin biosynthesis is mediated by phytochrome and cryptochrome[67] and appears to occur independently of photosynthesis.[68] Increase in photon fluence rates of white light over the range of 5 to 80 μmol m^{-2} s produces a decrease in anthocyanin production. Rates of production and amounts of accumulation of anthocyanin in both shoots and roots have been reported to vary with the age of the seedlings at the time of exposure to light.[69] Endogenously produced and exogenously supplied ethylene, as well as auxins, cytokinins, GA$_3$, and CO$_2$, have been found to markedly suppress anthocyanin synthesis in maize seedlings.[70] This is generally contrary to what has been found in most fruit (see Chapters 2 through 7) and is probably due to differences in the physiology of the tissues examined. Phosphorus-deficient maize plants appear to accu-mulate cyanidin 3-rhamnoside in their leaves[71] rather than cyanidin 3-glu-coside, but this remains to be confirmed. Knoll[72] and Knoll et al.[73] noted that a low P level together with low temperature in the root zone were probably two factors which stimulated anthocyanin production in corn. Other researchers[74,75] have investigated the effects of mineral nutrition on the pro-duction of anthocyanins in seedlings of 'ESI' corn, and they have found that nutrient deficiencies generally lead to increased production of anthocyanins.

The anthocyanins of corn have a very long history of use as colors, apparently having been used by the Incas.[76] Recently, the interest in corn anthocyanins has resurged, and several patents[77-82] describing various preparation and application processes or for colorants from the purple corn have been granted. Most processes use the red 'Morado' and 'Kulli' corn and an aqueous solvent extraction to obtain an anthocyanin extract which is then used to color candies and sugared syrups. A process patented by the Sugiyama Chemical Institute of Tokyo[81,82] also uses 'Morado' corn as the raw material but obtains the pigments by enzymatic digestion of the ground corn followed by spray drying.

IV. OATS

In oats (*Avena sativa* L.) the formation of anthocyanins is genetically regulated by a single recessive gene (*P*) which requires red light for its activation.[83] Nittler[84] reported that phosphate deficiency caused anthocyanins to develop in seedlings of 'Russell' oats. To our knowledge, no free or glycosylated anthocyanidins have been characterized in oats. However, many phenolic compounds are found in oats. They include: kaempferol, quercetin, luteolin, tricin, vitexin, isovitexin, the flavanone homoeriodictyol and its related chalcone, and a variety of phenolic acids.[25,85-89] The structure, occurrence, and function of oat phenolics have been thoroughly reviewed by Collins.[88]

V. RICE

Rice, *Oryza* spp., is one of the most important food crops. About half of the world population eats this valuable grain as its main food. The cultivated species are *O. glaberrima*, which is restricted to some African areas; and *O. sativa*, the most important cultivated species, which is divided into the varieties *indica* (widespread throughout the tropical and subtropical regions), *japonica*, and *javanica*.

The first studies on the inheritance of anthocyanins in rice were conducted in the first decades of this century.[90-94] In the 1940s and 1950s, several studies dealt with the distribution of anthocyanin pigments in various parts of the rice plant,[95-97] and several authors[98-100] symbolize the different genes which control the anthocyanin pigmentation. As knowledge of the genetics of rice increased, however, their naming system became inadequate, and in 1959, under the direction of the International Rice Research Institute (IRRI), new genetic symbols were developed.[101]

Rice has more mutants than corn, barley, or tomato. Therefore, the inheritance of various traits is much more complex in rice than in other plants. There are many complementary and inhibitory loci. An example of the complicated inheritance in rice is the interaction of different loci that control the pigmentation in the various plant parts. Rice is pigmented in the pericarp, lemma, pula, outer glume, apicule, beard, stigma, antherial suture, leaf sheath, leaf lamina, leaf margin, ligule, auricle, "pulvinus", collar joint, node, internode (culm), coleoptile, and even in the root.[98-104] Normally, the color of the seed coat is white, but in some varieties it is red, purplish-red, purple, or dark purple.[104]

Variations in color intensity observed in various parts of the plant suggest the existence of different genes which control the pigmentation in those different organs; Ramiah[105] and Misro et al.[106] observed in extensive varietal collections that the pigment distribution was never uniform and, in some, appeared to be an independent characteristic. In segregating progenies, the ratio between colored and uncolored is different in different parts.[107,108]

According to Nagao et al.,[109] two complementary allele series at the *C* and *A* loci determine the anthocyanin formation. The base gene for "chromogene" production is *C*; the conversion of chromogene into anthocyanin is controlled by *A*. There is a series of multiple alleles at the two loci; six alleles at locus *C* and four at locus *A*. Five other loci control the anthocyanin distribution in the other plant parts: *P* controls the distribution of anthocyanin precursors in the apiculate; *Pr* controls the color in the glume and rachilla; *Pl* with three alleles conditions the distribution of the pigment in the leaf lamina, leaf sheath, pulvinus, auricle, ligule, node, internode, and rachilla; *Pn* governs the color in the leaf apex and margins; and *Ps* controls the localization of the pigment in the stigma.[102,104]

In 1973, Setty and Misro[110] showed that gene *P* was made up of three complementary genes (Pa-Pb-Pc), and as a result proposed a modification to the genetic (C-A-P) system. Hadagal et al.[111] added two additional genes to the list of genes responsible for anthocyanin pigmentation in rice. The inhibitor gene — *I-Pi* — suppresses the effect of *Pi*,[109,112] and the inhibitor genes — *I-P* and *IPla* — influence the color of apicals and leaf apex, respectively.[112,113]

The genes for anthocyanin pigmentation in the different plant organs are found on 7 chromosomes.[114] Genes that control the anthocyanin pigmentation are present in all the associative groups according to the following scheme:

- Group I (WX) — color of the apiculate *P* with *wx*, waxy or glutinous endosperm
- Group II (PI) — purple leaf (PI) with absence of ligule and phenol reaction
- Group III (*A*) — anthocyanin activator A with red pericarp (Pn)
- Group IV (Pin) — pigmentation in the internode
- Group V (Pj, Prp, Lsp) — purple joint, purple pericarp, and purple leaf sheath, respectively
- Group VI (Ntv) — purple internode
- Group VII (Psh, Aj, Ps)[115]
- Group VIII — colored glume with colored stigma[116]

TABLE 3
Anthocyanins of Rice (*Oryza sativa* L.)

Anthocyanin	Plant part	Relative conc
Cyanidin 3-glucoside	Seed	+ + + +
	Leaves	+ + + +
	Internode	+ + + +
	Flower (stigma)	+ + + +
Cyanidin 3-rhamnoside	Seed	+
	Leaves	+ + +
	Internodes	+ + +
	Flower	+ + +
Cyanidin 3,5-diglucoside	Leaves	+
	Internodes	+ +
Malvidin 3-galactoside	Seed	+ + +

From Nagai, I. et al.,[102] Suzushino et al.,[131] Nagai et al.,[132] and Gangulee.[128,129]

Recent studies have shown an even more complex genetic inheritance of anthocyanin in rice and revealed the presence of a greater number of genes than previously reported.[117-122] In group III, the total number of genes has reached 16.[123]

One of the first reports on the role of anthocyanins in rice was by Kosaka,[124] who studied the relationship between the presence of anthocyanins and assimilation of nutrients during plant growth. Hayashi[125-127] identified the pigment in *O. sativa* var. *atropurpurea* as a monoglucoside of cyanidin. In the purple leaves of variety 'Meghat Patnai', Gangulee[128,129] found a monoside of cyanidin together with a free cyanidin and a diglucoside of cyanidin. Sharma and Seshadri[130] identified cyanidin diglucoside and delphinidin monoglucoside in different types of rice. According to Suzushiro et al.[131] and Nagai et al.,[102,132] the anthocyanin composition of rice is made up of four pigments: cyanidin 3-glucoside, cyanidin 3-rhamnoglucoside, cyanidin 3,5-diglucoside, and malvidin 3-galactoside. These pigments are distributed differently in the various cultivars examined and are not found in all plant parts. Cyanidin 3-glucoside and cyanidin 3-rutinoside are the most diffuse, while malvidin 3-galactoside is found particularly in the multicolored seed coats (Table 3). Nagai et al.[102] characterized 3-glucoside, 3-rhamnoside, and 3,5-diglucosides of cyanidin as the anthocyanins of the internodes; cyanidin 3-glucoside and malvidin 3-galactoside as the predominant red pigments of the rice grains; and cyanidin 3-glucoside and cyanidin 3-rhamnoside as the anthocyanins of the rice leaves and stigma.

During late maturity, the total content of anthocyanin in the leaves of rice increases from 5 to 20 mg/100 g. During the senescence phase, there is a highly significant negative correlation between the chlorophyll content and total anthocyanins.[133] The presence of tungro virus in the purple varieties of rice causes an extensive breakdown in the chlorophyll pigment and an increase in the anthocyanin pigmentation.[134]

In recent times, the anthocyanins of rice were extracted from various red-pigmented *Oryza* spp. and from the anthocyanin-containing husks[135] and used in the production of colored sake[136,137] and daikon.[138] The process patented by Katayama et al.[138] uses red leaves and stems as the raw material for production of anthocyanins, while the process of Fukui and Wagi[139] uses red rice bran and enzymes together with ethanol for the extraction of the anthocyanins. Purification of the extract is by ion-exchange resin.

VI. RYE

Rye, *Secale cereale* L., is an important cereal crop in the cool climates of northern Europe, Asia, and North America. The world produces an average of 28.6 million ton and uses the crop primarily for human food. Distillers use malt made from rye for rye whiskey and gin.

The presence of anthocyanins, chlorophyll, and carotenoids provides the color in the rye grain.[140] Seedlings germinated in alternating light and dark conditions produce two anthocyanin pigments which were identified as derivatives of cyanidin and pelargonidin by Metche and Gay.[141] In the purple leaves of rye, Metche[142] found cyanidin 3-glucoside, cyanidin 3-rhamnosyl-glucoside, and cyanidin 3-rhamnosyldiglucoside.

In seeds of 'Prolific' rye, Dedio et al.[143] identified delphinidin 3-rutinoside as the aleurone anthocyanin; cyanidin 3-rutinoside was detected in the coleoptiles and cyanidin 3-glucoside, in the first leaves. In the pericarp of purple-seeded rye cultivars, Dedio et al.[144] found cyanidin 3-glucoside, peonidin 3-glucoside, and the acylated forms of these glycosides as the major anthocyanins; and trace amounts of cyanidin 3-rutinoside, peonidin 3-rutinoside, and their acylated derivatives. Only cyanidin 3-rutinoside is present in the coleoptiles, and the first leaves contain only cyanidin 3-glucoside.[144] The proportion of the acylated forms and the ratio of the aglycones varies between cultivars. More recently, Busch et al.,[145] using UV spectroscopy, hydrolysis, and cochromatography, identified cyanidin 3-gentiobioside in the primary leaves of *S. cereale*.

The presence of anthocyanins in rye has been used as an aid in understanding the genetics of this plant. Thus, several authors have reported on the heritability of anthocyanins in rye. According to Watkins[146] and Watkins and White,[147] the occurrence of anthocyanins in 'Prolific' rye is genetically controlled by complementation of two genes, *A* and *B* for the seed color and by *A* and *R* for the red color in the coleoptile. In the plant type *virescens*, two independent, dominant genes — *A1* and *A2* — control the anthocyanin production.[148,149] The presence of anthocyanins in the auricle, the ligule, and the nodes appears to be a dominant character governed by two complementary genes, *R1* and *R2*.[150-152]

Melz et al.[153] isolated the gene for the lack of anthocyanins in chromosome 7R while the gene for pigmented grain, coleoptile, and aleuron is on chromosome 3[154] and is controlled by either a single dominant gene or by multiple, closely linked genes.[155]

VII. SORGHUM AND MILLET

A. SORGHUM

Sorghum (*Sorghum bicolor* [L.] Moench.) is a major staple grain crop on the African and Asian continents used for feed, food, shelter, and fuel production.[156] The pigmentation in sorghum grain is due to the presence of polyphenols in the seed coat, where they accumulate to form a continuous pigmented layer.[157] In brown, high-tannin sorghum with a dominant ''spreader'' gene, polyphenols are also present in the epicarp.[158] The major polyphenol components of sorghum grain are the flavonoids and their polymers, the condensed tannins; their amount varies widely in different cultivars.[156,159] Cultivars with increased levels of these and other phenolic compounds exhibit disease, insect, mold, bird, and postharvest sprouting resistance, as well as nutrient deficiencies.[157-159]

The anthocyanins of sorghum are derivatives of cyanidin, pelargonidin, apigeninidin, luteolinidin, and fisetinidin (Table 4). The mesocotyl (first internode) of *S. vulgare* produces red cyanidin 3-glucoside acylated with an unknown aliphatic acid when exposed to light after being grown in the dark for 4 days. After a long period in complete darkness, yellow apigeninidin (3-deoxypelargonidin) and orange luteolinidin (3-deoxycyanidin) are produced, each as the aglycone, 5-glucoside, and a further unidentified stable acid.[160,168] The unidentified orange pigment observed in whole kernels of red varieties of grain sorghum[166] is believed to be luteolinidin, since both luteolinidin and apigeninidin were subsequently found even in the seed of white varieties.[174] Glumes of low-polyphenol sorghum contain the 3-deoxyanthocyanins luteolinidin and apigeninidin in a 55:45 ratio.[163] Recently, it has been found that infection of sorghum mesocotyls with either a pathogenic or nonpathogenic fungus resulted in a rapid accumulation of apigeninidin and luteolinidin; cyanidin was not detected.[162] In tissue culture, sorghum cells often produce large amounts of apigeninidin and luteolinidin, but no detectable cyanidin or other anthocyanin.[159] Seed and leaf tissue of some cultivars have relatively high levels of the flavan-4-ols, apiforol and luteoforol (Table 4), which have hydroxylation patterns similar to apigeninidin and luteolinidin; and have no detectable flavan-3-ols, catechin or epicatechin, which correspond to the hydroxylation pattern of cyanidin.[159,172] The proanthocyanidins or polymeric flavonoids of sorghum, on the other hand, have been found to have the hydroxylation pattern of cyanidin, not that of the 3-deoxyanthocyanidins.

In the absence of light, ethylene promotes synthesis of apigeninidin and luteolinidin in mesocotyls of sorghum.[175-176] However, ethylene has no influence on the synthesis of cyanidin which requires light for its formation.[177,178]

TABLE 4
Anthocyanins and Other Flavonoids of Sorghum
(Sorghum bicolor [L.] Moench.)

Compound	Plant part	Ref.
Cyanidin	Mesocotyl	160
Cyanidin glycosides	Mesocotyl, roots, and shoots	161
Apigeninidin	Mesocotyl, glumes	160,162
Apigeninidin glycoside	Mesocotyl, roots, shoots, grain	161
	Glumes	163,164
Luteolinidin	Mesocotyl, glumes	160,162
Luteolinidin glycoside	Mesocotyl, roots, shoots, grain, and glumes	163,164, 165,166
Pelargonidin glycoside	Grain	163
Fisetinidin	Grain	167
Apigenin glycoside	Shoots	168
Luteolin	Mesocotyl	160
Luteolin glycoside	Shoots	168
Luteolin, 7-*O*-methyl	Glumes	163
Naringenin	Grain	169
Eriodictyol	Grain	169,170
(+)-Catechin	Grain	169
Luteoforol	Grain	171
	Several	164
Apiforol	Grain, leaf	172
	Leaf	173
Kaemperol	Grain	174
(+)-Taxifolin	Grain	169
Naringeninchalcone	Grain	169

The presence of anthocyanins in sorghum is linked to its susceptibility to rust *(Puccinia purpurea)*,[179] and to its resistance to downy mildew *(Peronosclespora sorghi)*.[180] The herbicide Alachlor inhibits the formation of anthocyanin in etiolated sorghum mesocotyls by interfering with the synthesis of phenylpropionic acid.[181]

The genes *R, Y, B1, B2,* and *S* control the color, quality, and quantity of polyphenols in the pericarp and testa of sorghum seed. The dominant genes, *R* and *Y,* strongly increase the degree of anthocyanin pigmentation.[182] In white sorghum, the pigment is made up of three chromogenes (I, II, and III) which give eriodictyol and pelargonidin by hydrolysis; polymers are also present.[170] Kambal and Bate-Smith[183] identified the yellow pigment as eriodictyol chalcone, while luteolinidin was obtained from the red seeds by hydrolysis. This deoxyanthocyanidin was derived from a pentahydroxyflavone — luteoforol — a reduction product of eriodictyol, by the genetic factor *R;* flavonoids were formed by the action of gene *Y* which regulates the biosynthesis.

Flour made from seeds of sorghum contains 11% protein, 3% lipids, and 71% starch.[184] The presence of polyphenols, however, makes the flour astringent and reduces its palatability and digestibility.[159,184,185] Pigment-free cultivars which have a higher carotenoid content in the endosperm and fewer tannins are most suited for processing into food products.[156,184] The bran produced from sweet sorghum contains cyanidin and pelargonidin glucosides[186] and has been suggested for use as a food colorant for cookies, candies, gelatins, and other products.[187] According to Olifson et al.,[187] this colorant is nontoxic and, when fed to guinea pigs for 6 months, produced no morphological changes in the liver, pancreas, kidneys, or thyroid; and no glycogens were found in the blood.

B. MILLET

Pearl millet, *Pennisetum americanum* (L.) Leeke (syn. *P. typhoides*), is a very important food crop in many developing countries, and is widely grown in most semiarid regions of Africa and Asia.[188]

Color of pearl millet varies from yellow to bluish gray. Polyphenols in the grain are responsible for a gray pigmentation which some populations in the world find undesirable. Traditionally, the bluish-gray color of milled grain has been removed by soaking overnight in water containing sour milk or tamarind pods which decreases the pH and removes of pH-dependent pigments.[189] Reichert et al.[188] suggested that *C*-glycosylflavonoids were responsible for their natural gray color. In the leaves of 545 individuals of cultivated foxtail millet (*Setaria italica* [L.] Beauv.) (109 pure lines from a world collection), Gluchoff-Fiasson et al.[190] identified 30 vitexin (8-*C*-glucosylapigenin) and 24 orientin (8-*C*-glucosylluteolin) derivatives.

Recently, in purple and sun-red varieties of *P. americanum*, Raju et al.[191] identified derivatives of cyanidin, delphinidin, and pelargonidin, with the proportion of aglycones varying among the cultivars. In *P. japonicum*, Shibata and Sakai[192] identified cyanidin 3-glucoside; and in a related species (*Panicum melinis*) originating from Brazil, Bobbio et al.[193] found peonidin 3-arabinosylglucoside.

VIII. WHEAT

Wheat (*Triticum* spp.) is the world's most important grain crop. Wheat kernels are milled into flour to make bread and other products. These products are the main food of hundreds of millions of people worldwide. As a result, wheat covers more of the earth's surface than any other food crop.

The first studies on anthocyanin pigmentation in wheat were conducted by Nillson-Ehle.[194,195] Recent reviews provide much information about the flavones, carotenoids, and xanthophylls in wheat, but almost nothing is reported on anthocyanins.[196]

It is known that when wheat sprouts are strongly illuminated they become more or less colored, depending on the variety, from the formation of anthocyanins. This was suggested as a means for the identification of cultivars.[197] According to Tusnjakova,[198] the cultivars with red-pigmented seeds are better adapted to the rigid conditions of northern regions. The presence of antho-cyanins in the stele is a characteristic associated with rust resistance.[199] Jonard[200] observed that the red-pigmented cultivars germinate more slowly than the white ones. Tandon et al.[201] found that in *T. aestivum* the red pigmentation in the seeds is associated with a higher protein content. The red-seeded cultivars had about 1.8% more protein than the amber types.

In the dark-red, light-red, and white cultivars, there is a relationship between the catechin and tannin contents and the degree of pigmentation. The pigment is contained almost exclusively in the seed coat.[202]

In *Triticum* spp., different organs of the plant are subject to anthocyanin pigmentation. The color of the coleoptile has been used as a genetic marker in the study of mutants obtained by irradiation or by chemical means.[202] Bhowal and Jha[204] studied the inheritance of the purple pigmentation of the auricle and found it to be controlled by genes on the 7D chromosome. The chromosomes containing the genes which control anthocyanin synthesis in the coleoptile of *T. aestivum* are 7A and 7B.[205] There are three different alleles in homologous loci which participate in the biosynthetic cycle producing cyanidin and peonidin and their acylated derivatives.[206] From crosses between various species of *Triticum*, Raut et al.[207] showed that the anthocyanin pig-mentation of the auricle, stem, and basal ring of the rachis was determined by a single dominant gene. According to Sheopuria and Singh,[208] the character for red pigmentation of the glume is dominant over white and is controlled by a single gene pair. In *T. durum*, the purple pigmentation of the pericarp and the red color of the auricle are inherited as dominant monogenic char-acters.[209,210]

The pigmentation intensity in *T. durum* is determined by genetic factors and is also strongly influenced by environmental conditions, especially light intensity and temperature.[111] Paquet[112] observed that the intensity of red pig-ment depended not only on genetic factors but also on the physical composition of the endosperm: if fleshy, the red tends to be lighter and if vitreous, the red darkens.

Anthocyanins of wheat which have been reasonably well characterized are listed in Table 5. Van Bragt et al.[213] found cyanidin and peonidin in coleoptiles from irradiated wheat sprouts. Gale and Flavell[205] reported the presence of cyanidin and peonidin glucosides and at least three other deriv-atives of the two aglycones with two of them being acylated. In the coleoptiles of 13 cultivars of winter and summer wheat from Switzerland, Vogel[214] found peonidin 3-glucoside (the most abundant), cyanidin 3-gentiobioside, and cyan-idin 3-glucoside. This author also showed that it was impossible to characterize cultivars based on the qualitative composition of these three anthocyanins because all cultivars analyzed contained the same three pigments.

In the pericarp of purple-seeded wheat *(T. aestivum)*, cultivar UM606a (Hard Federation/Chinese Spring/Nero/3/3* Pitic 62), Dedio et al.[215] identified cyanidin 3-glucoside, peonidin 3-glucoside, and trace amounts of the corre-

TABLE 5
Relative Concentration of Wheat
(*Triticum* spp.) Anthocyanins

	Grain	Coleoptile	
Anthocyanin	**a**	**a**	**b**
Cyanidin 3-glucoside	+ + + +	+	+ + +
Peonidin 3-glucoside	+ + +	−	+ + + +
Acylated cyanidin glucoside	+ +	+ + +	−
Acylated peonidin glucoside	+ +	−	−
Cyanidin 3-rutinoside	+	+ + + +	−
Peonidin 3-rutinoside	+	+ +	−
Acylated cyanidin rutinoside	−	+ +	−
Acylated peonidin rutinoside	−	+	−
Cyanidin 3-gentiobioside	−	−	+ +

Note: a, From Dedio et al.;[215] b, from Vogel.[214]

sponding rutinosides. Acylated derivatives of the cyanidin and peonidin glucosides were also detected. The coleoptiles contained mostly cyanidin 3-rutinoside, acylated cyanidin glucoside, and acylated cyanidin rutinoside with smaller amounts of the corresponding peonidin derivatives. There were also two acylated glycosides of cyanidin which were not characterized. The acyl groups of the acylated pigments were not aromatic in character.

Bolton[216] found cyanidin and delphinidin derivatives in the grains of lines Blue A and PBB whereas purple grains of NY1 and NY3 had malvidin derivatives and those of ND2, delphinidin and malvidin derivatives. According to Zeven,[217] purple grain color is caused by anthocyanins in the pericarp whereas blue color is caused by anthocyanins in the aleurone layer. Purple grains occur in tetraploid wheats from Ethiopia and in one bread wheat accession apparently native to China. Purple pericarp and blue aleurone grains have been studied primarily for their use as markers for special purple (feed) wheats to enable their distinction from those used for human consumption.[217,218] Both purple- and blue-grain, however, have also been suggested for use as markers in the production of hybrid seeds and in studies to measure the degree of natural cross pollination and of gamete transmission.[217] The genetics of purple pericarp and blue aleurone characters is discussed in detail in the recent review of Zeven.[217]

In addition to anthocyanins, wheat contains a variety of other flavonoids whose presence may have significant effects on the physiological and technological properties of the grain and processed products. Commercial wheat bran has recently been found to contain low levels of catechin and di-, tri-, and oligomeric proanthocyanidins,[219] in addition to 6,8-di-*C*-glycosylapigenins and their sinapic and ferulic acid esters.[220]

IX. WILD RICE

The wild rice or Indian rice, *Zizania aquatica* L., is from a botanical point of view quite far from ordinary rice (*Oryza* spp.), although both species belong to the Oryzaceae family.

The plant grows along lakeshores in Canada and the Great Lakes region of the United States. It is called wild rice because of its growth habits; and it is used in the same manner as ordinary rice, but it is not a true rice.

The anthocyanins of the leaf sheath and staminate florets of wild rice from Manitoba were identified by Gutek et al.[221] as cyanidin 3-glucoside and cyanidin 3-rhamnoglucoside which accounts for 75 and 25% of the total anthocyanin, respectively. At least a simple dominant gene (designated *A*) has been shown to control the anthocyanin production in wild rice.[221] However, the occurrence of red progeny from a self-pollinated white plant suggests the presence of a more complex control mechanism which remains to be elucidated.

REFERENCES

1. **Mullick, D. B.,** Biochemical genetics of the anthocyanins of barley (*Hordeum vulgare* L.), *Diss. Abstr. B,* 27, 2247, 1967.
2. **Mullick, D. B. and Brink, V. C.,** Localization of anthocyanins by peeling the investments of the barley caryopsis, *Crop Sci.,* 6, 204, 1966.
3. **Gassner, G. and Straib, W.,** Untersuchungen ueber den Einfluss der Mineralsalzernaehrung auf die Anthocyanbildung an jugen Gerstenpflanzen, *Angew. Bot.,* 19, 225, 1937.
4. **Hamy, A.,** Effect of phosphorus deficiency on barley leaf pigments, *C.R. Sciences Acad. Agric. Fr.,* 69(12), 935, 1983.
5. **Takahashi, R. and Yamamoto, J.,** Studies on the classification and geographical distribution of barley varieties. XII. On the pigmentation of the plant parts, *Rep. Ohara Agric. Res.,* 39, 25, 1950 (PBA 1957, 1758).
6. **Narula, P. N. and Mathur, V. S.,** Genetic studies in barley. II. Inheritance of pigmentation, *J. Indian Bot. Soc.,* 45, 24, 1966 (PBA 1967, 4113).
7. **Takahashi, R.,** The origin and evolution of cultivated barley, *Adv. Genet.,* 7, 227, 1955.
8. **Mullick, D. B., Faris, D. G., Brink, D. G., and Acheson, R. M.,** Anthocyanins and anthocyanidins of the barley pericarp and aleurone tissues, *Can. J. Plant Sci.,* 38, 445, 1958.
9. **McFarlane, W. D., Wye, E., and Grant, H. L.,** *Proc. Eur. Brew. Conf., Baden-Baden,* 1955, 298.
10. **Jende-Strid, B.,** Mutation affecting flavonoid synthesis in barley, *Carlsberg Res. Commun.,* 43, 265, 1978.
11. **Metche, M. and Urion, E.,** Isolation and identification of anthocyanin derivatives in the husk of barley, *C.R. Acad. Sci. Fr.,* 252, 356, 1961.
12. **Woodward, R. W. and Thieret, J. W.,** A genetic study of complementary genes for purple lemma, palea, and pericarp in barley (*Hordeum vulgare* L.), *Agron. J.,* 45, 182, 1953.

13. **Mayler, J. L. and Stanford, E. H.,** Color inheritance in barley, *J. Am. Soc. Agric.,* 34, 427, 1942.
14. **Litzenberger, S. C. and Green, J. M.,** Inheritance of awns in barley, *Agron. J.,* 43, 117, 1951.
15. **Takahashi, R., Yamamoto, J., Yasuda, S., and Itano, Y.,** Inheritance and linkage studies in barley, *Ber. Ohara Inst. Landwirtsch. Forsch.,* 10, 29, 1953.
16. **Mutry, G. S. and Jain, K. B. L.,** Genetic studies in barley. I. Inheritance of pigmentation in various plant parts, *J. Indian Bot. Soc.,* 38, 561, 1959.
17. **Devreux, A.,** Development of anthocyanins during brewing, *Brass. Malt. Belg.,* 11, 137, 1961.
18. **Jende-Strid, B. and Lundqvist, U.,** Diallelic tests of anthocyanin-deficient mutants, *Barley Genet. Newsl.,* 8, 57, 1978.
19. **Jende-Strid, B.,** Phenolic acids in grains of wild-type barley and proanthocyanidin-free mutants, *Carlsberg Res. Commun.,* 50, 1, 1985.
20. **Nilan, R. A., Kleinhofs, A., and Warner, R. L.,** Use of induced mutants of genes controlling nitrate reductase, starch deposition, and anthocyanin synthesis in barley, *Induced Mutations — A Tool in Plant Research,* Proc. Int. Symp., IAEA, Vienna, 1981, 183.
21. **Jende-Strid, B.,** Characterization of mutants in barley affecting flavonoid synthesis, *Hoppe-Seyler's Z. Physiol. Chem.,* 362, 12, 1981.
22. **Bhatia, I. S., Kanshal, G. P., and Bajaj, K. L.,** Chrysoeriol from barley seeds, *Phytochemistry,* 11, 1867, 1972.
23. **Van Sumere, C. F., Cottenie, J., de Greef, J., and Kint, J.,** Biochemical studies in relation to the possible germination regulatory role of naturally occurring coumarin and phenolics, *Recent Adv. Phytochem.,* 4, 166, 1972.
24. **Slominski, B. A.,** Phenolic acids in the meal of developing and stored barley seeds, *J. Sci. Food Agric.,* 31, 1007, 1980.
25. **Sosulski, F., Krygier, K., and Hogge, L.,** Free, esterified and insoluble-bound phenolic acids. III. Composition of phenolic acids in cereal and potato flours, *J. Agric. Food Chem.,* 30, 337, 1982.
26. **Marshall, H. G. and Pomeranz, Y.,** Buckwheat description, breeding, production and utilization, in *Advances in Cereal Science and Technology,* Pomeranz, Y., Ed., American Association of Cereal Chemistry, St. Paul, MN, 1982, 157.
27. **Mazza, G.,** Buckwheat *(Fagopyrum esculentum),* the crop and its importance, in *Encylopedia of Food Science, Food Technology and Nutrition,* MacRae, R., Ed., Academic Press, London, 1992, 534.
28. **Kaerstens, W. K. H.,** Anthocyanin and anthocyanin formation in seedlings of *Fagopyrum esculentum* Moench., *Rec. Trav. Bot. Neerl.,* 36, 85, 1939 (CA 33, 8687).
29. **Inoue, K., Hosokama, Y., and Shimadate, T.,** Anthocyanin pigments of buckwheat hypocotyls, *Kenkyu Kiyo Nihon Daigaku Bunrigakubu Shizen Kagaku Kenkyusho,* 17, 26, 1982 (CA 97, 88690m).
30. **Troyer, J. R.,** Anthocyanin pigments of buckwheat hypocotyls, *Ohio J. Sci.,* 58, 187, 1958 (CA 52, 12090h).
31. **Troyer, J.,** Anthocyanin formation in excised segments of buckwheat seedling hypocotyls, *Plant Physiol.,* 39, 907, 1964.
32. **Harraschain, H. and Mohr, H.,** Effect of visible light on the flavonoid synthesis and morphogenesis of buckwheat seedlings *(Fagopyrum esculentum).* II. Flavonol synthesis and hypocotyl growth, *Z. Bot.,* 51, 277, 1963.
33. **Amrhein, N.,** Biosynthesis of cyanidin in buckwheat hypocotyls, *Phytochemistry,* 18(4), 585, 1979.
34. **Margna, U., Laanest, L., Margna, E., Otter, M., and Vainjarv, T.,** Nitrogen-induced changes in the accumulation of buckwheat seedling flavonoids, *Eesti NSV Tead. Akad. Toim., Biol. Seer.,* 23(4), 298, 1974 (CA 82, 72078).

35. **Margna, U. and Vainjarv, T.,** Buckwheat seedling flavonoids do not undergo rapid turnover, *Biochem. Physiol. Pflanz.,* 176(1), 44, 1981.
36. **Shropshire, W., Jr. and Mohr, H.,** Gradient formation of anthocyanin in seedlings of *Fagopyrum* and *Sinapis* unilaterally exposed to red and far red light, *Photochem. Photobiol.,* 12, 145, 1970.
37. **Tohver, A.,** Effect of exogenous phenylalanine on the light sensitivity of anthocyanin accumulation in buckwheat seedlings, *Regul. Rosta Pitan. Rast.,* 1976, 33, 1976 (CA 87, 16465).
38. **Tohver, A.,** Anthocyanin accumulation in buckwheat hypocotyl under intermittent red and far red light, *Fiziol. Rast.,* 34, 742, 1987 (CA 107, 131053v).
39. **Troyer, J. R.,** The distribution of rutin and other flavonoid substances in buckwheat seedlings, *Plant Physiol.,* 30, 168, 1955.
40. **Troyer, J. R.,** Distribution of rutin and other flavonoid substances in mature buckwheat plants, *Lloydia,* 19, 216, 1956.
41. **Durkee, A. B.,** Polyphenols of the bran-aleurone fraction of buckwheat seed *Fagopyrum sagittatum* Gilib., *J. Agric. Food Chem.,* 25, 287, 1977.
42. **Margna, U., Margna, E., and Paluteder, A.,** Localization and distribution of flavonoids in buckwheat seedling cotyledons, *J. Plant Physiol.,* 136, 166, 1990.
43. **Emerson, R. A.,** The genetic relations of plant colors in maize, *Cornell Univ. Agric. Exp. Stn.,* 39, 3, 1921.
44. **Stadler, L. J.,** Spontaneous mutation of the *R* locus in maize. I. The aleurone-color and plant-color effects, *Genetics,* 31, 377, 1946.
45. **Stadler, L. J.,** Spontaneous mutation at the *R* locus in maize. II. Race differences in mutation rate, *Am. Nat.,* 82, 289, 1948.
46. **Rhoades, M. M.,** Effect of the bronze locus on the anthocyanin formation in maize, *Proc. Natl. Acad. Sci. U.S.A.,* 31, 91, 1945.
47. **McClintock, B.,** The control of gene action in maize, *Brookhaven Symp. Biol.,* 18, 162, 1965.
48. **Dellaporta, S. L., Greenblatt, I., Kermicle, J. L., Hicks, J. B., and Wessler, S. R.,** Molecular cloning of the maize *R-nj* allele by transposon tagging with *Ac,* in *Chromosome Structure and Function: Impact of New Concepts,* Gustafson, J. P. and Appels, R., Eds., 18th Stadler Genetics Symposium, Plenum Press, New York, 1987, 263.
49. **Taylor, L. P. and Briggs, W. R.,** Genetic regulation and photocontrol of anthocyanin accumulation in maize seedlings, *Plant Cell,* 2, 115, 1990.
50. **Wienand, U., Weydemann, U., Niesbach-Klösgen, U., Peterson, P. A., and Saedler, H.,** Molecular cloning of the *c2* locus of *Zea mays,* the gene coding for chalcone synthase, *Mol. Gen. Genet.,* 203, 202, 1986.
51. **O'Reilly, C., Shepherd, N. S., Pereira, A., Schwartz-Sommer, Z., Bertram, I., Robertson, D. S., Peterson, P. A., and Saedler, H.,** Molecular cloning of the *al* locus of *Zea mays* using the transposable elements *En* and *Mul, EMBO J.,* 4, 877, 1985.
52. **Fedoroff, N. V., Furtek, D. B., and Nelson, O. E., Jr.,** Cloning of the *bronze* locus in maize by a simple and generalizable procedure using the transposable controlling element *Activator (Ac), Proc. Natl. Acad. Sci. U.S.A.,* 81, 3825, 1984.
53. **McLaughlin, M. and Walbot, V.,** Cloning of a mutable *bz2* allele of maize by transposon tagging and differential hybridization, *Genetics,* 117, 771, 1987.
54. **Theres, N., Scheele, T., and Starlinger, P.,** Cloning of the *Bz2* locus of *Zea mays* using the transposable element *Ds* as a gene tag, *Mol. Gen. Genet.,* 209, 193, 1987.
55. **Paz-Ares, J., Weinand, U., Peterson, P., and Saedler, H.,** Molecular cloning of the *c* locus of *Zea mays:* a locus regulating the anthocyanin pathway, *EMBO J.,* 5, 829, 1986.
56. **Ludwig, S. R., Ledare, H. F., Dellaporta, S. L., and Wessler, S. R.,** *Lc,* a member of the maize *R* gene family responsible for tissue-specific anthocyanin production, encodes a protein similar to transcriptional activators and contains the *myc*-homology region, *Proc. Natl. Acad. Sci. U.S.A.,* 86, 7092, 1989.

57. **Chandler, V. L., Radicella, J. P., Robbins, T. P., Chen, J., and Turks, D.,** Two regulatory genes of the maize anthocyanin pathway are homologous: isolation of *B* utilizing *R* genomic sequences, *Plant Cell*, 1, 1175, 1989.

58. **Straus, J.,** Anthocyanin synthesis in maize endosperm tissue cultures. I. Identity of pigments and general factors, *Plant Physiol.*, 34, 536, 1959.

59. **Harborne, J. B. and Gavazzi, G.,** Effect of *Pr* and *pr* alleles on anthocyanin biosynthesis in *Zea mays*, *Phytochemistry*, 8, 999, 1969.

60. **Lawanson, A. O. and Osude, B. A.,** Identification of the anthocyanin pigments of *Zea mays* var. E.S.I., *Z. Pflanzenphysiol.*, 67, 460, 1972.

61. **Baraud, J., Genevois, L., and Panart, J. P.,** Anthocyanins of corn, *J. Agric. Trop. Bot. Appl.*, 11, 55, 1974 (CA 62, 3062c).

62. **Nakatani, N., Fukuda, H., and Fuwa, H.,** Major anthocyanin of Bolivian purple corn (*Zea mays* L.), *Agric. Biol. Chem.*, 43, 389, 1979.

63. **Seah, K. T. and Styles, E. D.,** Anthocyanins in maize, *Can. J. Genet. Cytol.*, 11, 482, 1969.

64. **Styles, E. D. and Ceska, O.,** Flavonoid pigments in genetic strains of maize, *Phytochemistry*, 11, 3019, 1972.

65. **Reddy, G. M. and Reddy, A. R.,** Action of *bronze-1* locus in anthocyanin synthesis in maize, *Proc. Symp. Contr. Mech. Cell Processes*, 1973, 397, 1973.

66. **Harborne, J. B. and Self, R.,** Malonated cyanidin 3-glucosides in *Zea mays* and other grasses, *Phytochemistry*, 26, 2417, 1987.

67. **Duke, S. O. and Naylor, A. W.,** Light control of anthocyanin biosynthesis in *Zea* seedlings, *Physiol. Plant.*, 37, 62, 1976.

68. **Duke, S. O., Fox, S. B., and Naylor, A. W.,** Photosynthetic independence of light induced anthocyanin formation in *Zea* seedlings, *Plant Physiol.*, 57, 192, 1976.

69. **Rengel, Z. and Kordan, H. A.,** Age, fluence rate, and anthocyanin production in *Zea mays* seedlings, *Ann. Bot. (London)*, 59, 41, 1987.

70. **Rengel, Z. and Kordan, H. A.,** Effect of growth regulators on light-dependent anthocyanin production in *Zea mays* seedlings, *Physiol. Plant.*, 69, 511, 1987.

71. **Bhatla, S. C. and Pant, R. C.,** Isolation and characterization of anthocyanins from P-deficient maize plant, *Curr. Sci.*, 46, 700, 1977 (CA 88, 19014).

72. **Knoll, H. A.,** Influence of soil temperature on the growth and contents of phosphorus and anthocyanins in maize, *Diss. Abstr. Int.*, B, 24, 3038, 1963 (JSFA 1964, 78ii).

73. **Knoll, H. A., Lathwell, D. J., and Brady, N. C.,** Influence of root zone temperature on the growth and contents of phosphorous and anthocyanin in maize, *Proc. Soil Sci. Soc. Am.*, 28, 400, 1964.

74. **Lawanson, A. O., Ojeniyi, G. A., Nduka, C. E., and Osueke, S. O.,** Contribution of cyanidin 3-galactoside and pelargonidin 3-glucoside to the pattern of anthocyanin accumulation during deficiencies of nitrogen, phosphorous and potassium in seedlings of *Zea mays*, *Phyton (Buenos Aires)*, 33, 183, 1975 (CA 84, 161827).

75. **Lawanson, A. O., Ojeniyi, A., Nduka, C. E., and Osueke, S. O.,** Distribution of cyanidin 3-galactoside and pelargonidin 3-glucoside in mineral-deficient maize seedlings, *Phyton (Buenos Aires)*, 33, 187, 1975 (CA 84, 147800).

76. **Paris, R. and Pereira-Alarcon, A.,** Flavonoids of some Peruvian dye plants, *Plant. Med. Phytother.*, 2, 90, 1968 (CA 69, 74492).

77. **Sanei Chemical Industries Ltd.,** Adjustable Colour, Japanese Patent, 4,838,154, 1973 (FSTA 6, 8T424).

78. **Sanei Chemical Industries Ltd.,** Food Colouring, Japanese Patent, 5,133,805, 1976 (FSTA 9, 4T212).

79. **Sanei Chemical Industries Ltd.,** Red Dye, Japanese Patent, 5,221,586, 1977 (FSTA 10, 4T90).

80. **Sanei Chemical Industries Ltd.,** Red Coloring, Japanese Patent, 5,649,544, 1981 (FSTA 14, 10T567).

81. **Sugiyama Chemical Institute,** Anthocyanin Food Coloring Agent from Purple Corn, Japanese Patent, 77,130,824, 1977 (CA 88, 150953).

82. **Sugiyama Chemical Institute,** Colouring Agent, Japanese Patent, 5,503,379, 1980 (FSTA 13, 8T394).

83. **Tegenkamp, T. R.,** Phenomena associated with hereditary albinism in *Avena, Diss. Abstr.,* 22, 3824, 1962 (PBA 1963, 1332).

84. **Nittler, L. W.,** Varietal differences among phosphorous-deficient oat seedlings, *Crop Sci.,* 8, 393, 1968.

85. **Taketa, F. J.,** Studies of Various Nutritional Factors in Relation to Dental Caries in the Rat, Ph.D. thesis, University of Wisconsin, Madison, 1957.

86. **Vogel, J. J.,** Studies on the Diet in Relation to Dental Caries in the Cotton Rat, Ph.D. thesis, University of Wisconsin, Madison, 1961.

87. **Collins, F. W.,** Oat phenolics: avenanthramides, novel substituted *N*-cinnamoylanthranilate alkaloids from oat groats and hulls, *J. Agric. Food Chem.,* 37, 60, 1989.

88. **Collins, F. W.,** Oat phenolics: structure, occurrence and function, in *Oats: Chemistry and Technology,* Webster, F. H., Ed., American Association of Cereal Chemistry, St. Paul, MN, 1986, 227.

89. **Collins, F. W., MacLachlan, D. C., and Blackwell, B. A.,** Oat phenolics: avenalunic acids, a new group of bound phenolic acids from oat groats and hulls, *Cereal Chem.,* 68, 184, 1991.

90. **Van der Stok, J. E.,** On the inheritance of grain color in rice, *Teysmannia,* 65, 5, 1908.

91. **Hector, G. P.,** Observations on the inheritance of anthocyanin pigment in paddy varieties, *Mem. Dep. Agric. India Bot. Ser.,* 8(2), 89, 1916.

92. **Hector, G. P.,** Correlation of colour characters in rice, *Mem. Dep. Agric. India Bot. Ser.,* 11(7), 153, 1922.

93. **Parnell, F. R., Ayyangar, G. N. R., and Ramiah, K.,** Inheritance of characters in rice, *Mem. Dep. Agric. India Bot. Ser.,* 9, 75, 1917.

94. **Jones, J. W.,** Inheritance of anthocyanin pigments in rice, *J. Agric. Res.,* 40, 1105, 1930.

95. **Dave, B. B.,** Inheritance of colour in Central Province's rice, *Nagpur Coll. Agric. Mag.,* 23, 20, 1948.

96. **Ramiah, K.,** *Rice Breeding and Genetics,* I.C.A.R., New Delhi, India, 1953.

97. **Ghose, R. L. M., Ghadge, M. B., and Subrahmanyan, V.,** *Rice in India,* I.C.A.R., New Delhi, India, 1956.

98. **Kadam, B. S. and Ramiah, K.,** Symbolization of genes in rice, *Indian J. Genet. Plant Breed.,* 3, 7, 1943.

99. **Nagao, S.,** Genetic analysis and linkage relationship of characters in rice, *Adv. Genet.,* 4, 181, 1951.

100. **Jodon, N. E.,** Present status of rice genetics, *J. Agric. Assoc. China,* 10, 5, 1955.

101. **Anon.,** Genetic symbols for rice recommended by the International Rice Commission, *I.R.C. Newsl.,* 8, 1, 1959.

102. **Nagai, I., Suzushino, G., and Tsuboki, Y.,** Genetic variation of anthocyanins in *Oryza sativa, Idengaku Zasshi,* 31, 441, 1962 (PBA 1965, 518).

103. **Shafi, M. and Khan, S.,** The inheritance of anthocyanin pigment in leafsheath, leafblade and stigma of paddy, *Pak. J. Sci. Res.,* 10, 34, 1958 (PBA 1958, 4239).

104. **Takahashi, M.,** Analysis of apiculus color genes essential to anthocyanin coloration in rice, *J. Fac. Agric. Hokkaido Univ.,* 50, 266, 1957 (PBA 1958, 4238).

105. **Ramiah, K.,** Anthocyanin genetics of cotton and rice, *Indian J. Genet. Plant Breed.,* 5, 1, 1945.

106. **Misro, B., Seetharaman, R., and Richharia, R. H.,** World genetic stock of rice. I. Pattern of anthocyanin pigment distribution, *Indian J. Genet. Plant Breed.,* 20, 113, 1960 (CA 58, 12868d).

107. **Nair, N. R.,** Genetics of rice (*Oryza sativa* L.). The Inheritance of Hairiness, Clustering, Beaked Habit, Anthocyanin Pigmentation and Interrelationships of Genes Governing These Characters, thesis, I.A.R.I., New Delhi, India, 1958.

108. **Iyer, K. N.,** Genetics of Rice (*Oryza sativa* L.). The Inheritance of Awning, Auricle, Ligule, and Junctura and Anthocyanin Pigmentation and the Interrelationship of Genes Governing These Characters, thesis I.A.R.I., New Delhi, India, 1959.

109. **Nagao, S., Takahashi, M., and Kinoshita, T.,** Genetical studies on rice plant. XXVI. Mode of inheritance and causal genes for one type of anthocyanin color character in foreign rice varieties, *J. Fac. Agric. Hokkaido Univ.*, 52, 20, 1962 (PBA 1963, 4545).

110. **Setty, M. V. N. and Misro, B.,** Complementary genic complex for anthocyanin pigmentation in the apiculus of rice (*Oryza sativa*), *Can. J. Genet. Cytol.*, 15(4), 779, 1973.

111. **Hadagal, B. N., Manjunath, A., and Goud, J. V.,** Linkage of genes for anthocyanin pigmentation in rice (*Oryza sativa* L.), *Euphytica*, 30, 747, 1981.

112. **Kondo, A.,** Fundamental studies on breeding rice by hybridization between Japanese and foreign rices. VII. Identification of the genes controlling anthocyanin pigmentation in Japanese and foreign rices, *Ikushugaku Zasshi*, 13, 92, 1963 (PBA 1965, 4133).

113. **Kondo, A.,** Fundamental studies on breeding rice by hybridization between Japanese and foreign rices. VIII. On some anthocyanin-distributing genes and their inhibitors, *Ikushugaku Zasshi*, 13, 241, 1963 (PBA 1964, 5864).

114. **Misro, B. R. H.,** Linkage studies in rice (*Oryza sativa* L.), *Oryza*, 18, 185, 1981 (PBA 1983, 438).

115. **Shafi, M. and Aziz, M. A.,** The inheritance of anthocyanin pigment in the outer glume and apiculus of rice, *Agric. Pak.*, 10, 217, 1959 (PBA 1960, 420).

116. **D'Cruz, R. and Dhulappanavar, C. V.,** Inheritance of pigmentation in rice, *Indian J. Genet.*, 23, 3, 1963 (PBA 1964, 2260).

117. **Kadam, B. S.,** Patterns of anthocyanin inheritance in rice. II. Japan-1 × T-433, *Res. J. Mahatma Phule Agric. Univ.*, 4, 32, 1973 (PBA 1974, 8666).

118. **Kadam, B. S.,** Patterns of anthocyanin inheritance in rice. IV. Two cases of six gene segregation, *Indian J. Genet. Plant. Breed.*, 36, 69, 1976 (PBA 1977, 8490).

119. **Kadam, B. S.,** Patterns of anthocyanin inheritance in rice. VII. Japanese rices, *Indian J. Genet. Plant. Breed.*, 41, 103, 1981 (PBA 1982, 3927).

120. **Yadav, L. N. and Tomar, N. S.,** Inheritance of anthocyanin pigmentation in stigma, outer glumes and apiculus of rice (*Oryza sativa* L.), *J. Indian Bot. Soc.*, 63, 190, 1984 (PBA 1985, 3513).

121. **Kolhe, C. L. and Bhat, N. R.,** Linkage of five anthocyanin genes in rice, *Indian J. Genet. Plant. Breed.*, 42, 92, 1982 (CA 27, 4535).

122. **Maekawa, M. and Kita, F.,** New gametophyte genes located in the third linkage group (chromosome 3) of rice, *Jpn. J. Breed.*, 35, 25, 1985 (PBA 1985, 6876).

123. **Thimmappaiah,** Inheritance and interrelationship of genes governing few characters in rice (*Oryza sativa* L.), *Mysore J. Agric. Sci.*, 10, 721, 1976 (PBA 1978, 4371).

124. **Kosaka, H.,** The relationship between different physiological aspects of plants and the appearance of pigments in the different parts of plants. IV. The relationship between the existence of anthocyanin pigments and the degree of ability to assimilate in some cultivated plants, *J. Dep. Agric. Kyushu Imp. Univ.*, 3, 251, 1933 (CA 27, 4535).

125. **Hayashi, K.,** Anthocyanins. XIII. Several anthocyanins containing cyanidin as the aglycone, *Acta Phytochim.*, 14, 55, 1944 (CA 45, 4786).

126. **Hayashi, K.,** Coloring matter of *Oryza sativa*, *Sigenkagaku Kekkyusyo Iho*, 9, 1, 1946 (CA 42, 3034e).

127. **Hayashi, K.,** Pigment of purple rice, *Misc. Rep. Res. Inst. Nat. Resour., Jpn.*, 9, 1, 1946 (CA 44, 9012i).

128. **Gangulee, H. C.,** Rice anthocyanin genetics. I. Review of anthocyanin genetics in general, *Port. Acta Biol., Ser. A*, 4, 99, 1955 (PBA 1956, 3361).

129. **Gangulee, H. C.,** Rice anthocyanin genetics. II. Some experimental results, *Bull. Bot. Soc. Bengal*, 9, 121, 1955 (CA 53, 22273f).

130. **Sharma, J. N. and Seshadri, T. R.,** Survey of anthocyanins from Indian sources, *J. Sci. Ind. Res.*, 14B, 211, 1955 (CA 50, 5846g).

131. **Suzushino, G., Hatanaka, A., Suzuki, Y., and Nagai, I.,** Anthocyanins in rice, *Nihon Daigaku Nojuigakubu Gakujutsu Kenkyu Hokoku*, 8, 65, 1957 (CA 55, 702c).

132. **Nagai, I., Suzushino, G., and Suzuki, Y.,** Anthoxanthins and anthocyanins in *Oryzaceae*, *Jpn. J. Breed.*, 10, 247, 1960.

133. **Kim, J. H. and Kwon, Y. W.,** Composition of chloroplast pigments, its change and anthocyanin development after synthesis in the remote rice cultivars, *Nonghak Yongu (Soul Taehakkyo)*, 2, 15, 1977 (CA 92, 194631).

134. **Subba Rao, B. L., Ghosh, A., and John, V. T.,** Effect of rice tungro virus on chlorophyll and anthocyanin pigments in two rice cultivars, *Phytopathol. Z.*, 94, 367, 1978 (CA 91, 52919).

135. **Bishonen Shuzo, K. K.,** Manufacture of Red Sake, Japanese Patent, 58, 51, 885 (CA 99, 51854).

136. **Bishonoen Shuzo, K. K.,** Food Dyes from Red Rice, Japanese Patent, 58, 179, 460 (CA 100, 66971).

137. **Yoshinaga, K., Takahashi, K., and Yoshizawa, K.,** Liqueur with pigments of red rice, *Nippon Jozo Kyokai Zasshi*, 81, 337, 1986 (CA 105, 113526f).

138. **Katayama, M., Onoe, A., Egawa, H., Imaki, T., and Nakano, J.,** Red Colors for Acidic Foods, Japanese Patent, 61,158,762, 1986 (CA 105, 207826).

139. **Fukui, K. and Wagi, M.,** Anthocyanin-Type Red Food Color and Its Manufacture from Red Rice, Japanese Patent, 2,248,465, 1990 (CA 114, 246180).

140. **Matz, S. S.,** *The Chemistry and Technology of Cereals as Food and Feed*, AVI Publishing, Westport, CT, 1959, 732.

141. **Metche, M. and Gay, R.,** Biogenesis of anthocyanin pigments, *Brauwissenschaft*, 17, 347, 1964.

142. **Metche, M.,** Anthocyanin pigments in some vegetables, *Brasserie*, 22, 69, 1967 (CA 67, 29835).

143. **Dedio, W., Kaltisikes, P. J., and Larter, E. N.,** The anthocyanins of *Secale cereale* (Rye), *Phytochemistry*, 8, 2351, 1969.

144. **Dedio, W., Hill, R. D., and Evans, L. E.,** Anthocyanin in the pericarp and coleoptiles of purple-seeded rye, *Can. J. Plant. Sci.*, 52, 981, 1972.

145. **Busch, E., Strack, D., and Weissenböck, G.,** Cyanidin 3-gentobioside from primary leaves of rye (*Secale cereale* L.), *Z. Naturforsch., Teil C*, 41(1/2), 485, 1986 (FCA 1986, 4940).

146. **Watkins, R.,** The inheritance of anthocyanin pigments in rye (*Secale cereale* L.), *Diss. Abstr. Int. B*, 26, 4986, 1966 (PBA 1967, 4080).

147. **Watkins, R. and White, W. J.,** The inheritance of anthocyanins in rye (*Secale cereale* L.), *Can. J. Genet. Cytol.*, 6, 403, 1964.

148. **Dumon, A. G.,** Analysis of heritable properties of rye (*Secale cereale* L.), *Studkring. Plantveredl., Wageningen Versl. N.*, 70, 966, 1961 (PBA 1962, 3217).

149. **Dumon, A. G. and Laeremans, R.,** Contribution a l'analyse genetique de l'anthocyane du seigle (*Secale cereale* L.), *Bull. Jard. Bot. Bruxelles*, 27, 507, 1957 (PBA 1958, 368).

150. **Fedorov, V. S. and Smirnov, V. G.,** Genetics of rye (*Secale cereale* L.), *Genetika, Moskow N.*, 2, 96, 1967 (PBA 1969, 506).

151. **Ruebenbauer, T. and Ruebenbauer, K.,** An attempt to clarify the mode of inheritance of anthocyanin coloration of the ligule and auricles in hybrids of rye inbred line, *Genet. Pol.*, 18, 193, 1977 (PBA 1978, 5322).

152. **Ruebenbauer, T., Kubera-Szpunar, L., and Kaleta, S.,** Genes governing qualitative characters in inbred lines of rye (*Secale cereale* L.), *Genet. Pol.*, 21, 377, 1980 (PBA 1981, 10419).

153. **Melz, G., Neumann, H., Muller, H. W., and Sturm, W.,** Genetical analysis of rye (*Secale cereale* L.). I. Results of gene localization on rye chromosomes using primary trisomics, *Genet. Pol.*, 25, 111, 1984 (PBA 1985, 7692).

154. **Sturm, W., Muller, H. W., and Neumann, H.,** Preliminary results on the use of primary triscomics to locate genes in rye, *Tagungsber., Akad. Landwirtschaftswiss. Dtsch. Demokr. Rep. N.*, 198, 153, 1981 (PBA 1982, 10221).

155. **Sturm, W., Neumann, H., and Melz, G.,** Trisomic analysis for the trait anthocyanin coloration in *Secale cereale* L., *Ark. Zuechtungsforsch.*, 11, 49, 1981 (PBA 1981, 7016).

156. **Creelman, R. A., Rooney, L. W., and Miller, F. R.,** Sorghum, in *Cereals: A Renewable Resource — Theory and Practice*, Pomeranz, Y. and Munck, L., Eds., American Association of Cereal Chemistry, St. Paul, MN, 1982, 395.

157. **Earp, C. F.,** Microscopy of the mature and developing caryopsis of *Sorghum bicolor* (L.) Moench., *Diss. Abstr. Int. B*, 45, 3673, 1984.

158. **Rooney, L. W. and Miller, F. R.,** Variation in the structure and kernel characteristics of sorghum, in *Proc. Int. Symp. Sorghum Grain Quality*, Rooney, L. W. and Murty, D. S., Eds., ICRISAT, Patancheru, Andhra Pradesh, India, 1982, 143.

159. **Butler, L. G.,** Sorghum polyphenols, in *Toxicants of Plant Origin*, Vol. IV, *Phenolics*, Cheeke, P. R., Ed., CRC Press, Boca Raton, FL, 1989, 95.

160. **Stafford, H. A.,** Flavonoids and related phenolic compounds produced in the first internode of *Sorghum vulgare* Pers. in darkness and in light, *Plant Physiol.*, 40, 130, 1965.

161. **McGrath, R. M., Kaluza, W. Z., Daiber, K. H., van der Riet, W. B., and Glennie, C. W.,** Polyphenols of sorghum grain, their changes during malting, and their inhibitory nature, *J. Agric. Food Chem.*, 30, 450, 1982.

162. **Nicholson, R. L., Kollipara, S. S., Vincent, J. R., Lyons, P. C., and Cadena-Gomez, G.,** Phytoalexin synthesis by the sorghum mesocotyl in response to infection by pathogenic and nonpathogenic fungi, *Proc. Natl. Acad. Sci. U.S.A.*, 1987, 5520.

163. **Misra, K. and Seshadri, T. R.,** Chemical components of *Sorghum durra* glumes, *Indian J. Chem.*, 5, 409, 1967 (CA 68, 51145).

164. **Bate-Smith, E. C.,** Luteoforol (3',4,4',5,7-pentahydroxyflavan) in *Sorghum vulgare*, *Phytochemistry*, 8, 1803, 1969.

165. **Gupta, R. K. and Haslam, E.,** Plant proanthocyanidins. V. Sorghum polyphenols, *J. Chem. Soc. Perkin Trans. 1*, 892, 1978.

166. **Nip, W. K. and Burns, E. E.,** Pigment characterization in grain sorghum. II. White varieties, *Cereal Chem.*, 48, 74, 1971.

167. **Blessin, G. W., VanEtten, C. H., and Dimler, R. J.,** An examination of anthocyanogens in grain sorghum, *Cereal Chem.*, 40, 201, 1963.

168. **Stafford, H. A.,** Changes in phenolic compounds and related enzymes in young plants of sorghum, *Phytochemistry*, 8, 743, 1969.

169. **Gujer, R., Magnolato, D., and Self, R.,** Glycosylated flavonoids and other phenolic compounds from sorghum, *Phytochemistry*, 25, 1431, 1986.

170. **Yasumatsu, K., Nakayama, T. O. M., and Chichester, C. O.,** Flavonoids of sorghum, *J. Food Sci.*, 30, 663, 1965.

171. **Bate-Smith, E. C. and Rasper, V.,** Tannins of grain sorghum: Luteoforol (leucoluteolinidin), 3',4,4',5,7-pentahydroxyflavan, *J. Food Sci.*, 34, 203, 1969.

172. **Watterson, J. J. and Butler, L. G.,** Occurrence of an unusual leucoanthocyanidin and absence of proanthocyanidins in sorghum leaves, *J. Agric. Food Chem.*, 31, 41, 1983.

173. **Haskins, F. A. and Gorz, H. J.,** Inheritance of leucoanthocyanidin content in sorghum leaves, *Crop Sci.*, 26, 286, 1986.

174. **Nip, W. K. and Burns, E. E.,** Pigment characterization in grain sorghum. I. Red varieties, *Cereal Chem.*, 46, 490, 1969.

175. **Cracker, L. E., Standley, L. A., and Starbuck, M. J.,** Ethylene control of anthocyanin synthesis in sorghum, *Plant. Physiol.*, 48, 349, 1971.

176. **Cracker, L. E.,** Ethylene, light and anthocyanin synthesis, *Plant Physiol.*, 51, 436, 1973.

177. **Downs, R. J. and Siegelman, H. W.,** Photocontrol of anthocyanin synthesis in milo seedlings, *Plant Physiol.*, 38, 25, 1963.

178. **Vince, D.**, Growth and anthocyanin synthesis in excised sorghum internodes. I. Effect of growth regulating substances, *Planta*, 82, 261, 1968.

179. **Indira, S., Rana, B. S., Rao, V. J. M., and Rao, N. G. P.**, Host plant resistance to rust in sorghum, *Indian J. Genet. Plant. Breed.*, 43, 193, 1983 (PBA 1985, 3487).

180. **Bhat, M. G., Gowda, B. T. S., Anahosur, K. H., and Goud, J. V.**, Inheritance of plant pigmentation and downy mildew resistance in sorghum, *SABRAO J.*, 14, 53, 1982 (PBA 1984, 5085).

181. **Molin, W. T., Anderson, E. J., and Porter, C. A.**, Effects of alachlor on anthocyanin and lignin synthesis in etiolated sorghum (*Sorghum bicolor* (L.) Moench.) mesocotyls, *Pestic. Biochem. Physiol.*, 25, 105, 1986.

182. **Hahn, D. H.**, Phenols of sorghum and maize: the effect of genotype and alkali processing, *Diss. Abstr. Int. B*, 45, 3674, 1984.

183. **Kambal, A. E. and Bate-Smith, E. C.**, A genetic and biochemical study on pericarp pigments in a cross between two cultivars of grain sorghum, *Sorghum bicolor, Heredity*, 37, 413, 1976.

184. **Hulse, J. H., Laing, E. M., and Pearson, O. E.**, *Sorghum and the Millets: Their Composition and Nutritive Value*, Academic Press, London, 1980, 28.

185. **Hahn, D. H., Rooney, L. W., and Earp, C. F.**, Tannins and phenols of sorghum, *Cereal Foods World*, 29, 776, 1984.

186. **Olifson, L. E., Nechaev, A. P., Osadchaya, N. L., and Mikhailova, L. F.**, Chemical nature of pigment from the integuments of sorghum grain, *Izv. Vyssh. Uchebn. Zaved., Pishch. Tekhnol.*, 1, 39, 1971 (CA 75, 22922).

187. **Olifson, L. E., Nechaev, A. P., Osadchaya, N. L., and Mikhailova, L. F.**, A new food dye made from sorghum grain and its toxicological characteristics, *Vopr. Pitan.*, 1, 76, 1978 (CA 88, 119568).

188. **Reichert, R. D., Young, C. G., and Christensen, D. A.**, Polyphenols in *Pennisetum* millet, in *Polyphenols in Cereals and Legumes*, Hulse, J. H., Ed., IDRC, Ottawa, Ont., Canada, 1980, 50.

189. **Hoseney, R. C. and Varriano,-Marston, E.**, Pearl millet: its chemistry and utilization, in *Cereals for Food and Beverages*, Inglett, G. E. and Munck, L., Eds., Academic Press, New York, 1980, 461.

190. **Gluchoff-Fiasson, K., Jay, M., and Viricel, M.-R.**, Detection of new flavonoid patterns in foxtail millet (*Setaria italica* (L.) Beauv.): comparison between pure lines and hybrids, *Biochem. Syst. Ecol.*, 18, 221, 1990.

191. **Raju, N. S. N., Rao, Y. S., Subba Rao, M. V., and Manga, V.**, Anthocyanidins of purple and sun-red pigmentation in pearl millet, (*Pennisetum americanum*), *Indian J. Bot.*, 8, 185, 1985 (CA 105, 187656).

192. **Shibata, M. and Sakai, E.**, Anthocyanin in the hair of spike or Japanese pearl millet (*Pennisetum japonicum*), *Bot. Mag. (Tokyo)*, 71, 193, 1958.

193. **Bobbio, F. O., Bobbio, P. A., and De Souza, S. C.**, Comparative stability of anthocyanins from *Panicum melinis* and *Solanum americanum, Bull. Liaison Groupe Polyphénols*, 14, 138, 1988.

194. **Nillson-Ehle, H.**, Kreuzungsuntersuchung an Hafer und Weizen, *Lunds Univ. Arsskr. Avd. 2*, 7(6), 3, 1911.

195. **Nillson-Ehle, H.**, Zur Kenntnis der mit der Keimungsphysiologie des Weizens in Zusammenhang stehenden inneren Faktoren, *Pflanzenzuechtung*, 2, 153, 1914.

196. **Fortmann, K. L. and Joiner, R. R.**, Wheat pigments and flour color, in *Wheat — Chemistry and Technology*, Pomeranz, Y., Ed., American Assocation of Cereal Chemistry, St. Paul, Minn., Vol. 3, Monogr. Ser., 1971.

197. **Terning, P. E.**, Anthocyanin colouring of the germ plants as a variety of character in wheat and a method for intensifying of the colour, *Medd. Statens Centr. Froekontrollanstalt*, 22, Stockholm, Sweden, 1947.

198. **Tusnjakova, M. M.**, Red-grained and red-eared forms of some wheat varieties, *Tr. Inst. Genet.*, 23, 145, 1956 (PBA 1957, 2786).

199. **Priilinn, O.**, A study of mutation variability in spring wheat on using chemical mutagens, *Industirov. Mutagenez Rast., Tallin, Estonia,* 1972, 284, 1972 (PBA 1973, 3444).

200. **Jonard, P.**, Relation entre la couleur du grain de diverses varietes de blé et leur aptitude a germer en moyettes, *Selectioneur,* 2, 41, 1933.

201. **Tandon, J. P., Dhillon, S. S., and Goswami, A. K.**, An association between locus for grain colour and protein content in wheat *(Triticum aestivum), Can. J. Genet. Cytol.,* 12, 28, 1970.

202. **Miyamoto, T. and Everson, E. H.**, Wheat seed pigmentation, *Agron. J.,* 50, 733, 1958.

203. **Rana, R. S. and Mathur, H. C.**, Enhanced outcrossing following mutagenic treatments in einkorn wheat, *Indian J. Genet.,* 27, 452, 1967 (PBA 1970, 7132).

204. **Bhowal, J. G. and Jha, M. P.**, An inhibitor of leaf and glume pigment in wheat, *Curr. Sci.,* 37, 568, 1968 (PBA 1969, 2094).

205. **Gale, M. D. and Flavell, R. B.**, The genetic control of anthocyanin biosynthesis by homoeologous chromosomes in wheat, *Genet. Res.,* 18, 237, 1971.

206. **Anon.,** Wheat, *Annu. Rep. Plant Breed. Inst., Cambridge, U.K., 1968 — Part IV,* p. 34, 1968.

207. **Raut, V. M., Kulkarni, V. S., Patil, V. P., and Deodokar, G. B.**, Genetic studies in tetraploid wheats. IV. Inheritance of anthocyanin pigmentation, *Indian J. Genet. Plant. Breed.,* 38, 199, 1978 (PBA 1979, 8810).

208. **Sheopuria, R. R. and Singh, S. P.**, Genetics of glume and seed colour of Newthatch wheat, *JNKVV Res. J.,* 4(1/2), 45, 1970 (PBA 1972, 4415).

209. **Sharman, B. C.**, Purple pericarp: a monofactorial dominant in tetraploid wheats, *Nature (London),* 181, 929, 1958.

210. **Villanueva Novoa, R.**, Inheritance of colour of auricles in wheat, *Bol. Trimestr. Exp. Agropec., Progr. Coop. Exp. Agropec., Lima,* 5(2), 25, 1956 (PBA 1957, 3918).

211. **McIntosh, R. A. and Baker, E. P.**, Inheritance of purple pericarp in wheat, *Proc. Linn. Soc. N.S.W.,* 92, 204, 1967.

212. **Paquet, J.**, Interaction de la coloration du tegument et de la texture de l'albumine dans la determination de l'intensité de la couleur du grain de blé, *Ann. Inst. Natl. Rech. Agron., Paris Ser. B,* 6, 27, 1956.

213. **Van Bragt, J., Brouner, J. B., and Zeven, A. C.**, The colour of the coleoptile of wheat. I. Anthocyanins of the coleoptiles of some *Triticinae, Wheat Inf. Serv.,* 25, 2, 1967.

214. **Vogel, F.**, Identification of anthocyanin pigments in wheat coleoptiles, *Schweiz. Land-wirtsch. Forsch.,* 9, 97, 1970 (FCA 1971, 127).

215. **Dedio, W., Hill, R. D., and Evans, L. E.**, Anthocyanin in the pericarp and coleoptiles of purple wheat, *Can. J. Plant Sci.,* 52, 977, 1972.

216. **Bolton, F. E.**, Inheritance of blue aleurone and purple pericarp in hexaploid wheat, *Diss. Abstr. Int., B,* 844, 19, 1968 (PBA 1970, 2684).

217. **Zeven, A. C.**, Wheats with purple and blue grains: a review, *Euphytica,* 56, 243, 1991.

218. **Gilchrist, J. A. and Sorrells, M. E.**, Inheritance of kernel colour in 'Charcoal' wheat, *J. Hered.,* 73, 457, 1982.

219. **McCullum, J. A. and Walker, J. R. L.**, Proanthocyanidins in wheat bran, *Cereal Chem.,* 67, 282, 1990.

220. **Feng, Y. and McDonald, C. E.**, Comparison of flavonoids in brain of flour classes of wheat, *Cereal Chem.,* 66, 516, 1989.

221. **Gutek, L. H., Woods, D. L., and Clark, K. W.**, Identification and inheritance of pigments of wild rice, *Crop. Sci.,* 21, 79, 1981.

Chapter 9

LEGUMES

I. BEAN

Von Portheim and Scholl[1] classified the pigment in *Phaseolus multiflorus* L. as an anthocyanin. Depending on the cultivar, beans show a large variety in the flavonoid content and color of the seed coat. Four colors are most common: black, brown, red, and white. These colors are correlated with the color of the hypocotyl, cotyledon, and leaf veins. For example, the black seed has a red hypocotyl.[2]

Baldovinos and Echegaray[3] isolated the pigments from the pericarp of *Ph. vulgaris* and measured the maxima of absorption between 542 and 576 nm. Recent patents[4,5] proposed the extraction of anthocyanins from black bean seeds for use as a colorant in foods and pharmaceuticals. The presence of anthocyanins in the hypocotyl of the plant is also a factor of resistance against attacks by *Macrophomina phaseolina*.[6]

The anthocyanin pigmentation in the hypocotyl of *Ph. vulgaris* requires three pairs of genes besides the base genetic factor *P*.[7] Anthocyanins may be produced in the leaves of beans by the action of ozone present in the atmosphere.[8] Robinson[9] attributed the origin of anthocyanin pigments to a precursor. The genotype determines the presence of the pigment. In some cases, more complementary genes are necessary, while in others a single dominant gene is sufficient to allow for the formation of the anthocyanin pigments.

In beans of *Ph. multiflorus* cv. Blue Coco, malvidin 3,5-diglucoside has been reported (Table 1).[10] The anthocyanins identified in *Ph. vulgaris* include delphinidin 3-glucoside, petunidin 3-glucoside, and malvidin 3-glucoside.[11,12] In 'Canadian Wonder' beans, Stanton and Francis[13] found pelargonidin 3-glucoside; cyanidin 3-glucoside; and some pelargonidin 3,5-diglucoside, delphinidin 3-glucoside, and cyanidin 3,5-diglucoside. The total flavonoid content was about 5% of the dry weight. These authors examined the bacteriostatic characteristics of the major anthocyanin of the seed coat, pelargonidin 3-glucoside, and found it to inhibit the growth of the yeast *Hansenula mrakii* but to stimulate the growth of two *Rhizobium* species.

Four anthocyanin pigments are responsible for the color in the cultivar Kintoki: pelargonidin 3-glucoside, pelargonidin 3,5-diglucoside, cyanidin 3-glucoside, and cyanidin 3,5-diglucoside.[14] In 'Kurodanekinugasa' beans, delphinidin 3-glucoside, petunidin 3-glucoside, malvidin 3-glucoside, and malvidin 3,5-diglucoside have been reported.[15]

Imrie and Hutton[16] identified malvidin, delphinidin, and petunidin as the anthocyanins of *Ph. atropurpureus* (= *Macroptilium atropurpureum*) flowers. The presence of anthocyanins in purple flowers of *Ph. acutifolius* is

TABLE 1
Anthocyanins of Selected Beans (*Phaseolus* spp.)

Species	Cultivar	Anthocyanin	Refs.
Ph. multiflorus	Blue Cuco	Malvidin 3,5-diglucoside	10
Ph. vulgaris	Several	Delphinidin 3-glucoside	11, 12
		Petunidin 3-glucoside	
		Malvidin 3-glucoside	
	Canadian Wonder	Pelargonidin 3-glucoside	13
		Pelargonidin 3,5-diglucoside	
		Cyanidin 3-glucoside	
		Cyanidin 3,5-diglucoside	
		Delphinidin 3-glucoside	
	Kintoki	Pelargonidin 3-glucoside	14
		Pelargonidin 3,5-diglucoside	
		Cyanidin 3-glucoside	
		Cyanidin 3,5-diglucoside	
	Kurodanekinugasa	Delphinidin 3-glucoside	15
		Petunidin 3-glucoside	
		Malvidin 3-glucoside	
		Malvidin 3,5-diglucoside	
Ph. atropurpureus	Unknown	Delphinidin glycoside	16
		Petunidin glycoside	
		Malvidin glycoside	
Ph. coccineus	Unknown	Malvidin 3,5-glucoside	27

controlled by a single dominant gene.[17] Okonkwo[18] studied the heritability of pigmentation of the pod and flowers of *Ph. vulgaris*; and Taware et el.[19] studied the genetics of hypocotyl, first leaf vein, petals, and seed coat color. Genetic studies on the pigmentation of the pod and its relationship to the pigmentation of other parts of the plant were also conducted by Vieira and co-workers.[20-22] Mutants of *Ph. vulgaris*, *Ph. trilobus*, and *Ph. aconitifolius* containing anthocyanins in the pod were obtained by irradiation treatments.[23-25]

Nozzolillo[26] studied the composition and biosynthesis of anthocyanins in legume seedlings and found malvidin in five of the seven species of *Phaseolus* investigated. Delphinidin was present in the other two species. In *Ph. vulgaris* and *Ph. coccineus*, the major pigment is malvidin 3,5-diglucoside.[27] Lima bean *(Ph. lunatus)* may contain anthocyanins in the hypocotyl, seed coat, and flowers; and their occurrence is determined by the dominant genetic factors *C*, *P*, and *R*.[28-30] The anthocyanidin present in the epidermal layer of seedlings grown in the light is malvidin.[27]

Anthocyanin pigmentation can provide characters useful to the taxonomy of *Phaeseolus-Vigna* complex. These include type of anthocyanidin present (malvidin in *Phaseolus* and cyanidin and/or delphinidin in *Vigna*, their location in the tissue, their rate of synthesis, and the presence or absence of anthocyanoplasts, but do not include intensity of coloration.[31] Recently, Hungria et al.[32] found the 3-glycosides of delphinidin, petunidin, malvidin, myricetin, quercetin, and kaempferol from seeds of black seeded *Ph. vulgaris* to act as nod gene-inducing factors in beans.

II. COWPEA AND MUNGBEAN

Cowpea, *Vigna unguiculata* (L.) Walp., is cultivated for its pods and seeds in the tropics and subtropics. In African countries, it is consumed as a boiled vegetable, as an ingredient in soups, and as a paste in steamed and fried dishes. In the U.S., immature seeds are used for canning. In India, it is consumed mostly as cooked green immature pods or cooked whole seeds in the form of curry with rice and other cereals. The seeds vary markedly in size, shape, and color. They are 2 to 12 mm long, globular or kidney shaped; smooth or wrinkled; white, green, buff, red, brown, or black; and variously speckled, mottled, blotched, or eyed.[33] The seed pigmentation is due to anthocyanins located in cells arranged in layers. In the colored pod there is only one layer of pigmented cells, but these pigmented cells are inner cells and retain their color during canning.[34,35]

The concentration of anthocyanin in cowpeas can be relatively high; however, the accumulation of these pigments does not appear to be related to the biosynthesis of proteins.[36] The principal pigments of cowpeas are glycosides of cyanidin with some delphinidin and malvidin present in some varieties.[31,36] The anthocyanin pigmentation is present in almost all the cultivars of cowpeas. Of the 1070 varieties examined by Mehra et al.[37], 99% had anthocyanins in at least one of the five plant parts examined. It also appears that the presence of anthocyanins in the seed of cowpeas is related to geographic distribution of this minor food legume.[38]

Smith[39] distinguished three dominant genes, *W* (Watson), *H* (Holstein), and *B* (Blood) that regulate the color of the seed coat; while Saunders[40] found that the purple was the dominant color over all the others, including black. Purple in the pod is, however, caused by two complementary genes.[41] According to Sen and Bhowal,[42] the purple pigmentation in the pods and the violet color in the flowers are produced by the same genetic factors which also occur in *V. catjang* and *V. sesquipedalis*. A cross between purple and green varieties of cowpeas gives a dominant red color.[43] Based on the anthocyanin pigmentation in the hypocotyl, Miyazaki et al.[44] and Miyazaki and Kawakami[45] distinguished two taxa, *V. radiata* var. *sublobata* which is probably the progenitor of *V. radiata* and *V. mungo*.

Louis and Kadambavan[46] obtained a mutant of cowpea lacking, totally or partially, anthocyanin pigmentation in the eye as a function of the amount of γ-radiation, while Jana and Rao[47] had the same results with *V. mungo* by combining irradiation and chemical treatment.

Mungbean, *V. mungo* (L.) Hepper (formerly called *Phaseolus mungo* L.), is a tropical plant grown in all the Asiatic countries for food use. In the U.S.,

TABLE 2
Anthocyanins in Plant Parts of Selected *Vigna* spp.

Species	Plant part	Pigment	Refs.
V. unguiculata	Seed coat	Cyanidin glycoside	31, 35
		Delphinidin glycoside	
		Malvidin glycoside	
V. mungo	Hypocotyl	Delphinidin 3-(*p*-coumaryl)-glucoside	49
	Seed coat	Delphinidin 3-glucoside	49, 51, 52
		Cyanidin 3-glucoside	
V. radiata	Hypocotyl	Delphinidin 3-(*p*-coumaryl)-glucoside	31, 49
	Cotyledon	Delphinidin 3,5-diglucoside	50
	Seed coat	Delphinidin 3-glucoside	31, 49, 51, 52
		Cyanidin 3-glucoside	
V. angularis	Seed coat	Delphinidin 3-glucoside	60

it is used primarily as bean sprouts in salads.[48] In various varieties of *V. mungo* and *V. radiata*, Ishikura et al.[49] identified delphinidin 3-(*p*-coumaryl) glucoside as the predominant pigment in the red/purple hypocotyl, while cyanidin 3-glucoside is the pigment in the hypocotyl of *V. mungo*. Delphinidin 3-glucoside is found in all the plants, particularly in the black seed coat of both *V. mungo* and *V. radiata* (Table 2).

The first complete identification of the anthocyanins in *V. radiata* (L). Wilczek or greengram was carried out using the sprouts. According to Gevers,[50] the pigment is delphinidin. The cotyledons contain delphinidin 3,5-digluco-side, possibly acylated with *p*- coumaric acid,[51] while the hypocotyl contains mainly delphinidin 3-glucoside.[31] Pandey et al.[52] identified variation in anthocyanins in the seed coat and hypocotyl of *V. radiata* and *V. mungo*. Delphinidin and cyanidin 3-glucosides plus other two anthocyanins were found.

The synthesis of anthocyanins in *V. mungo* is controlled by the phytochromes and is inhibited when treated with Methyrapone (SU-4885). A dose of 200 ppm inhibits 90% of the synthesis.[53] Some herbicides have also been found to inhibit the synthesis of anthocyanin in mung bean hypocotyls.[54] In *V. radiata*, Dumortier and Vendrig[55,56] used a chemical treatment (dimethyl-formamide and dimethylsulfoxide) to stimulate the production of anthocyanins.

The anthocyanin pigmentation in mung bean seedlings *(V. radiata)* is governed by two polymeric dominant genes.[57] Watt[58] studied the heritability of the seed coat color in many beans while Viswanatha[59] studied the heritability of anthocyanin pigmentation in *V. unguiculata*. In adzuki beans (*Ph. angularis* = *V. angularis*), Sasanuma et al.[60] extracted and identified delphinidin 3-glucoside from the black seed coat. In 1934, Kuroda and Wada[61] identified phloroglucin and protocatechuic acid as the products of degradation of anthocyanins in adzuki beans, and Fujiwara and Kajita[62] studied the color characteristics of processed products made from the adzuki bean.

III. FABABEAN

The anthocyanin pigmentation of faba flowers, *Vicia faba* (L.), is characterized by a spotted arrangement of the pigment (almost black in color), called anthophaein by Moebius[63] and vicin by Karrer and Widmer,[64] who characterized it as a mixture of glucosides and rhamnosides of delphinidin.

The presence of anthocyanins is correlated with a low sugar content in the flower nectar[65] and is regulated, according to Goyal,[66] by two genes *a1* and *a2* that control the formation in the flower and fruit. According to Rowlands and Corner,[67] a third gene seems to be required, and this serves to complement the two basic genes that control the synthesis of anthocyanins. Other genes apparently regulate the intensity and distribution of the anthocyanins in fababeans.[68]

The seed color can be violet, green, brown, or black. The violet pigment which develops only at the end of ripening is due to anthocyanins and is localized in the upper half of the cellular cavity in the palisade layer; the brown pigment is due to tannins and is localized in the cuticle of the inner and outer parts of the palisade and of the parenchyma, while the green color is due to the presence of chloroplasts.[69] According to Picard,[70] the green and red pigmentation of the seed coat is recessive, while violet, marbled, and black colors are dominant.

Crofts et al.[71] used anthocyanins as genetic markers for the selection of fababean cultivars free of tannins. Picard[72] also found that plants lacking anthocyanin pigmentation in the flowers produced seeds lacking tannins. Similar observations were made by Moreno Yanguela and Cubero Salmeron.[73] Harbone[10] reported that all species of *Vicia* contain delphinidin, petunidin, and malvidin 3-rhamnoside-5-glucosides. Pecket[74] observed that flavonol glycosides play a role in *Vicia* flower color by their action as copigments of the anthocyanins. Recently, the flavonoids rutin, quercitrin, hyperoside, quercetin 3-arabinoside, kaempferol 3-rutinoside, and robinin (kaempferol-3-robinoside-7-rhamnoside) were identified in leaves of six *Vicia* species, and their pattern was utilized to draw taxonomic relationships between *Vicia* species.[75]

IV. PEA

Statham et al.[76] isolated and characterized the 3-sophoroside-5-glucosides and the 3-sambubioside-5-glucosides of cyanidin and delphinidin in the purple pod of *Pisum sativum* L. Other authors have reported 3-rhamnoglucoside-5-glucosides of delphinidin, petunidin, and malvidin.[10]

In pea pods, Nakatani et al.[77] extracted mainly cyanidin and delphinidin glycosides, with minor quantities of malvidin and pelargonidin derivatives responsible for the various color shades ranging from orange to red, blue, and purple. By crossing, Yaeger and Meader[78] also obtained peas with purple pods.

Genetic studies of the pigments of *P. sativum* have shown that one gene, *B*, controls the hydroxylation at the 5' position of the anthocyanidin.[79] Another gene, *Cr,* controls methylation and glycosylation.[10] Other genes involved in pea pigmentation determine the color of the testa, seed, cotyledon,[80] and pod.[81] The flowers contain other pigments including 3-sambubiosides rather than 3-sophorosides.[82] Color expression in flowers is controlled by the *A* gene, whereas plants homozygous for the *a* allele do not produce anthocyanins. According to Hrazdina and Weeden,[83] however, this locus does not seem to control chalcone synthase, the first enzyme of the flavonoid pathway. A white-petaled mutant contains malvidin 3-rhamnoglucoside in the uncolored pseudobase form.[84]

Nowacki[85] observed that at the precursor stage anthocyanins and flavonols compete for the pigmentation of the flower. The anthocyanins in pea sprouts tend to preferentially form in the basal part of the internodes in response to genetic and environmental factors. The anthocyanins are primarily derivatives of delphinidin followed by cyanidin.[86] There is no genetic relationship between the presence of anthocyanins and stem length.[87] Anthocyanins in germinated peas provide a resistance factor against attacks by *Pythium ultimum*,[88,89] anthracnoses, *Ascochyta pisi*,[90] and pathogenic fungi such as *Mycosphaerella pinodes*.[91] They also may influence the tannin content of the seed coat.[92] Anthocyanins also provide a valid index for the genetic selection of cultivars with a high canning quality[93] or with low levels of condensed tannins.[94]

Hrazdina et al.[95] found anthocyanins and flavonol glycosides to be localized only in the vacuoles of epidermal cells of A 681-230 peas, but the enzymes involved in their biosynthesis were found in both the epidermal and the parenchyma cells.

V. SOYBEAN

Soybeans, *Glycine max* Benth., are a most important source of edible plant protein and are covered by a pigmented layer. The anthocyanins that cause this coloration are located in the epidermis of the palisade layer that makes up the testa of the soybean.[96]

Anthocyanins were first described in soybeans by Nagai.[97] Kuroda and Wada[98-100] and Manabe et al.[101] found that cyanidin 3-glucoside is the major anthocyanin of the seed coat of soybean seeds. Yoshikura and Hamaguchi[102] identified delphinidin 3-glucoside and cyanidin 3-glucoside in the black-seeded variety. Taylor[103] identified pelargonidin 3-glucoside in the reddish buff seed coats of 'T236' soybeans.

The dark pigmentation in the seed coat of black cultivars is due to the presence of a pleiotropic pubescence color allele *T* that controls the synthesis of cyanidin 3-glucoside. The flower color is due to allele *W* that is believed to be responsible for the trihydroxylation of delphinidin 3-glucoside.[104] The instability of a mutable allele at the W_4 locus has been attributed to a transposable element that produces a line chimeric for anthocyanin pigmentation.[105,106]

The infection by soy mosaic virus (SMV)[107] and Cd pollution[108] are among the factors that increase pigmentation. In a study on the precursors of phenolic compounds of the soybean, Hoagland[109] found that anthocyanin production was increased by amino methyl phosphoric acid (AMPA) and was depressed by glyphosate and glyphosine.

Cyanidin 3-glucoside is believed to serve as an antioxidant in toshi oil, a fermented soy sauce obtained by *Aspergillus soyae* and used in Chinese cooking to garnish fish and pork dishes.[110] This anthocyanin was recently separated from the black soybean using isotachoelectrophoresis.[111]

Malvidin is the purple pigment in the stem of some wild varieties, and apparently is responsible for the resistance of these varieties to bean flies.[112]

The anthocyanins which form by photoinduction in soybean sprouts are glycosides of malvidin.[27] Using HPLC, delphinidin, malvidin, and petunidin can be separated from the purple and bronze hypocotyls of soybeans. Malvidin is the most frequently occurring anthocyanidin, and this pigment is five times more concentrated in the purple than in the bronze form.[113]

Soybean flowers range in color from white to dark purple. Hartwig and Hinson[114] reported that the purple color of the hypocotyl is dominant over green, purple petal color is dominant over white, and black color of the hyle is dominant over brown color. These colors are regulated by complementary genes.[115] According to recent reports,[104-106,116] anthocyanin synthesis in flowers is not dependent upon a single gene, *W*, as was proposed by Bhatt and Torrie;[117] instead its expression is determined by other complementary genes which control the anthocyanin distribution in the various plant parts.

VI. SWEET AND GRASS PEA

The interest in *Lathyrus* pigments arises from the studies by Beale et al.[118] on genetics of flower color in the sweet pea (*L. odoratus* L.). Previously, Karrer and Widmer[64] isolated and partially characterized delphinidin 3-rhamnoside in the same species. Since 1960, however, the pigments of *Lathyrus* species have been fully elucidated. Harborne[119,120] identified malvidin 3-rhamnoside in petals of *L. sativus* or grass pea, and 19 anthocyanins in petals of *L. odoratus*. The pigments in petals of mauve and purple forms of sweet peas are pelargonidin, cyanidin, peonidin, delphinidin, petunidin, and malvidin 3-rhamnoside-5-glucosides; pelargonidin, cyanidin, and peonidin 3-galactosides; pelargonidin and peonidin 3-galactoside-5-glucosides; pelargonidin, cyanidin, and peonidin 3-(2^G-xylosylgalactosides) or 3-lathyrosides. More recently, Reichel and Hiller[122] isolated 11 pigments in petals of 'Valencia' sweet peas and characterized: cyanidin 3-glucoside, peonidin 3-glucoside, peonidin 3,5-diglucoside, peonidin triglucoside, pelargonidin 3-glucoside, pelargonidin 3,5-diglucoside, and two pelargonidin triglucosides.

The color of flowers reflects the anthocyanin and flavonol pattern. When the color is red, there are delphinidin, petunidin, malvidin, kaempferol, and quercetin in the petals; when the color is pink, cyanidin and peonidin together with quercetin and kaempferol are present; pelargonidin and kaempferol are present in salmon flowers.[119,123-125]

Arkatov et al.[126] studied the anthocyanin composition of flowers in 112 sweet pea varieties and concluded that all forms examined could be classified in three main classes containing 1, 2, and 3 anthocyanidins, respectively: (1) pelargonidin; (2) cyanidin and peonidin; (3) delphinidin, petunidin, and malvidin.

Pecket[127] found that color variation during flower senescence in *L. hirsutus* was due to a change in flavonol:anthocyanin ratio with a loss of the latter. Sarkar and Banerjee[128] identified cyanidin and pelargonidin 3,5-diglucosides in the seeds of *L. sativus*.

VII. OTHER LEGUMES

A. CHICKPEA

Balasubrahmanyam[129] studied the heritability of pink, blue, and white pigmentation in the flower petals of *Cicer arietinum* L. It was found that it is controlled by three factors: *C* and *B* combined give a blue color and *P* associated with *B* and *C* gives red color. When *B* or *C* is lacking, white results. A factor *A* gives a blue-brown color to the testa that becomes red-brown with the effect of gene *R*.

The blue color of the flower appears to be associated with sterility, while pink is not.[130] The color of the corolla (pink, blue, salmon, and white) is, however, controlled by two genes.[131] In the cultivar Chikodi, variation in the color of petals is controlled by gene *P*, which controls the violet and red color of the petals as well as the pink and violet color of the corolla.[132]

The anthocyanins responsible for the color of the petals of *C. pinnatifolia* are delphinidin, petunidin, and malvidin 3-rhamnoside-5-glucosides; and this glycosidic pattern is typical of the Viciaceae.[10] The flavonol pigments of *C. arietinum* leaves and stems are the 3-*O*-glucosides of kaempferol, quercetin, and isorhamnetin; kaempferol 3-*O*-apiosylglucoside; kaempferol 3-(malonylglucoside); and kaempferol 3-(apiosylmalonylglucoside).[133]

B. LENTIL

Lentils (*Lens culinaris* Medic.) with red and yellow cotyledons can be easily distinguished by examining the seed coat under a UV light. In a cross between the two, red is dominant over yellow.[134] Plantlets of the high-yield variety, Laird, are pigmented with anthocyanins.[135] Nozzolillo[27] identified delphinidin as the aglycone of the subepidermal pigment in the stem of *L. culinaris*.

D'Arcy and Jay[136] confirmed that the anthocyanin in the lentil seeds is a diglucoside of delphinidin. It is present in such a small quantity (about 60 ppm) that it is impossible to make a more precise identification. Tricetin, luteolin, and two proanthocyanidins are also present. The cotyledon, however, contains primarily kaempferol and 5-deoxykaempferol.[136]

C. LUPINE

Lupine, *Lupinus* spp. *Genistae*, is a widely distributed legume with a long history of use as a human food.[137] Harbone[10] reported that the glucosidic arrangement of the anthocyanins in *Lupinus* spp. was quite simple, 3-glucosides or 3,5-diglucosides as in all the Papilionatae. Baeyer[138,139] identified pelargonidin 3,5-diglucoside in the blue flowers of *L. polyphemus*.

An earlier chromatographic analysis by Bragdø[140] showed the presence of pelargonidin, cyanidin, and delphinidin glycosides, in which the degree of hydroxylation of the anthocyanin was regulated by the gene *B*. The genetics of anthocyanic pigmentation in lupine flowers is complex. Al-Bassam[141] studied the disappearance of pigmentation by mutation in *L. polyphyllus*, while Forbes et al.[142] were interested in the relationship between anthocyanin pigmentation in *L. angustifolius* L. and resistance to parasitic attack. Wozny et al.[143] reported a minor quantity of anthocyanins in *L. luteus* plants that had been treated with lead salts.

D. WINGED BEAN

The winged bean, *Psophocarpus tetragonolobus* (L.) DC, is a less-known high protein plant from the family Leguminosae. This plant seems to have originated in Madagascar, and it is currently diffused throughout Asia, tropical Africa, the Caribbean, and tropical America. The importance of this crop is due to the fact that all the plant organs (seeds, mature and immature pods, tubers, leaves, flowers) are edible.[144]

The color of the testa in winged bean varies from white to black by way of green, tan, brown, and purple.[145] The flower color varies from white to blue to violet. The variety Tp-1 has light blue flowers with blue stripes, while the variety Chimbu has creamy to brown flowers and creamy to purple seeds.[146] In samples from Java, Sastrapradja et al.[147] observed that plants with violet flowers produced seeds with violet seed coats, whereas plants with blue flowers produced seeds with tan to brown seed coats. Erskine[148] found that purple stem color was dominant over green. Le Hing et al.[149] found 11 flavonoids in the seed coat of the cultivar Binh Minh, and identified apigenin-7,4'-*O*-rutinoside, apigenin-6-*C*-glycoside, apigenin-8-*C*-glycoside, isovitexin, apigenin 4'-*O*-rhamnoside, and rutin. In the variety Chimbu, Lachman et al.[150] identified malvidin 3-*O*-rhamnoside-5-*O*-glucoside as the main phenolic compound in the peel of winged beans.

REFERENCES

1. **Von Portheim, L. and Scholl, E.,** An investigation concerning the formation and chemistry of anthocyanin, *Ber. Bot. Ges.,* 26, 480, 1908 (CA 3, 1537).

2. **Anon.,** Developing bean varieties for Latin America needs, *Activ. Turrialba,* 2, 3, 1974 (PBA 1975, 7860).

3. **Baldovinos, G. and Echegaray, A. A.,** Possible determining factors in the pigmentation of the pericarp of *Phaseolus vulgaris, Biol. Tec. Soc. Agron. Mex.,* 1, 39, 1954 (CA 49, 4093b).

4. **Hoffmann, P.,** Natural Food Colorant, *PCT Int. Appl.,* WO 820035, 1982 (CA 96, 121213).

5. **Cai, C., Dong, H., Dong, X., and Dong, T.,** Extraction method for polyphenol edible natural pigment, *Faming Zhuanli Shenquing Gankai Shuomingshu,* 1, 031, 549, 1989 (CA 113, 76888).

6. **Gonzalez-Avila, M., Aladro, R., and Marrero, H.,** Correspondence between the hypocotyl anthocyanin content in different types of French beans and infections by *Macrophomina phaseolina, Cienc. Agric., Havana,* 9, 79, 1981 (PBA 1983, 1722).

7. **Nakayama, R.,** Genetical studies on kidney beans *(Phaseolus vulgaris).* II. On the inheritance of hypocotyl color. I. *Hirosaki Daigaku Nogakubu Gakujutsu Hokoku,* 4, 80, 1958 (PBA 1959, 3248).

8. **Howell, R. K. and Kremer, D. F.,** Chemistry and physiology of pigmentation in leaves injured by air pollution, *J. Environ. Qual.,* 2, 434, 1973 (CA 81, 10222).

9. **Robinson, R.,** Formation of anthocyanins in plants, *Nature (London),* 137, 172, 1936.

10. **Harborne, J. B.,** *Comparative Biochemistry of Flavonoid Compounds,* Academic Press, London, 1967.

11. **Feenstra, W. J.,** Chemical aspects of the action of three seed coat color genes of *Phaseolus vulgaris, K. Ned. Akad. Wet. Proc.,* 62C, 119, 1959 (CA 53, 16286f).

12. **Feenstra, W. J.,** Biochemical aspects of seedcoat colour inheritance in *Phaseolus vulgaris* L., *Meded. Landbouwhogesch., Wageningen,* 60, 53, 1960 (CA 58, 751c).

13. **Stanton, W. R. and Francis, B. J.,** Ecological significance of anthocyanins in the seed coats of the Phaseolae, *Nature (London),* 211, 970, 1966.

14. **Yoshikura, K. and Hamaguchi, Y.,** Anthocyanins of kidney beans (Kintoki), *Eiyo To Shokuryo,* 24, 275, 1971 (CA 75, 148494).

15. **Okita, C., Suwa, H. K., Yoshikura, K., and Hamaguchi, Y.,** Anthocyanins of *Phaseolus vulgaris,* cv. Kurodanekinugasa, *Eiyo To Shokuryo,* 25, 427, 1972 (CA 77, 137394).

16. **Imrie, B. C. and Hutton, E. M.,** The inheritance of flower color in *Macroptilium atropurpureum, J. Hered.,* 69, 54, 1978.

17. **Thomas, C. V.,** Genetic, morphological and physiological studies of drought and heat resistance in tepary beans *(Phaseolus acutifolius* A. Gray) and common beans *(P. vulgaris* L.), *Diss. Abstr. Int. B,* 44, 410, 1983 (PBA 1984, 1963).

18. **Okonkwo, C. A.,** Genetics of flower and pod color in *Phaseolus vulgaris* L. and studies of natural and artificial pollination in *P. vulgaris* and *P. coccineus* L., *Diss. Abstr. Int. B,* 44, 3651, 1984 (PBA 1985, 3006).

19. **Taware, S. P., Varghese, P., and Patil, V. P.,** Genetic studies on French bean, *Phaseolus vulgaris* L., *Biovigyanam,* 13, 70, 1987 (PBA 1989, 8003).

20. **Sanchez, A. L. and Vieira, C.,** Inheritance of pod colour in *Phaseolus vulgaris* L., *Ceres, Minas Gerais,* 12, 106, 1964 (PBA 1965, 7613).

21. **De Moraes, C. F. and Vieira, C.,** Inheritance of pod colour in *Phaseolus vulgaris* L., *Ceres, Minas Gerais,* 15, 199, 1968 (PBA 1970, 4069).

22. **Vieira, C.,** Inheritance of pod colour in *Phaseolus vulgaris* L., *Ceres, Minas Gerais,* 16, 63, 1969 (PBA 1970, 6369).

23. **Gutenmaher, P. and Priadcencu, A.,** Valuable Frenchbean mutants by irradiation, *Lucr. Stiint., Inst. Agron. Bucuresti, Ser. A,* 16, 9, 1973 (PBA 1976, 9710).

24. **Rangaswami Ayyanagar, K. and Subramanian, D.,** Gamma radiation induced mutants in *Phaseolus trilobus* Ait. (panyppayir) and *P. aconitifolius* Jacq. (mat bean), *Symp. Use Radiations Radioisot. Studies Plant Productivity, India,* 1974, 27 (PBA 1975, 995).

25. **Klein, W. H., Withrow, R. B., Elstad, V., and Proce, L.,** Photocontrol of growth and pigment synthesis in bean seedlings as related to irradiance and wavelength, *Am. J. Bot.,* 44, 15, 1957.

26. **Nozzolillo, C.,** A study of anthocyanins in seedling legumes, *Can. J. Bot.,* 51, 911, 1973.

27. **Nozzolillo, C.,** Anthocyanin pigments in bean seedlings, *Phytochemistry,* 10, 2552, 1971.

28. **Allard, R. W.,** Inheritance of some seed-coat colors and patterns in lima beans, *Hilgardia,* 22, 167, 1953.

29. **Allard, R. W.,** Genes modifying the *Cc* and *Rr* loci in lima beans, *Proc. Am. Soc. Hortic. Sci.,* 68, 386, 1956.

30. **Bemis, W. P.,** Inheritance of a base seed-coat color factor in lima beans, *J. Hered.,* 48, 124, 1957.

31. **Nozzolillo, C. and McNeill, J.,** Anthocyanin pigmentation in seedlings of selected species of *Phaseolus* and *Vigna (Fabaceae), Can. J. Bot.,* 63, 1066, 1985.

32. **Hungria, M., Joseph, C. M., and Phillips, D. A.,** Anthocyanidins and flavonols, major nod gene inducers from seeds of a black-seeded common bean (*Phaseolus vulgaris* L.), *Plant Physiol.,* 97, 751, 1991.

33. **Chavon, J. K., Kadam, S. S., and Salunkhe, D. K.,** Cowpea, in *Handbook of World Food Legumes,* Salunkhe, D. K. and Kadam, S. S., Eds., CRC Press, Boca Raton, FL, 1989, 1.

34. **Burns, E. E. and Winzer, J. W.,** Pigment characteristics of the southern pea, *Vigna sinensis, Proc. Am. Soc. Hortic. Sci.,* 80, 449, 1962.

35. **Culver, W. H. and Cain, R. F.,** Nature, causes, and correction of discoloration of canned blackeye and purple-hull peas (field peas), *Tex. Agric. Exp. Stn. Bull.,* 748, 5, 1952 (CA 47, 16291b).

36. **Bhatt, J. P.,** Protein and Amino Acid Contents of Southern Peas *(Vigna sinensis)* and Their Relation to Anthocyanin Development, Ph.D. thesis, Texas A & M University, College Station, TX, 1971 (CA 75, 95394).

37. **Mehra, K. L., Singh, C. B., Kohli, K. S., and Magoon, M. L.,** Divergence and distribution of pigmentation on plant parts in a world collection of cowpea, *Vigna sinensis, Genet. Iber.,* 26/27, 79, 1975 (PBA 1976, 9726).

38. **Anon.,** Centre Agricultural Research Surinam, 1970 Annu. Rep., *CELOS Bull.,* 13, 50 pp. (PBA 1972, 7005).

39. **Smith, F. L.,** Inheritance of three seed-coat color genes in *Vigna sinensis* Savi., *Hilgardia,* 24, 279, 1956.

40. **Saunders, A. R.,** Inheritance in the cowpea (*Vigna sinensis* Endb.). I. Colour of the seed coat, *S. Afr. J. Agric. Sci.,* 2, 285, 1959 (PBA 1960, 4442).

41. **Saunders, A. R.,** Inheritance in the cowpea (*Vigna sinensis* Endb.). II. Seed coat colour pattern: flower, plant, and pod colour, *S. Afri. J. Agric. Sci.,* 3, 141, 1960 (PBA 1961, 2749).

42. **Sen, N. K. and Bhowal, J. G.,** Genetics of *Vigna sinensis* (L.) Savi, *Genetics,* 32, 247, 1961 (PBA 1960, 5598).

43. **Venugopal, R. and Goud, J. V.,** Inheritance of pigmentation in cowpea, *Curr. Sci.,* 46, 277, 1977 (PBA 1977, 12196).

44. **Miyazaki, S., Kawakami, J., and Ishikura, N.,** Phytogenetic relationship and classification of *Vigna radiata-mungo* complex, *JARQ,* 17, 225, 1984.

45. **Miyazaki, S. and Kawakani, J.,** Differences in several characters among mungbean, black gram and their wild derivatives, *Jpn. J. Trop. Agric.,* 25, 1, 1981 (PBA 1982, 5382).

46. **Louis, I. H. and Kadambavan, S. M.,** Induction of white eye mutant in cowpea (*Vigna sinensis* (L.) Savi, *Madras Agric. J.,* 62, 95, 1975 (PBA 1976, 10710).

47. **Jana, M. K. and Rao, S. A.,** Anthocyaninless mutant of black gram *(Phaseolus mungo),*
    Mutat. Breed. Newslett., 3, 11, 1974 (PBA 1975, 7859).
48. **Duke, J. A.,** *Handbook of Legumes of World Economic Importance,* Plenum Press, New
    York, 1981.
49. **Ishikura, N., Iwata, M., and Miyazaki, S.,** Flavonoids of some *Vigna* plants in Leg-
    uminosae, *Bot. Mag. (Tokyo),* 94, 197, 1981 (CA 95, 183936).
50. **Gevers, R.,** Interakties tussen phytochroom en groeirregulatoren, *Licentiaatsverandeling
    K.U. Leuven, Belg.,* 63, 1973.
51. **Proctor, J. T. A.,** Photocontrol of anthocyanins synthesis in the apple and in the mung-
    bean, *Diss. Abstr.,* 1970.
52. **Pandey, R. N., Pawar, S. E., Chintalwar, G. J., and Bhatia, D. R.,** Seed coat and
    hypocotyl pigments in greengram and blackgram, *Proc. Indian Acad. Sci., Plant Sci.,*
    99, 301, 1989 (CA 112, 155392).
53. **Geuns, J. M. C.,** Methyrapone and the inhibition of growth, anthocyanin, carotenoid
    and chlorophyll biosynthesis in mung bean seedlings, *Biochem. Physiol. Pflanz.,* 171,
    435, 1977 (CA 87, 178368).
54. **Mito, N., Yoshida, R., and Oshio, H.,** Effects of hormone-type herbicides on proton
    extinction and anthocyanin synthesis by mung bean hypocotyl sections, *Zasso Kenkyu,*
    35, 332, 1990 (CA 114, 223468).
55. **Dumortier, F. M. and Vendrig, J. C.,** Effect of 'penetrant-carriers' in relation to the
    photocontrol of anthocyanin synthesis in mung beans *(Phaseolus aureus* Roxb.), *Z.
    Pflanzenphysiol.,* 87, 313, 1978 (CA 89, 1016).
56. **Dumortier, F. M. and Vendrig, J. C.,** Excretion of a natural inhibitor of anthocyanin
    synthesis during germination of mung bean seedlings, *Can. J. Bot.,* 61, 3279, 1983.
57. **Yadav, R. D. S.,** Genetics of anthocyanin pigmentation in mungbean *(Vigna radiata*
    (L.) Wilczek, *Indian J. Genet. Plant Breed.,* 47, 297, 1987 (PBA 1990, 1989).
58. **Watt, E. E.,** The composition and origin of the texture layer and the inheritance of colour
    and luster of mungbean seeds, *Diss. Abstr. Int. B,* 36, 3163, 1976 (CA 84, 118526).
59. **Viswanatha, K. P.,** Inheritance studies of a few qualitative characters in cowpea *(Vigna
    unguiculata* (L.) Walp.), *Thesis Abstr.,* 6, 241, 1980 (PBA 1982, 916).
60. **Sasanuma, S., Takeda, K., and Hayashi, K.,** Anthocyanins. LV. Black red pigment
    of adzuki bean, *Bot. Mag. (Tokyo),* 79, 807, 1966 (CA 67, 51019).
61. **Kuroda, C. and Wada, M.,** The colouring matter of adzuki bean, *Proc. Imp. Acad.
    (Tokyo),* 10, 472, 1934 (CA 29, 1134).
62. **Fujiwara, T. and Kajita, T.,** Studies on the color tone of ann (bean jam). 1. Analysis
    of adzuki husk pigments and effect of water quality on the color tone of adzuki ann,
    Kaisegaku Kenkyu, 34, 1, 1987 (CA 108, 36480).
63. **Moebius, M.,** Die Anthophaein, der braune Bluethenfarbstoff, *Ber. Dtsch. Bot. Ges.,*
    18, 341, 1900.
64. **Karrer, P. and Widmer, R.,** Untersuchungen über die Pflanzenstoffe, *Helv. Chim.
    Acta,* 10, 67, 1927.
65. **Smaragdova, N. P., Stepanova, L. N., and Borodina, L. N.,** On the biology of fodder
    broad beans *(Vicia faba* L.) in the Chernozem belt, in *Biology and Cultivation of Agri-
    cultural Crops,* Moskow Univ., Moskow, U.S.S.R., 1966, 101 (PBA 1969, 6083).
66. **Goyal, R. D.,** Breeding behaviour of anthocyanin pigmentation in flower and stem of
    broad bean *(Vicia faba* Linn.), *Sci. Cult.,* 31, 147, 1965 (PBA 1966, 1183).
67. **Rowlands, D. G. and Corner, J. J.,** Genetics of pigmentation in broad beans, *Proc.
    1st Int. Congr. Food Sci. Technol.,* London, II, 1962, 151.
68. **Clark, E. M. and Donnelly, E. D.,** *Vicia:* inheritance of plant color and recessive
    epistasis in flower, *Crop Sci.,* 4, 661, 1964.
69. **Erith, A. G.,** The inheritance of color, size, form of seed and flower color in *Vicia faba*
    L., *Genetics,* 13, 477, 1930.
70. **Picard, J.,** La coloration des teguments du grain chez la fevérole *(Vicia faba* L.). Etude
    de l'heredité des differentes colorations, *Ann. Amelior. Plant.,* 13, 97, 1963 (PBA 1964,
    3173).

71. **Crofts, H. J., Evans, L. E., and McVetty, P. B. E.**, Inheritance, characterization and selection of tannin-free fababeans (*Vicia faba* L.), *Can. J. Plant Sci.*, 60, 1135, 1980.

72. **Picard, J.**, Aperçu sur l' heredité du caractère absence de tannins dans les grains de fevérole (*Vicia faba* L.), *Ann. Amelior. Plant.*, 26, 101, 1976 (PBA 1976, 6667).

73. **Moreno Yanguela, M. T. and Cubero Salmeron, J. I.**, The inheritance of qualitative characters in *Vicia faba* L., *An. Inst. Nac. Invest. Agrar. (Spain) Prod. Ser. Veg.*, 1, 61, 1971 (PBA 1973, 3201).

74. **Pecket, R. C.**, Variation in flower color in *Vicia, Nature (London)*, 213, 1240, 1967.

75. **Perrino, P., Maruca, G., Linsalata, V., Bianco, V. V., Lester, R. N., and Lattanzio, V.**, Flavonoid taxonomic analysis of *Vicia* species of section *Faba, Can. J. Bot.*, 67, 3529, 1989.

76. **Statham, C. M., Crowden, R. K., and Harborne, J. B.**, Biochemical genetics of pigmentation in *Pisum sativum, Phytochemistry*, 11, 1083, 1972.

77. **Nakatani, M., Yamamoto, Y., and Hori, H.**, Extraction of Anthocyanin Pigments from Pea, Japanese Patent, 01,09,271, 1989 (CA 111, 150765).

78. **Yeager, A. F. and Meader, E. M.**, Breeding new vegetable varieties, *Bull. N.H. Agric. Exp. Stn.*, 440, 33p., 1957 (PBA 1858, 3453).

79. **Dodds, K. S. and Harborne, J. B.**, *Annu. Rep. John Innes Hortic. Inst.*, 1964, 34.

80. **Lamprecht, H.**, The inheritance of colours of *a* seeds of *Pisum, Agric. Hortic. Genet., Landskrona*, 17, 1, 1959 (PBA 1959, 4650).

81. **Lamprecht, H.**, Anthocyanin-coloured stripe along the nerve and anthocyanin-coloured speckling of *Pisum* pods, and the inheritance of these characters, *Agric. Hortic. Genet., Landskrona*, 21, 149, 1963 (PBA 1965, 1317).

82. **Statham, C. M. and Crowden, R. K.**, Anthocyanin biosynthesis in *Pisum*. Sequence studies in pigment production, *Phytochemistry*, 13, 1835, 1974.

83. **Hrazdina, G. and Weeden, N. F.**, Enzymic control of anthocyanin expression in the flowers of pea *(Pisum sativum)* mutants, *Biochem. Genet.*, 24, 309, 1986.

84. **Crowden, R. K.**, Pseudobase of malvidin 3-rhamnoside 5-glucoside in *am* mutants of *Pisum sativum, Phytochemistry*, 21, 2989, 1982.

85. **Nowacki, E.**, Some negative correlation in biochemical characters inheritance, *Genet. Pol.*, 10, 76, 1968 (CA 74, 95575).

86. **Nozzolillo, C.**, On the role of anthocyanins in vegetative tissues, a relationship with stem length in pea seedlings, *Can. J. Bot.*, 57, 2554, 1979.

87. **Nozzolillo, C.**, Anthocyanin pigments in pea seedlings: genetically controlled and environmentally influenced, *Can. J. Bot.*, 56, 2890, 1978.

88. **Ewing, E. E.**, Factors for resistance to post-emergence damping-off in pea (*Pisum sativum* L.) incited by *Pythium ultimum* Trow., *Diss. Abstr.*, 20, 1518, 1959 (HA 1960, 3834).

89. **Stasz, T. E., Harman, G. E., and Marx, G. A.**, Time and site of infection of resistant and susceptible germinating pea seeds by *Pythium ultimum, Phytopathology*, 70, 730, 1980.

90. **Janyska, A.**, The question of the inheritance of resistance to anthrachnose in the pea, *Bull. Vysk. Ustavu Zalinarsk Olomuc*, 7, 49, 1963 (PBA 1964, 7003).

91. **Sorgel, G.**, The question of ideas up to the present on resistance to fungal pathogens: illustrated by the example of foot rot and leaf and pot spot in peas, *Sitzungsber. Dtsch. Akad. Landw. Wiss. Berlin*, 5, 20, 1956 (PBA 1957, 3600).

92. **Schneider, A.**, Tannin-type condensation products of anthocyanidins in the seed coat of *Pisum arvense, Naturwissenschaften*, 39, 452, 1952 (CA 47, 8838a).

93. **Monti, L. M. and Scarascia-Mugnozza, G. T.**, Impiego di una mutazione per precocità in un programma d'incrocio per il miglioramento genetico del pisello da industria, *Genet. Agrar., Pavia*, 24, 195, 1970 (PBA 1971, 9317).

94. **Stickland, R. G.**, Condensed tannins of pea seeds, *Plant Sci. Lett.*, 34, 403, 1984.

95. **Hrazdina, G., Marx, G. A., and Hoch, H. C.**, Distribution of secondary plant metabolites and their biosynthetic enzymes in pea (*Pisum sativum* L.) leaves. Anthocyanins and flavonol glycosides, *Plant Physiol.*, 70, 745, 1982.

96. Alexsandrova, V. G. and Aleksandrova, O. G., The distribution of pigments in the testa of some varieties of soybean, *Glycine hispida* Maxim., *Bull. Appl. Bot. Genet. Plant Breed. U.S.S.R.*, 3, 4, 1934 (CA 29, 5149).

97. Nagai, I., A genetico-physiological study of the formation of anthocyanins and brown pigments in plants, *Tokyo Univ. Coll. Agric. J.*, 8, 1, 1921.

98. Kuroda, C. and Wada, M., The colouring matter of kuro-mame, *Proc. Imp. Acad. (Tokyo)*, 9, 17, 1933 (CA 27, 2448).

99. Kuroda, C. and Wada, M., Kuromamin, the colouring matter of kuro-mame, *Proc. Imp. Acad. (Tokyo)*, 9, 517, 1933 (CA 28, 1701).

100. Kuroda, C. and Wada, M., Kuromamin, the colouring matter of kuro-mame, *Proc. Imp. Acad. (Tokyo)*, 11, 189, 1935 (CA 29, 5843).

101. Manabe, T., Kubo, S., Kodama, M., and Bessho, Y., Prevention of discoloration in red parts of white peach. III. Some properties of an anthocyanin pigment isolated from black soybean peels, *Nippon Shokuhin Kogyo Gakkai-Shi*, 12, 472, 1965.

102. Yoshikura, K. and Hamaguchi, Y., Anthocyanins of the black soybean, *Eiyo To Shokuryo*, 22, 367, 1969 (CA 72, 63600).

103. Taylor, B. H., Environmental and Chemical Evaluation of Variations in Hilum and Seedcoat Colors in Soybean, MSc thesis, University of Arkansas, Fayetteville, 1976.

104. Buzzell, R. I., Buttery, B. R., and MacTavish, D. C., Biochemical genetics of black pigmentation of soybean seed, *J. Hered.*, 78, 53, 1987.

105. Groose, R. W., Schulte, S. M., and Palmer, R. G., Germinal reversion of an unstable mutation for anthocyanin pigmentation in soybean, *Theor. Appl. Genet.*, 79, 161, 1990.

106. Groose, R. W., Weigelt, H. D., and Palmer, R. G., Somatic analysis of an unstable mutation for anthocyanin pigmentation in soybean, *J. Hered.*, 79, 263, 1988.

107. Tu, J. C., Localization of infectious soybean mosaic virus in mottled soybean seeds, *Microbios*, 14, 151, 1975 (CA 85, 90328).

108. Ogawa, H. and Naito, S., Effect of cadmium ion on soybeans, *Kanagawa-Ken Eisei Kenkyusho Nempo*, 23, 141, 1974 (CA 85, 41679).

109. Hoagland, R. E., Effects of glyphosate on metabolism of phenolic compounds. VI. Effects of glyphosine and glyphosate metabolites on phenylalanine ammonia-lyase activity, growth, and protein, chlorophyll, and anthocyanin levels in soybean *(Glycine max)* seedlings, *Weed Sci.*, 28, 393, 1980.

110. Liu, T. Y., Oil from fermented black soybean, *Glycine max* (L.) Merrill var. 'O'Tau' (toshi) and its antioxidative activity, *Chung Kuo Nung Yeh Hua Hsueh Hui Chih*, 22, 240, 1984 (CA 103, 86749).

111. Tsuda, T. and Fukuba, H., Separation and identification of anthocyanins in food by isotacoelectrophoresis, *Nippon Eiyo Shokuryo Gakkai-Shi*, 42, 79, 1989 (CA 110, 211110).

112. Chiang, H. S. and Norris, D. M., Purple stem, a new indicator of soybean stem resistance to bean flies (Diptera: Agromyzidae), *J. Econ. Entomol.*, 77, 121, 1984.

113. Peters, D. W., Wilcox, J. R., and Vorst, J. J., Hypocotyl pigments in soybeans, *Crop Sci.*, 24, 237, 1984.

114. Hartwig, E. E. and Hinson, K., Inheritance of flower color of soybeans, *Crop Sci.*, 2, 152, 1962.

115. Raut, V. M., Halwanker, G. B., and Patil, V. P., Genetic studies in soybean. I. Inheritance of anthocyanin pigmentation, *J. Marahashtra Agric. Univ.*, 9, 40, 1984 (PBA 1985, 3036).

116. Groose, R. W. and Palmer, R. G., Gene action governing anthocyanin pigmentation in soybean, *J. Hered.*, 82, 498, 1991.

117. Bhatt, G. M. and Torrie, J. H., Inheritance of pigment color in the soybean, *Crop Sci.*, 8, 617, 1968.

118. Beale, G. H., Robinson, G. M., Robinson, R., and Scott-Moncrieff, R., Genetics and chemistry of flower color variations, *J. Genet.*, 37, 375, 1939.

119. Harborne, J. B., Flavonoid pigments of *Lathyrus odoratus*, *Nature (London)*, 187, 240, 1960.

120. **Harborne, J. B.,** Plant polyphenols. IX. The glycosidic pattern of anthocyanin pigments, *Phytochemistry,* 2, 85, 1963.

121. **Pecket, R. C.,** The variation of flower color in the genus *Lathyrus, New Phytol.,* 59, 138, 1960.

122. **Reichel, L. and Hiller, W.,** Ueber die Farbstoffe der Wicke, *Naturwissenschaften,* 47, 83, 1983 (CZ 1960, 14408).

123. **Sakata, Y. and Uemoto, S.,** The flower colors of sweet pea. 1. Anthocyanidin pigmentation in flowers of spring-flowering sweet pea, *Engei Gakkai Zasshi,* 45, 181, 1976 (CA 86, 185907).

124. **Sakata, Y. and Arisumi, K.,** The flower color of sweet pea. II. The constitution of the anthocyanin and flavonol pigments in the various strains of the garden sweet pea, *Kagoshima Daigaku Nogakubu Gakujutsu Hokoku,* 33, 13, 1983 (CA 99, 85276).

125. **Ratkin, A. V., Zaprometov, M. N., Andreev, V. S., and Evdokimova, L. I.,** Study of the biosynthesis of anthocyanidins and flavonols in blossoms of the sweet pea, *L. odoratus* L., *Z. Obshch. Biol.,* 41, 685, 1980 (CA 93, 201081).

126. **Arkatov, V. V., Andreev, V. S., and Ratkin, A. V.,** Genetic control of flower color formation in sweet pea, *Lathyrus odoratus* L. Communication I. Characteristics of anthocyanin pigment in sweet pea flowers, *Genetika,* 12(8), 30, 1976 (CA 86, 2454).

127. **Pecket, R. C.,** Color changes in flowers of *L. hirsutus* during senescence, *Nature (London),* 211, 1251, 1966.

128. **Sarkar, J. K. and Banerjee, P. K.,** Chromatographic examination of anthocyanin in *Lathyrus sativus* L., *Curr. Sci.,* 46, 62, 1977 (CA 86, 86152).

129. **Balasubrahmanyam, R.,** Inheritance of seed-coat color in gram *(Cicer arietinum), Madras Agric. J.,* 37, 379, 1950 (PBA 1956, 797).

130. **Singh, D. and Syam, R.,** Ovule sterility in gram, *Cicer arietinum, Curr. Sci.,* 28, 294, 1959 (PBA 1960, 917).

131. **More, D. C. and D'Cruz, R.,** Genetic studies in Bengal gram (*Cicer arietinum* L.). III. T-54-A × D-70-10, *Poona Agric. Coll. Mag.,* 60, 27, 1970 (PBA 1971, 9351).

132. **Pawar, A. M. and Patil, J. A.,** Genetic studies in gram, *J. Mahashtra Agric. Univ.,* 7, 261, 1982 (PBA 1983, 7752).

133. **Börger, G. and Barz, W.,** Malonated flavonol 3-glucosides in *Cicer arietinum, Phytochemistry,* 27, 3714, 1988.

134. **Wilson, V. E., Law, A. G., and Warner, R. L.,** Inheritance of cotyledon color in *Lens culinaris* (Medic.), *Crop Sci.,* 10, 205, 1970.

135. **Slinkard, A. E. and Bhatty, R. S.,** Laird lentil, *Can. J. Plant Sci.,* 59, 503, 1979.

136. **D'Arcy, A. and Jay, M.,** Les flavonoides des grains de *Lens culinaris, Phytochemistry,* 17, 826, 1978.

137. **Kadam, S. S., Chougule, B. A., and Salunkhe, D. K.,** Lupine, in *Handbook of World Food Legumes,* Salunkhe, D. K. and Kadam, S. S., Eds., CRC Press, Boca Raton, FL, 1989, 163.

138. **Baeyer, E.,** Farbstoffe der roten, violetten und blauen Lupinenblueten, *Chem. Ber.,* 92, 1062, 1959 (CA 53, 14242d).

139. **Baeyer, E.,** In der Natur vorkommende Metallkomplexe, *Chimia,* 16, 333, 1962.

140. **Bragdø, M.,** Interspecific crosses in *Lupinus,* cytology and inheritance of flower colour, *Hereditas,* 43, 338, 1957 (PBA 1958, 618).

141. **Al-Bassam, H.,** Biochemical changes associated with mutations of loss of anthocyanin pigmentation of flowers in some plants, *Dokl. Akad. Nauk SSSR,* 168, 1405, 1966 (CA 65, 10977b).

142. **Forbes, I., Jr., Wells, H. D., Edwardson, J. R., and Ostazeski, S. A.,** Inheritance of resistance to gray leaf spot disease and of an anthocyanin inhibitor in blue lupines, *Lupinus angustifolius* L., *Crop Sci.,* 1, 184, 1961.

143. **Wozny, A., Zatorska, B., and Mlodzianowski, F.,** Effect of lead on the development of lupine seedlings and ultrastructural localization of lead in the roots, *Acta Soc. Bot. Pol.,* 51, 345, 1982 (CA 99, 117389).

144. **Hymowitz, T. and Boyd, J.,** Origin, ethnobotany, and agricultural potential of the winged bean *(Psophocarpus tetragonolobus)*, *Econ. Bot.*, 31, 180, 1977.

145. **Rajendran, R., Satyanaryana, A., Selvaraj, Y., and Bhargava, B. S.,** Some investigations on the indigenous and exotic collections of winged bean in India, in *1st Int. Symp. Dev. Potentials Winged Bean*, Los Baños, Laguna, Philippines, January 1978, 1980, 71, (HA 1981, 8632).

146. **Bistocchi, N., Cagiotti, M. R., and Crosta, G.,** Indagine preliminare sulla biologia del fagiolo alato *(Psophocarpus tetragonolobus* (L.) DC.), *Ann. Fac. Agrar. Univ. Perugia*, 33, 389, 1979.

147. **Sastrapradja, S., Aminah Lubis, S. H., Lubis, I., and Sastrapradja, D.,** A survey of variation in *Psophocarpus tetragonolobus* (L.) DC with reference to the Javanese samples, *Ann. Bogor.*, 6, 221, 1978.

148. **Erskine, W.,** Measurements of the cross-pollination of winged bean in Papua, New Guinea, *SABRAO J.*, 12, 11, 1980 (HA 1982, 6242).

149. **Le Hing, T., Hubacek, J., Lachman, J., Pivec, V., Borek, V., and Rehakova, V.,** Flavonoid substances in the seed coat of the Binh Minh cultivar of winged beans, *Psophocarpus tetragonolobus* (L.) DC, *Agric. Trop. Subtrop.*, 19, 193, 1986 (HA 1986, 3420).

150. **Lachman, J., Rehakova, V., Hubacek, J., and Pivec, V.** Flavonoid substances and saccharides in the seeds of winged bean *Psophocarpus tetragonolobus* (L.) DC., *Sci. Agric. Bohemosl.*, 14, 265, 1982.

Chapter 10

ROOTS, TUBERS, AND BULBS

I. CARROT

The color of carrots (*Daucus carota* L.) may be white, yellow, orange, red, and even black. Cultivars possessing cyanic color contain anthocyanins which may occur in trace amounts in pink cultivars to 1750 mg/kg in black carrots.[1,2] Several authors[1-7] have reported on the anthocyanin composition of carrot roots and flowers. Krishnamoorty and Seshadri[3] reported the first partial characterization of the anthocyanins in the purple-black-rooted *D. carota* ssp. *sativa* in which they identified only a cyanidin derivative. Harborne[1] provided the first complete identification of the anthocyanins in leaves, flowers, stems, and roots of *Daucus* species. Three anthocyanins were separated and identified as cyanidin glycosides. The first pigment was the rare cyanidin 3-(2^G-xylosylgalactoside) or 3-lathyroside, a sugar previously found in *Lathyrus odoratus* petals (see Chapter 9, Section VI). The second pigment was cyanidin 3-xylosylglucosylgalactoside, and the third was cyanidin 3-feruloylxylosylglucosylgalactoside. In the subspecies *D. carota maritimus*, sinapic and ferulic acid derivatives of cyanidin 3-glucosylgalactoside were found, instead of the triglycosides. More recently, Harborne and Grayer[5] reported that unpublished FAB-MS studies by their group have indicated that in cyanidin 3-sinapylxylosylglucosylgalactoside the sinapic acid is located on the glucose residue.

The inheritance of anthocyanins in the petioles of *D. carota* var. *sativa* was studied by Angell and Gabelman.[6] It appears that the absence of anthocyanins in white flowers can be attributed to the lack of the enzyme chalcone synthase, which is active in flowers containing anthocyanins. This characteristic of wild carrot flowers has been used for the study of the flavonoid metabolism which involves phenylalanine-ammonia lyase, chalcone isomerase as well as chalcone synthase.[8]

A cell culture of carrot retains its ability to synthesize anthocyanins even after storage at −140°C.[9] In cambial tissue, the biosynthesis of these pigments *in vitro* is related to the presence of auxins, such as indolacetic acid.[10] Hopp et al.[11] isolated intact vacuoles which contained anthocyanins and the enzymes responsible for their biosynthesis. Dougall et al.[12] isolated single-cell clones of wild carrots which differed 40-fold in the level of anthocyanin produced when measured under standard conditions. Alfermann and Reinhard[13] described cell lines capable of producing anthocyanins in the dark and cell lines which require light and auxin.

A partial identification of the anthocyanins produced in cell culture of carrot was first presented by Schmitz and Seitz,[14] who reported the occurrence of cyanidin 3-glucosylxyloside. Later, Hemingson and Collins[15] found four nonacylated pigments: cyanidin 3-glucosylgalactoside, cyanidin 3,5-digalactoside, cyanidin 3-glucoside, and cyanidin 3-galactoside. This finding, however, was discredited by Harborne et al.,[16] who identified the major pigment of carrot tissue culture as cyanidin 3-sinapylxylosylglucosyl galactoside, the same pigment already identified in the leaves and flowers.[1] The erroneous report of Hemingson and Collins[15] arose mainly from the fact that these authors inadvertently partially hydrolyzed the pigment during purification.

II. CELERY

Celery, *Apium graveolens* L., is a popular vegetable related to parsley and parsnip. People eat the crisp leafstalks either raw or cooked. In cultivated white celery, red pigmentation is an undesirable characteristic because the pink color is generally associated with fungal infection (pink rot).[17]

Anthocyanins have been found in celery *(A. graveolens)* from the Old World temperate regions and in austral celery *(A. australe)*, native to the New World.[18] Both species contain the cyanidin aglycone and a linear trisaccharide (xylose-glucose-galactose). However, only *A. graveolens* contains the ferulyl, sinapyl, and *p*-coumaryl esters of cyanidin 3-xylosylglucosyl-galactoside. *A. australe* lacks the *p*-coumaryl derivatives.[1]

According to Arus and Orton,[19] the red pigmentation trait in the petioles of celery is dominant over the green coloration and is controlled by a single gene.

III. CHICORY

Chicory, *Cichorium intybus* L., is a minor crop whose roots are used as a coffee substitute and as a source of inulin. The leaves and young shoots may be used in salads or as cooked vegetables.

In salad chicory, the presence of anthocyanins in leaves is controlled by a genetic factor which is independent of those that govern the coloration of the venation in mature leaves.[20] Because of this, it is possible through plant breeding to obtain cultivars of chicory with specific pigmentation.[20,21]

As early as 1939, Lawrence et al.[22] identified delphinidin as the anthocyanidin in the flowers of *C. intybus*, and Harborne[23] noted that the glycoside of delphinidin had the chromatographic characteristics of the rare 3-rhamnoside-5-glucoside. These flowers have the peculiarity already noted by Kastle and Haden[24] that during the day they lose color according to the amount of light. This phenomenon was investigated by several authors, including Todt[25] and Proctor and Creasy,[26] who found that this effect was due to the presence of an enzymatic system that causes the decoloration of the anthocyanins.

Timberlake et al.[27] analyzed the anthocyanins in the red leaves of 'Rossa di Verona' chicory and found three pigments which, after hydrolysis, gave cyanidin and glucose. The main pigment was identified as cyanidin 3-glucoside; the other two had the characteristics of aliphatic acyl derivatives. Cyanidin and delphinidin were confirmed as the aglycones of chicory by Cappelletti and Caniato,[28] who also reported that cyanidin is the aglycone present in the greatest quantity.

Using FAB-MS and NMR spectroscopy, Bridle et al.[29] identified cyanidin 3-*O*-β-(6-*O*-malonyl)-D-glucopyranoside in the leaves of 'Rossa di Verona' chicory. Later work by Takeda et al.[30] using HPLC and FAB-MS found delphinidin 3-(6″-malonylglucoside) and delphinidin 3-(6″-malonylglucoside)-5-malonylglucoside in freshly picked flowers of chicory.

IV. GARLIC

The pigmentation of garlic (*Allium sativum* L.) is due to the presence of carotenoids, anthocyanins, and chlorophylls. The anthocyanins are localized in the inner scale leaves of the bulb and tend to disappear with storage. Du and Francis[31] separated three major and four minor pigment bands by paper chromatography from an extract of the inner scales of garlic. The three major anthocyanins were identified as cyanidin 3-glucoside (the most representative) followed by cyanidin 3-glucoside monoacylated and cyanidin 3-glucoside triacylated with an aliphatic acid. Other investigations on garlic anthocyanins have been reported by Boscher,[32] Borukh and Demkevich,[33] Kenmochi and Katayama,[34] and Hilal et al.[35] All have confirmed the presence of cyanidin 3-glucoside; and in *A. vineale*, Boscher[32] found petunidin glucosides. In the flowers of *A. aschersonianum*, Hilal et al.[35] also found acylated derivatives of cyanidin 3-glucoside, which they did not fully characterize. Further work using modern techniques such as HPLC, NMR, and FAB-MS is needed to fully characterize the anthocyanins of garlic and allied crops. In a recent review of the literature on the chemical composition of onions and allied crops, Fenwick and Hanley[36] also emphasize this point.

V. GINSENG

Ginseng is a perennial herb of eastern Asia *(Panax ginseng)* and eastern North America *(P. quinquefolius)*. It has a long, fleshy root used as medicine in a number of countries. Many people believe it can cure various illnesses, but its medicinal value has not been proven.

The principal anthocyanin found in the skin of red ginseng roots is pelargonidin 3-glucoside.[37] According to Lee and Park,[37] *P. ginseng* also contains an additional unidentified pigment not found in the species from North America, and this may provide a means for differentiating the two species. The anthocyanins of the ginseng family, Araliaceae, are characteristically glycosylated with 3-xyloside-galactoside;[38] and, therefore, it would be expected that this sugar also be present in the pigment of ginseng. Recently, Li and Zhu[39] separated four pigments from a cell culture of *P. ginseng*, and, using UV and NMR, identified the main pigment as cyanidin 3-*O*-β-D-xylopyranyl-(1-2)-β-D-glucopyranoside.

VI. JERUSALEM ARTICHOKE

Jerusalem artichoke (*Helianthus tuberosus* L.) is a tuber-producing plant closely related to the sunflower. It is native to North America, and it has long been grown in Europe for animal feed. The tubers, which are rich in carbohydrates and proteins,[40] have a number of uses including: food and feed, high-fructose syrup, ethyl alcohol, protein, fodder yeast, brandy, lactic acid, propionic acid, mannitol, pectic substances, and as substrate for the production of acetone and butanol.[41-48]

Most cultivars of Jerusalem artichoke produce tubers with white flesh and white skin, a few cultivars produce white-fleshed and purple-skinned tubers, and a very few genetic accessions produce tubers with purple skin and purple flesh.[49] The purple color in the skin and/or flesh of tubers is due to anthocyanins.[50] In the tissues of white-fleshed tubers treated with growth regulators, Swanson et al.[51] reported the formation of a red pigment which appeared to be an anthocyanin. Ibrahim et al.[52] identified the major anthocyanin in callus tissue culture of the Jerusalem artichoke, treated with kinetin and indolacetic acid, as a glucoside of cyanidin.

VII. ONION

Depending on its genetic makeup, the onion plant (*Allium cepa* L.) can produce red, yellow, or white bulbs. The color of red onions is due to the presence of anthocyanins which are found in crystalline form in the epidermal cells.[53]

Cyanidin 3-pentoseglucoside was the first anthocyanin identified in onion.[54] Later, Fouassin[55] used paper chromatography to separate three cyanidin glycosides, including a monoside and a diglucoside. The report[56] that peonidin 3-arabinoside was present in membranes of 'Southport Red Globe' onions was shown to be incorrect.[57] Using improved paper chromatographic techniques, Fuleki[57-60] separated eight anthocyanins in extracts of 'Ruby' and 'Southport Red Globe' onions. Seven of the eight pigments contained cyanidin as aglycone and one contained peonidin. The peonidin derivative was identified as the 3-glucoside and the two most abundant anthocyanins were cyanidin 3-glucoside and cyanidin 3-diglucoside. The latter compound was neither the 3-sophoroside (β,1-2 linkage) nor the 3-gentobioside (β,1-6 linkage); and it was Du et al.[61] who, using paper chromatography coupled with spectral analysis and degradation studies, showed it to be the novel 3-laminariobioside, a diglucoside possessing β,1-3 glucosylglucose linkage. Kenmochi and Katayama[34] confirmed cyanidin 3-glucoside as the major pigment of red onion and found cyanidin 3-laminarioside in samples from two cultivars.

An extensive search for the presence and nature of acylated anthocyanins in onions was carried out by Moore and co-workers.[62-64] Their studies emphasized the importance of the extracting solvent system for the isolation of acylated pigments which are especially sensitive to the presence of mineral acid in the extractant. On the basis of chromatographic and spectral analyses using UV-Vis, infrared radiation (IR), and NMR, these authors suggested that many of the anthocyanins of the onion may be acylated but that this does not involve cinnamic acid groupings.[64] Recently, Terahara and Yamaguchi[65] reported that the major anthocyanin in red onion is cyanidin 3-glucoside acylated with malonic acid.

Data on anthocyanin content of red onions were provided by Feldman et al.[66] who found between 8.7 and 20.7 mg/100 g in three cultivars examined. Irradiation with 250 krd of γ-rays causes rapid breakdown of the anthocyanins immediately after treatment or during storage;[67] the mechanism of the breakdown and the nature/fate of the end products remains, however, unknown.

Rieman[68] suggested the presence of three allelomorphic genes: W, which controls the synthesis of red pigment; Wy, which regulates the synthesis of the yellow pigment; and recessive w, which produces a white bulb. Davis[69] proposed that the dominant white genotype could be distinguished from the recessive yellow by the lack of a gene, R, responsible for red coloration by exposing the fresh scales to ammonia vapor. Clarke et al.[70] postulated the presence of a single gene, C, required for the expression of all pigmentation, in combination with the gene, R, for the red scales and the recessive gene r for yellow scales. Jones and Peterson[71] found that crosses between some yellow-golden cultivars gave light-red F1 progenies, suggesting that inheritance of color in onion was more complex than previously believed. Later work by El-Shafie and Davis[72] showed that five genes — I, C, G, L, and R — are responsible for the white, yellow, red, and brown colors of onion; and that these five genes work in a specific order in the biosynthesis of pigments. According to Patil et al.,[73] four genes are necessary for the expression of red color: $C1$; $C2$; and two duplicated complementary factors, $R1$ and $R2$.

In addition to anthocyanins, red onions also contain considerably more flavonols, especially quercetin, which may range from 2.5 to 6.5 g/100 g dry weight in colored onions but in trace amounts ($\sim$1 mg/100 g) in the skins of white onion cultivars.[74] Similarly, the total content of flavonols in edible portions of a red onion is reportedly double that of yellow cultivars.[75]

VIII. POTATO

The potato (*Solanum tuberosum* L.), with an annual production of nearly 300 million metric tons, ranks as the fourth major food crop of the world, exceeded only by wheat, rice, and maize.[76] The flesh color of potatoes is generally white or yellow. A few cultivars with red or purple flesh are, however, available. The degree of tuber flesh pigmentation can vary from just a slight pigmentation of the vascular ring to a complete pigmentation of the entire tuber.[77] The presence of anthocyanins in the periderm or peripheral cortical cells forms the basis of the skin pigmentation.[78,79]

The first detailed examination of the anthocyanins of the cultivated potato was carried out by Chmielewska,[80,81] who studied the pigments present in the skin and flesh of a purple-black variety called Negresse. A malvidin 3-rhamnosylglucoside, acylated with *p*-coumaric acid (negretein), and a 3-monoglucoside (tuberin) of a supposedly unknown anthocyanidin (tuberinidin) were reported. In 1960, Harborne[82] described the isolation and characterization of ten anthocyanins in the tubers or flowers of various colored forms of potatoes. Dependent on the cultivar and red or purple color, the anthocyanins which may be found in potatoes are the 3-rhamnosylglucoside (or 3-rutinoside) of pelargonidin, delphinidin, and petunidin; and the 3-(*p*-coumarylrutinoside)-5-glucosides of pelargonidin, cyanidin, delphinidin, peonidin, petunidin, and malvidin.

In pink and purple pigmented potatoes, Howard et al.[83] found single anthocyanins which were identified as the 3-(*p*-coumarylrutinoside-5-glucosides of pelargonidin and peonidin, respectively. In 'Desirée' and 'Urgenta' potatoes, Sachse,[84] however, identified pelargonidin 3-(*p*-coumarylrutinoside-5-glucoside and peonidin ferulyl-3-rutinoside-5-glucoside. The full structure of petanin was recently elucidated by Andersen et al.[85] as petunidin 3-*O*-(6-*O*-*p*-coumaroyl-α-L-rhamnoside-β-D-glucoside)-5-*O*-β-D-glucoside.

The inheritance of anthocyanin pigmentation has been studied by numerous authors and reviewed by Von Rathlef and Siebeneick,[86] Swaminathan and Howard,[87] Howard,[88,89] and recently by De Jong.[90] The designation and function of the genes responsible for anthocyanin pigmentation in cultivated potatoes are summarized in Table 1. Depending on whether the cultivated diploids (2n = 2x = 24) or tetraploids (2n = 4x = 48) are considered, the number and functions of the known genes are considerably different. In his recent review, De Jong,[90] however, points out that the differences between the two groups are due more to differences in alleles which have thus far been identified rather than differences in the presence or absence of different loci. In both groups, the production of anthocyanins is controlled by three genes: *P* controls the synthesis of trihydroxylated pigments (delphinidin and petunidin), *R* regulates the production of cyanidin and peonidin, and *Ac* controls the acylation process.[87,91-93] According to Harborne,[93] gene *Ac* also regulates the degree of glycosylation and methoxylation of anthocyanins. Both *Ac* and *acac* genotypes occur among the cultivated diploids. Thus far, at least in tetraploids, all anthocyanins described are acylated.[87] The reason for this apparent difference between tetraploids and cultivated diploids is not clear, but it would appear that the *ac* allele is either absent or present at a very low frequency in tetraploids.

According to Swaminathan and Howard,[87] the gene *D* is a basic factor in tetraploid potatoes; and it is necessary for the development of pigmentation

TABLE 1
Summary of Genes for Anthocyanin Pigmentation in Cultivated Potatoes

Function		Diploid		Tetraploid
I. Production of pigments	P:	Blue or purple[91]	P:	Blue or purple[87,92]
	R:	Red; several alleles[91]	R:	Red[87,92]
II. Chemical alteration of pigments	Ac:	Acylation[93]		
III. Distribution of pigments	I:	Distributes pigment to tuber skin[94]	D:	Basic gene for distribution to many plant tissues[87,92]
	B:	Series of alleles controlling distribution to floral abscission layer, eyebrow, embryo pot and nodal band[94,95]		
	F:	Distributes pigment to flower[86,94]	F:	Factor for flower color in the presence of D[87,96]
			E:	Distributes pigment to eye in the presence of D[87,96]
			M:	Functions only in the presence of D; restricts pigmentation of tuber periderm to areas around the eye[87,97]
	Ow:	Distributes pigment to ovary wall[98]		
	Pf:	Distributes pigment to tuber flesh[99]		
	Pw:	Distributes pigment to whorl[100]		
	Ul:	Distributes pigments to underside of leaf[100]		
IV. Function not (yet) clearly defined (may be distribution genes)	Pd:	Pigmentation of dorsal side of leaf[101]		
	Pv:	Pigmentation of ventral side of leaf[101]		
	PSC:	Purple skin pigmentation[102]		

From De Jong, H., *Am. Potato J.*, 68, 585, 1991. With permission.

in various plant parts. Genotypes with a D---R--- constitution produce red pigment in the phelloderm of the tuber skin. In diploids, the gene *I* has been assigned a similar function.[94] According to Dodds and Long,[94] the effect of *I* is limited to the tuber, whereas in the tetraploids, *D* is considered to have an effect throughout the whole plant. Recently, De Jong[99] identified a gene for pigmented flesh *(Pf)* in cultivated diploids which is closely linked to the *I* gene.

The *F* gene acts as an intensifier of pigmentation in potato flowers and requires *D* and *R* (or *P*) genes for expression.[94-96] Garg et al.[101] studied the inheritance of leaf pigmentation in cultivated diploids for which they assigned the gene symbols *Pd* (pigmentation of dorsal side of leaf) and *Pv* (pigmentation of ventral side of leaf). They concluded that *Pd* and *Pv* are linked. It is conceivable that *Pd* and *Pv* are similar to *Ul* (underleaf pigmentation) and *Pw* (pigmented whorl) of Kessel and Rowe.[100] More recently, Gebhardt et al.,[102] using restriction fragment length polymorphism (RFLP) analysis, identified a gene *PSC* (purple skin color) in cultivated diploids.

Foliar applications of 2,4-D have been used commercially to promote the production of anthocyanin in the periderm of red-skinned potatoes.[103] Recently, however, Fritz et al.[104] reported that 2,4-D and ethephon foliarly applied to red-skinned 'Norland' potatoes reduced tuber yields and had little effect on periderm color intensity at harvest or throughout storage.

In addition to anthocyanins, a variety of other phenolic compounds are present in potato tubers. These are monohydric phenols, coumarins, flavonols, flavones, and polyphenols. The flavonol glycosides are quercetin 3-glucoside and quercetin 3-rutinoside.[23] The tannins are mostly localized in suberized tissue of the potato and impart tan coloration to the skin. Tyrosine, a monohydric phenol, constitutes 0.1 to 0.3% of the dry weight of the tuber, whereas chlorogenic acid constitutes 0.025 to 0.15% of the dry weight of potatoes.[105,106]

IX. SWEET POTATO

The sweet potato is the edible tuber of *Ipomoea batatas* L. The presence of anthocyanins in *Ipomoea* species was noted in studies by Smith,[107,108] who examined the variations of flower colors in *I. leerii*, and by Almeida Cousin,[109] who investigated the anthocyanins of *I. roxa* flowers and their use as pH indicators.

Rangaswami Ayyanagar and Sampathkumar[110] studied the relationship between leaf form and anthocyanin pigmentation, and compared how these two characteristics were controlled by genes. In *I. batatas* the anthocyanins are contained in the skin, while the carotenoids are found in the flesh of the tuber. The biosynthesis of these pigments is genetically regulated and is independent of chlorophyll synthesis.[111] According to Mikell et al.,[112] the pigmentation of the skin and pulp is controlled by complementary genetic

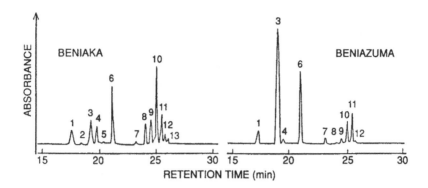

FIGURE 1. Analytical HPLC of anthocyanins from the periderm of 'Beniaka' and 'Beniazuma' sweet potatoes. Peak 1: cyanidin-3GG-5G; 2: cyanidin-3G-5G; 3: peonidin-3GG-5G; 4: *p*-hydroxybenzoyl-cyanidin-3GG-5G; 5: peonidin-3G-5G; 6: *p*-hydroxybenzoyl-peonidin-3GG-5G; 7: feruloyl-peonidin-3GG-5G; 8: caffeoyl-*p*-hydroxybenzoyl-cyanidin-3GG-5G; 9: caffeoyl-cyanidin-3GG-5G; 10: caffeoyl-*p*-hydroxybenzoyl-peonidin-3GG-5G; 11: caffeoyl-peonidin-3GG-5G;12: caffeoyl-feryloyl-peonidin-3GG-5G; 13: caffeoyl-peonidin-3G-5G (GG: glucosylglucoside, G: glucoside). (From Miyazaki, T., Tsuzuki, W., and Suzuki, T., *J. Jpn. Soc. Hortic. Sci.*, 60(1), 217, 1991. With permission.)

factors; and Hernandez et al.[113] observed mutagenic changes in the pigments as a result of irradiation.

The use of sweet potato pigments as a source of natural food color was first proposed by Kazui and Umaki.[114] Cascon et al.[115] described a process for the extraction of the pigments using enzymatic degradation of starch from the tuber with α-amylase and amyloglycosidase followed by concentration of the product to 72° Brix and spray-drying or freeze-drying of the concentrate to a final anthocyanin content of 0.12%. Bassa and Francis[116] studied the stability of a colorant produced by extraction with 1% HCl in water followed by purification on a CG-50 resin, in a model beverage. When compared with cyanidin 3-glucoside from blackberries, a commercial source of anthocyanin, and a deacylated preparation from sweet potatoes, the original sweet potato extract was the most stable.

The anthocyanins of sweet potatoes were reported to be dicaffeylpeonidin-3-sophoroside-5-glucoside by Imbert et al.[117] and acylated cyanidin 3-glucosylfructoside-5-xyloside by Tsukui et al.[118] Recently, using absorption spectra, HPLC, ^{1}H-NMR, and FAB-MS, Miyazaki et al.[119] isolated and characterized 13 anthocyanins from the periderm of *I. batatas* (Figure 1). The anthocyanins found in the periderm of 'Beniaka' and 'Beniazuma' sweet potatoes are acylated with caffeic and *p*-hydroxybenzoic acids. Those contained in the flesh of 'Yamagawamurasaki' and 'Tanegashimamurasaki' sweet potatoes are acylated with cinnamic acid derivatives. According to a very recent report by Odake et al.,[120] the two major anthocyanins of 'Yamagawamurasaki' sweet potatoes are the 3-caffeylferulylsophoroside-5-glucosides of cyanidin and peonidin. In the flowers of the congeneric species *I. cairica*, Pomilio and Sproviero[120] identified cyanidin 3-(*p*-coumaroyl-caffeoyl)-so-phoroside-5-glucoside, cyanidin 3-(dicaffeoyl)-sophoroside-5-glucoside, and cyanidin 3-(caffeyl)-sophoroside-5-glucoside. In the flowers of *I. congesta*, there are anthocyanins with an acylated glucosidic pattern analogous to *I. cairica*. They are acylated derivatives of peonidin, three of which are isomers of peonidin 3-(caffeyl)-sophoroside-5-glucoside, peonidin 3-(*p*-coumaryl-caf-feyl)-sophoroside-5-glucoside, and peonidin 3-(*p*-coumaryl)-sophoroside-5-glucoside.[122] The flowers of *I. fistulosa* contain peonidin 3-arabinosylglu-coside and kaempferol.[123]

Cell cultures of sweet potatoes have also been suggested for use as a potential source of anthocyanins,[123-126] and Nozue et al.[126] patented a process for the production of pigments from callus cultures of sweet potatoes. The process involves the use of potassium nitrate and ammonium sulfate as com-ponents of the medium in combination with illumination at 1,500 to 10,000 lx.

Treatment of sweet potatoes with phosphate preparations[127] and ethephon[128] is known to increase the production of anthocyanins in the skin of the roots, and it has been proposed as a means of increasing the color intensity of the roots.[127,128]

X. TARO

Taro, *Colocasia esculenta* (L.) Schott. (syn. *C. antiquorum* L.), is a tropical plant of the *Arum* family grown for its starchy rootlike stem or corm. It is a popular food source in southeast Asian and Middle East countries where it is processed into a variety of products including a paste called poi.[129]

In the 'Lehua Maoli' taro, Chan et al.[130] identified three anthocyanins: pelargonidin 3-glucoside, cyanidin 3-glucoside, and cyanidin 3-rutinoside. The highest anthocyanin concentration is found in a film that covers the corm (16 mg/100 g) while inside the corm and in the pedicel it is only 4.9 mg/100 g. Anthocyanins are not present in the leaves of taro.

Compared to other plants, taro can be considered low in anthocyanin content. Nonetheless, the presence of red pigments gives poi a bluish-gray color which turns pink following lactic fermentation, decreasing the pH of the product from 6.3 to 4.5. Strauss et al.[131] reported on the inheritance of anthocyanin pigmentation in taro.

XI. YAM

Yams belong to the family Dioscoreaceae. The edible yam is genus *Dios-corea* and a common species is *D. alata*. It is a plant that has thick roots much like those of sweet potato. These roots are a major food source for the inhabitants of many tropical countries. Although yam is considered primarily a source of carbohydrates, it is also an excellent source of vitamins, especially vitamin C. Most aspects of the production, processing, and products of yams were reviewed by Oke.[132]

Many varieties/cultivars of yams produce roots that are pigmented yellow by xanthophylls and/or red by anthocyanins. The tubers of *D. bulbifera* are pigmented yellow by lutein, violaxanthin, zeaxanthin, auroxanthin, and cryptoxanthin but also contain chlorophyll and an unidentified anthocyanin.[133]

D. alata or large yam may have white or pigmented pulp depending on the cultivar. Rasper and Coursey[134] were the first to identify the anthocyanins in this plant. They isolated and identified the principal pigment as cyanidin 3,5-diglucoside. Based on the chromatographic characteristics, two other minor pigments appear to be cyanidin 3-glucoside and cyanidin 3-rhamnoglucoside.

Imbert and Seaforth[135] reported that the anthocyanins of *D. alata*, cultivar St. Vincent, are acylated pigments. According to these authors, the main pigment is cyanidin 3-gentobioside acylated with ferulic acid, while the other two are glucosides of cyanidin acylated with ferulic acid. In yams from India, Karnick[136] identified a variety of phenolic compounds including: caffeic, coumaric, ferulic, and sinapic acids; and the flavonoids kaempferol, quercetin, and cyanidin.

Tsukui et al.[137] identified three pigments in purple yam, *D. batatas*. After precipitation with lead acetate, the anthocyanins were separated by thin-layer chromatography and identified by classical physicochemical techniques as cyanidin 3-gentobioside acylated with *p*-coumaric acid, malvidin 3-gentobioside-5-rhamnoglucoside acylated with *p*-coumaric acid, and malvidin 3-gentobioside-5-glucoside acylated with *p*-coumaric acid.

According to Rhodes and Martin,[138,139] the anthocyanin pigmentation of yams is an important factor in varietal characterization because they were able to classify *D. alata* into three groups based on the correlation between anthocyanin pigmentation and other characteristics.

The presence of anthocyanins is a negative attribute in yams, and the less pigmented cultivars seem to be preferred.[140] Removal of the anthocyanins from the skin of the roots would allow for a better food use of *Dioscorea*[140,141] and, at the same time, produce a natural color for foods, pharmaceuticals, and cosmetics.

REFERENCES

1. **Harborne, J. B.,** A unique pattern of anthocyanins in *Daucus carota* and other Umbelliferae, *Biochem. Syst. Ecol.*, 4, 31, 1976.
2. **Canbas, A.,** Study on the pigment of black carrot, *Doga Bilim Derg.*, Ser. D2, 9, 394, 1985 (CA 104, 17688).
3. **Kirshnamoorty, V. and Seshadri, T. R.,** Survey of anthocyanin from Indian sources. III., *J. Sci. Ind. Res. (India)*, 21B, 591, 1962 (CA 58, 8235e).
4. **Darmanyan, E. B. and Dudkin, M. S.,** Phenolic compounds of industrial carrot varieties, *Izv. Vyssh. Uchebn. Zaved., Pishch. Tekhnol.*, 5, 33, 1976 (CA 86, 84065).

5. **Harborne, J. B. and Grayer, R. J.,** The anthocyanins, in *The Flavonoids: Advances in Research Since 1980,* Harborne, J. B., Ed., Chapman & Hall, London, 1988, 1.

6. **Angell, F. F. and Gabelman, W. H.,** Inheritance of purple petiole in carrot, *Daucus carota* var. *sativa, HortSci.,* 5, 175, 1970.

7. **Braun, G. and Seitz, U.,** Accumulation of caffeic, ferulic and chlorogenic acid in relation to the accumulation of cyanidin in two lines of *Daucus carota, Biochem. Physiol. Pflanz.,* 168, 93, 1975 (CA 84, 2335).

8. **Hinderer, W., Noe, W., and Seitz, H. U.,** Differentiation of metabolic pathways in the umbel of *Daucus carota, Phytochemistry,* 22, 2417, 1983.

9. **Dougall, D. K. and Whitten, G. H.,** The ability of wild carrot cell cultures to retain their capacity for anthocyanin synthesis after storage at −140°C, *Planta Med.,* 1980 (Suppl.), 129, 1980 (CA 94, 27479).

10. **Sugano, N. and Hayashi, K.,** Anthocyanins. LVII. Dynamic interrelation of cellular ingredients relevant to the biosynthesis of anthocyanin during tissue culture of carrot aggregen, *Bot. Mag. (Tokyo),* 80, 440, 1967 (CA 68, 85007).

11. **Hopp, W., Hinderer, W., Petersen, M., and Seitz, H. U.,** Anthocyanin-containing vacuoles isolated from protoplasts of *Daucus carota* cell cultures, *Physiol. Prop. Plant Protoplasts,* 1985, 122, 1985 (CA 103, 68280).

12. **Dougall, D. K., Johnson, J. M., and Whitten, G. H.,** A clonal analysis of anthocyanin accumulation by cell cultures of wild carrot, *Planta,* 149, 292, 1980 (CA 93, 146602).

13. **Alfermann, W. and Reinhard, E.,** Isolation of anthocyanin-producing and nonproducing tissue culture of *Daucus carota:* influence of auxin on anthocyanin production, *Experientia,* 27, 353, 1971 (CA 74, 108105).

14. **Schmitz, M. and Seitz, U.,** Inhibition of anthocyanin synthesis by gibberellic acid in callus cells of *Daucus carota, Z. Pflanzenphysiol.,* 68, 259, 1972 (CA 78, 39215).

15. **Hemingson, J. C. and Collins, R. P.,** Anthocyanins present in cell cultures of *Daucus carota, J. Nat. Prod.,* 45, 385, 1982.

16. **Harborne, J. B., Mayer, A. M., and Bar-Nun, N.,** Identification of the major anthocyanin of carrot cells in tissue cultures as cyanidin 3-(sinapoylxylosylglucosylgalactoside), *Z. Naturforsch., C,* 38, 1055, 1983 (CA 100, 82764).

17. **Potter, H. S., Cloninger, C. K., and Drost, A.,** Foliar and soil application of chemicals for control of pink rot of celery, *Mich. Agric. Exp. Stn. Q. Bull.,* 40, 734, 1958 (JSFA 1959, 122ii).

18. **Timberlake, C. F. and Bridle, P.,** Anthocyanins in fruit and vegetables, in *Recent Advances in the Biochemistry of Fruits and Vegetables,* Friend, J. and Rhodes, M. J. C., Eds., Academic Press, London, 1981, 221.

19. **Arus, P. and Orton, T. J.,** Inheritance patterns and linkage relationship of eight genes of celery, *J. Hered.,* 75(1), 11, 1984 (PBA 1984, 4701).

20. **Kiss, P. A. D.,** Inheritance of red pigmentation in salad chicory, *Zuechter,* 35, 297, 1965 (PBA 1966, 5178).

21. **Kiss, P. A. D.,** Une chicoree de Bruxelles a feuilles rouges, *Rev. Hortic., Paris,* 135, 542, 1963 (HA 1964, 2871).

22. **Lawrence, W. J. C., Price, J. R., Robinson, G. M., and Robinson, R.,** Distribution of anthocyanins in flowers, fruits and leaves, *Phil. Trans. R. Soc. (London), Ser. B,* 230, 149, 1939 (CA 33, 8689).

23. **Harborne, J. B.,** *Comparative Biochemistry of the Flavonoids,* Academic Press, London, 1967, 227.

24. **Kastle, J. H. and Haden, R. L.,** The color changes occurring in the blue flowers of wild chicory, *Cichorium intybus, Am. Chem. J.,* 46, 315, 1911 (CA 5, 3839).

25. **Todt, D.,** Opening and anthocyanin content variations in blossoms of *Cichorium intybus* in light-dark cycle and under constant conditions, *Z. Bot.,* 50, 1, 1962.

26. **Proctor, J. T. A. and Creasy, L. L.,** An anthocyanin-decolorizing system in florets of *Cichorium intybus, Phytochemistry,* 8, 1401, 1969.

27. **Timberlake, C. F., Bridle, P., and Tanchev, S. S.**, Some unusual anthocyanins occurring naturally or as an artifact, *Phytochemistry*, 10, 165, 1971.

28. **Cappelletti, E. M. and Caniato, R.**, Leaves of some *Cichorium intybus* L. cultivars as a source of anthocyanin pigments, *Plant. Med. Phytother.*, 18(1), 3, 1984 (CA 101, 207628).

29. **Bridle, P., Loeffler, R. S. T., Timberlake, C. F., and Self, R.**, Cyanidin 3-malonylglucoside in *Cichorium intybus*, *Phytochemistry*, 23, 2968, 1984.

30. **Takeda, K., Harborne, J. B., and Self, R.**, Identification and distribution of malonated anthocyanins in plants of the Compositae, *Phytochemistry*, 25, 1337, 1986.

31. **Du, C. T. and Francis, F. J.**, Anthocyanins in garlic (*Allium sativum* L.), *J. Food Sci.*, 40, 1101, 1975.

32. **Boscher, J.**, Composition of anthocyanin pigments present in inflorescent bulblets of *Allium vineale*, *C.R. Acad. Sci. Fr., Ser. D*, 271, 584, 1970 (CA 74, 1002).

33. **Borukh, I. F. and Demkevich, L. I.**, The pigments in some varieties of West Ukraine garlic, *Izv. Vyssh. Uchebn. Zaved., Pishch, Teckhnol.*, 3, 47, 1976 (CA 85, 199571).

34. **Kenmochi, K. and Katayama, O.**, Studies on the utilization of plant pigments. I. Anthocyanin pigments of red garlic *(Allium sativum)* and red onion *(A. cepa)*, *Nippon Shokuhin Kogyo Gakkai-Shi*, 22, 598, 1975 (HA 1976, 11214).

35. **Hilal, S. H., Shabana, M. M., and Ibrahim, A. S.**, Flavonoids and anthocyanins of the flowers of *Allium aschersonianum* Barb. and *Muscari comosum* Mill., *Egypt. J. Pharm. Sci.*, 24, 67, 1983 (CA 104, 145491).

36. **Fenwick, G. R. and Hanley, A. B.**, The genus *Allium*. II., *CRC Crit. Rev. Food Sci. Nutr.*, 22, 273, 1985.

37. **Lee, Q. H. P. and Park, H.**, Studies on the anthocyanin pigments in fruits of *Panax* species. I. Identification of major pigment, *J. Korean Agric. Chem. Soc.*, 23(4), 242, 1980 (CA 94, 205450).

38. **Ishikura, N.**, A further survey of anthocyanins and other phenolics in *Ilex* and *Euonymus*, *Phytochemistry*, 14, 1439, 1975.

39. **Li, S. and Zhu, W.**, Studies on pigment cell culture of *Panax ginseng*, *Zhiwu Xuebao*, 32, 103, 1990 (CA 113, 129397).

40. **Mazza, G.**, Distribution of sugars, dry matter and protein in Jerusalem artichoke tubers, *Can. Inst. Food Sci. Technol. J.*, 18, 263, 1985.

41. **Mazza, G.**, Sorption isotherms and drying rates of Jerusalem artichoke (*Helianthus tuberosus* L.), *J. Food Sci.*, 49, 384, 1984.

42. **Hoehn, E.**, Food potential of Jerusalem artichoke, *Proc. Annu. Conf. Manitoba Agron.*, p. 128, 1982.

43. **Fleming, S. E. and Groot Wassink, J. W. D.**, Preparation of high-fructose syrup from the tubers of Jerusalem artichoke (*Helianthus tuberosus* L.), *CRC Crit. Rev. Food Sci. Nutr.*, 12, 1, 1979.

44. **Guiraud, J. P., Caillaud, J. M., and Galzy, P.**, Optimization of alcohol production from Jerusalem artichoke, *Eur. J. Appl. Microbiol. Biotechnol.*, 14, 81, 1982.

45. **Apaire, V., Guiraud, J. P., and Galzy, P.**, Selection of yeasts for single cell protein production on media based on Jerusalem artichoke extracts, *Z. Allg. Mikrobiol.*, 23, 211, 1983.

46. **Arrazola, J. M.**, New raw material for production of fodder yeast, *Proc. 4th Int. Congr. Microbiol.*, 26, p. 555, 1947.

47. **Benk, E., Koeding, G. V., Treiber, H., and Bielecki, F.**, Jerusalem artichoke brandy. III. Results of investigation of laboratory produced Jerusalem artichoke brandy, *Alkohol-Ind.*, 83, 463, 1970.

48. **Kosaric, N., Wieczorek, A., Costantino, G. P., and Duvnjak, Z.**, Industrial processing and products from Jerusalem artichoke, in *Advances in Biochemical Engineering*, Springer-Verlag, Berlin, Germany, 1985, 1.

49. **Kiehn, F. and Chubey, B. B.,** Variability in agronomic and compositional characteristics of Jerusalem artichoke, paper presented at the Int. Congr. Food Non-Food Appl. Inulin Inulin-Containing Crops, Wageningen, The Netherlands, February 18–21, 1991.

50. **Miniati, E.,** Anthocyanin content of Jerusalem artichoke, unpublished results, 1986.

51. **Swanson, C. R., Hendricks, S. B., Toole, V. K., and Hagen, C. E.,** Effect of 2,4-dichlorophenoxyacetic acid and other growth regulators on the formation of a red pigment in Jerusalem artichoke tuber tissue, *Plant Physiol.,* 31, 315, 1956 (CA 50, 17296b).

52. **Ibrahim, R. K., Thakur, M. L., and Permanand, B.,** Formation of anthocyanins in callus tissue culture, *Lloydia,* 34, 175, 1971 (CA 75, 106030).

53. **Schorr, L.,** Anthocyanin in protoplasm, *Z. Wiss. Mikrosk.,* 52, 369, 1936 (CA 33, 4627).

54. **Robinson, G. M. and Robinson, R.,** A survey of anthocyanins. II. *Biochem. J.,* 26, 1647, 1932.

55. **Fouassin, A.,** Identification par chromatographie des pigments anthocyaniques des fruits et des legumes, *Rev. Ferment. Ind. Aliment.,* 11, 173, 1956.

56. **Brandwein, B. J.,** Pigments in three cultivars of the common onion *(Allium cepa), J. Food Sci.,* 30, 680, 1965.

57. **Fuleki, T.,** An improved chromatographic method for the separation and detection of anthocyanins, *Rep. Hortic. Res. Inst. Ont.,* 1968, 91, 1969.

58. **Fuleki, T.,** The anthocyanins of strawberry, rhubarb, radish and onion, *J. Food Sci.,* 34, 365, 1969.

59. **Fuleki, T.,** Pigments responsible for the color of red onion, *Rep. Hortic. Res. Inst. Ont.,* 1969, 124, 1970 (HA 1970, 7639).

60. **Fuleki, T.,** Anthocyanins in red onion, *Allium cepa, J. Food Sci.,* 36, 101, 1971.

61. **Du, C. T., Wang, P. L., and Francis, F. J.,** Cyanidin 3-laminariobioside in spanish red onion *(Allium cepa), J. Food Sci.,* 39, 1265, 1974.

62. **Moore, A. B.,** Acylated Anthocyanins in Red Onions, Ph.D. thesis, University of Massachusetts, Amherst, 1981 (CA 95, 95740).

63. **Moore, A. B., Francis, F. J., and Clydesdale, F. M.,** Changes in chromatographic profile of anthocyanins of red onion during extraction, *J. Food Prot.,* 45, 738, 1982.

64. **Moore, A. B., Francis, F. J., and Jason, M. E.,** Acylated anthocyanins in red onions, *J. Food Prot.,* 45, 590, 1982.

65. **Terahara, N. and Yamaguchi, M.,** Anthocyanins in a red onion, *Bull. Fac. Hortic.* Minamikyusyu University, Takanabe, Miyazaki, Japan, 15, 59, 1985.

66. **Feldman, A. L., Gusar, Z. D., and Girkovakaya, E. B.,** Biochemical characteristics of varieties of southern Ukrainan onions, *Konservn. Ovoshchesuch. Promst.,* 4, 19, 1973 (CA 79, 15877).

67. **Cachin, K. and Kurosaki, T.,** Effects of gamma radiation on sprout inhibition, growth of microorganisms and chemical composition of Shinan Red onions, *J. Jpn. Soc. Hortic. Sci.,* 40, 91, 1971 (HA 1972, 3928).

68. **Rieman, G. H.,** Genetic factors for pigmentation in the onion and their relation to disease resistance, *J. Agric. Res.,* 42, 251, 1931.

69. **Davis, E. W.,** Rapid identification of recessive white onion bulbs by use of ammonia fumes, *J. Hered.,* 45, 122, 1954 (PBA 1955, 590).

70. **Clarke, A. E., Jones, H. A., and Little, T. M.,** Inheritance of bulb color in the onion, *Genetics,* 29, 569, 1944.

71. **Jones, H. A. and Peterson, C. E.,** Complementary factors for light red bulb color in onions, *Proc. Am. Soc. Hortic. Sci.,* 59, 457, 1952.

72. **El-Shafie, M. W. A. and Davis, G. N.,** Inheritance of bulb color in the onion *(Allium cepa* L.), *Hilgardia,* 38, 607, 1967 (PBA 1968, 7068).

73. **Patil, J. A., Deokar, A. B., and Maslekar, S. R.,** Inheritance of bulb colour in onion, *Res. J. Mahatma Phule Agric. Univ.,* 2(1), 92, 1971 (PBA 1973, 6352).

74. **Hermann, K.,** On the contents of localization of phenolics in vegetables, *Qual. Plant. Plant Foods Hum. Nutr.,* 25, 231, 1976.

75. **Kiviranta, J., Huovinen, K., and Hiltunen, R.,** Variation of flavonoids in *Allium cepa, Planta Med.,* 517, 1986.

76. **Talburt, W. F. and Smith, D.,** *Potato Processing,* AVI/Van Nostrand Reinhold, New York, 1987.

77. **De Jong, H.,** Inheritance of pigmented tuber flesh in cultivated diploid potatoes, *Am. Potato J.,* 64, 337, 1987.

78. **Gray, D. and Hughes, J. C.,** The tuber quality, in *The Potato Crop: The Scientific Basis for Improvement,* Harris, P. M., Ed., Chapman & Hall, New York, 1978, 504.

79. **Vidner, J.,** Colour of the flesh and skin of tubers in a world collection of potatoes, *Genet. Shlechteni,* 8, 51, 1972 (PBA 1972, 8264).

80. **Chmielewska, I.,** Pigments of violet potatoes, *Rocz. Chem.,* 15, 491, 1935 (CA 30, 2964).

81. **Chmielewska, I.,** Pigments of violet potatoes, *Bull. Soc. Chim. Fr.,* 3, 1575, 1936 (CA 30, 7579).

82. **Harborne, J. B.,** Plant polyphenols. I. Anthocyanin production in the cultivated potato, *Biochem. J.,* 74, 262, 1960.

83. **Howard, H. W., Kukimura, H., and Whitmore, E. T.,** The anthocyanin pigments of the tubers and sprouts of *tuberosum* potatoes, *Potato Res.,* 13, 142, 1970 (CA 73, 106335).

84. **Sachse, J.,** Anthocyane in Kartoffelsorten Urgenta und Desirée (*Solanum tuberosum* L.), *Z. Lebensm. Unters. Forsch.,* 153, 294, 1973 (CA 80, 69376).

85. **Andersen, O. M., Opheim, S., Aksnes, D. W., and Froeystein, N. A.,** Structure of petanin, an acylated anthocyanin isolated from *Solanum tuberosum,,* using homo- and hetero-nuclear two-dimensional nuclear magnetic resonance techniques, *Phytochem. Anal.,* 2, 230, 1991.

86. **Von Rathlef, H. and Siebeneick, H.,** On some crosses of Peruvian cultivars of *S. andigenum* Juz. et Buk. with Richters Jubel and the genetics of skin color, tuber color, flesh color, flower color and tuber shape in the potato, *Genetica,* 16, 153, 1934.

87. **Swaminathan, M. S. and Howard, H. W.,** The cytology and genetics of the potato (*Solanum tuberosum*) and related species, *Bibliographia Genet.,* 16, 1, 1953.

88. **Howard, H. W.,** Potato cytology and genetics 1952–1959, *Bibliogr. Genet.,* 19, 87, 1960.

89. **Howard, H. W.,** *Genetics of the Potato, Solanum tuberosum,* Springer-Verlag, New York, 1970, 126.

90. **De Jong, H.,** Inheritance of anthocyanin pigmentation in the cultivated potato: a critical review, *Am. Potato J.,* 68, 585, 1991.

91. **Dodds, K. S. and Long, D. H.,** The inheritance of colour in diploid potatoes. I. Types of anthocyanidins and their genetic loci, *J. Genet.,* 53, 136, 1955.

92. **Salaman, R. N.,** The inheritance of colour and other characters in the potato, *J. Genet.,* 1, 6, 1911.

93. **Harborne, J. B.,** Phenolic compounds of the potato, *Proc. 1st Congr. Food Sci. Technol.,* London, 1962, 423.

94. **Dodds, K. S. and Long, D. H.,** The inheritance of colour in diploid potatoes. II. A three-factor linkage group, *J. Genet.,* 54, 27, 1956.

95. **Hermsen, J. G. Th. and Verdenius, J.,** Selection from *Solanum tuberosum* Group Phureja of genotypes combining high-frequency haploid induction with homozygosity for embryo-spot, *Euphytica,* 22, 244, 1973.

96. **Lunden, A. P.,** Some more evidence of autotetraploid inheritance in the potato (*Solanum tuberosum*), *Euphytica,* 9, 225, 1960.

97. **Kelly, J. P.,** Seed progeny of a potato with faintly coloured tubers, *J. Genet.,* 14, 197, 1924.

98. **De Jong, H. and Rowe, P. R.,** Genetic markers in inbred clones of cultivated diploid potatoes, *Potato Res.,* 1, 200, 1972.

99. **De Jong, H.,** Inheritance of pigmented tuber flesh in cultivated diploid potatoes, *Am. Potato J.,* 64, 337, 1987.

100. **Kessel, R. and Rowe, P. R.,** Inheritance of two qualitative traits and a proposed genetic map for their linkage group in diploid potatoes, *Potato Res.,* 17, 283, 1974.

101. **Garg, K. C., Tiwari, S. P., and Sharma, K. P.,** Inheritance of leaf pigmentation in dihalpoid-Phureja hybrids of potato, *J. Indian Potato Assoc.,* 8, 31, 1981.

102. **Gebhardt, C., Ritter, E., Debener, T., Schachtschabel, U., Walkemeier, B., Uhrig, H., and Salamini, F.,** RFLP analysis and linkage mapping in *Solanum tuberosum, Theor. Appl. Genet.,* 78, 65, 1989.

103. **Nelson, D. C. and Bristol, D. W.,** 2,4-D on potatoes, Agric. Ext. Serv. Univ. Minn. N.D. State Univ., Red River Valley Potato Facts No. 5, 1975.

104. **Fritz, V. A., Hebel, J. B., Borowski, A. M., and Hung, P. E.,** Ethephon and 2,4-D do not improve periderm color and may decrease yield of red-skinned 'Norland' potato, *HortSci.,* 26, 553, 1991.

105. **Reeve, R. M., Hautala, E., and Weaver, M. L.,** Anatomy and compositional variation within potatoes. II. Phenolics, enzymes and other minor components, *Am. Potato J.,* 46, 347, 1969.

106. **Craft, C. and Audia, W.,** Phenolic substances and barrier formation in potato, *Bot. Gaz. (Chicago),* 123, 211, 1962.

107. **Smith, E. P.,** Flower colors as natural indicators, *Trans. Proc. Bot. Soc. Edinburgh,* 30, 230, 1931 (BA 1932, 24621).

108. **Smith, E. P.,** The calibration of flower color indicators, *Protoplasma,* 18, 112, 1933 (CA 27, 3236).

109. **Almeida Cousin, J. C.,** The colouring matter of *Ipomea roxa, Rev. Quim. Farm. (Santiago),* 1, 298, 1936 (CA 33, 4627).

110. **Rangaswami Ayyanagar, K. and Sampathkumar, R.,** Inheritance of leaf shape and colour in sweet potato, *Indian J. Genet. Plant Breed.,* 38, 262, 1978 (PBA 1977, 9300).

111. **Kehr, A. E., Ting, Y. C., and Miller, J. C.,** Site of carotenoids and anthocyanin synthesis in sweet potato, *Proc. Am. Soc. Hortic. Sci.,* 65, 396, 1955.

112. **Mikell, J. J., Hernandez, T. P., and Miller, J. C.,** Preliminary studies on the inheritance of skin and flesh color of the sweet potato, *Proc. 52nd Annu. Conv. Assoc. South. Agric. Workers,* p. 113, 1955 (PBA 1956, 222).

113. **Hernandez, T. P., Hernandez, T., and Miller, J. C.,** Effect of different rates of irradiation on sweet potato roots and tomato seeds, *Proc. 56th Annu. Conv. Assoc. South. Agric. Workers,* p. 157, 1959 (PBA 1960, 3703).

114. **Kazui, T. and Umaki, T.,** Anthocyanin Dye from Sweet Potatoes, Japanese Patent, 177,097, 1948 (CA 45, 5416h).

115. **Cascon, S. C., Carvalho, M. P. M., Moura, L. L., Guimaraes, I. S. S., and Philip, T.,** Natural colorants from purple sweet potato for use in foods, *Bol. Pesqui. — EMBRAPA, Cent. Tecnol. Agric. Aliment.,* 9, 25, 1984 (CA 104, 49979).

116. **Bassa, I. A. and Francis, F. J.,** Stability of anthocyanins from sweet potatoes in a model beverage, *J. Food Sci.,* 52, 1753, 1987.

117. **Imbert, M. P., Seaforth, C. E., and Williams, D. B.,** Anthocyanin pigment of the sweet potato *Ipomoea batatas, Proc. Am. Soc. Hortic. Sci.,* 88, 481, 1966 (CA 65, 14098h).

118. **Tsukui, A., Kuwano, K., and Mitamura, T.,** Anthocyanin pigment isolated from purple root of sweet potato, *Kaisegaku Zasshi,* 34, 153, 1983 (CA 99, 102276).

119. **Miyazaki, T., Tsuzuki, W., and Suzuki, T.,** Composition and structure of anthocyanins in the periderm and flesh of sweet potatoes, *J. Jpn. Soc. Hortic. Sci.,* 60(1), 217, 1991.

120. **Odake, K., Terahara, N., Saito, N., Toki, K., and Honda, T.,** Chemical structures of two anthocyanins from purple sweet potato, *Ipomea batatos, Phytochemistry,* 31, 2127, 1992.

121. **Pomilio, A. B. and Sproviero, J. F.**, Acylated anthocyanins from *Ipomoea cairica*, *Phytochemistry*, 11, 1125, 1972.

122. **Pomilio, A. B. and Sproviero, J. F.**, Acylated anthocyanins. II. Complex anthocyanins from *Ipomoea cairica*, *Phytochemistry*, 11, 2323, 1972.

123. **Gupta, O. C. D., Gupta, R., and Gupta, P. D.**, Chemical examination of flowers of *Ipomoea fistulosa*, *Planta Med.*, 38, 147, 1980 (CA 93, 3905).

124. **Nishimaki, T. and Nozue, M.**, Isolation and culture of protoplasts from high anthocyanin-producing callus of sweet potato, *Plant Cell Rep.*, 4, 248, 1985.

125. **Nozue, M. and Yasuda, H.**, Occurrence of anthocyanoplasts in cell suspension cultures of sweet potato, *Plant Cell Rep.*, 4, 252, 1986.

126. **Nozue, M., Kikuma, M., Miyamoto, Y., Fukuzaki, E., Matsumura, T., and Hashimoto, Y.**, Red Anthocyanin Pigment and Its Manufacture with Callus Culture of *Ipomoea batatas*, Japanese Patent, 63,233,993, 1988 (CA 111, 132628f).

127. **Yamazaki, Y. T., Owari, K., Kawasaki, Y., Yamada, T., and Yoshihara, K.**, Red color development of sweet potato epidermis by treatment with phosphate preparations, *Eisei Shikensho Hokoku*, 104, 154, 1986.

128. **Vyas, S. P., Bohra, S. P., and Sankhla, N.**, Antagonism between morphactin and ethylene in root-coiling of *Ipomoea pentaphylla*, *Z. Pflanzenphysiol.*, 69, 185, 1973 (CA 78, 132630).

129. **Greenwell, A. B. H.**, Taro with special reference to its cultural and uses in Hawaii, *Econ. Bot.*, 1, 276, 1947.

130. **Chan, H. T., Jr., Kao-Jao, T. H. C., and Nakayama, T. O. M.**, Anthocyanin composition of taro, *J. Food Sci.*, 42, 19, 1977.

131. **Strauss, M. S., Stephens, G. C., Gonzales, C. J., and Arditti, J.**, Genetic variability in taro, *Colocasia esculenta* (L.) Schott (Araceae), *Ann. Bot. (London)*, 45, 429, 1980.

132. **Oke, O. L.**, Yam, a valuable source of food and drugs, *World Rev. Nutr. Diet.*, 15, 156, 1972 (FSTA 7, 11J1698).

133. **Martin, F. W., Telek, L., and Ruberte, R. M.**, Yellow pigments of *Dioscorea bulbifera*, *J. Agric. Food Chem.*, 22, 335, 1974.

134. **Rasper, V. and Coursey, D. G.**, Anthocyanins of *Dioscorea alata* L., *Experientia*, 23, 601, 1967 (CA 67, 97638).

135. **Imbert, M. P. and Seaforth, C.**, Anthocyanins of *Dioscorea alata* L., *Experientia*, 24, 417, 1968 (CA 69, 675).

136. **Karnick, C. R.**, Phytochemical investigation of some *Dioscorea* species and varieties found in India, *Q. J. Crude Drug Res.*, 11, 1761, 1971.

137. **Tsukui, A., Kuwano, K., Mitamura, T., and Tanimura, W.**, The anthocyanin pigments of powdered Philippine purple yam, *Nippon Nogei Kagaku Kaishi*, 51(8), 5, 1977 (CA 87, 180667).

138. **Martin, F. W. and Rhodes, A. M.**, Correlation among greater yam (*Dioscorea alata*, L.) cultivars, *Trop. Agric.*, 50, 183, 1973 (FSTA 5, 10J1653).

139. **Rhodes, A. M. and Martin, F. W.**, Multivariate studies of variations in yams (*Dioscorea alata* L.), *J. Am. Soc. Hortic. Sci.*, 97, 685, 1972.

140. **Martin, F. W., Cabanillas, E., and Guadalupe, R.**, Selected varieties of *Dioscorea alata* L., the Asian greater yam, *J. Agric. Univ. P.R.*, 59, 165, 1975 (FSTA 8, 2J305).

141. **Carreno-Diaz, R. and Grau, N.**, Anthocyanin pigments in *Dioscorea tryphida* L., *J. Food Sci.*, 42, 615, 1977.

Chapter 11

COLE CROPS

I. CABBAGES

Knowledge of the red pigmentation in cabbage, *Brassica oleracea* var. *capitata* forma *rubra* L., dates back to the 19th century.[1] Rubrobrassicin was the name given to the anthocyanin pigment of red cabbage which Chmielewska[2-4] identified as a glycoside of cyanidin esterified with sinapic acid. Later chromatographic studies, however, showed that the pigment composition was much more complex than first believed. In fact, Fouassin[5] separated six red, violet, and blue cyanidin derivatives from cabbage. Their Rf values showed that they were acylated heterosides or biosides.

The composition of rubrobrassicin was further elucidated by Stroh[6] who showed that it was composed of cyanidin and D-glucose in a 1:3 ratio. Later chemical and chromatographic tests[7] revealed that its structure is cyanidin 3-sophoroside-5-glucoside. Using paper chromatography, Metche[8] described three pigments: cyanidin 3-sophoroside-5-glucoside and two cyanidin derivatives, one containing sinapic acid and one acylated with two ferulic residues. Harborne[9] noted that the sinapic acid may have been present as an impurity in the form of esters. Later analyses by Tanchev and Timberlake[10] confirmed two sinapic esters of cyanidin, cyanidin 3-sophoroside-5-glucoside acylated with sinapic acid, and cyanidin 3-sophoroside-5-glucoside diacylated with sinapic acid. These authors also characterized cyanidin 3-sophoroside-5-glucoside and cyanidin 3,5-diglucoside. The esterification with cinnamic acids was further confirmed by Lanzarini and Morselli[11] who identified four of the ten pigments separated by paper chromatography: cyanidin 3-sophoroside-5-glucoside acylated with ferulic acid, cyanidin 3-sophoroside-5-glucoside acylated with *p*-coumaric acid, and the already known cyanidin 3,5-diglucoside and cyanidin 3-sophoroside-5-glucoside.

More recent studies by Hrazdina et al.[12] with 'Red Danish' cabbage comfirm the results of the previously cited works and add to the list of identified pigments cyanidin 3-sophoroside-5-glucoside acylated with malonic acid and cyanidin 3-sophoroside-5-glucoside acylated with ferulic acid, bringing the total of identified anthocyanin pigments to eight.

In 1987, using HPLC, Idaka[13] separated more than 15 anthocyanins from *B. oleracea* and fully characterized 11 of these as 3-*O*-(β-D-glucopyranosyl)-5-*O*-(β-D-glucopyranosyl)cyanidin, 3-*O*-(6-*O*-*E*-ferulyl-β-D-glucopyranosyl)-5-*O*-(β-D-glucopyranosyl)cyanidin, 3-*O*-(2-*O*-(β-D-glucopyranosyl)-β-D-glucopyranosyl)-5-*O*-(β-D-glucopyranosyl)cyanidin, 3-*O*-(6-*O*-*E*-*p*-coumaryl-2-*O*-(β-D-glucopyranosyl)-β-D-glucopyranosyl)-5-*O*-(β-D-glucopyranosyl) cyanidin, 3-*O*-(6-*O*-*E*-ferulyl-2-*O*-(β-D-glucopyranosyl)-β-D-glucopyranosyl)

-5-*O*-(β-D-glucopyranosyl)cyanidin, 3-*O*-(6-*O*-*E*-sinapyl-2-*O*-(β-D-glu-co-pyranosyl)-β-D-glucopyranosyl)-5-*O*-(β-D-glucopyranosyl)cyanidin, 3-*O*-(6-*O*-(4-*O*-(β-D-glucopyranosyl) *E*-*p*-coumaryl)-2-*O*-(β-D-glucopyranosyl)-β-D-glucopyranosyl)-5-*O*-(β-D-glucopyranosyl) cyanidin, 3-*O*-(6-*O*-(4-*O*-(β-D-glucopyranosyl) -*E*-ferulyl)-2-*O*-(β-D-glucopyranosyl) -β-D-glucopyrano-syl)-5-*O*-(β-D-glucopyranosyl)cyanidin, 3-*O*-(6-*O*-*E* -*p*-coumaryl-2- *O*-(2-*O* -*E*-sinapyl-β-D-glucopyranosyl)-β-D-glucopyranosyl)-5-*O* -(β-D-glucopy-ranosyl)cyanidin, 3-*O*-(6-*O*-*E*-ferulyl-2-*O*-(2-*O*-*E*-sinapyl-β-D-glucopyran-osyl)-β-D-glucopyranosyl)-5-*O*-(β-D-glucopyranosyl)cyanidin, 3-*O*-(6-*O*-*E*-sinapyl-2-*O* -(2-*O*-*E* -sinapyl-β-D-glucopyranosyl-β-D-glucopyranosyl)-5-*O*-(β-D-glucopyranosyl)cyanidin (Figure 1).[13-17] Nakatani et al.[18] and Ikeda et al.[19] also reported the same diacylates.

In vivo, the anthocyanins of red cabbage are dissolved in the cellular sap and are synthesized in intracellular organelles called anthocyanoplasts.[20] Small and Pecket[21] presented evidence that the anthocyanoplast in red cabbage is bound by a single tripartite membrane approximately 10 nm in thickness. Neumann,[22] however, did not find vacuolar organelles but only osmiophilic globules which lacked membranes and internal structures.

Several genetic, physiological, and agronomic factors influence the pigmentation of red cabbage. Pecket and Small[20] observed that cabbage shoots which have more sugar form anthocyanins more easily, though they did not find a quantitative relationships between the level of sugars and anthocyanins. The accumulation of anthocyanins in the leaves of red cabbage grown on saline substrate containing isosmotic solutions of Na_2SO_4 or NaCl were interpreted by Strogonof and Dostanova[23] and Dostanova[24] as a specific and protective reaction of the plant to the toxicity of the salts. Anthocyanins were also produced in response to UV irradiation.[25]

When red cabbage is grafted onto white varieties, the characteristics of both species appear in the successive generation.[26] However, the anthocyanin concentration in the progeny is much lower.[27] Bubarova[28] studied the effects of crossing radish with cabbage and kohlrabi on the anthocyanin pattern and concentration. The results revealed no qualitative changes, because cyanidin remained the only anthocyanidin present in the hybrids. Also, the hybrid from the radish × kohlrabi crosses inherited the glycosidic arrangement of the radish anthocyanins, but with higher levels of pigmentation. Eenink[29] observed that the genotypes for anthocyanin pigmentation in crosses between *B. oleracea* and *Raphanus sativus* or *Eruca sativa* resulted in matromorphic plants of heterozygotic type. Cabbage and mustard seed capacity of germination, which is higher in darker seeds, has been correlated with the presence of anthocyanin in the seed coat.[30,31] Several studies have addressed the inheritance of anthocyanin pigmentation in cabbage[32,33] and used red cabbage to investigate the mechanism of phytochrome action in the control of biosynthesis of anthocyanins in *Brassica*.[34-36]

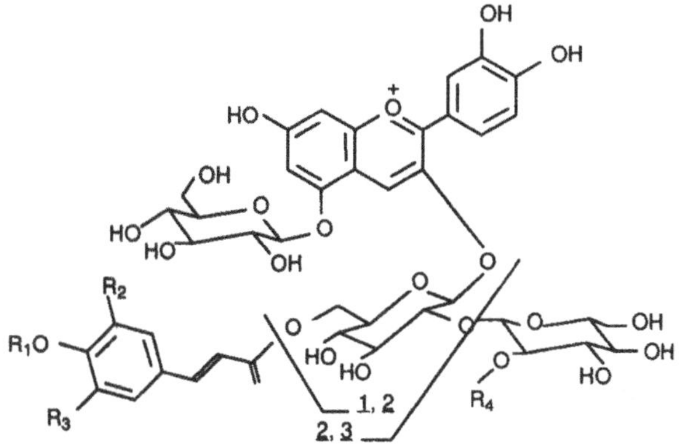

		FABMS (m/z)
1	R_4 = H	773
2		611
3 R_1 = R_3 = H R_2 = OCH$_3$		787
4 R_1 = R_2 = R_3 = H	R_4 = H	919
5 R_1 = R_3 = H R_2 = OCH$_3$	R_4 = H	949
6 R_1 = H R_2 = R_3 = OCH$_3$	R_4 = H	979
7 R_1 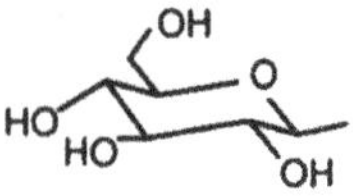 =	R_2 = R_3 = H R_4 = H	1081
8 R_1 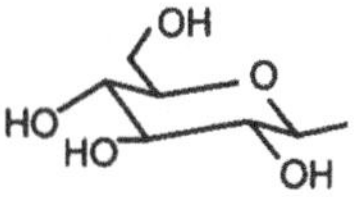 =	R_2 = OCH$_3$ R_3 = R_4 = H	1111
9 R_1 = R_2 = R_3 = H	R_4 = sinapyl	1125
10 R_1 = R_3 = H R_2 = OCH$_3$	R_4 = sinapyl	1155
11 R_1 = H R_2 = R_3 = OCH$_3$	R_4 = sinapyl	1185

FIGURE 1. Structures of anthocyanins from red cabbage, *Brassica oleracea*. (Adapted from Idaka, E., Japanese Patents.[13-15] With permission.)

A variety of factors are known to influence the biosynthesis of anthocyanins in red cabbage: treatment with indolacetic acid (IAA) leads to a markedly lower anthocyanin content.[37] Application of herbicides may either inhibit or stimulate anthocyanin production.[38,39] Cabbage plantlets grown in the dark accumulate low levels of anthocyanins (about 0.35 nmol/g) which increase over 14 times to 5 nmol/g after 6 d in the light. The synthesis is linked to the phenylalanine ammonia-lyase (PAL) activity;[41] and the most effective light is that in the red, blue, and UV regions of the spectrum.[40] There is almost no benefit in exposing the plant to light in the far red region.[40,41] In cabbage, there is only one photoreceptor, a copper-containing flavoprotein which is most active at 690 and 450 nm.[42] Kinetin stimulates the nonphotosynthetic formation of anthocyanins, and its action is enhanced by the shikimic and cinnamic acids as a result of its effect on the cycle of these acids though not by stimulating the PAL action. The effect is probably due to an increased permeability of the cellular membranes caused by the kinetin.[42]

Over the years, considerable interest has been shown in the use of red cabbage pigments as a pH indicator.[43-48] This is because cabbage anthocyanins, like most other anthocyanins, assume a vast range of colors ranging from red to blue to green to yellow as the pH varies from acidic to alkaline. Grube et al.[45] suggested that cabbage pigment could be used for a restricted pH range, 8.5 to 10.0. After a spectrophotometric study of the coloration at varying pH, Wolf[46] concluded that it could be used as a universal pH indicator. Currently, however, red cabbage pigment extract is used as a pH indicator only in teaching exercises.[47,48]

Since the mid-1970s, there has been an increased interest in red cabbage pigments as a possible source of natural colors for beverages and other products. Shewfelt and Ahmed[49] reported a successful procedure to isolate the pigment using 350 ppm SO_2 in water. They employed an ion exchange procedure to purify the pigment and prepared freeze-dried powders suitable for coloring dry beverage mixes.[50] These authors also demonstrated that the pigment preparations obtained from red cabbage imparted a color superior to that obtained with Red No. 40. Color stability was satisfactory both in the dry and in reconstituted beverage mixes during refrigerated storage. Sapers[51] described a process for the preparation of a deodorized colorant, prepared with the use of a recyclable odor adsorbent resin, Amberlite XAD-7. The anthocyanins are later eluted from the adsorbent with acidified ethanol. The resulting colorant possesses greater heat and storage stability than colorants based on grapes, red beets, and cranberries[52] (Figures 2 and 3).

Several patents describing the preparation and application of red food colorants from red cabbage were granted to Japanese investigators in the 1980s.[53-64] One of these patents[53] describes preparation methods involving aqueous or ethanolic extraction followed by treatment with potassium sulfate for 20 h at 35°C. The extract is filtered, acidified with sulfuric and citric acids, concentrated, neutralized with potassium hydroxide, and diluted with a 3% ethanol solution. Shibata and co-workers patented the use of pigments from red cabbage in soft drinks[54] and candies.[55] Variations of the extraction process with water or alcohol containing SO_2 (followed by purification with a cationic resin) to eliminate the odors and extraneous materials, as well as the application of ultrafiltration to the manufacturing of colorants were recently patented.[56,57] Idaka[58] patented the use of chromatography for removal of acids from acidic anthocyanin pigments. Novel purification methods aimed at removing the odor associated with the sulfur compounds have also been

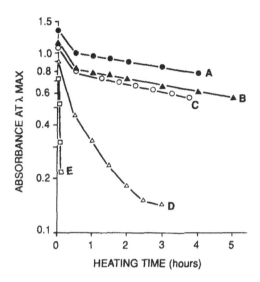

FIGURE 2.   Effects of heating at 100°C on the absorbance of pH 3 buffer solutions of colorants: (A) 'Meteor' cabbage; (B) 'Resistant Red Acre' cabbage; (C) 'Red Head' cabbage; (D) cranberry concentrate; and (E) red beet. (From Sapers, G. M., Taffer, I., and Ross, L. R., *J. Food Sci.*, 46, 105, 1981. With permission.)

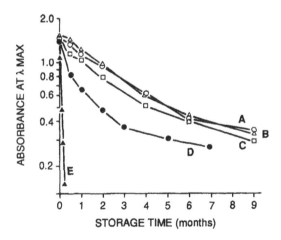

FIGURE 3.   Effects of storage in light at 25°C on the absorbance of pH 3 buffer solutions of colorants: (A) 'Meteor' cabbage; (B) 'Red Head' cabbage; (C) 'Resistant Red Acre' cabbage; (D) cranberry concentrate; and (E) red beet. (From Sapers, G. M., Taffer, I., and Ross, L. R., *J. Food Sci.*, 46, 105, 1981. With permission.)

developed.[59-63] Precipitation of impurities from pigment extracts with sodium titanate was proposed by Koda and Yasuda.[59] Similarly, Tadamasa and Yasuda[61] suggested the use of sodium polyphosphate, and Yasuda and Kotake[62] patented a process for the removal of proteins with tannins. The use of supercritical CO_2 for the production of a high purity product, as well as production of pigments from tissue culture of cabbage has also been proposed.[63,64]

Presently, a commercially available red cabbage color is marketed under the name San Red RC.[65-68] This product provides a deep blue-to-red color at pH 3.0 or below and becomes more blue in tone as the pH is increased to 4.0 and above. The manufacturer recommends the use of this colorant in beverages, chewing gums, candies, sherberts, dressings, yogurt, and other fermented products.[65,66]

II. RADISH

The anthocyanins of the red-skinned radish (*Raphanus sativus* L.) possess the same basic glycosidic pattern as those of red cabbage, although the radish also contains pelargonidin derivatives.[69] The first investigation on the anthocyanins of radish were published in 1910, when Sacher[70,71] and Schwertschlager[72] reported on the potential of these pigments as an acid-base indicator. Fouassin[5] conducted the first chromatographic separation and reported the presence of five pigments, probably acylated derivatives of pelargonidin. Harborne[73] found that the pigmentation of the roots and corolla of red *R. sativus* is made up of pelargonidin derivatives, while the purple forms contain cyanidin as the aglycone. Harborne and Paxman[74] attributed the presence of cyanidin in *R. sativus* to gene *H*, which controls hydroxylation of anthocyanins in radish. As a result, the genetically recessive scarlet form of radish contains cyanidin derivatives.[74,75] Earlier, Harborne and Sherratt[76] had attributed the pigmentation of 'Scarlet Globe' red radish to pelargonidin 3-diglucoside-5-glucoside acylated with *p*-coumaric or ferulic acid, and called it raphanin.

Using chromatographic and electrophoretic analyses, Ishikura and Hayashi[77] separated five derivatives of pelargonidin 3-diglucoside-5-glucoside acylated with *p*-coumaric, caffeic, and ferulic acids. Later work by these authors[78] led to the characterization of a triglucoside of cyanidin esterified with three molecules of cinnamic acid in a purple hybrid of *R. sativus*. This triglucoside of cyanidin is structurally analogous to the pelargonidin derivative encountered in the red variety.[79]

Using paper chromatography, Fuleki[80] resolved and characterized 13 pigments from extracts of the epidermal layer of red-skinned radish roots (Table 1). Pelargonidin 3-sophoroside-5-glucosides acylated with *p*-coumaric, ferulic, or caffeic acid predominated. The other major anthocyanin was pelargonidin 3-sophoroside-5-glucoside. Pelargonidin 3-sophoroside-5-glucoside acylated with *p*-coumaric, ferulic, and caffeic acids was a minor component. However, in *R. caudatus*, Lele[81] found malvidin 3,5-diglucoside.

TABLE 1
Anthocyanins Identified in Radish Roots *(Raphanus sativus)*

Cultivar	Anthocyanin	Relative conc	Ref.
Scarlet Globe	Pelargonidin 3-diglucoside-5-glucoside acylated with *p*-coumaric or ferulic acid		76
Iwakumi-aka	Pelargonidin 3-diglucoside-5-glucoside acylated with *p*-coumaric or ferulic and caffeic acids		77
Purple var.	Cyanidin 3-sophoroside-5-glucoside acylated with *p*-coumaric or ferulic acid		73
Red var.	Pelargonidin 3-sophoroside-5-glucoside acylated with *p*-coumaric or ferulic acid		73
Purple hybrid	Cyanidin 3-diglucoside-5-glucoside acylated with *p*-coumaric and/or ferulic and/or caffeic acids		78
French Breakfast, Scarlet White Tip, and two unknown red cultivars	Pelargonidin 3-sophoroside-5-glucoside acylated with *p*-coumaric, ferulic, and caffeic acids	+	80
	Pelargonidin 3-sophoroside-5-glucoside acylated with *p*-coumaric and ferulic acids	+ +	
	Pelargonidin glycoside	+	
	Pelargonidin 3-sophoroside-5-glucoside acylated with *p*-coumaric acid	+ + + +	
	Pelargonidin 3-sophoroside-5-glucoside acylated with ferulic acid	+ + + +	
	Pelargonidin 3-sophoroside-5-glucoside acylated with caffeic acid	+ + + +	
	Pelargonidin glycoside	+ + +	
	Pelargonidin glycoside	+	
	Pelargonidin glycoside	+ +	
	Pelargonidin glycoside	+	
	Pelargonidin 3-sophoroside-5-glucoside	+ + +	
	Pelargonidin glycoside	+	
	Pelargonidin glycoside	+	

In radish, anthocyanin production may be stimulated by treatments with leucine, phenylalanine, and valine, as well as with aspartic acid and threonin. The reason for this increased production following treatment with amino acids is that amino acids, such as phenylalanine and leucin, are believed to be precursors in the synthesis of anthocyanins, and also to be incorporated into the B-ring of the anthocyanin molecule.[82-85]

Genetically, the qualitative composition of anthocyanins apparently remains unchanged with an increase in chromosome number. However, in tetraploid genotypes anthocyanin concentration increases by 20 to 70% compared to the diploids.[86] Heterosis affects anthocyanin composition and may lead to the production of malvidin and peonidin glycosides in F1 hybrids in addition to cyanidin and pelargonidin derivatives.[87]

The anthocyanins in *R. sativus* seedlings are primarily located in the cells of the subepidermal layer of the cotyledons and the hypocotyl axis.[88] Their production involves the phytochrome mechanisms; and light intensity, quality, and duration are the predominant factors affecting the extent of accumulation.[89] The biosynthesis of the pigments by way of the shikimic acid cycle is intensified by treating the *R. sativus* seedlings with *H*-indol-3-acetic acid, gibberellic acid, and kinetin.[90,91] The latter favors the formation of anthocyanins in low light conditions rather than in bright light.[92] Other phytohormones that may enhance or inhibit anthocyanin production include: chorocholine, gibberellic acid, abscissic acid, ethanol, and cycloheximide.[93-96] Grisafi and Venturella[97] obtained an increased anthocyanin production in seedlings grown on 10^{-4} *M* chloramphenicol. Antibiotics such as terramycin, chloramphenicol, streptomycin, and aureomycin also stimulate the production of anthocyanins in radish seedings; and this unique characteristic is used as a biological method for measuring the antibiotic activity of these substances.[98-100]

III. TURNIP

Turnip *(Brassica rapa)* is a cool-season, fast-growing vegetable of the mustard family, Cruciferae. Many of its varieties contain anthocyanins, often found together with chlorophylls.

Grill and Vince[101-103] and Grill[104] found that in turnip seedlings, anthocyanins are synthesized in the cotyledons, and light is required for their synthesis. Shibutani and Okamura[105] classified the Japanese varieties of *B. rapa* into two groups based on the presence of cyanidin or pelargonidin derivatives. Aoba[106-108] studied the distribution and heritability of the pigmentation in the hypocotyl, petiole, and roots of turnip varieties from Japan. According to Lein,[109] the anthocyanin pigmentation is the dominant characteristic and is controlled by the presence of a single gene. The synthesis of anthocyanins in seedlings depends on the phytochromes[110] or on their response to chemical treatment.[111] Hoshi[112] studied the pigmentation of turnip roots in which the purple pelargonidin 3-diglucoside-5-glucoside is present along with the flavonol isorhamnetin; cyanidin 3-diglucoside-5-glucoside is found in the red roots with the corresponding flavonol kaempferol, and only flavonols are found in the white roots.

In red turnip *(Brassica campestris* L.), Igarashi et al.[113] isolated and identified in root peel cyanidin 3-diglucoside-5-glucoside, cyanidin 3,5-diglucoside, and cyanidin 3-glucoside. These authors also investigated the effects of these anthocyanins on serum cholesterol level in rats and found a significant decrease in atherogenic cholesterol. The genetics of anthocyanin pigmentation in *B. campestris* was studied by Cours and Williams[114] and

Gugnani et al.[30] who reported a significant relationship between seed germinability and presence of anthocyanins. In pickled red turnips, anthocyanin diffusion from the skin to the inner white flesh can be prevented by packing the vegetable under pH-controlled conditions.[115]

IV. MUSTARD

Black mustard is *Brassica nigra* Koch (*Sinapis nigra* L.), oriental mustard is *B. juncea* Coss., and yellow or white mustard is *B. hirta* Moench (*B. alba*, Rabenh., *Sinapis alba* L.). In seedlings of yellow mustard grown in the light, Takeda et al.[116] identified cyanidin 3-sambubioside-5-glycoside acylated with malonic, sinapic, and *p*-coumaric acids ($C_{55}H_{69}O_{29}$); cyanidin 3-sambubioside-5-glucoside acylated with malonic, sinapic, and ferulic acids ($C_{56}H_{59}O_{30}$); cyanidin 3-sambubioside-5-sophoroside acylated with malonic, sinapic, and *p*-coumaric acids ($C_{61}H_{67}O_{34}$); and cyanidin 3-sambubioside-5-sophoroside acylated with malonic, sinapic, and ferulic acids ($C_{62}H_{69}O_{35}$). Therefore, the anthocyanin composition of mustard is similar to that of other Cruciferae.

An unusual spatial distribution of flavonoids within the epidermis was reported in cotyledons of yellow mustard because the flavonol quercetin was found in the upper epidermis, and anthocyanins accumulated only in the lower epidermis.[117]

In *B. juncea*, the pigments were identified as peonidin 3-glucoside and peonidin 3-galactoside.[119] The quantity found in 290 mg/100 g; however, up to a concentration of 40 mg/100 g, the anthocyanins are not visible because their presence is masked by the other cell pigments.[120]

The genetics of pigmentation of *B. nigra* was the object of a study by Delwiche and Williams[121] while Bertsch and Mohr[122] studied the light dependence of the synthesis of anthocyanins in *B. hirta*.

V. OTHERS

Using paper chromatography, Hoshi and Hosoda[123] found cyanidin 3-sophoroside-5-glucoside in *B. nigra* (black mustard), *B. rapa* (turnip), *B. oleracea* (red cabbage), *B. napus* (rape), *B. juncea* (oriental mustard), and *R. sativus* (radish); and pelargonidin 3-sophoroside-5-glucoside in *R. sativus* and *B. rapa*. Hoshi and Hosoda[124] observed the presence of two anthocyanins and five yellow flavonols in the leaves and petals of *B. campestris*, *B. nigra*, *B. oleracea*, *B. juncea*, *B. napus*, and *B. carinata*. From the distribution of these seven flavonoids, it is possible to distinguish all these species of *Brassica*, with the exception of *B. campestris* which seems to include numerous subspecies.

REFERENCES

1. **Wheldale, M.,** *The Anthocyanin Pigments of Plants,* Cambridge University Press, Cambridge, 1916, 318.
2. **Chmielewska, I.,** Untersuchungen ueber das Rotkohlfarbstoffe, *Rocz. Chem.,* 13, 725, 1933 (ZA 1936, I, 2361).
3. **Chmielewska, I.,** The coloring matter of red cabbage. II, *Rocz. Chem.,* 16, 384, 1936.
4. **Chmielewska, I., Smardzewska, I., and Kulesza, J.,** Red coloring matter of cabbage *(Brassica oleracea).* III, *Rocz. Chem.,* 18, 176, 1938 (CA 33, 2561).
5. **Fouassin, A.,** Identification par chromatographie d'anthocyanines de fruits et de legumes, *Rev. Ferment. Ind. Aliment.,* 11, 173, 1956.
6. **Stroh, H. H.,** Ueber die Anthocyane des Rotkohls. I. Zur Konstitution des Rubrobrassinchlorid, *Z. Naturforsch., Teil B,* 699, 1959 (CA 54, 15555b).
7. **Stroh, H .H. and Seidel, H.,** Red cabbage anthocyans. III. Structure of the sinapic ester of rubrobrassin, *Z. Naturforsch., Teil B,* 20, 39, 1965 (CA 62, 14987f).
8. **Metche, M.,** Anthocyanin pigments in some vegetables, *Brasserie,* 22, 69, 1967 (CA 67, 29835).
9. **Harborne, J. B.,** Plant polyphenols. XI. The structure of acylated anthocyanins, *Phytochemistry,* 3, 151, 1964.
10. **Tanchev, S. S. and Timberlake, C. F.,** The anthocyanins of red cabbage *(Brassica oleracea), Phytochemistry,* 8, 1825, 1969.
11. **Lanzarini, G. and Morselli, L.,** Gli antociani del cavolo rosso, *Ind. Conserve,* 49, 16, 1974 (CA 81, 62341).
12. **Hrazdina, G., Iredale, H., and Mattick, L. R.,** Anthocyanin composition of *Brassica oleracea* cv. Red Danish, *Phytochemistry,* 16, 297, 1977.
13. **Idaka, E.,** Acylated Anthocyanins from Red Cabbage, Japanese Patent, 62,209,173, 1987 (CA 108, 35008).
14. **Idaka, E.,** Acylated Anthocyanins from Plants and Elucidation of Their Structures, Japanese Patent, 63,113,078, 1988 (CA 110, 8589).
15. **Idaka, E.,** Acylated Anthocyanins from Red Cabbage, Japanese Patent, 1,74,271, 1989.
16. **Idaka, E., Suzuki, K., Yamakita, H., Ogawa, T., Kondo, T., and Goto, T.,** Structure of monoacylated anthocyanin isolated from red cabbage, *Brassica oleracea, Chem. Lett.,* 1, 145, 1987 (CA 106, 135274v).
17. **Idaka, E., Yamakita, H., Ogawa, T., Kondo, T., Yamamoto, M., and Goto, T.,** Structure of three diacylated anthocyanins isolated from red cabbage, *Brassica oleracea, Chem. Lett.,* 6, 1213, 1987 (CA 107, 130902j).
18. **Nakatani, N., Ikeda, K., Nakamura, M., and Kikuzaki, H.,** Structure of diacylated anthocyanins from red cabbage *(Brassica oleracea), Chem. Express,* 2, 555, 1987 (CA 107, 233170).
19. **Ikeda, K., Kikuzaki, H., Nakamura, M., and Nakatani, N.,** Structure of two acylated anthocyanins from red cabbage *(Brassica oleracea), Chem. Express,* 2, 563, 1987 (CA 108, 3432).
20. **Pecket, R. C. and Small, C. J.,** Occurrence, location and development of anthocyanoplasts, *Phytochemistry,* 19, 2571, 1980.
21. **Small, C. J. and Pecket, R. C.,** The ultrastructure of anthocyanoplasts in red cabbage, *Planta,* 154, 97, 1982.
22. **Neumann, D.,** Subcellular localization of anthocyanins in red cabbage seedlings, *Biochem. Physiol. Pflanz.,* 178, 405, 1983 (HA 1985, 132).
23. **Strogonof, B. P. and Dostanova, R. K.,** Dependence of anthocyanin and leucoanthocyanin content of plants on type of salination, *Fiziol. Rast.,* 13, 509, 166 (CA 65, 11280g).
24. **Dostanova, R. H.,** The effect of Na_2SO_4 and NaCl on the metabolism of plastid pigments in plants, *Fiziol. Rast.,* 13, 614, 1966 (HA 1967, 4801).

25. **Sbrana, C. and Lercari, B.**, Due risposte adattative indotte dall'UV-A in *Brassica oleracea* var. *capitata*, *Agric. Mediterr.*, 119, 276, 1989 (HA 1990, 1724).

26. **Glushchenko, I. E., Sokolova, L. K., Kruzilin, A. S., and Shvedskaya, Z. M.**, Changes in pigment composition of chimera plants of cabbage and in their seed progenies, *Skh. Biol.*, 1, 213, 1966 (CA 65, 15805c).

27. **Sokolova, L. K., Glushchenko, I. E., Kruzilin, A. S., and Shvedskaya, Z. M.**, A change in the pigment composition of vegetative and sexual hybrids of cabbage, in *Nutrition and Manurial Fertilization of Plants*, Abstr. 12.55.67, Nauk. Dumka, Kiev, Ukraine, 1966, 68, (PBA 1969, 7727).

28. **Bubarova, M.**, Anthocyanin pigments in intergenus hybrids obtained by crossing radishes *(Raphanus sativus)* with common cabbage (*Brassica oleracea* var. *capitata*) and radishes with kohlrabi *(Brassica caulorapa)* and in the parental forms, *Gradinar. Lozar. Nauka*, 11, 88, 1974 (CA 82, 14088).

29. **Eenink, A. H.**, Matromorphy in *Brassica oleracea* L. IV. Formation of homozygous and heterozygous diploid products of gametogenesis and qualitative genetical research on matromorphic plants, *Euphytica*, 23, 719, 1974.

30. **Gugnani, D., Banerjee, S. K., and Singh, D.**, Germination capacity in relation to seed coat color in cabbage and mustard, *Seed Sci. Technol.*, 3, 375, 1975 (PBA 1976, 5829).

31. **Szweykowska, A.**, Anthocyanin and the effect of light on the development of cabbage seedlings, *Acta Soc. Bot. Pol.*, 26, 349, 1957 (CA 53, 7322f).

32. **Sampson, D. R.**, New light on the complexities of anthocyanin inheritance in *Brassica oleracea*, *Can. J. Genet. Cytol.*, 9, 352, 1967.

33. **Baggett, J. R.**, Inheritance of internal pigmentation in green cabbage (*Brassica oleracea* var. *capitata* L.), *Euphytica*, 27, 593, 1978.

34. **Pecket, R. C. and Hathout-Bassim, T. A.**, Mechanism of phytochrome action in the control of biosynthesis of anthocyanin in *Brassica oleracea*, *Phytochemistry*, 13, 815, 1974.

35. **Pecket, R. C. and Hathout-Bassim, T. A.**, Effect of kinetin in relation to photocontrol of anthocyanin biosynthesis in *Brassica oleracea*, *Phytochemistry*, 13, 1395, 1974.

36. **Hathout-Bassim, T. A. and Pecket, R. C.**, Effect of membrane stabilizers on phytochrome-controlled anthocyanin biosynthesis in *Brassica oleracea*, *Phytochemistry*, 14, 731, 1975.

37. **Kang, B. G. and Burg, S. P.**, Role of ethylene in phytochrome-induced anthocyanin synthesis, *Planta*, 110, 227, 1973.

38. **Grebinskii, S. O. and Khimil, M. V.**, The effect of chlormequat chloride on biosynthesis of anthocyanins in red cabbage and *Tradescantia* leaves, *Fiziol. Biokhim. Kult. Rast.*, 12, 409, 1980 (HA 1981, 1195).

39. **Hartmann, H. D.**, Chemical weed control and the root system of cabbage, *Angew. Bot.*, 43, 47, 1969 (HA 1970, 913).

40. **Ku, P. K. and Mancinelli, A. L.**, Photocontrol of anthocyanin synthesis. I. Action of short, prolonged and intermittent irradiations on the formation of anthocyanins in cabbage, mustard and turnip seedlings, *Plant Physiol.*, 49, 212, 1972.

41. **Mancinelli, A. L. and Rabino, I.**, Photoregulation of anthocyanin synthesis. X. Dependence of photosynthesis of high irradiance response anthocyanin synthesis in *Brassica oleracea* leaf disk and *Spirodela polyrrhiza*, *Plant Cell Physiol.*, 25, 1153, 1984.

42. **Hrazdina, G. and Creasy, L. L.**, Light induced changes in anthocyanin concentration, activity of phenylalanine ammonia-lyase and flavanone synthase and some of their properties in *Brassica oleracea*, *Phytochemistry*, 18, 581, 1979.

43. **Matula, V. H.**, Anthocyanin as an indicator in acidimetry, *Chem. Ztg.*, 48, 305, 1924 (CA 18, 2481).

44. **Matula, V. H. and Macek, C. B.**, The anthocyanin as indicator in neutralization analysis, *Chem. Obz.*, 11, 84, 1936 (CA 30, 7059).

45. **Grube, O., Dieckmann, K., and Gundermann, R.**, The pigment of red cabbage as an indicator in the pH range 8.5–10, *Chem. Ztg.*, 67, 34, 1943 (CA 37, 5671).

46. **Wolf, F. T.,** Absorption spectra of the anthocyanin pigment of red cabbage, a natural wide-range pH indicator, *Physiol. Plant,* 9, 559, 1956.

47. **Furihata, K., Hori, T., and Takamura, I.,** Color changes of red cabbage indicator in acidity and alkalinity, *Tokyo Gakugei Daigaku Kiyo, Dai-4-Bu,* 31, 287, 1979 (CA 93, 6340).

48. **Forster, M.,** Plant pigments as acid-base indicators — an exercise for the junior high school, *J. Chem. Educ.,* 55, 107, 1978 (CA 88, 135680).

49. **Shewfelt, R. L. and Ahmed, E. M.,** Anthocyanin extracted from red cabbage shows promise as coloring for dry beverage mixes, *Food Prod. Dev.,* 11, 4, 52, 1977.

50. **Shewfelt, R. L. and Ahmed, E. M.,** Enhancement of powdered soft drink mixes with anthocyanin extracts, *J. Food Sci.,* 43, 435, 1978.

51. **Sapers, G. M.,** Deodorization of a colorant prepared from red cabbage, *J. Food Sci.,* 47, 972, 1982.

52. **Sapers, G. M., Taffer, I., and Ross, L. R.,** Functional properties of a food colorant prepared from red cabbage, *J. Food Sci.,* 46, 105, 1981.

53. **Wakabayashi, S., Kotake, K., and Yasuda, S.,** Red Food Coloring Agent from Purple Cabbage, Japanese Patent, 8,025,460, 1980 (CA 93, 24809s).

54. **Shibata, M., Wakabayashi, M., and Yasuda, S.,** Coloring Agent for Soft Drinks, Japanese Patent, 8,026,816, 1980 (CA 93, 6375t).

55. **Shibata, M., Wakabayashi, M., and Yasuda, S.,** Coloring Agent for Candies, Japanese Patent, 8,026,817, 1980 (CA 93, 6374s).

56. **Mano, S., Tezuka, T., and Tomikanehara, T.,** Anthocyanin Red Pigment Extraction from Plants for Coloring Foods, Japanese Patent, 63,270,766, 1988 (CA 111, 150763).

57. **Yasuda, A., Koto, T., and Obata, M.,** Prevention of Color Changes of Red Cabbage Pigments, Japanese Patent, 61,282,032, 1986 (CA 106, 83282).

58. **Idaka, E.,** Removal of Acids from Acidic Anthocyanin Pigments by Chromatography, Japanese Patent, 63,278,971, 1988 (CA 112, 4902).

59. **Koda, T. and Yasuda, A.,** Anthocyanin Colorants from Red Cabbage, Japanese Patent, 61,101,560, 1986 (CA 105, 189724q).

60. **Inoue, T. and Yasuda, A.,** Pigments from Red Cabbage as Food Colorant, Japanese Patent, 6, 197, 361, 1986 (CA 197, 76432).

61. **Tadamasa, H. and Yasuda, A.,** Anthocyanin Food Colors from Red Cabbage, Japanese Patent, 6, 197, 362, 1986 (CA 105, 96252).

62. **Yasuda, A. and Kotake, K.,** Manufacture of Hydrophilic Food Colors from Vegetable Materials, Japanese Patent, 60, 176, 562, 1985 (CA 104, 147466).

63. **Yasuda, A., Kotake, K., and Obata, M.,** Production of Stable, Odorless Red Cabbage Pigment, Japanese Patent, 6, 185, 166, 1986 (CA 105, 41538m).

64. **Koda, T.,** Manufacture of Red Anthocyanin Pigments from *Brassica* Tissue Culture, Japanese Patent, 62, 181, 796, 1987 (CA 109, 21693).

65. **Murai, K. and Wilkins, D.,** Natural red color derived from red cabbage, *Food Technol.,* 44, 131, 1990.

66. **Anon.,** San Red RC Food Colors, Information Bull., San-Ei Chemical Industries, Osaka, 561, Japan, 2, 1990.

67. **Anon.,** Sun Red RC Red Cabbage Color, Tech. Rep., San-Ei Chemical Industries, Osaka, 561, Japan, 2, 1990.

68. **Anon.,** Natural Colors, Inf. Bull., San-Ei Chemical Industries, Osaka, 561, Japan, 3, 1990.

69. **Timberlake, C. F. and Bridle, P.,** Distribution of anthocyanins in food plants, in *Anthocyanins as Food Colors,* Markakis, P., Ed., Academic Press, New York, 1982, 125.

70. **Sacher, J. F.,** A very sensitive indicator, *Chem. Ztg.,* 34, 1192, 1910.

71. **Sacher, J. F.,** The coloring matter of red radish, *Chem. Ztg.,* 34, 1333, 1910 (CA 5, 1723).

72. **Schwertschlager, J.,** The coloring matter of red radish, *Chem. Ztg.,* 34, 1257, 1910 (CA 5, 1723).

73. **Harborne, J. B.,** Plant polyphenols. IX. The glycosidic pattern of anthocyanin pigments, *Phytochemistry,* 2, 85, 1963.

74. **Harborne, J. B. and Paxman, G. J.,** Genetics of anthocyanin production in the radish, *Heredity,* 19, 505, 1964.

75. **Hoshi, T., Takemura, E., and Hayashi, K.,** Anthocyanins. XLII. Genetic modification of hydroxylation pattern in radish anthocyanins, *Bot. Mag. (Tokyo),* 76, 431, 1963 (CA 61, 13642e).

76. **Harborne, J. B. and Sherratt, H. S. A.,** Variations in the glycosidic pattern of anthocyanins. II, *Experientia,* 13, 486, 1957.

77. **Ishikura, N. and Hayashi, K.,** Studies on anthocyanins. XXXVIII. Chromatographic separation and characterization of the component anthocyanins in radish root, *Bot. Mag. (Tokyo),* 76, 6, 1963 (CA 60, 14580d).

78. **Ishikura, N., Hoshi, T., and Hayashi, K.,** Anthocyanins. XLV. Crystallization and characterization of the basic triglucoside common to all component in purple pigment of hybrid radish, *Bot. Mag. (Tokyo),* 78, 8, 1965 (CA 64, 2411f).

79. **Ishikura, N. and Hayashi, K.,** Biogenetic interrelation between anthocyanin and some of the concomitant substances in radish and turnip. Studies on anthocyanins, *Bot. Mag. (Tokyo),* 78, 481, 1965 (CA 65, 10959h).

80. **Fuleki, T.,** The anthocyanins of strawberry, rhubarb, radish and onion, *J. Food Sci.,* 34, 365, 1969.

81. **Lele, S. S.,** Das pigment von *Raphanus caudatus* L., *J. Sci. Ind. Res. (India), Sect. B,* 18, 243, 1959 (CZ 1963, 10142).

82. **Ishikura, N. and Hayashi, K.,** Studies on anthocyanins. XXVII. Anthocyanins in red roots of radish, *Bot. Mag. (Tokyo),* 75, 28, 1962 (CA 57, 3874h).

83. **Ishikura, N. and Hayashi, K.,** Anthocyanins. XLVI. Separation and identification of the complex anthocyanins in purple radish, *Bot. Mag. (Tokyo),* 78, 91, 1965 (CA 64, 2411f).

84. **Ishikura, N. and Hayashi, K.,** Studies on anthocyanins. LII. Fate of 14C-labeled amino acids administered to the seedlings of red radish with special regard to anthocyanin biosynthesis, *Bot. Mag. (Tokyo),* 79, 156, 1966 (CA 63, 17373c).

85. **Ishikura, N. and Hayashi, K.,** Anthocyanins. LIII. Incorporation of leucine-14C into the molecular framework of anthocyanin appearing in red radish, *Bot. Mag. (Tokyo),* 79, 308, 1966 (CA 66, 73306).

86. **Sovoshkin, I. P. and Demina, T. G.,** Effect of the doubling of chromosomes on the qualitative composition and content of anthocyanins in radish roots, *Genetika,* 7(11), 21, 1971 (CA 76, 56745).

87. **Nasir, F., Ahmed, J., and Kahn, M. I.,** Expression of heterosis for biochemical traits in *Raphanus sativus* L., *Z. Acker Pflanzenbau,* 155, 159, 1985 (CA 104, 106447).

88. **Politis, J.,** Genes producing anthocyanin in seedling of *Raphanus sativus* and other Cruciferae, *C.R. Acad. Sci. Fr.,* 225, 256, 1947 (CA 42, 1336).

89. **Bellini, E. and Martelli, M.,** Anthocyanin synthesis in radish seedlings. Effect of continuous far-red irradiation and phytochrome transformations, *Z. Pflanzenphysiol.,* 70, 12, 1973.

90. **Straub, V. and Lichtenthaler, H. K.,** Effect of 1H-indol-3-acetic acid on the formation of chloroplast pigments, plastic quinones, and anthocyanins in *Raphanus* seedlings, *Z. Pflanzenphysiol.,* 70, 34, 1973.

91. **Straub, V. and Lichtenthaler, H. K.,** Effect of gibberellic acid A3 and kinetin on the formation of photosynthetic pigments, lipoquinones, and anthocyanins in *Raphanus* seedlings, *Z. Pflanzenphysiol.,* 70, 308, 1973.

92. **Buschmann, C. and Lichtenthaler, H. K.,** The effect of cytokinins on pigment accumulation and Hill-activity of radish seedlings, *Photosynth. Plant Prod., Jt. Meet. O.E.C.D. Studienzent. Weikersheim,* 1981, 172 (CA 99, 191891).

93. **Jain, V. K. and Guruprasad, K. N.,** Effect of chorocholine and gibberellic acid on the anthocyanin synthesis in radish seedlings, *Physiol. Plant,* 75, 233, 1989 (CA 111, 4310).

94. **Guruprasad, K. N. and Laloraya, M. M.,** Effect of pigment precursors on the inhibition of anthocyanin biosynthesis by GA and ABA, *Plant Sci. Lett.,* 19, 73, 1980 (HA 1981, 8570).

95. **Lichtenthaler, H. K. and Thiess, D. E.,** Inhibition of the light-induced chlorophyll, carotenoid and anthocyanin synthesis by ethanol, *Physiol. Plant.,* 30, 260, 1974.

96. **Billington, R. W. and Heyes, J. K.,** Effect of inhibitors on light-stimulated synthesis in radish hypocotyls, *Nature (London),* 227, 858, 1970.

97. **Grisafi, F. and Venturella, G.,** Anthocyanins in hypocotyl and cotyledons of *Raphanus sativus* L. seedlings, *G. Bot. Ital.,* 116, 63, 1982 (CA 100, 188087).

98. **Cortesi, V. R. and Girard, R.,** Influence of adrenocorticotropin (ACTH) on the culture and morphogenesis of two fanerogames, radish and bean, *Bull. Soc. Pharm. Bordeaux,* 92, 23, 1954 (CA 49, 12609a).

99. **Netien, G. and Lacharme, J.,** Action of terramycin on formation of pigments in radish plantlets, *Bull. Soc. Chim. Biol.,* 37, 643, 1955 (CA 50, 2745f).

100. **Netien, G. and Lacharme, J.,** The anthocyanin content of radish cotyledons as a measure of antibiotic activity, *C.R. Acad. Sci. Fr.,* 240, 692, 1955 (CA 49, 7659b).

101. **Grill, R. and Vince, O.,** Anthocyanin formation in turnip seedlings *(Brassica rapa),* evidence for two light steps in the biosynthetic pathway, *Planta,* 63, 1, 1964.

102. **Grill, R. and Vince, O.,** Photocontrol of anthocyanin synthesis in turnip seedlings. II. The possible role of phytochrome in the response to prolonged irradiation with far-red or blue light, *Planta,* 67, 122, 1965.

103. **Grill, R. and Vince, O.,** Photocontrol of anthocyanin synthesis in turnip seedlings. III. The photoreceptors involved in responses to prolonged irradiation, *Planta,* 70, 1, 1965.

104. **Grill, R.,** Photocontrol of anthocyanin synthesis in turnip seedlings. VI. The effect of feeding precursors, *Planta,* 76, 11, 1967.

105. **Shibutani, S. and Okamura, T.,** On the mode of coloration of coloured turnip varieties. II. The anthocyanins of coloured turnip varieties, *Engei Gakkai Zasshi,* 25, 111, 1956 (PBA 1957, 1912).

106. **Aoba, T.,** Taxonomic relations and routes of introduction of the turnip varieties distributed in eastern Japan, *Nogyo Oyobi Engei,* 35, 1729, 1960 (PBA 1962, 4985).

107. **Aoba, T.,** Studies on the classification and geographical distribution of Japanese vegetable varieties. II. On the character of F1 hybrids of turnip from eastern Japan, *Engei Gakkai Zasshi,* 30, 147, 1962 (PBA 1962, 3511).

108. **Aoba, T.,** On the genetics of testa colour in the turnip, *Ikushugaku Zasshi,* 20, 173, 1970 (PBA 1971, 8106).

109. **Lein, K. A.,** Genetische und Physiologische Untersuchungen zur Bildung von Glucosinolaten in Rapssamen. I. Zur Vererbung der Glucosinolatarmut, *Z. Pflanzenphysiol.,* 67, 243, 1972 (PBA 1973, 4506).

110. **Grill, R.,** Effect of ethanol treatment on apparent phytochrome synthesis and anthocyanin formation in turnip seedlings *(Brassica rapa* L.), *Z. Pflanzenphysiol.,* 101, 361, 1981.

111. **Choudri, R. S. and Bhatnagar, V. B.,** Effect of maleic hydrazide on the keeping quality of turnip *(Brassica rapa),* *Indian J. Hortic.,* 12, 1, 1955 (CA 50, 5222i).

112. **Hoshi, T.,** Genetical studies on anthocyanins in *Brassicaceae.* II. Genetical studies on the formation of anthocyanins and flavonols in turnip varieties, *Bot. Mag. (Tokyo),* 88, 249, 1975 (CA 84, 118522).

113. **Igarashi, K., Abe, S., and Satoh, J.,** Effects of atsumi-kabu (red turnip, *Brassica campestris* L.) anthocyanin on serum cholesterol in cholesterol-fed rats, *Agric. Biol. Chem.,* 54, 171, 1990.

114. **Cours, B. J. and Williams, P. H.,** Genetic studies in *Brassica campestris, Cruciferae Newsl.,* 2, 38, 1977 (PBA 1981, 1533).

115. **Abe, T.,** Pickling of Red Turnip under pH-Controlled Conditions, Japanese Patent, 2,163,037, 1990 (CA 113, 170762).

116. **Takeda, K., Fischer, D., and Grisebach, H.,** Anthocyanin composition of *Sinapsis alba*, light induction of enzymes and biosynthesis, *Phytochemistry*, 27, 1351, 1988.

117. **Beggs, C. J., Kuhn, K., Boecker, R., and Wellman, E.,** Phytochrome-induced flavonoid biosynthesis in mustard (*Sinapis alba* L.) cotyledons. Enzymic control and differential regulation of anthocyanin and quercetin formation, *Planta*, 172, 121, 1987.

118. **Mohr, H.,** Influence of monochromatic radiation on longitudinal growth of hypocotyl and anthocyanin synthesis in seedlings of *Sinapis alba* L. (= *Brassica alba* Boiss.), *Planta*, 49, 389, 1957.

119. **Park, K. H.,** Studies on the anthocyanins of *Brassica juncea*. I. Identification of anthocyanins, *Hanguk Nonghwahakhoe Chi*, 22, 33, 1979 (CA 92, 57035).

120. **Park, K. H.,** Studies on the anthocyanins in *Brassica juncea*. II. Quantitative determination of anthocyanins, *Hanguk Nonghwahakhoe Chi*, 22, 39, 1979 (CA 92, 57036).

121. **Delwiche, P. A. and Williams, P. H.,** Thirteen marker genes in *Brassica nigra*, *J. Hered.*, 72, 289, 1981.

122. **Bertsch, W. and Mohr, H.,** Interpretation of the far-red band effective in the high-energy reaction of photomorphogenesis. Light-dependent synthesis of anthocyanins in mustard seedlings *(Brassica hira)*, *Planta*, 65, 245, 1965.

123. **Hoshi, T. and Hosoda, T.,** Genetical studies on anthocyanin in Brassicaceae crops, *Jpn. J. Breed.*, 17, 47, 1967 (PBA 1968, 1491).

124. **Hoshi, T. and Hosoda, T.,** Interrelation between the genome and flavonoid composition in the genus *Brassica*, *Jpn. J. Breed.*, 28, 137, 1978 (PBA 1979, 759).

Chapter 12

OTHER CROPS

I. ASPARAGUS

The presence of anthocyanins in *Asparagus officinalis* L. is evident from the red color of the tips of the edible bracts. Generally, the cultivars high in anthocyanin content are, however, judged to be of lower quality and poor yielders. Among 14 European and American cultivars tested in 1980, Spaganiva and its selection SH72, both lacking anthocyanins, were found most acceptable.[1] Chen and Shen[2] and Chen[3] reported that in the cultivar Connover Colossal only 40% of the plants were colored. According to Hepler et al.,[4] the lack of anthocyanin in a selection of asparagus void of purple pigmentation was due to a single dominant gene, *Ip*, which inhibited anthocyanin production.

The presence of anthocyanins and phenolics is believed to be associated with a low soluble carbohydrate content, which explains the lower sensory quality of red-pigmented asparagus. Formation of anthocyanins is favored by darkness and low temperatures.[5]

Robinson and Robinson[6] reported that the red pigment of asparagus was a cyanidin diglucoside, but Wann and Thompson[7] were the first to isolate and characterize the two main pigments as cyanidin 3-glucoside and cyanidin 3,5-diglucoside, possibly acylated with carboxylic acids. Later, Francis[8] identified cyanidin 3-glucosylrutinoside as the main pigment, followed by cyanidin 3-rutinoside, in the reddish bracts of 'Mary Washington' asparagus (Table 1). Two other anthocyanins detected were thought to be peonidin 3-glucosylrutinoside and peonidin 3-rutinoside.

The presence of anthocyanins in asparagus has no detrimental effect on storability of the canned product. Rather, quercetin causes problems because it reacts with tin ions to form a yellow lacquer.[9]

Not all *Asparagus* varieties contain anthocyanins. *A. persicus* lacks anthocyanins but contains colorless flavonoids.[10] In the flowers of *A. gonocladus*, Tiwari et al.[11] identified malvidin 3-glucoside.

II. *CLITORIA TERNATEA*

Clitoria ternatea L. or butterfly pea is found throughout India, and its flowers are used for medical and food purposes. The entire plant is used as an antidote for snake poison. Raman[12] studied the blue color of *C. ternatea* flower *in vivo*, while Misra and Gyani[13] proposed that it be used as an acid/base indicator.

Ranaganayaki and Singh[14] found cyanin, kaempferol, and an unknown anthocyanin in the blue flowers while only kaempferol was present in the

TABLE 1
Anthocyanins of Asparagus Spears
(*Asparagus officinalis* L.)

Anthocyanin	Relative conc
Cyanidin 3-glucosylrutinoside	+ + + +
Cyanidin 3-rutinoside	+ +
Peonidin 3-glucosylrutinoside	+
Peonidin 3-rutinoside	+

Adapted from Francis, F. J., *J. Food Sci.*, 32, 430, 1967.

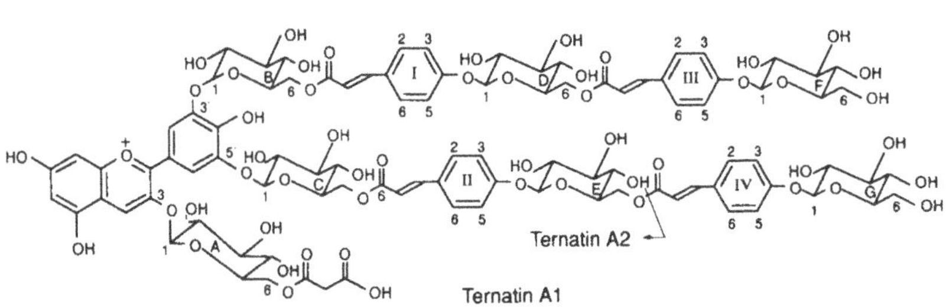

FIGURE 1. Structure of ternatin A. (From Terahara, N., Saito, N., Honda, T., Toki, K., and Osajima, Y., *Heterocycles*, 31, 1773, 1990. With permission.)

white flowers. Lowry and Chew[15] reported the presence of delphinidin 3,5-diglucoside in the flowers of this plant from Malaysia and suggested using the flower extracts as a food colorant. Srivastava and Pande[16] identified monoglucosides of malvidin, delphinidin, and petunidin in the flowers.

More recently, Saito et al.[17] isolated six polyacylated anthocyanins from *C. ternatea* flowers, all based on delphinidin 3,3'5'-triglucoside. The structure of the deacylated anthocyanins has been further characterized as delphinidin 3,3',5'-tri-*O*-β-D-glucopyranoside.[18] The polyacylated pigments have been named ternatin A1, A2, B1, B2, D1, and D2,[19,20] and have been found to be stable in neutral solution for several months.[18] Ternatin A1, the largest anthocyanin known to date, is composed of delphinidin with seven molecules of D-glucose (G), four molecules of *p*-coumaric acid (C), and one molecule of malonic acid (Figure 1). Ternatin A2 has a molecular weight of 1799 and consists of delphinidin, six molecules of D-glucose, three of *p*-coumaric acid, and one of malonic acid.[20] In the ternatins B1 and B2, the 3'- and 5'-side chains consist of -CGCG or -CGC and -CGC or CG, respectively (see Figure 2 in Chapter 1).[21] The blue flowers of *C. ternatea* also contain five flavonols identified as kaempferol, kaempferol 3-glucoside, robinin, quercetin, and quercetin 3-glucoside.[17]

III. EGGPLANT

The presence of anthocyanins in the purple skin of eggplant or aubergine (*Solanum melongena* L.) was first reported by Kuroda and Wada,[22-24] who isolated the principal pigment and called it nasunin after the Japanese name for the plant, nasukon. They suggested that the structure of this anthocyanin was delphinidin 3-diglucoside acidified with *p*-coumaric acid, and they also noted the presence of delphinidin 3-glucoside. The structure of nasunin was further characterized by Takeda et al.[25] and Sakamura et al.[26] The first group of authors showed that it was delphinidin *p*-coumaryl-monorhamnoside-diglucoside, identical to violanin (the pigment extracted from the petals of Giant pansy). These authors, however, were uncertain about the position of the sugars in the molecule. Sakamura et al.[26] determined the position of the sugars and identified the structure of nasunin as delphinidin 3-(-*p*-coumaryl-L-rhamnoside)-D-glucoside-5-glucoside chloride. The rhamnose-glucose bond is in the 1–6 position, and the *p*-coumaric acid is attached at the 4 position of the rhamnose.[27]

Using chromatography on a cellulose column, Sakamura and Obata[28] separated three delphinidin derivatives from the fruit skin of *S. melongena*. Using similar techniques, Nagashima and Taira[29] also found three pigments which they identified as delphinidin 3-diglucosyl-*p*-coumaric, delphinidin 3-glucoside, and delphinidin 3,5-diglucoside. Casoli and Dall'Aglio[30] used paper chromatography to separate four pigments. The major pigment was delphinidin 3-rutinoside acylated with an unidentified acid. The two other pigments were believed to be delphinidin 5-glucoside-3-rutinoside and delphinidin 3-glucoside. Later work by Pifferi and Zamorani[31] established the presence of six pigments, and identified: delphinidin 3-glucoside, delphinidin 3-diglucoside-5-glucoside, delphinidin 5-glucoside-3-diglucoside acylated with an unidentified acid, and delphinidin 5-glucoside-3-diglucoside-caffeic acid.

In Bulgarian eggplant, Tanchev et al.[32] found only delphinidin 3-rutinoside and small quantities of delphinidin 3-rutinoside-5-glucoside. They found no acylated pigments, and attributed the lack of sugar acylation to genetic differences in eggplant with respect to the place of origin. Similarly, Hocking[33] found no nasunin in Australian eggplants. Ramaswamy and Rege[34] isolated seven pigments in eggplants from India and all had delphinidin as the aglycone. Among the sugars where rhamnose and glucose, and they noted the presence of *p*-coumaric and caffeic acids as acyl residues.

Abe and Gotoh[35-37] used genetic analysis to show the presence of acylated anthocyanins in the Burma cultivar and their absence in 'Black Beauty' eggplants. According to these authors, the acylation is governed by a single dominant gene. The pigment in 'Burma' is a hydroxycinnamic derivative of delphinidin 3,5-diglucoside while in 'Black Beauty' it is delphinidin 3-rutinoside.

The genetics of eggplant coloration were further studied by Topoleski[38] and Janick and Topoleski,[39] and a guide chart for color combinations in hybrids of eggplants was described by Sambandam.[40] *Ac* is the gene which determines the anthocyanin structure while *Puc* controls the nonphotodependent synthesis of the fruit pigment.[41,42] The purple color of the fruit is dominant over yellow.[43] The purple color in the hypocotyl is a dominant character controlled by a single gene.[44] The anthocyanic pigmentation of the leaf veins, stem, and petioles is genetically linked.[45] Sambandam[46] found that there are several interconnected genes which control the pigmentation of the hypocotyl *(Ph)*, epicotyl *(Al)*, leaves *(Pb)*, leaf margin *(Pm)*, superior and inferior parts of the calyx *(Pu and Pl)*, suture of the anther *(St)*, and the skin *(C)*. Patil and More,[47-49] however, described a more complex genetic arrangement with two complementary genes controlling the leaf and leaf vein pigmentation, while the skin and fruit color are believed to be under the control of three genes, *Pfa, Pfb1*, and *Pfb2*. The coloration of the fruit is associated with its form, and in leaf veins, purple color is always dominant over green.

Consumers prefer uniformly colored eggplants with dark skin varying from dark purple to black. The plant, however, is not very well adapted to low temperatures; therefore, to obtain intensely and uniformly pigmented fruit, temperature must be controlled, even by warming the soil.[50] In fact, cool seasons can produce a variety of abnormalities, including nonuniform pigmentation of the fruit.[51] In general, anthocyanin content of the fruit is a good quality indicator of eggplant cultivars[52-54] and has been correlated with polyphenol,[55] glycoalkaloid,[56] and chlorophyll[57] contents of the fruit. Thus, from a comparison in solanin and anthocyanin contents of 21 cultivars of eggplants, Bajaj et al.[56] concluded that the most acceptable cultivars were those rich in anthocyanins and low in glycoalkaloids. Similarly, Sidhu et al.[57] reported that there is a direct correlation between anthocyanin content and chlorophyll. Recently, it was reported that the Hunter color parameters are highly correlated with anthocyanin content[58] and that the chilling injuries, causing a loss in color of the fruit skin, are negatively correlated with anthocyanin content.[59] Also, when eggplant fruits are harvested overripe, the color changes toward yellowish.[60]

In addition to anthocyanins, eggplants are rich in chlorogenic acid which is the major phenolic compound in the seeds and their surrounding parenchyma.[61] In harvested fruit, this acid is believed to stimulate the activity of polyphenoloxidase (PPO) which causes color loss in eggplant anthocyanins.[62,63] According to Sakamura et al.,[64] the maximum activity of eggplant PPO is at pH 6 and 35°C, and it affects the various anthocyanins differently. It reacts with the anthocyanin structure in the following order: pelargonidin = peonidin > cyanidin > delphinidin.

The separation of enzymatic preparations obtained from *S. melongena* on a DEAE-cellulose column gives two different PPO (A and B).[64] Both oxidize the orthodiphenols, but they have different substrate and pH specificities. The

Michaelis constant of enzyme a for chlorogenic acid is $1.6 \times 10^{-3} M$, while for catechol it is $0.4 \times 10^{-3} M$; for enzyme B the values are $0.6 \times 10^{-3} M$ and $4.0 \times 10^{-3} M$, respectively. The mechanism does not imply sugar hydrolysis even though it is an oxidative-type reaction because from the oxidation of catechol to orthoquinone, insoluble brownish polymers form.[65]

IV. FENNEL

Fennel, *Foeniculum vulgare* Miller, is indigenous to Italy. The lower part of the plant, a type of false bulb, is the edible portion, while the upper part and inflorescence are used as spices. The anthocyanin pigments present in the stem are cyanidin 3-ferulylxylosylglucosylgalactoside and cyanidin 3-sinapyl-xylosylglucosylgalactoside.[66] This glucosidic arrangement is typical of the Umbelliferae.[67,68] Other flavonoids identified in fennel are kaempferol; quercetin; 3'-OMe quercetin; and *p*-hydroxybenzoic, gentisic, vanillic, syringic, protocatechuic, and ferulic acids.[69] Related to *F. vulgare* or common fennel is the sea fennel or *Crithmum maritimum* L., which is native to sea cliffs, sand, or shingle along the coastal reefs of Italy, France, and the British Isles. The fleshy leaves are aromatic and can be salted, boiled, and pickled or used as an herb. The plant is characterized by an anthocyanin pigment in the leaf, cyanidin 3-sinaplylxylosyl-glucosylgalactoside.[67]

V. GLOBE ARTICHOKE

The globe artichoke, *Cynara scolymus* L., is a gray-green vigorous plant that looks like a thistle. It is 60 to 90 cm tall, with large, prickly leaves with fleshy bases around the flower head or choke. These fleshy leaf bases are the parts usually eaten baked, fried, stuffed, or served with sauces. There are purple and green cultivars.

The anthocyanins in the bracts and flowers of purple cultivars are reportedly cyanidin derivatives.[70,71] Foury and Aubert[70] also observed that the purple cultivars are rich in orthodiphenols, and therefore, not suitable for processing because they develop brown pigments when heated.

A more detailed identification of the anthocyanins of globe artichoke was reported by Pifferi and Vaccari.[72] These authors, using spectrophotometric and chromatographic methods of analysis, isolated seven anthocyanins in four purple cultivars. Three of the pigments were identified as cyanidin 3-caffeylglucoside, cyanidin 3-caffeylsophoroside, and cyanidin 3-dicaffeyl-sophoroside. The other four pigments, because of their low concentration, could not be fully characterized; available information, however, indicates that these minor pigments contain cyanidin, caffeic acid, sugar residues, and a hydroxycinnamic acid. The presence of small quantities of cyanidin 3-glucoside in the inner part of artichoke bracts has been confirmed recently by El-Negoumy et al.[73] Other phenolic compounds occurring in globe artichokes include: luteolin 7-rutinoside, luteolin 7-glucoside, and luteolin 7-rutinoside-4'-glucoside; free apigenin and luteolin; and caffeic, vanillic, syringic, *p*-coumaric, and ferulic acids.[74,75]

VI. HOLLYHOCK

Hollyhock flowers, *Althaea rosea* Cav. var. *nigra*, have been used as medicine and colorants throughout recorded time. In 1892, Glan[76] characterized the red pigment as a glycoside derived from protocatechuic acid. Willstätter and Zollinger[77] called the red pigment of hollyhock altaein and identified it as a monoglucoside of mirtillidin. Karrer et al.,[78] however, reported that the anthocyanins of hollyhock flowers were glycosides of delphinidin, petunidin, and malvidin.

In 1956, Fouassin[79] separated seven pigments from the black, mauve, and red varieties of *A. rosea*: three derivatives of delphinidin, two of malvidin, and two of petunidin. The pigments of *A. rosea* were also studied by Nair et al.[80] who identified cyanidin 3-glucoside and cyanidin 3-rutinoside. More recently, Kohlmunzer et al.[81] identified six pigments: three glucosides of delphinidin, cyanidin, and malvidin; and 3,5-diglucosides of the same three anthocyanidins.

According to Kasperskaya et al.,[82] the anthocyanins of hollyhock flowers are an excellent colorant for gelatins, jams, and candies, and they are relatively stable over time. There is about a 20% loss in color after 3 months of storage.[82] For use as colorants or pharmaceutical products, the pigments are extracted with aqueous SO_2 solutions acidified with citric acid and then concentrated. The concentrated product is purified by precipitating the pectins and other impurities with ethyl alcohol, added in a 1:1:5 ratio, and reconcentrating it under reduced pressure to obtain a product containing about 50 g of anthocyanins per liter of product.[83] Hollyhock flower pigments can also be extracted with hot acetone (50 to 55°C) mixed in a 1:1 ratio with the petals and then recovering, under reduced pressure, the extracted colorant and the solvent, which can be reused.[84]

The anthocyanins of *Althaea rosea* var. *nigra* have been proposed for use as pharmacologicals;[81,85] and according to Kohlmunzer et al.,[81] their protective activity on blood capillaries is comparable to that provided by the anthocyanins of *Vaccinium myrtillus* and superior to that of rutin.

VII. PEANUT

The peanut is the fruit of the peanut plant, *Arachis hypogaea* L., and is a kind of pea, not a nut. Like other legumes, peanuts bear seeds in pods (shells). The peanut plant is unusual because its pods develop underground. For this reason, the peanuts are also called groundnuts.

Investigations on the pigmentation that characterized the seed coat have shown that different genes are involved in the inheritance of the white, purple, red, or pink color of the seed coat or testa of peanuts.[86-88] However, no correlation between color of the seed coat and anthocyanic composition has been found.[89-92] Halevy and Ashri[93] identified the aglycone in the seed coat of various peanuts as cyanidin, pelargonidin, and peonidin as well as glucose, which is present in the anthocyanins of three peanut cultivars. The pink line 123 contains a glucoside of cyanidin; the red Congo cultivar has a glucoside of cyanidin and one of pelargonidin; and the blue-black Pearl Black cultivar contains glucosides of pelargonidin, cyanidin, and peonidin.

Bilquez and Lecomte[94] reported that the normal yellow pigmentation of *A. hypogaea* flower is due to a flavone, while the pink appears to be due to an anthocyanin. The anthocyanin pigmentation of the flowers takes on a semicircular or half-moon shape whose inheritance is correlated with the absence of anthocyanins in the stem and with white testa of the seed.[95] According to Sathiamoorthy and Natarajan,[96] two genes are involved in the formation of anthocyanins in the purple pegs of peanuts; to our knowledge, however, the identity of these anthocyanins has not been reported.

VIII. PEPPER

The color of ripe *Capsicum annuum* L. — peppers, chile peppers, or paprika — is due to a complex mixture of carotenoids (capsanthin, capsarubin, cryptocapsin, and xanthophylls), and therefore, it is rather rare that the deep purple unripe fruit of *C. annuum* or *C. frutescens* contains anthocyanins. However, as peppers ripen, they become red and may contain anthocyanins.[97] In the cultivar Bighart, the chlorophyll decreases during ripening while the anthocyanins, ascorbic acid, and carotenoids increase.[98] Sharma and Seshadri[99] reported the presence of an unusual anthocyanin (petunidin diglucoside) in four cultivars of chile peppers grown at the Indian Agricultural Research Institute, New Delhi. The anthocyanins were, for the most part, confined to the pericarp. Recently, Saga and Kikuchi[100] characterized delphinidin 3-(p-coumaroylrhamnoside)-glucoside-5-glucoside or nasunin in *C. annuum* Goshiki. Endo[101] studied the inheritance of anthocyanins in crosses between the varieties Goshiki and Takanotsume, and established the minimum number of genes to be 1.18. Rashid Khan and Munir[102] investigated the inheritance of anthocyanins in various parts of stems and flowers of *C. annuum* and postulated a color factor D, the effects of which are intensified by genes A and B. Neither of these produces color when present singly but together they produce color in the absence of D. These authors also found that node color is completely linked with anther color as follows: purple anthers — light purple nodes; blue anthers — reddish blue nodes; and yellow anthers — green nodes.

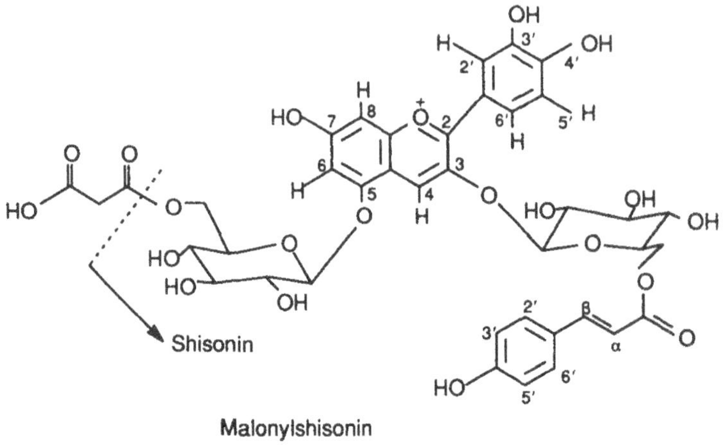

FIGURE 2.   Structures of malonylshisonin and shisonin. (From Kondo, T., Tamura, H., Yoshida, K., and Goto, T., *Agric. Biol. Chem.*, 53, 797, 1989. With permission.)

IX. PERILLA

Perilla — *Perilla frutescens* var. *crispa*, Benth. (*P. ocymoides* var. *crispa*) — is an herb, native to the Himalayas, China, and Japan, grown for its colored foliage. In Japan, the purple leaves of perilla (Japanese name, shiso) have been used for coloring traditional Japanese pickles such as umeboshi (pickled Japanese plum) and benishōga (pickled ginger).[103,104] Recently, the leaf extracts have been evaluated for use as a natural food color.[104]

The first chemical studies on the red pigments of perilla were conducted by Kondo[105] and Kuroda and Wada,[24,106] who showed that the pigment, which they called shisonin, was cyanidin 3,5-diglucoside and its *p*-coumaric esters. Later, Hayashi and Abe[107] and Takeda and Hayashi[108] confirmed that shisonin was cyanidin 3-(*p*-coumarylglucoside)-5-glucoside. The position of acylation was determined by Watanabe et al.,[27] who established the structure as 3-(6-*p*-coumaryl-D-glucopyranoside)-5-D-glucopyranoside. Later NMR studies by Goto et al.[109] confirmed this stereostructure. In *P. nankinensis*, Jadot and Niebes[110] also found cyanidin 3-(6-*p*-coumaryl-D-glucoside)-5-D-glucoside to be the principal anthocyanin present.

Recently, however, Kondo et al.[103] reported that shisonin is actually an artifact and that the genuine pigment of perilla leaves is the malonylated anthocyanin malonylshisonin (Figure 2). Ishikura[111] reported the existence of four anthocyanins in leaves of perilla which he characterized as cyanidin-3-(*p*-coumarylglucoside)-5-glucoside, cyanidin-3-ferulylglucoside-5-glucoside, cyanidin-3-glucoside-5-glucoside-(*p*-coumaric + caffeic acids), and cyanidin-3-glucoside-5-glucoside-(organic acid). Later, Yoshida et al.[104] isolated and fully characterized nine pigments: malonylshisonin, shisonin, caffeylmalonylcyanin, malonyl-*cis*-shisonin, caffeylcyanin, *cis*-shisonin, *cis*-caffeylmalonylcyanin, *cis*-caffeylcyanin, and cyanin. Yoshida et al.[104] also examined the color stability of malonylshisonin, shisonin, caffeylmalonylcyanin, malonyl-*cis*-shisonin, *cis*-caffeylcyanin in an acidic aqueous solution ($5 \times 10^{-5} M$, pH 2.0, 50°C); they found that malonylshisonin and malonyl-*cis*-shisonin possessed the same color stability and were considerably more stable than shisonin and cyanin. Color stability studies with shisonin were also conducted by Miyauchi[112] and Suyama et al.,[113] who found that in an acidic environment the degradation of this anthocyanin followed a first-order kinetics and that addition of fructose, saccharose, Cu and Fe ions, H_2O_2, and O_2 had little effect on color stability.

In the light, the perilla pigment is relatively stable,[114] and its stability can be further improved by mixing the pigment with carboxymethylcellulose.[115] γ-Irradiation of shisonin (up to 150 krd) seems to discolor the pigment especially when dissolved in methanol.[116]

Perilla pigment, for use as food colorant, can be purified with ion-exchange resins and elution with dilute alkali or organic solvent/water mixtures[117] or by ultrafiltration and reverse osmosis.[118,119] Extraction with 10% citric acid solution followed by ultrafiltration with a membrane having a molecular weight cut-off of 6000 resulted in a 60% recovery of anthocyanin.[119] Also, when perilla leaves are preserved by pickling, the anthocyanins leach out of the leaves into the brine. The spent brine, which may contain up to 1.98 g of anthocyanin/100 g of dry leaves used, has been recommended for use as a source of pigment suitable as food colorant.[119-124] Relatively pure anthocyanin pigment can be obtained by ultrafiltration of the brine with a membrane of a <2000 cut off followed by electrodialysis to remove the NaCl salt.[123] The resulting anthocyanin extract can be used in a variety of food and pharmaceutical applications including manufacturing of a red alcoholic liqueur made by mixing the pigment of perilla leaves with sake.[124]

Recently, anthocyanin pigments have also been produced by cell/tissue cultures of perilla.[125-129] A high amount of anthocyanin pigments, 3 g/l, was obtained in a bubble column bioreactor after 10 d of cultivation at an aeration rate of 0.1 vvm with light irradiation at 27.2 W/m². Likewise, 2.9 g anthocyanins per liter of medium was produced using an aerated and agitated bioreactor at an agitation speed of 130 rpm, an aeration rate of 0.1 vvm, and a light irradiation intensity of 27.2 W/m².[125]

X. PISTACHIO

Pistachio, *Pistacia vera* L., is a shrub of the Mediterranean area and the Middle East region that produces edible nuts used as food and flavoring.[130,131] Red color in the pistachio nut is an important quality factor because seeds with a brilliant purple color in their upper part and a greenish silvery color in the lower part seem to be most desirable.[132] In addition, the change in color of the nut is an accepted maturity index.[133,134]

It is known that the reddish color of the skin covering the kernel of the pistachio gives the characteristic reactions of the anthocyanins.[135] By precipitation with lead acetate, Oliai[136] separated the anthocyanins from total phenolic fraction in *P. vera*, but provided no further information on the nature of these flavonoids. The occurrence of myricetin in the genus *Pistacia* was reported by Perkin and Wood,[137] and the presence of proanthocyanidins in *P. lentiscus* was reported by Freudenberg et al.[138] Catechin and epicatechin were later reported occurring in pistachio nuts by Feucht and Nachit.[139] Other flavonoids were identified by Hiroi et al.[140] in the wood of *P. chinensis*. The phenolics of *P. terebinthus* and *P. lentiscus* were studied by Koco,[141] and reported to possess antimicrobial activity.[142] Cyanidin 3-galactoside has been identified as the main anthocyanin pigment of the pistachio nut by chromatographic, chemical, and spectral analyses.[143]

XI. RHUBARB

Rhubarb (*Rheum* spp.) contains anthocyanins in the epidermal layer of the stalk, particularly at the base, and in the roots. *R. palmatum* is used for pharmaceutical purposes; and *R. rabarbarum* and *R. rhaponticum* are eaten fresh and are processed into extracts used to make liqueurs and syrups.[144]

From an agronomic point of view, the absence of natural or artificial light is essential for maximum development of anthocyanins in rhubarb.[145,146] Canning of rhubarb using minimum preparation steps provides a product with superior organoleptic characteristics and with minimum color loss.[147]

The anthocyanins of rhubarb stalks were first identified by Gallop[148] as cyanidin 3-glucoside and cyanidin 3-rutinoside. The same two pigments were confirmed by Blundstone and Crean[149] and by Hetmanski and Nybom.[150] In 'Canada Red' rhubarb, Wrolstad and Heatherbell[151] also found cyanidin 3-glucoside and cyanidin 3-rutinoside present in the relative amounts of 87 and 13%, respectively. A third anthocyanin (present in trace amounts) was also reported and later identified as a chloride of cyanidin, an artifact from the extraction and purification steps. In 'German Wine', 'Sunrise', 'Sutton's Seedless', and 'Valentine' rhubarb, Fuleki[152] found three anthocyanins. Two of these pigments were the 3-glucoside and 3-rutinoside of cyanidin and accounted for 33 to 58% and 66 to 41%, respectively (Table 2). The third pigment was identified as a bioside of cyanidin and was present in small quantities in all the examined cultivars.

Later work by Wrolstad and Struthers[153] using polyvinylpyrrolidone (PVP) column chromatography, in order to avoid artifacts, found four anthocyanins: the two principal pigments already noted, a third anthocyanin which seemed to be a derivative of delphinidin glycosylated at position 5, and a fourth pigment which appeared to be a furanosic form of one of the two major pigments.

In a study on the distribution and evolution of flavonoids in petioles and limbs of rhubarb during development, Tronchet[154] found that the predominant

TABLE 2
Relative Quantity of Individual Anthocyanins in
Rhubarb Cultivars

	Cultivar				
Anthocyanin	German Wine	Sunrise	Sutton's Seedless	Valentine	Canada Red
Cyanidin 3-glucoside	33	48	48	58	87
Cyanidin 3-rutinoside	66	51	50	41	13
Others	1	1	2	1	t

Note: t = Trace.

Adapted from Fuleki, T., *J. Food Sci.*, 34, 365, 1969.

pigments were anthocyanins and anthraquinones. Monomeric anthraquinones, their glycosides, and dimeric anthrone glycosides (sennosides) were also identified by Ohshima et al.[155] in methanol extracts of rhubarb. In the seeds and leaves of ornamental rhubarb, *R. tataricum*, cyanidin 3-glucoside and cyanidin 3,5-diglucoside have been reported.[156,157]

XII. ROSELLE

Roselle, *Hibiscus sabdariffa* L., is a tropical plant grown for its edible red calyx and for a fiber (rosella hemp) made from its stem. The calyxes contain anthocyanins and are excellent for making a transparent, brilliant red gelatin. They are also used to make jams, jellies, and refreshing fruity beverages.

Forsyth and Simmonds[158] were the first to characterize the anthocyanidins of roselle as cyanidin and delphinidin. Thin-layer and paper chromatographic analyses of red sepals by Du and Francis[159] revealed the presence of four anthocyanins: delphinidin 3-sambubioside (delphinidin 3-xylosylglucoside) or hybiscin and cyanidin 3-sambubioside (cyanidin 3-xylosylglucoside) or gossypicyanin as the major pigments and delphinidin 3-glucoside and cyanidin 3-glucoside as the minor ones. These authors also reported the total anthocyanin content expressed as delphinidin 3-glucoside at 1.5 g/100 g of dry calyx. Recently, Pouget et al.[160] (using HPLC analysis) reported the relative proportion of delphinidin and cyanidin 3-sambubiosides at 70.9 and 29.1%, respectively.

Since the early 1970s, roselle has received considerable attention as a potential source of natural red colorant for foods, pharmaceuticals, and cosmetics. Esselen and Sammy[161,162] reported on the extraction and stability of 'Early Dwarf' roselle pigment as a colorant. According to these authors, when the right proportions of roselle calyxes or a concentrated aqueous extract were added to apple gelatin or pectin, the color retention was very good even after

12 months of storage in the dark at room temperature. After 18 months, it was still acceptable. High temperature, however, had a negative effect. After 6 months of storage at 37.7°C, the product had largely lost its color. Desirable color characteristics of the product were retained up to 18 months with storage at temperatures less than 1.6°C. Esselen and Sammy[162] also reported that savory drinks, prepared by first steeping roselle calyxes in hot water and then adding sugar and spices (cinnamon and cloves) to it, retained their color for 6 weeks at 37.7°C. However, after 3 months, there was an almost complete loss of red color, and it turned brownish and was accompanied by a noticeable deterioration of the aroma. Storage of this drink at room temperature even up to 18 months did not show marked alteration, and there was no loss in color or aroma when stored at 1.6°C. The presence of spices had no effect on color stability. Similar color stability observations were made with carbonated drinks prepared using roselle concentrate.

Clydesdale et al.[163] prepared a spray-dried powder from roselle concentrate. They added the colorant to a dry beverage mix and a dry gelatin dessert mix, and studied the pigment color stability under a number of conditions. During a 16-week period, the color loss was greater than that of the synthetic colorant, Red Dye No. 2, but it remained highly acceptable for the first 4 weeks at room temperature. Similar findings were reported by Saeed and Ahmed,[164] who showed a correlation between color and aroma degradation in a carbonated beverage stored at room temperature for 90 d.

Osman[165] and Pouget et al.[166] studied the extraction and stability of pigment extracts but reported somewhat contradictory results. Osman[165] found that the color intensity of the extract decreased with addition of bisulfite, peroxide, ethylenediaminetetraacetic acid (EDTA), and some amino acids; and increased with stannous chloride, calcium sulfate, glucono-δ-lactone, and Pectinal R-10. Pouget et al.,[166] on the other hand, reported that addition of sulfite or metabisulfite markedly improved the color of aqueous solutions containing 1% roselle anthocyanin extract. These seemingly contradictory findings are probably due to differences in pH and temperature at which the color stability experiments were conducted.

Recently, Zeng et al.[167-169] described an extraction technique for roselle anthocyanins which uses water at 50°C, an extraction period of 4 h, and a calyx/solvent ratio of 1:10. According to the authors, pigment extracted by this technique follows a first-order kinetics of thermal degradation which is rapid at temperatures above 100°C and practically instantaneous at 165 to 170°C. When this hot water extract of roselle anthocyanins is added to agarose, it produces candied fruit which has an intense red color and which is stable and attractive. The concentration of pigment in the candies varies from 0.3 to 0.6% for the hard candies and 0.1 to 0.2% for the soft candies.

Roselle anthocyanins have also been produced by tissue and callus culture,[171,172] and their synthesis has been increased by treating the plants with gibberellic acid and benzyladenine.[173] A recent unconfirmed report,[174] however, suggests that the roselle pigment may act as a mutagen of *Salmonella typhimurium* strains.[174]

XIII. SAFFRON

Crocus sativus L., saffron, is widely used for the production of the prized yellow colorant obtained from the pistils. The sepals may find application as a source of red pigment, because they contain flavonols and anthocyanins.[175]

Malvidins and cyanidins were found in *C. sativus* by Maroto[176] and in *C. vernus* by Fouassin.[79] In various *Crocus* species, Price et al.[177] found the anthocyanins to be mainly diglucosides of delphinidin accompanied in some species by malvidin. Lawrence et al.[178] identified delphinidin 3,5-diglucoside as the main pigment in six of the eight species examined. The presence of delphinidin 3,5-diglucoside was confirmed in *C. sativus* by Saito et al.[179] Thus, the most common pigment in *Crocus* spp. is delphinidin, followed by petunidin and malvidin, depending upon the species.[180] As observed by Harborne,[181] however, it is somewhat surprising that malvidin is not the predominant pigment of *Crocus* and that no mutations to cyanidin or pelargonidin forms have been observed, even though *Crocus* has been in cultivation for a very long time and hybridization is common in the genus.

XIV. SUGARCANE

Sugarcane, *Saccharum officinarum* L., is a tall grass that grows in tropical and semitropical countries. It produces sturdy stalks 2 to 5 m high and about 5 cm in diameter. These stalks are used to produce sugar. The outer covering of the stalk may be reddish-brown, dark-red, or purple in color, and the entire stalk is often colored internally as well.

Sankaranarayan and Narasimhan[182] found anthocyanins in 311 of the 435 clones studied. When only one organ was pigmented, it was generally the internode, root primordia, or dewlap. In six clones, pigmentation was observed in seven organs.

Using chromatographic techniques and physicochemical tests, Misra and Dubey[183] characterized the anthocyanin of sugarcane as peonidin 3-galactoside. In the cultivar NCo 310, Smith and Hall[184] identified the anthocyanin pigment as the 3-deoxyanthocyanidin, luteolinidin, occurring in the growing point of the cane plant.

Factors which influence the degree of pigmentation of sugarcane include: sulfur deficiency in the soil and presence of disease, both of which promote the formation of red pigments.[185,186]

Pigmented sugarcane varieties are diffuse throughout Asia, and the presence of anthocyanins in the stalks can cause problems with the purification of sugar.[187] According to Goodacre and Coombs,[188] flavonols and anthocyanins of sugarcane can undergo enzymatic oxidation which lead to the development of a brown color in sugar syrup. Also, some anthocyanins can remain with the sugar fraction and appear in the raw sugar crystals,[189] thereby lowering the sugar quality.

XV. SUNFLOWER

Sunflower (*Helianthus annuus*, L.) is one of the world's leading oilseed crops, second only to soybean in world oil production.[190] A sunflower cultivar with red pigment in the corolla and in other parts of the plant was first reported in the early 1900s; the first studies on color inheritance were also conducted during this period.[191-193] The anthocyanins in sunflower have a definite distribution throughout various parts of the plant.[194,195] Plants with pigmented stems, however, do not necessarily produce pigmented seeds because pigmentation in different parts of the plant is regulated by different genes.[196] In fact, the anthocyanin pigmentation in the plant is regulated by a dominant gene, T, while that of the seed is under the control of a complementary gene, I.[197] The genes G and T_1, together with the gene T, condition the presence of anthocyanins in the ligule of the flower and the hypoderm of the pericarp.[198] Luczkiewicz[199,200] reported that the anthocyanins in the stem and leaf veins are regulated by a single dominant gene. Mosjidis[201] showed the relationship between the anthocyanin pigmentation in six parts of the flower.

Some types of male-sterile sunflowers can be grouped according to the presence of anthocyanins in the hypocotyl[202-204] and are, therefore, useful in plant selection.[205] Anthocyanins are also used as genetic markers in the production of hybrids.[206] Plant hormones induce the synthesis of anthocyanins in sunflower cotyledons while their synthesis is inhibited by actinomycin D.[207]

The presence of anthocyanins in purple-hulled cultivars of the sunflower has recently been implicated to provide resistance to damage by redwing blackbirds *(Agelaius phoeniceus)*.[208-212] This claim, although well founded, has not been fully proven. Other factors, such as seed morphology and/or the presence of unknown bird repellents in purple-hulled seeds, may be responsible for the apparent lack of preference of purple-hulled seeds by the redwing blackbirds.

Anthocyanins were first noted in the corolla of the sunflower. Sando[213] reported the presence of a monoglycoside of cyanidin which he considered to be a product of flavonol reduction. Vaccari et al.[214] separated nine anthocyanins from purple sunflower hulls and identified with certainty the two major glycosides as peonidin 3-divanillyl-sambubioside and cyanidin 3-vanillyl-sambubioside. Two other pigments were tentatively identified as cyanidin 3-vanillyl-glucosylarabinoside and an acylated diglucoside of malvidin. Only a few sunflower cultivars have hulls containing anthocyanins, and of these, very few are pigmented by the glucosides present at a concentration of 2.2 g/100 g dry hulls.[214] The oilcake from purple-hulled seeds provides an anthocyanin-rich material that can be used to produce a red colorant. Depending on the cultivar, when the oilcake is extracted with acidified aqueous ethanol and the extract is concentrated, a dark red product is obtained. This product can be used to color food and pharmaceutical and cosmetic products.[215,216] An extraction process which maximizes the recovery of pigment from hulls has been patented.[217]

The purple hulls from the cultivar Neagra de Cluj are particularly rich in pigments characterized as cyanidin 3-glucoside (approximately 35%), an acylated form of cyanidin 3-glucoside (15%), cyanidin 3-xyloside (14%), an acylated form of cyanidin 3-xyloside (29%), and less than 10% of two unidentified compounds.[211] Holm[218] also reported that for maximum extraction and minimal degradation, sunflower hull anthocyanin pigment should be extracted with 18 to 28% ethanol and 2 to 9% tartaric acid at 8°C for approximately 3 min. According to Mok and Hettiarachchy,[219] however, extraction of anthocyanins from sunflower hulls is most effective when carried out at room temperature using 1 part ground hulls to 20 parts of aqueous SO_2 solution at a concentration of 1000 ppm.

When a sunflower hull pigment solution is heated, the red color gradually fades while the brown pigment increases. Degradation of anthocyanin upon heating follows first order reaction kinetics.[219] The rate constants are reportedly lower at pH 3.0 than at pH 1.0 and 5.0,[219] indicating that the sunflower hull anthocyanins are more stable at pH 3.0.

The economic feasibility of producing a natural red colorant from purple-hulled sunflowers depends on the market price for the product and on its cost of production. According to Wiesenborn et al.,[220] the total plant investment to produce an annual volume of 355,000 kg of extract is $5.18 million, excluding working capital requirements. Thus, the immediate prospects for commercial production of anthocyanins from sunflower hulls are not very promising.

XVI. TAMARILLO

Tamarillo (*Cyphomandra betacea* [Cav.] Sendt.), or tomato tree, is native to Peru and grows well throughout the subtropical regions. The fruit is eaten fresh or is processed into juice, jams, or jellies, and condiments. Some cultivars produce fruit with red skin, a yellow-orange pulp, and purple gelatin surrounding the seed. The quantity of pigment in the skin increases as it ripens over a 15-week period, and the harvest of unripe fruit stops color development.[221] Also, as fruit ripening proceeds, the anthocyanin and organic acid contents decrease in the pulp while the anthocyanins increase in the skin.[222]

Dawes[223] reported the presence of at least eight anthocyanins in tamarillos, the main one being delphinidin based. Wrolstad and Heatherbell[224] reported on the identification of anthocyanins and distribution of flavonols in the fruit. According to these authors, the pulp contains very few anthocyanins, while it is rich in anthoxanthin and procyanidin. Delphinidin 3-rutinoside is the principal pigment in the gelatin near the seed while cyanidin 3-rutinoside is the primary pigment in the skin, which contains 9.7×10^{-4} mM/g fresh weight (FW). There are more monoglucosides in the juice than in the skin. The juice contains six anthocyanins: delphinidin 3-rutinoside, pelargonidin 3-rutinoside, delphinidin 3-glucoside, pelargonidin 3-glucoside, cyanidin 3-glucoside, and cyanidin 3-rutinoside. The highest anthocyanin concentration, however, is around the seed; there a value of 6.4×10^{-3} mM/g FW expressed as delphinidin 3-rutinoside is found with small quantities of two other pigments, one having the characteristic of pelargonidin 3,5-diglucoside and the other as a diglucoside of cyanidin or delphinidin. In the skin, there also appears to be a small quantity of anthocyanins acylated with *p*-coumaric, caffeic, and ferulic acids.[224]

In a Brasilian variety of tamarillo, Bobbio et al.[225] identified an anthocyanin arrangement which was significantly different from that reported by Wrolstad and Heatherbell;[224] this was attributed to misidentification of the species used by the American researchers. Bobbio et al.[225] found pelargonidin 3-glucosylglucose in the seed and skin of the fruit and peonidin 3-glucosylglucose and malvidin 3-glucosylglucose in the seeds. In addition to differences in the anthocyanidins, the diglucoside present in the three pigments identified by Bobbio et al.[225] was maltose, which has not been reported in other anthocyanins.[226] In fact, the disaccharides found to date have been rutinose, sambubiose, lathyrose, and sophorose.[227]

XVII. TAMARIND

Tamarind (*Tamarindus indica* L.) is widely distributed throughout southeast Asia (India, Java, the Philippines), and it is grown commercially in most tropical regions.[228]

Annually, India produced 250,000 tonnes of tamarind from both cultivated and wild trees. About 200 kg of fruit can be harvested from a single tree, but only the pulp, which makes up 50% of the fruit, is used. It is low in nutritional value and has modest medicinal properties.[229] The pulp is eaten candied or in jams; but most commonly it is used to prepare a high quality, refreshing drink.

Lewis and Johar[230] reported that the anthocyanin pigment in the red tamarind fruit is cyanidin 3-glucoside. Lewis et al.[231] confirmed this identification in colored fruit that differs from the white fruit in which the pigment is leucoanthocyanin.

XVIII. TOMATO

Butler and Chang[232] identified pelargonidin as the anthocyanidin in the purple stem of *Lycopersicon esculentum* L., and reported that this pigmentation is associated with a lack of phosphorus. Von Wettstein-Knowles[233,234] identified the acylated anthocyanin petunidin 3-(*p*-coumaryl)rutinoside-5-glu-coside, and Hussey[235] found petunidin 3-glucoside in tomato plants grown in the greenhouse.

Anthocyanin synthesis in tomato seedlings is controlled by phytochrome.[236,237] and is influenced by the presence of hormones.[238,239] Lesley et al.[240] showed that the gene *Wo* (from a woolly mutant) influenced the expression of gene *aW* which is responsible for the lack of anthocyanins. Durand[241] reported that the gene *Wo* is a better genetic marker than *aa*, apparently also responsible for the absence of anthocyanins. Anais[242] used anthocyanins as a genetic marker to evaluate the efficiency of tomato pollination by the bee, *Exomalopsis*; and Rick et al.[243] found that lack of anthocyanins did not increase the susceptibility of the plants to attacks by *Epitrix hirtipennis*. The pink pigmentation at the base of the hypocotyl and its genetic relationship were studied by Ibrahim et al.,[244] and Koornneeff et al.[245] determined the effect of light on the biosynthesis of anthocyanins in tomato mutants.

There are several agronomic factors that promote the formation of anthocyanins in tomatoes, and of these, phosphorus deficiency is the best-recognized promoter of anthocyanin accumulation,[246-249] and certain herbicides are known to inhibit the synthesis of anthocyanins.[250]

REFERENCES

1. **Steiner, H.,** Green asparagus, testing international asparagus varieties, *Gemuese*, 16, 186, 1980 (HA 1981, 5501).
2. **Chen, M. S. and Shen, T. F.,** Cytological studies on the chromosomes of *Asparagus officinalis* L. I. Karyotype and hybridization of *Asparagus officinalis* L., *Taiwan Asparagus Res.*, 68, 22, 1980 (PBA 1983, 2566).
3. **Chen, M. S.,** Cytogenic studies on *Asparagus officinalis* L. II. Studies on artificially induced polyploid asparagus, *J. Agric. Res. China*, 31, 137, 1982 (PBA 1983, 1628).
4. **Hepler, P. R., Thompson, A. E., and McCollum, J. P.,** Inheritance of resistance to asparagus root, *Ill. Exp. Stn. Bull.*, 607, 47, 1957 (HA 1957, 4541).
5. **Hsue, J. C. and Lin, C. H.,** Study of the postharvest physiology of white asparagus stalks under different storage times and environment, *Sci. Res. Abstr. Rep. China, Taipei, Taiwan*, 1979, 314 (HA 1981, 1246).
6. **Robinson, R. and Robinson, G. M.,** A survey of anthocyanins. IV, *Biochem. J.*, 28, 1712, 1934.
7. **Wann, E. V. and Thompson, A. E.,** Anthocyanin pigments in asparagus, *Proc. Am. Soc. Hortic. Sci.*, 87, 270, 1965.
8. **Francis, F. J.,** Anthocyanins of asparagus, *J. Food Sci.*, 32, 430, 1967.
9. **Wu, S. M. and Wu, C. Y.,** Relationship between the development of the color and flavor and the detinning of canned asparagus, *Tech. Rep., Food Ind. Res. Dev. Inst., Taiwan N.*, 137, 23, 1979.
10. **Akhmedov, U. A. and Khalmatov, K. K.,** *Asparagus persicus* growing in Uzbekistan, *Med. Zh. Uzb.*, 6, 40, 1965 (CA 63, 18649a).
11. **Tiwari, K. P., Masood, M., and Minocha, P. K.,** Chemical examination of flowers of *Asparagus gonocladus* Baker, *J. Indian Chem. Soc.*, 55, 520, 1978 (CA 89, 160160).

12. **Raman, C. V.,** Floral colors and their origin, *Curr. Sci.,* 38, 451, 1969 (CA 71, 36377).

13. **Misra, R. and Gyani, B. P.,** Light absorption and approximate indicator constants of extracts of some Indian flowers and their indicator properties, *J. Indian Chem. Soc.,* 16, 167, 1953 (CA 49, 1886).

14. **Ranaganayaki, S. and Singh, A. K.,** Isolation and identification of pigments of the flowers of *Clitoria ternatea, J. Indian Chem. Soc.,* 56, 1037, 1979 (CA 92, 211799).

15. **Lowry, J. B. and Chew, L.,** Use of extracted anthocyanin as a food dye, *Econ. Bot.,* 28, 61, 1974.

16. **Srivastava, B. K. and Pande, C. S.,** Anthocyanins from the flowers of *Clitoria ternatea, Planta Med.,* 32, 138, 1977 (CA 87, 180749).

17. **Saito, N., Abe, K., Honda, T., Timberlake, C. F., and Bridle, P.,** Acylated delphinidin glucosides and flavonols from *Clitoria ternatea, Phytochemistry,* 24, 1583, 1985.

18. **Terahara, N., Saito, N., Honda, T., Toki, T., and Osajima, Y.,** Further structural elucidation of the anthocyanin, deacylternatin, from *Clitoria ternatea, Phytochemistry,* 29, 3686, 1990.

19. **Terahara, N., Saito, N., Honda, T., Kenjiro, T., and Osajima, Y.,** Structure of ternatin Al, the largest ternatin in the major blue anthocyanins from *Clitoria ternatea* flowers, *Tetrahedron Lett.,* 31, 2920, 1990.

20. **Terahara, N., Saito, N., Honda, T., Kenjiro, T., and Osajima, Y.,** Structure of ternatin A2, one of the *Clitoria ternatea* flower anthocyanins having the unsymmetrical side chains, *Heterocycles,* 31, 1773, 1990.

21. **Terahara, N., Saito, N., Honda, T., Toki, K., and Osajima, Y.,** Acylated anthocyanins of *Clitoria ternatea* flowers and their acyl moieties, *Phytochemistry,* 29, 949, 1990.

22. **Kuroda, C. and Wada, M.,** Coloring matter of eggplant (nasu), *Proc. Imp. Acad. (Tokyo),* 9, 51, 1933 (CA 27, 3215).

23. **Kuroda, C. and Wada, M.,** Coloring matter of eggplant (nasu), *Proc. Imp. Acad. (Tokyo),* 11, 235, 1935 (CA 29, 6598).

24. **Kuroda, D. and Wada, M.,** Constitution of natural colouring matters kuromamin, shisonin and nasunin, *Bull. Chem. Soc. Jpn.,* 11, 272, 1936 (CA 30, 5998).

25. **Takeda, K., Abe, Y., and Hayashi, K.,** Anthocyanins. XXXIX. Violanin as a complex triglucoside of delphinidin, and its occurrence in the variety Burma of *Solanum melongena, Proc. Jpn. Acad.,* 39, 225, 1963 (CA 59, 14298f).

26. **Sakamura, S., Shibusa, S., and Obata, Y.,** The structure of the major anthocyanin in eggplant, *Agric. Biol. Chem.,* 27, 663, 1963.

27. **Watanabe, S., Sakamura, S., and Obata, Y.,** Structure of acylated anthocyanins in eggplant and perilla; and the position of acylation, *Agric. Biol. Chem.,* 30, 420, 1966 (CA 65, 2621h).

28. **Sakamura, S. and Obata, Y.,** Anthocyanase and anthocyanins occurring in eggplant, *Solanum melongena* L., *Agric. Biol. Chem.,* 25, 750, 1961 (CA 56, 7705a).

29. **Nagashima, Y. and Taira, T.,** On the pigment of eggplant, *Eiyo To Shokuryo,* 17, 420, 1965 (CA 64, 10086).

30. **Casoli, U. and Dall'Aglio, G.,** Ricerche sugli antociani de melanzana, *Ind. Conserve,* 44, 18, 1969 (FSTA 2, 7J708).

31. **Pifferi, P. G. and Zamorani, A.,** Studi sui pigmenti naturali, Nota IV. Gli antociani del *Solanum melongena, Ind. Agrar.,* 7, 51, 1969 (FSTA 1, 8J708).

32. **Tanchev, S. S., Ruskov, P. J., and Timberlake, C. F.,** The anthocyanin of Bulgarian aubergine *(Solanum melongena), Phytochemistry,* 9, 1681, 1970.

33. **Hocking, P.,** Anthocyanins in the Australian eggplant, *Food Technol. Aust.,* 23, 285, 1971 (FSTA 4, 1J26).

34. **Ramaswamy, S. and Rege, D. V.,** Chromatographic identification of skin pigments of *Solanum melongena, J. Food Sci. Technol., India,* 12, 250, 1975 (FSTA 8, 11J1756).

35. **Abe, Y. and Gotoh, K.,** Genetic control of acylation of anthocyanin pigment in eggplant, *Annu. Rep. Natl. Inst. Genet., Jpn.,* 1955(6), 75, 1956 (PBA 1957, 3830).

36. **Abe, Y. and Gotoh, K.,** Anthocyanin in the fruit-coats of two varieties of eggplant, Burma and Black Beauty, *Annu. Rep. Natl. Inst. Genet., Jpn.,* 1956(7), 49, 1957 (PBA 1958, 3879).

37. **Abe, Y. and Gotoh, K.,** Biochemical and genetical studies on anthocyanins in eggplant, *Bot. Mag. (Tokyo),* 72, 432, 1959 (CA 55, 6616).

38. **Topoleski, L. D.,** Graft-induced alterations in *Solanum melongena, Diss. Abstr.,* 23, 3088, 1963 (PBA 1964, 3162).

39. **Janick, J. and Topoleski, L. D.,** Inheritance of fruit color in eggplant *(Solanum melongena), Proc. Am. Soc. Hortic. Sci.,* 83, 547, 1963.

40. **Sambandam, C. M.,** Guide chart for color combinations in hybrid eggplants, *Econ. Bot.,* 21, 309, 1967 (PBA 1968, 5556).

41. **Tigchelaar, E. R.,** Genetics and photocontrol of anthocyanin coloration in eggplant *(Solanum melongena* L.), *Diss. Abstr.,* 27, 3790, 1967 (PBA 1968, 7175).

42. **Tigchelaar, E. C., Janick, J., and Erickson, H. T.,** The genetics of anthocyanin coloration in eggplant *(Solanum melongena), Genetics,* 60, 475, 1968 (PBA 1970, 4054).

43. **Hagiwara, T. and Takeda, M.,** Is this not an instance of parthenogenesis in the eggplant?, *Nogyo Oyobi Engei,* 32, 376, 1957 (PBA 1959, 4608).

44. **Sambandam, C. N.,** Inheritance of hypocotyl colour in brinjal, *Indian J. Genet.,* 24, 175, 1964 (PBA 1966, 5250).

45. **Sinha, B. K., Prakesh, R., and Haque, M. F.,** Linkage studies in brinjal, *Indian J. Genet.,* 26, 223, 1966 (PBA 1967, 7126).

46. **Sambandam, C. M.,** Some studies on the genetics of the purple colour in eggplant, *Annamalai Univ. Agric. Res. Annu.,* 1, 1, 1969 (HA 1970, 6465).

47. **Patil, S. K. and More, D. C.,** Inheritance studies in brinjal, *J. Maharashtra Agric. Univ.,* 8, 43, 1983 (PBA 1984, 1913).

48. **Patil, S. K. and More, D. C.,** Inheritance studies of some characters in brinjal, *J. Maharashtra Agric. Univ.,* 8, 47, 1983 (PBA 1984, 1914).

49. **Patil, S. K. and More, D. C.,** Genetics of pigmentation in brinjal, *J. Maharashtra Agric. Univ.,* 8, 126, 1983 (PBA 1984, 1912).

50. **Nothmann, J., Rilski, I., and Spigelman, M.,** Effects of air and soil temperature on colour of eggplant fruits *(Solanum melongena), Exp. Agric.,* 14, 189, 1978 (FSTA 11, 6J965).

51. **Nothmann, J., Rilski, I., and Spigelman, M.,** Flowering pattern, fruit growth and color development of eggplant during the cool season in a subtropical climate, *Sci. Hortic.,* 11, 217, 1979 (FSTA 12, 7J1076).

52. **Awashti, C. P., Abidi, A. B., and Dixit, J.,** Biochemical constituents of promising cultivars of brinjal, *Indian J. Hortic.,* 44, 245, 1987 (CA 110, 74098).

53. **Chadha, M. L., Hedge, R. K., and Bajaj, K. L.,** Heterosis and combining ability studies of pigmentation in brinjal *(Solanum melongena* L.), *Veg. Sci.,* 15, 64, 1988 (PBA 1990, 4711).

54. **Chadha, M. L. and Sidhu, A. S.,** Punjab Neelam variety of brinjal *(Solanum melongena* L.), *J. Res., Punjab Agric. Univ.,* 25, 508, 1988 (FSTA 21, 7J117).

55. **Singh, D. K., Gautam, N. C., Awashti, C. P., and Singh, R. D.,** Biochemical composition of fruits of promising brinjal varieties and hybrids *(Solanum melongena* L.), *Veg. Sci.,* 15, 141, 1988 (HA 1990, 9003).

56. **Bajaj, K. L., Kaur, G., and Chadha, M. L.,** Glycoalkaloid content and other chemical constituents of the fruit of some eggplant *(Solanum melongena)* varieties, *J. Plant Foods,* 3, 163, 1979 (HA 1982, 2246).

57. **Sidhu, A. S., Kaur, G., and Bajaj, K. L.,** Biochemical constituents of varieties of eggplant, *Veg. Sci.,* 9, 112, 1982 (FSTA 17, 6J143).

58. **Molla, E., Esteban, R. M., and Lopez-Andreu, F. J.,** Determination of fruit color. Application to eggplant fuit, *Alimentaria,* 197, 55, 1988 (FSTA 22, 5J172).

59. **Lopez-Andreu, F. J., Molla, E., Fernandez, M., and Esteban, R. M.,** Cold storage of aubergines. Changes in quality, *Alimentaria,* 192, 31, 1988 (FSTA 22, 5J188).

60. **Aubert, S. and Pochard, E.,** Problemes de conservation en fraise l'aubergine (*Solanum melongena* L.). II. Essais preliminaires de mise au froid, *Rev. Hortic.,* 216, 35, 1981.

61. **Rhee, J. K. and Iwata, T.,** Histological observations on the chilling injury of eggplant fruits during cold storage, *J. Jpn. Soc. Hortic. Sci.,* 51, 237, 1982 (HA 1983, 1783).

62. **Sakamura, S. and Obata, Y.,** Anthocyanase and anthocyanins occurring in eggplant. II. Isolation and identification of chlorogenic and related compounds from eggplant, *Agric. Biol. Chem.,* 27, 121, 1963 (CA 58, 12857a).

63. **Sakamura, S., Watanabe, S., and Obata, Y.,** Anthocyanase and anthocyanins occurring in eggplant. III. Oxidative decoloration of the anthocyanin by polyphenoloxidase, *Agric. Biol. Chem.,* 29, 181, 1965.

64. **Sakamura, S., Shibusa, S., and Obata, Y.,** Separation of polyphenoloxidase for anthocyanin degradation in eggplant, *J. Food Sci.,* 31, 317, 1966.

65. **Casoli, U., Dall'Aglio, G., and Leoni, C.,** Azione di estratti enzimatici vegetali sui pigmenti antocianici delle melanzane, delle mele e del ribes, *Ind. Conserve,* 44, 193, 1969.

66. **Harborne, J. B.,** Anthocyanin in *D. carota* and other Umbelliferae, *Biochem. Syst. Ecol.,* 4, 31, 1976.

67. **Timberlake, C. F. and Bridle, P.,** Anthocyanins in fruits and vegetables, in *Recent Advances in the Biochemistry of Fruits and Vegetables,* Friend, J. and Rhodes, M. J. C., Eds., Academic Press, London, 1981, 221.

68. **Hrazdina, G.,** Anthocyanins, in *The Flavonoids, Advances in Research,* Harborne, J. B. and Mabry, T. J., Eds., Chapman & Hall, London, 1982, 135.

69. **Umadevi, L. and Daniel, M.,** Phenolics of some fruit species of the Apiaceae, *Natl. Acad. Sci. Lett.,* 13, 439, 1990.

70. **Foury, C. and Aubert, S.,** Observations preliminaires sur la presence et la repartition de pigments anthocyaniques dans un mutant d'artichaut (*Cynara scolymus* L.) a fleurs blanches, *Ann. Amelior. Plant.,* 27, 603, 1977 (CA 88, 117793).

71. **Aubert, S. and Foury, C.,** Couleur et pigmentation anthocyanique de l'artichaut (*Cynara scolymus* L.), in *3° Congr. Int. Stud. Carciofo,* 1979, 57, 1979 (HA 1982, 7877).

72. **Pifferi, P. G. and Vaccari, A.,** Studi sui pigmenti naturali. X. Gli antociani del carciofo (*Cynara scolymus* L.), *Ind. Conserve,* 53, 107, 1978.

73. **El-Negoumy, S. I., El-Sayed, N. H., and Saleh, M. A. M.,** Flavonoid glycosides of *Cynara scolymus, Fitoterapia,* 58, 178, 1987.

74. **Lattanzio, V. and Morone, I.,** Variations of the orthodiphenol content of *Cynara scolymus* L. during the plant growing season, *Experientia,* 35, 993, 1979.

75. **Lattanzio, V. and Van Sumere, C. F.,** Changes in phenolic compounds during the development and cold storage of artichoke (*Cynara scolymus* L.) heads, *Food Chem.,* 24, 37, 1987.

76. **Glan, R.,** Ueber den Farbstoff der Schwarzen Malve *(Althaea rosea),* Inaugural Diss., University of Erlangen, Erlangen, Germany, 1892.

77. **Willstätter, R., and Zollinger, E. H.,** Untersuchung ueber die Anthocyane, *Justus Liebigs Ann. Chem.,* 408, 110, 1915.

78. **Karrer, P., Widmer, R., Helfenstein, A., Hurliman, W., Nievergelt, O., and Montsarrat-Thoms, P.,** Ueber Pflanzenfarbstoffe. IV. Zur Kenntnis der Anthocyane und Anthocyanidine, *Helv. Chim. Acta,* 10, 729, 1927.

79. **Fouassin, A.,** Identification par chromatographie des pigments anthocyaniques des fruits et des legumes, *Rev. Ferment. Ind. Aliment.,* 11, 173, 1956.

80. **Nair, A. G. R., Nagarajan, S., and Subramanian, S. S.,** Flavonoids of the flowers of *Hibiscus abelmoschus* and *Althaea rosea, Curr. Sci.,* 33, 431, 1964 (PBA 1965, 4745).

81. **Kohlmunzer, S., Konska, G., and Wiatr, E.,** Anthocyanosides of *Alcea rosea* L. var. *nigra* as vasoprotective agents, *Herba Hung.,* 22, 13, 1983 (CA 101, 65981).

82. **Kasperskaya, T. V., Viktorova, G. K., Smirnov, E. V., and Salikhov, S. A.,** New colorant for confectionery products, *Klebopek. Kondit. Prom.,* 1, 41, 1982 (FSTA 15, 9L632).

83. **Rakhimkanov, Z. B., Karimdzhanov, A. K., Ismailov, A. I., and Mukhamedova, F. K.**, Red food colorings from plants, *Uzb. Khim. Zh.*, 5, 46, 1983 (CA 100, 66793).

84. **Kasumov, M. A., Kerimov, Y. B., Babaev, R. A., and Isaev, N. Y.**, Method of Obtaining a Red Food Colorant, U.S.S.R. Patent, 704, 971, 1979 (FSTA 12, 8T408).

85. **Turowska, I., Kohlmunzer, S., Skwara, J., Grzybek, J., and Wiatr, E.**, Anthocyanin and anthocyanin-free form and mucilage level in flowers of *Alcea rosea*. II. Viscosimetric analysis, anthocyanosides and other active substances, *Herba Pol.*, 20, 20, 1974 (CA 81, 132806).

86. **Hammons, R. O.**, White testa inheritance in the peanut, *J. Hered.*, 54, 139, 1963.

87. **Rigoni, V. A. and Krapovickas, A.**, Observations on the inheritance of colour in the testa of groundnut (*A. hypogaea* L.), *Arch. Fitotec. Urug.*, 5, 406, 1953 (PBA 1955, 857).

88. **Harvey, J. E.**, Testa Color Inheritance in Peanuts (*Arachis hypogaea* L.), Ph.D. thesis, University of Georgia, Athens, 1967 (PBA 1969, 3321).

89. **Surendra, P. and Srivastava, D. P.**, Inheritance of testa colour in groundnut (*Arachis hypogaea* L.), *Sci. Cult., Calcutta*, 33, 489, 1967 (PBA 1969, 1291).

90. **Ashri, A.**, Further evidence for a second red testa gene in peanut, *Arachis hypogaea* L., *Oleagineux*, 25, 393, 1970.

91. **Srivastava, A. N.**, A new genotype for red testa in *Arachis hypogaea* L., *Madras Agric. J.*, 59, 39, 1972.

92. **Srivastava, A. N.**, Supplementary factors that influence seedcoat colour in groundnut (*Arachis hypogaea* L.), *Indian J. Agric. Sci.*, 43, 911, 1973.

93. **Halevy, A. H. and Ashri, A.**, Partial identification of anthocyanin pigments in various testas of peanuts, *Arachis hypogaea, Oleagineux*, 26, 771, 1971 (CA 77, 4007).

94. **Bilquez, A. F. and Lecomte, J.**, Heredite de la coloration des fleurs chez l'arachide, *Oleagineux*, 24, 411, 1969.

95. **Srinivasalu, N. and Loganathan, N. S.**, Inheritance of the purple crescent on the standard petal of *Arachis hypogaea* L., *Curr. Sci.*, 28, 497, 1969 (PBA 1960, 3135).

96. **Sathiamoorty, M. R. and Natarajan, S. T.**, Inheritance of anthocyanin pigmentation in pegs of the groundnut plant (*Arachis hypogaea* L.), *J. Maharashtra Agric. Univ.*, 3, 264, 1978 (PBA 1981, 2440).

97. **Pruthi, J. S.**, Spices and condiments: chemistry, microbiology, technology, in *Advances in Food Research Suppl. 4*, Academic Press, New York, 1980, 150.

98. **Baslouny, F. M. and Biswas, P. K.**, Ascorbic acid, pigments and mineral element contents associated with growth and development of pimento pepper, *Proc. Fla. State Agric. Soc.*, 94, 268, 1981 (HA 1983, 3348).

99. **Sharma, J. N. and Seshadri, T. R.**, Survey of anthocyanins from Indian sources. II, *J. Sci. Ind. Res.*, 14, 211, 1955.

100. **Saga, K. and Kikuchi, J.**, Anthocyanin in the Goshiki pepper (*Capsicum annuum* L.) in relation to fruit growth, *Hirosaki Daigaku Nogakubu Gakujutsu Hokoku*, 40, 14, 1983 (CA 100, 188912).

101. **Endo, T.**, Inheritance of fruit color in *Capsicum, Annu. Rep. Natl. Inst. Genet., Jpn.*, 1952, 46, 1953 (PBA 1957, 233).

102. **Rashid Khan, A. and Munir, M.**, Inheritance studies in the common chillies *(Capsicum annuum)* in West Pakistan, *W. Pak. J. Agric. Res.*, 1, 124, 1962 (PBA 1964, 2715).

103. **Kondo, T., Tamuta, H., Yoshida, K., and Goto, T.**, Structure of malonylshisonin, a genuine pigment in purple leaves of *Perilla ocymoides* L. var. *crispa* Benth, *Agric. Biol. Chem.*, 53, 797, 1989 (CA 111, 54176).

104. **Yoshida, K., Kondo, T., Kameda, K., and Goto, T.**, Structures of anthocyanins isolated from purple leaves of *Perilla ocymoides* L. var. *crispa* Benth. and their isomerization by irradiation of light, *Agric. Biol. Chem.*, 54, 1745, 1990 (CA 113, 188025).

105. **Kondo, K.**, Untersuchung ueber Anthocyan und Anthocyanidin. V. Ueber den Farbstoff von *Perilla ocymoides* L. var. *crispa* Benth., *J. Pharm. Soc. Jpn.*, 51, 254, 1931.

106. **Kuroda, C. and Wada, M.,** Coloring matter of shiso, *Proc. Imp. Acad. (Tokyo),* 11, 28, 1935 (CA 29, 3338).

107. **Hayashi, K. and Abe, Y.,** Anthocyanins. XXV. Paper chromatographic investigation on anthocyanins occurring in the leaves of *Perilla* varieties, *Bot. Mag. (Tokyo),* 68, 71, 1955 (CA 49, 15176g).

108. **Takeda, K. and Hayashi, K.,** Oxidative degradation of acylated anthocyanins showing the presence of organic acid-sugar linkage in the 3-position of anthocyanidin. Experiments on ensatin, awobanin and shisonin, *Proc. Jpn. Acad.,* 40, 510, 1964.

109. **Goto, T., Takase, S., and Kondo, T.,** PMR spectra of natural acylated anthocyanins. Determination of stereostructure of awobanin, shisonin and violanin, *Tetrahedron Lett.,* 27, 2413, 1978 (CA 90, 38177).

110. **Jadot, J. and Niebes, P.,** Identification and characterization of acylated anthocyanins present in the leaves of *Perilla nankinensis, Bull. Soc. R. Sci. Liege,* 37, 593, 1968 (CA 70, 93912).

111. **Ishikura, N.,** Anthocyanins and flavones in leaves and seeds of Perilla plant, *Agric. Biol. Chem.,* 45, 1855, 1981 (CA 95, 165593).

112. **Miyauchi, T.,** On the extract of red pigment from *Perilla* leaf, *Shokuhin Kogyo,* 10, 62, 1975.

113. **Suyama, K., Tamate, M., and Adachi, S.,** Colour stability of shisonin, red pigment of a perilla (*Perilla ocymoides* L. var. *crispa* Benth.), *Food Chem.,* 10, 69, 1983.

114. **Ikawa, F.,** Studies on natural coloring agents. XII. Properties of anthocyanin pigments and use for foods, *Aichi-Ken Shokuhin Kogyo Shinkenjo Nempo,* 19, 15, 1978 (CA 92, 20825).

115. **Hayakawa, S., Nachigami, Y., and Yamaguchi, I.,** Stabilization of some naturally occurring pigments, betanine, crocin and shisonin as food additives, *Tokyo Kasei Daigaku Kenkyu Kiyo,* 18, 71, 1978 (CA 90, 85434a).

116. **Shiroshi, M., Tada, M., Hosoda, H., and Shiga, M.,** Effect of gamma irradiation on plant pigments, *Shokuhin Shosha,* 4, 63, 1969 (FSTA 4, 11J1818).

117. **San-Ei Chemical Industries,** Coloring Material from Beefsteak Plants, Japanese Patent, 151,767, 1981 (CA 96, 161159).

118. **Chiang, B. H., Chung, M. Y., and Hwang, L. S.,** Anthocyanin extraction from perilla leaves and concentration by ultrafiltration and reverse osmosis, Paper No. 396, presented at the IFT Ann. Meet. Food Expo, Atlanta, GA, June 9–12, 1985.

119. **Chung, M. Y., Hwang, L. S., and Chiang, B. H.,** Concentration of perilla anthocyanins by ultrafiltration, *J. Food Sci.,* 51, 1494, 1986.

120. **Wu, C. F. and Hwang, L. S.,** Anthocyanins in perilla leaves, *Food Sci. (China),* 7, 60, 1980.

121. **Huang, S. A. and Hwang, L. S.,** Studies on the natural red pigment of perilla: changes of anthocyanin content with variety and growing stage, *Food Sci. (China),* 7, 161, 1980.

122. **Hwang, L. S., Tarng, J., and Chiang, B. H.,** Extraction, concentration and stability of Perilla anthocyanins, paper presented at the 7th World Congr. Food Sci. Technol., Singapore, Malaysia, September 28–October 2, 1987.

123. **Lin, S. S., Chiang, B. H., and Hwang, L. S.,** Recovery of perilla anthocyanins from spent brine by diafiltration and electrodialysis, *J. Food Eng.,* 9, 21, 1989.

124. **Ohba, T., Hasuo, T., Akita, O., and Yamamoto, Y.,** A liqueur using the extract of *Perilla ocymoides* var. *crispa, Nippon Jozo Kyokai Zasshi,* 80, 287, 1985 (CA 103, 69749).

125. **Zhong, J.-J., Seki, T., Kinoshita, S. I., and Yoshida, T.,** Effect of light irradiation on anthocyanin production by suspended culture of *Perilla frutescens, Biotechnol. Bioeng.,* 38, 653, 1991.

126. **Ota, S.,** Perilla Pigment Production by Callus Cultivation, Japanese Patent, 61,195,688, 1985 (CA 196, 30211).

127. **Ota, S., Tamaoki, Y., Niwada, H., and Kawachi, M.,** Red Pigment Manufacture with Beefsteak Plant Callus Tissue Culture, Japanese Patent, 6,402,593, 1989 (CA 112, 230718).

128. **Ota, S., Yamaoki, Y., Niwada, H., and Kawachi, M.,** Preparation of Anthocyanin Red Pigments from *Perilla frutescens* Leaves, Japanese Patent, 63,156,865, 1988 (CA 111, 150762).

129. **Tamura, H., Fujiwara, M., and Sigisawa, H.,** Production of phenylpropanoids from cell cultured callus tissue of the leaves of Akachirimen-shiso (*Perilla* spp.), *Agric. Biol. Chem.*, 53, 1971, 1989.

130. **Woodroof, J. G.,** *Tree Nuts: Production, Processing, Products*, Vol. 2, AVI Publishing, Westport, CT, 1979, chap. 25.

131. **Pomini, L.,** Il pistacchio, *Riv. Ital., EPPOS*, 60, 659, 1978.

132. **Spina, P. and Pennisi, F.,** La coltura del pistacchio in Sicilia, *Frutticoltura*, 19, 533, 1957.

133. **Crane, J. C.,** Quality of pistachio nuts as affected by time of harvest, *J. Am. Soc. Hortic. Sci.*, 103, 332, 1978.

134. **Nimadzhanova, N. K. and Rafieva, M. G.,** Content of pigments in pistachio nut in relation to ripening, *Izv. Akad Nauk Tadzhikistan, Otd. Biol. Nauk*, 4, 70, 1978.

135. **Schormueller, H.,** *Handbuch der Lebensmittelchemie*, Vol. 5, Springer-Verlag, Berlin, 1969, 332.

136. **Oliai, M.,** Les pigments flavoniques de la graine de pistache (*Pistacia vera* L., Anacardiaceae), *Bull. Soc. Pharm. Bordeaux*, 109, 193, 1970.

137. **Perkin, A. G. and Wood, P. J.,** Yellow colouring principles contained in various tannin matters. V. *Pistacia lentiscus, P. terebinthus, Tamaris africana, T. gallica, Ailanthus glandulosa, J. Chem. Soc.*, 73, 374, 1898.

138. **Freudenberg, K., Fikentschler, H., Harder, M., and Schmidt, O.,** Die Umwaldung des Cyanidins in Catechin. XX. Mitteilung ueber Gerbstoffe und aehnliche Verbindungen, *Justus Liebigs Ann. Chem.*, 444, 135, 1925.

139. **Feucht, W. and Nachit, M.,** Untersuchungen ueber Flavane und Flavolane in Fruechten und Fruechteschalen, *Mitt. Rebe Wein*, 24, 293, 1974.

140. **Hiroi, T., Takahashi, T., and Imamura, H.,** Wood extractives. XV. Constituents of *Pistacia chinensis* wood, *Nippon Mokuzai Gakkaishi*, 12, 324, 1966.

141. **Koco, P.,** Chemical composition of *Pistacia lentiscus* and *Pistacia terebinthus*, *Bul. Univ. Shteteror Tiranes, Ser. Shkencat Natyrore*, 21, 119, 1967.

142. **Malekzadeh, F.,** An antimicrobial compound in two *Pistacia* species, *Micopathol. Mycol. Appl.*, 54, 73, 1974.

143. **Miniati, E.,** Anthocyanin pigment in the pistachio nut, *Fitoterapia*, 52, 267, 1981.

144. **Treptow, H.,** Rhabarber (Rheumarten) und seine Verwendung, *Ernaehrung*, 9, 179, 1985 (CA 104, 67571).

145. **Canham, A. E.,** Artificial light in rhubarb growing, *Shinfield Progr.*, 14, 26, 1969 (HA 1969, 6877).

146. **Kmiecik, W.,** Assessment of the suitability of five rhubarb varieties for forcing, *Acta Agrar. Silvestria, Agrar.*, 14, 35, 1974 (PBA 1975, 5006).

147. **Ruben, G. M. and Sanitskaya, A. F.,** Chemico-technological evaluation of rhubarb cultivars, *Konserv. Ovoschchesush. Prom.*, 11, 37, 1984 (FSTA 18, 2J86).

148. **Gallop, R. A.,** Variety, composition and colour in canned fruits, particularly rhubarb, Fruit. Veget. Canning Quick Freezing Research Assoc., *Chipping Campden Sci. Bull. N.*, 5, 1965.

149. **Blundstone, H. A. W. and Crean, D. E. C.,** The pigments of red fruits, Reseach Report Fruit Veget. Preserv. Research Assoc., Chipping Campden, U.K., 1966.

150. **Hetmanski, W. and Nybom, N.,** The anthocyanins of edible rhubarb, *Fruchtsaft-Ind.*, 13, 256, 1968 (FSTA 1, 7J567).

151. **Wrolstad, R. E. and Heatherbell, D. A.,** Anthocyanin pigments of rhubarb, Canada Red., *J. Food Sci.*, 33, 592, 1968.

152. **Fuleki, T.,** The anthocyanins of strawberry, rhubard, radish and onion, *J. Food Sci.*, 34, 365, 1969.

153. **Wrolstad, R. E. and Struthers, B. J.,** Polyvinylpyrrolidone (PVP) column chromatography of strawberry, rhubarb and raspberry anthocyanins, *J. Chromatogr.,* 55, 405, 1971.

154. **Tronchet, J.,** Flavonols of the leaves (petioles and limbs) of *Rheum,* their distribution and their evolution during development, *Ann. Sci. Univ. Besançon, Bot.,* 6, 29, 1969 (CA 74, 95439).

155. **Ohshima, Y., Ohno, Y., Kosiyama, K., and Takahashi, K.,** High-performance liquid chromatographic separation of rhubarb constituents, *J. Chromatogr.,* 360, 303, 1986.

156. **Chumbalov, T. K. and Nurgalieva, G. N.,** Anthocyanins from the seeds of *Rheum tataricum.* II, *Khim. Prir. Soedin,* 3, 59, 1967 (CA 67, 768).

157. **Nurgalieva, G. and Chumbalov, T. K.,** Anthocyanins of *Rheum tataricum, Fenolnye Soedin. Ikh. Biol. Funkts., Mater. Vses. Symp.,* 1966, 93 (CA 71, 10257).

158. **Forsyth, W. G. C. and Simmonds, N. V.,** A survey of the anthocyanins of some tropical plants, *Proc. R. Soc., Sect. B,* 142, 549, 1954.

159. **Du, C. T. and Francis, F. J.,** Anthocyanins of roselle *(Hibiscus sabdariffa), J. Food Sci.,* 38, 818, 1973.

160. **Pouget, M. P., Vennat, B., Lejeune, B., and Pourrat, A.,** Identification of anthocyanins of *Hibiscus sabdariffa* L., *Lebensm. Wiss. Technol.,* 23, 101, 1990.

161. **Esselen, W. B. and Sammy, G. M.,** Roselle — a natural red colorant for foods?, *Food Prod. Dev.,* 7(2), 80, 1973.

162. **Esselen, W. B. and Sammy, G. M.,** Application for roselle as a red food colorant, *Food Prod. Dev.,* 9(10), 37, 1975.

163. **Clydesdale, F. M., Main, J. H., and Francis, F. J.,** Roselle *(Hibiscus sabdariffa* L.) anthocyanins as food colorants for beverages and gelatin desserts, *J. Food Prot.,* 42, 204, 1979.

164. **Saeed, A. R. and Ahmed, M. O.,** Storage stability of carbonated beverage from roselle calyxes *(Hibiscus sabdariffa), Sudan J. Food Sci. Technol.,* 9, 78, 1977 (FSTA 11, 4H598).

165. **Osman, E. M.,** Effect of different factors on the extraction, concentration and stability of color in roselle juice — a suggested potential natural colorant for foods, *Diss. Abstr. Int. B,* 44, 3346, 1984 (FSTA 17, 9T12).

166. **Pouget, M. P., Lejeune, B., Vennat, B., and Pourrat, A.,** Extraction, analysis and study of the stability of *Hibiscus* anthocyanins, *Lebensm. Wiss. Technol.,* 23, 103, 1990.

167. **Zeng, H., Dong, Y., and Yang, J.,** Preparation and application of red pigment from *Hibiscus sabdariffa* L. I. Selection of the most optimal extraction conditions, *Shipin Yu Fajiao Gongye,* 5, 27, 1984 (CA 102, 23040).

168. **Zeng, H., Dong, Y., and Yang, J.,** Preparation and application of red pigment from *Hibiscus sabdariffa* L. II. Effect of heat on the red pigment from roselle, *Shipin Yu Fajiao Gongye,* 6, 31, 1984 (CA 102, 202765).

169. **Zeng, B., Huang, Z., Zheng, H., and Chen, Y.,** Stability of edible roselle red dyes, *Shipin Kexue (Beijing),* 69, 1, 1985 (CA 104, 108078).

170. **Stahl, E., Rau, G., and Carius, W.,** Aufschliessen und Aufblahen von Pflanzlichem Material durch CO_2-Hockdruckbehandlung. I. Leinsamen und Hibiscusblueten, *Z. Lebensm. Unters. Forsch.,* 182, 33, 1986.

171. **Mizukami, H., Tomita, K., Ohashi, H., and Hiraoka, N.,** Anthocyanin production in callus cultures of roselle *(Hibiscus sabdariffa* L.), *Plant Cell Rep.,* 7, 553, 1988 (CA 110, 93518).

172. **Oohashi, Y., Mizukami, H., Tomita, K., Hiraoka, N., and Fujimoto, K.,** Manufacture of Anthocyanin Pigments by Tissue Culture of *Hibiscus sabdariffa* calluses, Japanese Patent, 130,594, 1989 (CA 110, 210989).

173. **Shalaby, A. S. and Mostafa, H. A. M.,** Effects of gibberellic acid and B-nine on growth, anthocyanin, and acidity contents in roselle *(Hibiscus sabdariffa* L.), *Z. Acker-Pflanzenbau,* 153, 321, 1984 (CA 102, 108117).

174. **Asanoma, M., Miyabe, M., and Sakabe, Y.**, Mutagenicity of natural food additives in *Salmonella thyphimurium*. II, *Nagoyashi Eisei Kenkyushjoho*, 30, 53, 1984 (CA 102, 202766).

175. **Garrido, J. L., Diez De Bethencour, C., and Revilla, E.**, Flavonoid composition of hydrolized extracts of *Crocus sativus* L., *An. Bromatol.*, 39, 69, 1987 (CA 107, 216326).

176. **Maroto, A. L.**, Natural anthocyanin pigments, *Rev. R. Acad. Cienc. Exactas, Fis. Nat. Madrid*, 44, 79, 1950 (CA 46, 581h).

177. **Price, J. R., Robinson, G. M., and Robinson, R.**, Occurrence of kaempferol in *Crocus*, *J. Chem. Soc.*, 1938, 281, 1938 (CA 32, 3456).

178. **Lawrence, W. J. C., Price, J. R., Robinson, G. M., and Robinson, R.**, Distribution of anthocyanins in flowers, fruit and leaves, *Phil. Trans. R. Soc., B*, 230, 149, 1939 (CA 33, 8689).

179. **Saito, N., Mitsui, S., and Hayashi, K.**, Delphin, the anthocyanin of medicinal saffron and its identity with hyacin as shown by paper chromatography of partial hydrolizates, *Bot. Mag. (Tokyo)*, 73, 231, 1960 (CA 55, 15476a).

180. **Lokar, L. and Poldini, L.**, Pigmenti antocianici nelle specie di *Crocus* L. del Friuli-Venezia Giulia, *G. Bot. Ital.*, 111, 367, 1977 (HA 1978, 7460).

181. **Harborne, J. B.**, *Comparative Biochemistry of the Flavonoids*, Academic Press, London, 1967, 241.

182. **Sankaranarayan, P. and Narasimhan, R.**, Pattern distribution of anthocyanin pigment in world assembly of noble sugarcane (*Saccharum officinarum* L.), *Indian J. Agric. Sci.*, 41, 440, 1971 (PBA 1972, 5719).

183. **Misra, K. and Dubey, R. C.**, Anthocyanin of sugarcane, *Curr. Sci.*, 43, 544, 1974 (CA 82, 28504).

184. **Smith, P. and Hall, P.**, Sugar cane anthocyanins as colour precursors and phytoalexins, *Proc. 14th Congr. Int. Soc. Sugar Cane Technol.*, 1971, 1139.

185. **Dutt, A. K.**, Sulphur deficiency in sugarcane, *Emp. J. Exp. Agric.*, 10, 257, 1962.

186. **Hokama, K.**, Biochemical studies on sugarcane reddening, *Ryukyu Daigaku Nogakubu Gakujutsu Hokoku*, 20, 37, 1973 (CA 82, 28712).

187. **Smith, P. and Paton, N. H.**, Sugarcane flavonoids, *Sugar Technol. Rev.*, 12, 117, 1985 (FSTA 17, 12L11).

188. **Goodacre, B. C. and Coombs, J.**, Formation of colour in cane juice by enzyme-catalised reactions. II. Distribution of enzyme and colour precursors, *Int. Sugar J.*, 80, 323, 1978 (FSTA 13, 3L433).

189. **Tu, C. C. and Onna, K.**, The non-sugar constituents of Hawaiian raw sugarcane crystals, *Proc. 10th Congr. Inst. Soc. Sugar Cane Technol.*, 1959, 291.

190. **Mazza, G. and Jayas, D. S.**, Equilibrium moisture characteristics of sunflower seeds, hulls, and kernels, *Trans. ASAE*, 34, 534, 1991.

191. **Cockerell, T. D. A.**, The red sunflower, *Pop. Sci. Mon.*, 1912, 373.

192. **Cockerell, T. D. A.**, Specific and varietal characters in annual sunflower, *Am. Nat.*, 49, 609, 1915.

193. **Flek, G. N.**, Breeding and genetics, in *Sunflower Science and Technology*, Carter, J. F., Ed., American Society of Agronomy, Crop Science Society of America, Soil Science Society of America, Publishers, Madison, WI, 1978, 279.

194. **Bottazzi, G. B.**, Pigmentazioni antocianiche nel girasole, *Ann. Sper. Agrar. (Pavia)*, 2, 186, 1949.

195. **Bottazzi, G. B.**, Pigmentazioni antocianiche nel girasole, *Genet. Agrar.*, 2, 186, 1950.

196. **Leclerq, P.**, Heredite de quelques caracteres qualitatifs chez le tournesol, *Ann. Amelior. Plant*, 18, 307, 1968.

197. **Velkov, V. N.**, Inheritance of anthocyanin pigmentation in certain inbred lines of sunflower, *Genet. Sel.*, 3, 279, 1970 (PBA 1971, 5927).

198. **Demurin, YA. N. and Tolmachev, V. V.**, Inheritance of some marker characters in sunflowers, *Ref. Zh.*, 5, 65, 1988 (PBA 1989, 2191).

199. **Luczkiewicz, T.,** Genetic analysis of some traits in sunflower (*Helianthus annuus* L.), Proc. 6th Int. Sunflower Conf., *Genetics,* 1974, 259, 1975.
200. **Luczkiewicz, T.,** Inheritance of some characters and properties in sunflower (*Helianthus annuus* L.), *Genet. Pol.,* 16(2), 167, 1975.
201. **Mosjidis, J. A.,** Inheritance of color in the pericarp and corolla of the disc florets in sunflower, *J. Hered.,* 73, 461, 1982.
202. **Vranceanu, V. A. and Stoenescu, F. M.,** Sunflower single hybrids, a production prospect for the near future, *Probl. Agric., Bucuresti,* 21(10), 21, 1969.
203. **Pogorletskii, B. K.,** Male sterility in sunflower from the teratological point of view, in *Probl. Onkol. Teratol. Rast., Leningrad, Russia, Nauka,* 1975, 297, 1975 (PBA 1977, 10687).
204. **Burlov, V. V. and Kostyuk, S. V.,** Identification of closely linked genes for anthocyanin plant colour and male sterility in sunflower and the breeding of lines with nuclear male sterility, *NauchnoTekh. Byull. Vses. Sel. Inst.,* 1, 33, 1981 (PBA 1982, 9615).
205. **Vranceanu, V. A. and Stoenescu, F. M.,** Producing hybrid sunflower seed on the basis of genetic male sterility, *Probl. Agric., Bucuresti,* 24(3), 69, 1972.
206. **Caramangiu, P.,** Using an anthocyanin marker in the production of hybrid seed in sunflower, *Probl. Agric., Bucuresti,* 26(3), 30, 1974 (PBA 1974, 6145).
206. **Caramangiu, P.,** Using an anthocyanin marker in the production of hybrid seed in sunflower, *Probl. Agric., Bucuresti,* 26(3), 30, 1974 (PBA 6145).
207. **Servettaz, O., Castelli, D., and Longo, C. P.,** The effect of Benzyladenine on anthocyanin accumulation in excised sunflower cotyledons, *Plant Sci. Lett.,* 4, 361, 1975.
208. **Mah, J.,** Feeding Behavior of Redwinged Blackbirds on Sunflowers with Different Bird-Resistant Features, Ph.D. thesis, North Dakota State University, Fargo, 1988.
209. **Mason, J. R., Adams, M. A., Dolbeer, R. A., Stehn, R. A., Woronecki, P. P., and Fox, G. J.,** Contribution of seed hull characteristics to resistance of sunflower to blackbird damage, *N.D. Farm Res.,* 43, 16, 1986.
210. **Mason, J. R., Bullard, R. W., Bolbeer, R. A., and Woronecki, P. P.,** Red-winged blackbird (*Agelaius phoeniceus* L.) feeding response to oil and anthocyanin levels in sunflower meal, *Crop Prot.,* 8(6), 455, 1989.
211. **Bullard, R. W., Woronecki, P. P., Dolbeer, R. A., and Mason, J. R.,** Biochemical and morphological characteristics in maturing achenes from purple-hulled (Neagra de Cluj) and oilseed sunflower hybrid cultivars, *J. Agric. Food Chem.,* 37, 886, 1989.
212. **Mason, J. R. and Adams, M. A.,** Anthocyanin Bird Repellents, U.S. Patent, 4,888,173, 1989.
213. **Sando, C. E.,** Anthocyanin formation in *Helianthus annuus, J. Biol. Chem.,* 64, 71, 1925.
214. **Vaccari, A., Pifferi, P. G., and Zaccherini, G.,** Anthocyanins of sunflower *Helianthus annuus, J. Food Sci.,* 47, 40, 1982.
215. **Pifferi, P. G.,** Coloring substance by extracting sunflower seeds, *Fr. Demande,* 2,331,605, 1977.
216. **Pifferi, P. G. and Vaccari, A.,** The anthocyanins of sunflower. II. A study of the extraction process, *J. Food Sci.,* 18, 629, 1983.
217. **Pifferi, P. G.,** Process for Obtaining a Dye Substance of Vegetable Origin, U.S. Patent 4,156,077, 1979.
218. **Holm, E. T.,** Extraction and Characterization of the Pigment in Purple-Hulled Sunflower, Ph.D. thesis, University of Minnesota, St. Paul, MN, *Diss. Abstr. Int., B,* 48(8), 2160, 1988.
219. **Mok, D. and Hettiarachchy, N. S.,** Heat stability of sunflower-hull anthocyanin pigment, *J. Food Sci.,* 56, 553, 1991.
220. **Wiesenborn, D., Golz, J., Hanzel, J., Hegenson, D., Hettiarachchy, N., Holm, E., and Lindley, J.,** Red food colorant extract derived from purple-hulled sunflower, *N.D. Farm Res. J.,* 49, 19, 1991.

221. **Pratt, H. K. and Reid, M. S.,** The tamarillo: fruit growth and maturation, ripening, respiration, and the role of ethylene, *J. Sci. Food Agric.,* 27, 399, 1976.
222. **Heatherbell, D. A., Reid, M. S., and Wrolstad, R. E.,** The tamarillo: chemical composition, during growth and maturation, *N.Z. J. Sci.,* 25, 239, 1982 (HA 1983, 1352).
223. **Dawes, S. N.,** Processing potential and composition of New Zealand sub-tropical fruits, *Food Technol. N.Z.,* 7, 22, 1972.
224. **Wrolstad, R. E. and Heatherbell, D. A.,** Identification of anthocyanins and distribution of flavonoids in tamarillo fruit (*Cyphomandra betacea* (Cav.) Sendt.), *J. Sci. Food Agric.,* 25, 1221, 1974.
225. **Bobbio, F. O., Bobbio, P. A., and Rodriguez-Amaya, D. B.,** Anthocyanins of the Brazilian fruit *Cyphomandra betacea, Food Chem.,* 12, 189, 1983.
226. **Timberlake, C. F. and Bridle, P.,** Anthocyanins, in *The Flavonoids,* Harborne, J. B., Mabry, T. J., and Mabry, H., Eds., Chapman & Hall, London, 1975, 214.
227. **Timberlake, C. F.,** Anthocyanins — occurrence, extraction and chemistry, *Food Chem.,* 5, 69, 1980.
228. **Lefevre, J. D.,** Revue de la litterature sur le tamarindier, *Fruits,* 26, 687, 1971.
229. **Lewis, Y. S. and Neelakantan, S.,** The chemistry, biochemistry and technology of tamarind, *J. Sci. Ind. Res., India,* 23, 204, 1984.
230. **Lewis, Y. S. and Johar, D. S.,** Characterization of the pigment in red tamarind (*Tamarindus indica* L.), *Curr. Sci.,* 25, 325, 1956.
231. **Lewis, Y. S., Dwarakanath, C. T., and Johar, D. S.,** Further studies on red tamarind, *Food Sci., Mysore,* 7(2), 44, 1958.
232. **Butler, L. and Chang, L. O.,** Graft-relationships of the tomato anthocyanin mutants of the tomato, *Genetics,* 41, 636, 1956 (PBA 1957, 2385).
233. **Von Wettstein-Knowles, P.,** Mutations affecting anthocyanin synthesis in the tomato. I. Genetics, histology, and biochemistry, *Hereditas,* 60, 317, 1968 (PBA 1970, 4025).
234. **Von Wettstein-Knowles, P.,** Mutations affecting anthocyanin synthesis in the tomato. II. Physiology, *Hereditas,* 61, 255, 1969 (PBA 1970, 6307).
235. **Hussey, G. C.,** Growth and development in the young tomato. II. The effect of defoliation on the development of the shoot apex, *J. Exp. Bot.,* 14, 326, 1963 (HA 1963, 5277).
236. **Ayers, J. and Mancinelli, A. L.,** Phytochrome control of anthocyanin synthesis in tomato seedlings, *Plant Physiol.,* 44 (Suppl.), 19, 1969 (HA 1971, 1490).
237. **Mancinelli, A. L. and Schwartz, O. M.,** The photoregulation of anthocyanin synthesis. IX. The photosensitivity of the response in dark and light-grown tomato seedlings, *Plant Cell Physiol.,* 25, 93, 1984.
238. **Aung, L. H.,** Gibberellins and other hormones on the tomato hypocotyl assay, *Proc. Plant Growth Regul. Work. Group,* 1979, 58 (CA 92, 105751).
239. **Khan, M. I.,** Gibberellic acid bioassay based on the inhibition of anthocyanin production in tomato seedlings, *Biol. Plant.,* 22, 401, 1980 (HA 1981, 4266).
240. **Lesley, J. W., Lesley, M. M., and Soost, R. K.,** Variegation due to chromosome loss induced by a gene in the woolly mutant of the tomato, *J. Hered.,* 70, 103, 1979 (PBA 1980, 4530).
241. **Durand, Y.,** Relationships between the marker genes *aa* and *Wo* and the male-sterility gene *ms35,* in *Genetics and Breeding of Tomato,* Philouze, J., Ed., INRA, Versailles, France, 1981, 225 (PBA 1983, 1652).
242. **Anais, G.,** Study of the rate of natural outcrossing due to *Exomalopsis* in the tomato, *Lycopersicon esculentum,* Consequence for plant breeding, *Nouvelles Agron. Antilles Guyane,* 3(3/4), 626, 1977 (PBA 1981, 11044).
243. **Rick, C. M., Quiros, C. F., Lange, W. H., and Stevens, M. A.,** Monogenic control of resistance of the tomato to the tobacco flea beetle: probable repellence by foliage volatiles, *Euphytica,* 25, 521, 1976 (PBA 1977, 6877).

244. **Ibrahim, H., Shannon, S., and Robinson, R.,** High pigmentation *hp* gene and anthocyanin production in the hypocotyls of tomato seedlings, *Genetika*, 1(1), 87, 1969 (PBA 1973, 3999).

245. **Koornneeff, M., Cone, J. W., Dekens, R. G., O'Herne-Robers, E. G., Spruit, C. J. P., and Kendrick, R. E.,** Photomorphogenic responses of long hypocotyl mutants of tomato, *J. Plant Physiol.*, 120, 153, 1985 (PBA 1985, 8258).

246. **Wartenberg, H. and Blumhor, T.,** Investigation on hyperchlorophyllization and chloroplast structure in tomato plant suffering from P deficiency, *Phytopathol. Z.*, 55, 101, 1966 (HA 1966, 4913).

247. **Ulrychova, M. and Sosnova, V.,** Effect of P deficiency on anthocyanin content in tomato plants, *Biol. Plant.*, 12, 231, 1970 (HA 1971, 1468).

248. **Rodeia, N. and Borges, M.,** Alteration in P and anthocyanins related to the presence of mycoplasmas in tomato plants, *Port. Acta Biol., Ser. A*, 13(1-4), 72, 1974 (CA 85, 156670).

249. **Rodeia, N. T.,** Variations in the levels of P and anthocyanins in relation to the presence of mycoplasmas in *Lycopersicon esculentum*, Mill. *Cienc. Biol., Mol. Cell. Biol.*, 8, 81, 1983 (CA 101, 87579).

250. **Shirakawa, N.,** Study on the selective herbicide CMMP. IV. On the mode of action of CMMP, *J. Jpn. Soc. Hortic. Sci.*, 39, 193, 1970 (HA 1971, 6337).

INDEX

A

N

O

Printed and bound by CPI Group (UK) Ltd, Croydon, CR0 4YY

22/10/2024

01777632-0001